ENVIRONMENT
New Challenges

ENVIRONMENT
New Challenges

Editors
Professor (Dr.) Arvind Kumar
FLS (London), FASc (Swiss), FISEC, FSESc, FAZ, FZA (Gold Medalist)
Pro-vice Chancellor
S.K.M. University, Dumka – 814 101 (Jharkhand)
&
Dr. Preeti
Research Scholar
S.K.M. University, Dumka – 814 101 (Jharkhand)

2008
DAYA PUBLISHING HOUSE
Delhi - 110 035

ISBN10 81-7035-519-2
ISBN13 978-81-7035-519-9
EAN: 9788170355199

Published by : **Daya Publishing House**
1123/74, Deva Ram Park
Tri Nagar, Delhi - 110 035
Phone: 27383999
Fax: (011) 23244987
e-mail : dayabooks@vsnl.com
website : www.dayabooks.com

Showroom : 4760-61/23, Ansari Road, Darya Ganj,
New Delhi - 110 002
Phone: 23245578, 23244987

Laser Typesetting : **Classic Computer Services**
Delhi - 110 035

Printed at : **Chawla Offset Printers**
Delhi - 110 052

PRINTED IN INDIA

Preface

Sustainable environmental management ensures a sound balance between optimum environmental production and renewal of natural resources. It takes account of the merits and demerits of different environmental activities of past and present, so that the needs of the moments are satisfied as well as the carrying capacity of natural resources are kept in reserve for fulfilling the future requirements. This is possible, if the pragmatic principles of ecology are properly adhered to, in planning, management and developmental activities in the environment, giving due care to prevailing environmental conditions and long term socio-economic needs.

In order to meet these new challenges for better environment, the continuous monitoring of the environment is required. Environmental monitoring is meant for programmed process for measurement and collection of information about the existing status of environment and can be done by physico-chemical and biological methods. The physico-chemical monitoring deals with quantitative estimation of physical and chemical parameters of environment, on the other hand employing of biological parameters for assessing environmental perturbation is termed as bio-monitoring. To restore steady state condition of the same, the deteriorations of ecosystem are detected in terms of monitoring of limited number of abiotic and biotic factors. The present endeavour is an attempt to know the recent environmental challenges and to develop certain strategies for sustainable environment.

My special thanks and appreciation go to the scientists whose contributions have enriched this volume. I wish to express my sincere gratitude to Dr. P.C. Hembram, Hon'ble Vice Chancellor, S.K.M. University, Dumka, Professor M.C. Dash, Hon'ble former Vice Chancellor of Sambalpur University, Professor N.C. Datta of Calcutta University, Professor K.C. Pandey of Lucknow University, Professor S.K. Konar of Kalyani University, Professor D.K. Belsare of Bhopal University, Professor U.C. Goswami of Gauhati University, Professor P.C. Mishra of Sambalpur University, Professor P. Natarajan of Kerala University, Professor P. S. Murthy of Bangalore University, Professor Ajit Varma of JNU, Professor A.L. Bhatia of Jaipur University, Professor A.K. Mittal of BHU, Professor

S.P. Hosmani of Mysore University, Professor K. Kapoor of Udaipur University, Professor K.B. Reddy of Nagarjuna University, Professor M. Vikram Reddy of Pondicherry University, Professor B.K. Tiwari of NEHU, Professor G. Tripathi of Jodhpur University, Professor R. Ramalingam of Annamalai University, Professor Sharif U. Ahmad of Nagaland University, Professor G.K. Kulkarni of Aurangabad University, Professor S.U. Meshram of Nagpur University, Professor S.K. Battish of PAU, Professor B.M. Sharma of Manipur University, Professor B.D. Joshi of Hardwar University, Professor G.C. Pandey of Faizabad University, Professor K. C. Sharma of Ajmer University, Professor M. Raziuddin of Hazaribag University, Professor U.S. Bagde of Mumbai University, Professor Gurdeep Singh of I.S.M., Dhanbad, Dr. P.K. Goel of Karad, Professor S.P. Roy and Shri Tribhuwan Poddar of Bhagalpur University, Bhagalpur for encouragements.

I also express my deep sense of gratitude to my parents whose blessings have always prompted me to pursue academic activities deeply. I am also thankful to my wife, *Kumari Bimla* and my two lovely sons, *Kumar Pallav Shivshankaran* and *Kumar Prasun Ramakrishnan* whose natural smiles extended to me relief all through this tiresome endeavour.

Last but not the least, I am also thankful to Mr. Anil Mittal, Proprietor, Daya Publishing House, New Delhi for taking keen interest in bringing out of this book. Finally, I will always remain a debtor to all my well-wishers for their blessings, without which this book would not have come into existence.

Professor A. Kumar

Contents

Chapter 1

Environmental Challenges for Copepods

Anil Kumar Jha[1] and Arvind Kumar[2]
[1]Department of Zoology, A.S. College, Deoghar, Jharkhand, India
[2]Environmental Science Research Unit, Post Graduate Department of Zoology
S.K.M. University, Dumka, India

Copepods are critical components of marine food webs. They play fundamental roles in the transfer of organic matter. In estuaries and other coastal systems, they are an essential part of the diet of many commercially important fish and invertebrates species. Therefore, it is essential to consider how copepods are affected by stressors in the coastal environment. Despite the call for studies of the long-term chronic effects of exposure of zooplankton to hydrocarbons (Corner *et al.*, 1976), we still have a very poor understanding of the sub-lethal impacts of chemical contaminants on marine copepods, the dominant mesozooplankton in coastal waters. The problem is much more complex however. Since a species likely encounters multiple stressors in the environment and also interacts with other species as predator or prey, simply elucidating the impact of a single stressor on a single species has limited value in predicting ecosystem-level changes. While every effort should be made to reduce contamination of our coastal waters, evidence suggests that despite chronic exposure to certain stressors, some copepod species are able to maintain population densities through acclimation and adaptation.

During the last 3 decades, estuaries and other coastal zone systems around the globe have shown increasing evidence of degradation, primarily due to human activities. This is of great concern because these regions have enormous ecological and economic importance, *e.g.*, they are widely recognized as essential nursery grounds for a variety of commercially important fish and invertebrates. Planktonic copepods are important components of marine pelagic foods webs, especially in the rich productive estuaries of coastal regions. The nauplius stage is a key component of the diet of the many fish larvae that develop in inshore nursery habitats. While most of the life cycle of these copepods is spent in the water column, the eggs of some species sink to the seabed and remain dormant in the sediments for periods of days to years.

Scientific assessments of the biotic impacts of eutrophication and chemical pollution have usually focused on mortality of commercially important species. Moreover sice widespread kills of fish and invertebrates (*e.g.*, crabs and shrimps) are obvious to the eyes and nose, they have received considerable attention from the news media. Narrow attention to major mortality events of these taxa alone, however, likely misses subtle community and ecosystem level changes, which only become manifest after a long period of time. For instance it may take years for the sublethal impacts of stressors (*e.g.*, hypoxia and toxicants) on individual development, growth and reproductive rates to become evident at the community level. Moreover, eutrophication accompanied by changes in the species composition of the primary producers could lead to shifts in the zooplankton community. These bottom-up processes may eventually affect higher trophic level. Shifts in the planktonic copepod community structure could have major impacts on fish populations since copepods are a major prey item in the diets of the larval and juvenine fish.

This article focuses on our current understanding of the impact of eutrophication and chemical pollution on marine copepod species that reside in the water column as nauplii, copepodites, and adult. Since some of these species have benthic resting eggs, it is necessary to consider processes taking place not only in the water column, but also on the seabed. The studies that I described were selected to exemplify the effects of eutrophication and chemical pollution on copepods. This is not a comprehensive review of the available literature.

Problems of Eutrophication

Eutrophication may affect copepods via several different mechanisms. Increased organic matter in the form of phytoplankton may initially be beneficial as it provides more food for grazing copepods. This was observed for the Japanese brakish water system of lake Shinji, Ohashi river, lake Nakaumi and Sakai Strait. A comparison of the mesozooplankton community of these waters in 1995-1997 to the community assemblage present 40 years earlier revealed that an increased in biomass had occurred (Uye *et al.*, 2000). If on the other hand, zooplankton grazing cannot keep pace with increased phytoplankton growth or there is a shift in species composition, an excessive buildup of organic matter followed by sedimentation to the seafloor may occur. A comparison of the plankton community structure of Osaka Bay, Japan to neighbouring waters revealed that the primary production rate was highest in the inshore highly eutrophic area; yet net-zooplankton (which was dominated by copepods) was not abundant (Uye *et al.*, 1999). The authors concluded that the net-zooplankton could not keep up with the production of phytoplankton, perhaps because the species composition had shifted to a size too small to be consumed by the copepods that dominated the net-zooplankton community. Hypoxic bottom waters were evident in Osaka Bay during the period of study in the 1990s as would be expected when too much organic matter sinks to the seabed. The impact of hypoxia on copepods is addressed below.

Since eutrophication is a process that typically takes place over many years, long-term data sets are necessary to demonstrate community-level changes in composition. Freshwater systems have received a great deal of study in this regard, and there is evidence of a shift in prominence from calanoid to cyclopoid copepods as lakes undergo eutrophication (*e.g.*, Gannon and Stemberger, 1978; Majer, 1996); however I could only find a few "before and after" field studies that addressed these issues in coastal marine systems. Uye (1994) studied the structure of the summer copepod community in 2 eutrophic bays of Japan. A comparison of the results to data sets from 40 yr earlier revealed that the median body weight of the copepod in the direction of offshore, and that the value for this parameters was also lower in Tokyo Bay was considered more eutrophic because the nitrogen-loading rate was approximately 50 per cent higher than in Osaka Bay. It was suggested eutrophication, there had been a general shift to smaller-sized copepods with egg sacs, in particular the cyclopoid *Oithona davisae*. The prominence of this species was attributed to a change in the food spectrum (small flagellates became more abundant) and / or the fact that it does not spawn eggs into the water column. The recent behaviour work of Paffenhofer and Mazzocchi (2002) on *Oithona plumifera* supports the suggestion that small motile flagellates are the preferred prey of cyclopoids. The possession of egg sacs was cited as an advantage in regions that develop hypoxic / anoxic bottom waters as a result of eutrophication. Spawned eggs that fall to the sediments may be prevented from hatching due to reduced oxygen levels. Uye also suggested that the shift to smaller-sized copepods contributed to the decline in the bay's planktivorous fish populations due to "bottom-up" processes. However, it is also possible that a "top-down", mechanism was responsible since predatory jelly-fish had also become abundant. A study of another system however found no evidence of change in community structure accompanying eutrophication. A comparison of the mesozooplankton community of the Japanese brackish-water system of lake Shinji, Ohashi river, lake Nakaumi, and Sakai strait to the community assemblage present 40 years earlier showed no substantial change in the community structure (Uye *et al.*, 2000). According to the authors, these water bodies had been subjected to greater nutrient loading from domestic sewage and agriculture during the last few decades. Copepods were dominant members of the zooplankton community in all of the systems, but there was no evidence for a shift to smaller sized species with egg sacs even though benthic anoxia occurred in the system.

Studies of benthic communities have revealed that reduced diversity accompanies eutrophication, and there is some evidence that it also occurs in copepod communities. Siokou-Frangou *et al.* (1998) found a less-diverse zooplankton community in Elefsis bay, a semi-enclosed region in the Aegean sea, compared to the neighbouring waters of Saronikos Gulf. The dominant copepod species in the bay was *Acartia clause*. The bay region was considered eutrophic based on nutrient levels and chlorophyll concentrations, but it had also experienced hydrocarbon and heavy metal pollution. In an earlier paper (Siokou-Frangou *et al.*, 1995), it was reported that this bay experienced some hypoxia during summer months. However, there was no indication that copepod species with egg sacs were dominant in Elefsis Bay, as Uye (1994) had reported for the eutrophic bays of Japan. The authors concluded that eutrophication of Elefsis Bay contributed to the reduction in zooplankton faunal diversity, however since the bay also experienced chemical pollution, eutrophication may not have been the only factor.

According to Ingrid *et al.* (1996) one must consider the structure of the grazing community in order to predict how a system will respond to nutrient enrichment. They suggested that every system has a critical level in terms of nutrient supply, and that organic matter will build-up when that level is exceeded. Grazers affect that critical level through a variety of mechanisms. Grazers determine how much material passes through the grazing food chain versus that which sediments out to the bottom. In the process, they sequester nutrients and release nutrients through excretion and fecal pellet production. More food particles will avoid ingestion, accumulate, and settle to the bottom if the material is not appropriate for grazers. Prey selectivity is determined by many factors, but size is regarded as very important. Subsequent eutrophic and heterotrophic bacterial production will be determined by the rate at which the grazers return nutrients to the system and deliver organic matter to the seafloor in the form of fecal pellets. The example is given of a stable coastal system in which oligotrich ciliates compete with the smaller life stages of copepods for small food particles, and in turn are the prey of larger stages. Under limited and transient food conditions, the authors suggest that ciliates with their higher growth rates will out-compete the copepods. However, under high nutrient concentrations, copepods should fare better because the growth of larger diatoms, which are the preferred food of copepods, should be favoured. The authors suggest that *Acartia* spp. May be key elements in food webs because they act as both predators consuming ciliates and competitors consuming phytoplankton.

In addition to trophodynamic effects, eutrophication often leads to the occurrence of hypoxia/ anoxia. In coastal waters, the reduction in dissolved oxygen usually beings at the sediment-water interface and extends upwards into the water column. A classic example of eutrophication leading to the increased occurrence of hypoxia/anoxia is Chesapeake Bay on the north-east coast of the US Roman *et al.* (1993) reported that the vertical distribution of *Acartia tonsa* in Chesapeake bay was altered by the presence of benthic hypoxia. Subsequent studies by Stalder and Marcus (1997) showed that *A. tonsa* did not avoid hypoxia suggesting that increased mortality would occur. On the other hand, Decker *et al.* (2003) subsequently showed that *A. tonsa* from Chesapeake Bay avoided hypoxia whereas *A. tonsa* from Florida did not. Stalder and Marcus (1997) had used animals from Florida in their study. These results suggest that over time, *A. tonsa* responds to chronic hypoxia in the environment by altering its migratory behaviour. The appropriate experiments have not been done to clarify whether this response is due to physiological or genetic adjustments or both. Although this means that mortality might not increase, the distribution of the species may be compressed into upper oxygenated layers. This change in the vertical distribution might affect the susceptibility of the species to predation as shown by Breiberg *et al.* (1997). In addition, species that become sluggish due to hypoxia will be more susceptible to predation than those which can maintain avoidance behaviour (Breiberg *et al.*, 1997). Changes in predation rates will alter the flow of energy in stressed systems.

In addition to the potential lethal effect of hypoxia on copepods, it is also importance to recognize that sub-lethal responses can affect population dynamics and hence community structure over longer time periods. Work in progress in out laboratory is directed at clarifying the sub-lethal impacts of hypoxia in copepods. Our research shows that hypoxic dissolved oxygen concentrations that do not lead to mortality of *A. tonsa* nevertheless result in reduced development, growth, and reproductive rates.

Hypoxia/anoxia caused by eutrophication may also impact copepod population dynamics by preventing hatching and reducing the viability of benthic resting eggs. In addition, benthic stages have to cope with exposure to hydrogen sulfide that often accompanies anoxic conditions. Marcus *et al.* (1997) showed that the rates of non-diapause copepod eggs of *Acartia tonsa*, *Centropages hamatus*, and *Labidocera aestiva* that were subjected to anoxia and/or hydrogen sulfide were similar regardless of which stressor was tested. On the other hand, the species differeed in their sensitivity to stressors. *A tonsa* was most sensitive, surviving only about 2-3 wk. Diapause eggs are much more tolerant of exposure, however, as evidenced by no reduction in hatching success for the diapause eggs of *C. hamatus* that were exposed to sulfide and/or anoxia for 20 wk (Marcus and Lutz, 1998), whereas substantial mortality was observed for non-diapause eggs of *C. hamatus* exposed for only 4 wk (Marcus *et al.*, 1997).

Effect of Pollutants

Chemical pollutants may affect marine organisms through various pathways. Dissolved materials may be absorbed directly from seawater. Contaminated particles may be ingested. In both cases, the contaminants may be assimilated and passed up to higher food levels via trophic transfer or they may be eliminated in fecal pellets (McManus *et al.*, 1983). Benthic organisms in bays and estruaries are at particular risk of exposure because chemical pollutants are often associated with fine sediments that accumulate in low-energy environments. The biological impact of stressors occurs at various levels including the individual population, and community levels. Most studies of chemical stressors have focused on short-term impacts on individuals as evidenced by cellular, biochemical, physical, and behavioural responses as well as determing lethal dosages. Although many studies have speculated on the population and community level effects, few studies have actually documented zooplankton population and community level effects due to chemical pollution because long term study is required.

Generally the highest levels of chemical contaminants are found in estuaries and coastal waters adjacent to large urbanized and industrial areas compared to other portions of the oceans. Whereas many studies of the impact of chemical pollutants are done with organisms isolated from the field, the 1974 CEPEX (Controlled Environmental Pollution Experiment) project was a multidisciplinary study of the impact of chemical pollutants on marine systems. The approach used large enclosed mesocosms. Reeve *et al.* (1977) summarized the results of the CEPEX experiments that pertained to the effects of copper on copepods. The study revealed the lethal and sub-lethal effects of copper exposure. The ingestion and filtration rates of animals taken from control bags were compared to animals taken from bags with 5 or 10 μgL^{-1} of copper. When offered a standard food, the ingestion and filtration rates were higher for the control animals. In addition, egg production was much higher for the controls. Changes in the bags also revealed the potentials. Changes in the bags also revealed the potential impact of changes on the copepod community. With fewer copepods, due to higher mortality upon exposure to copper and possibly also reduced feeding, phytoplankton were also to bloom because they were no longer being cropped by the herbivore grazers.

Two and a half decades later, there is renewed interest in the effects of metals on copepods. While previous research focused primarily on the lethal effects of metals in the dissolved state, current research is considering sub-lethal effects, food web transfer, and biogeochemical cycling. Hook and Fisher (2001) determined the impact of Hg, Cd and Ag on *A. tonsa* and *A. clause*. The concentrations tested were at environmentally realistic levels such as those that occur in urbanized estuaries. They exposed adults to food that had been exposed to metals and also dissolved metals in the water. They found that exposure to contaminated food resulted in assimilation of the metals primarily into internal tissues. With direct exposure, the metals showed up in the exoskeleton, particularly in the case of Hg. The metals led to sub-lethal effects (*e.g.*, decreased egg production and hatching, ovarian development, and protein concentration in eggs) at concentrations 2-3 orders of magnitude less than lethal concentrations. That study emphasized the importance of considering uptake pathways (as pointed out by Corner *et al.*, 1976), cumulative body burdens and different toxins. The study also emphasized the need for investigations of the effects of long-term chronic exposure to metals. In a subsequent study, Hook and Fisher (2002) showed similar responses to Zn and Mn, although the concentrations that led to an effect differed among the various metals that were tested. The results of that study also showed that toxicity was associated with the degree of bringing to sulfur. The authors suggested that the metals interfere with sulfur binding to the enzymes associated with vitellogenesis, leading to decreased egg production.

Large numbers of copepod eggs have been found in the seabed sediments of bays and estuaries and it has been suggested that these resting eggs are critical to the perpetuation of planktonic populations. The resting egg essemblage most likely includes a mixture of quiescent non-diapause eggs and diapause eggs. Since the seabeds of many urbanized estuaries and bays are contaminated with various chemical toxicants, there is concern that the viability and hatching of resting eggs will be adversely affected in such areas. The impact of hydrogen sulfide on copepod egg viability was already mentioned above. However, those results suggest that diapause eggs should be more tolerant of introduced chemical toxicants than non-diapause eggs as is the case for the diapause and non-diapause stages of many other organisms. Lindley and co-workers examined the viability of copepod resting eggs from British estuaries in which the degree of chemical contamination varied. In their initial paper (Lindley *et al.*, 1998), the highest hatching success (92 per cent) was observed for resting eggs from the estuary having the lowest concentration of PAHs and the lowest hatching success (14 per cent) from the estuary with the highest concentration of PAHs. Although the concentration of PAHs was used as an indicator of contamination, it is certainly possible that other stressors contributed to the differences in hatching success. Given the design of the study, it is not possible to know if the eggs were diapause or non-diapause, though it seems likely that they were the latter. Subsequently, Lindley *et al.* (1999) examined the impact of 2 organochlorine compounds on copepod eggs of *Eurytemora affinis* and *A. bifilosa*. The eggs that they tested were most likely not diapause eggs since the control eggs hatched within a few days. Although the concentration tested were generally higher than natural values, the results indicated that exposing eggs to these compounds led to reductions in viability and hatching. On the other hand, no effect on hatching success was detected in resting eggs exposed to Orimulsion and fuel oil #6 (Suderman and Marcus, 2002). In that 3-mo microcosm study, intact (sediments from Tampa Bay, FL, USA were dosed with one of 5 hydrocarbon treatments via hydrocarbon-coated sand and compared to untreated controls. *Acartia tonsa* eggs were non-viable in all treatments after only a few weeks of incubation, as evidenced by a marked decline in the abundance of nauplii that hatched from a sub-sample of the sediments. However, there was no evidence that exposure to concentrations of 700 or even 7000 ppm (levels expected in the event of a spill) of either fuel led to significant increases in resting egg mortality as compared to the controls. The results indicate that regardless of environmental conditions, eggs of *A. tonsa* from Tempa Bay sediments are likely quiescent non-diapause eggs, and that these eggs do not remain viable in the seabed for extended periods of time.

There is evidence that copepod species may adapt to the presence of chemical stressors analogous to what Decker *et al.* (2003) observed for hypoxia. In a study of the effects of copper on A. tonsa from Bascayne Bay, USA, Reeve *et al.* (1976) observed considerable variation in the 24-h LC_{50} values, which he attributed to physiological and/or differentiation among seasonally separated population. He concluded that there was wide genetic variability within species in tolerance to toxic substances, such that populations that may be reduced dramatically in an area may return to their original population size over time. To substantiate this idea, Reeve *et al.* (1976) referred to an unpublished observation by John Steele of the copepods from Loch Ewe in Scotland. Copepods from Ewe were not affected by copper concentrations (no citation provided) comparable to those tested as part of the CEPEX experiments. Steele suggested that the ambient copper concentrations were higher in the Loch than in Saanich inlet where the CEPEX experiments were conducted, and as a result, the Scottish animals were more accustomed to higher copper levels and were thus more tolerant. However, Reeve *et al.* also pointed out that the chelating abilities of the waters may have differed such that the levels of biologically available copper were not the same.

References

Decker, M.B., Breitberg, D. and Marcus, N., 2003. Geographical differences in behavioural responses to hypoxia: local adaptation to an anthropogenic stressed? *Ecol. Apol.*, 13: 1104–1109.

Gannon, J.E. and Stemberger, R.S., 1977. Zooplankton (especially crustaceans and rotifers) as indicators of water quality. *Trans. Am. Microsc. Soc.*, 97: 16–35.

Hook, S.E. and Fisher, N.S., 2001. Reproductiotive toxicity of metals in calanoid copepods. *Mar. Biol.*, 138: 1131–1140.

Hook, S.E. and Fisher, N.S., 2002. Relating the reproductive toxicity of five ingested metals in calanoid copepods with sulfur affinity. *Mar. Environ. Res.*, 53: 161–174.

Ingrid, G.T., Andersen, T. and Vadstein, O., 1996. Pelagic food webs and eutrophication of coastal waters: Impact of grazers on algal communities. *Mar. Pollut. Bull.*, 33: 1–6.

Kennish, M., 1997. *Pollution Impacts on Marine Biotic Communities.* CRC Press, Boca Raton, FL, pp. 310.

Lindley, J.A., Donkin, P., Evans, S.V., George, C.L. and Uil, K.F., 1999. Effects of 2 organochlorine compounds on hatching and viability of calanoid copepod eggs. *J. Exp. Mar. Biol. Ecol.*, 242: 59–74.

Lindley, J.A., George, C.L., Evans, S.V. and Donkin, P., 1998. Viability of calanoid copepod eggs from intertidal sediments a comparison of 3 estuaries. *Mar. Ecol. Prog. Ser.*, 162: 183–190.

Maier, G., 1996. Copepod communities in lakes of varying trophic degree. Arch. Hydrobiol. 136: 455–465.

Marcus, N.H. and Lutz, R.V., 1998. Longevity of non-diapause and diapause eggs of *Centropages hamatus* (Copepoda : Calanoida) from the Northern Gulf of Mexico. *Mar. Biol.*, 131: 249–257.

Marcus, N.H., Lutz, R.V. and Chanton, J.P., 1997. Impact of anoxia and sulfide on the viability of eggs of three planktonic copepods. *Mar. Ecol. Prog. Ser.*, 146: 291–295.

McManus, G.B., Wyman, K.D., Peterson, W.T. an d Wurster, C.F., 1983. Factors affecting the elimination of PCB's in the marine copepod *Acartia tonsa. Estuar. Coast. Shelf*, 17: 421–430.

National Research Council, 2000. *Clean Coastal Waters: Understanding and Reducing the Effects of Nutrient Population.* National Academy Press, Washington DC, pp. 405.

Nixon, S.W., 1995. Coastal marine eutrophication: a definition, social causes and future concerns. *Ophelia*, 41: 199–219.

Paffenhofer, G.A. and Mazzocchi, M.G., 2002. On some aspects of the behaviour of *Oithona plumifera* (Copepoda : Cyclopoida). *J. Plankton. Res.*, 24: 129–136.

Reeve, M.R., Gamble, J.C. and Walter, M.A., 1977. Experimental observations on the effects of copper on copepods and other zooplankton: controlled ecosystem pollution experiment. *Bull. Mar. Sci.*, 27: 92–104.

Reeve, M.R., Grice, G.D., Gibson, V.R., Walter, M.A., Darcy, K. and Ilkeda, T., 1976. A controlled environmental pollution experiment (CEPEX) and its usefulness in the study of larger marine zooplankton under toxic stress. In: *Effects of Pollutants on Aquatic Organisms*, (Ed.) A.R.M. Lockwood. Cambridge University Press, Cambridge, UK, pp. 145–162.

Roman, M., Gauzens, A.L., Rhinehart, W.K. and White, J., 1993. Effects of low oxygen water on Chesapeake Bay zooplankton. *Limnol. Oceanogr.*, 38: 1603–1614.

Sinkou-Frangou, I., Papathanassiou, E., Lepretre, A. and Frontier, S., 1998. Zooplankton assemblages and influence of environmental parameters on them in a Mediterranean coastal area. *J. Plankton. Res.*, 20: 847–870.

Stalder, L. and Marcus, N.H., 1997. Zooplankton responses to hypoxia behavioural patterns and survival of three species of calanoid copepods. *Mar. Biol.*, 127: 599–608.

Suderman, B. and Marcus, N.H., 2002. The effects of Orimulsion and fuel, oil #6 on the hatching success of copepod resting eggs in the seabed of Tampa Bay, Florida. *Environ. Pollut.*, 120 : 787–795.

Chapter 2

Hydrobiological Survey of the Manipur River with Reference to Fish Migration

Laishram Kosygin*[1] and T. Romi Singh[2]
[1]Lake Monitoring and Research Laboratory, Loktak Development Authority, Ningthoukhong – 795 126 Manipur, India
[2]Department of Fisheries, Government of Manipur, Lamphel – 795 004, Manipur, India

ABSTRACT

A survey was conducted to study the migratory fishes in the downstream of Manipur River up to Myanmar boarder. A total of 20 fish species belonging to 16 genera and 8 families have been collected. Distribution pattern of fishes in the river stretch was affected by water quality. This low catch and poor diversity might be due to low dissolved oxygen level, high concentration of free CO_2 and turbidity of water. Fish diversity was relatively high at Section 2 and 3 where water quality is comparatively better. *A. bengalensis* was the only catadromous fish collected during the survey exhibiting long distance migration. Fishes like *Labeo dero, L. angra, L. bata,* and *Osteobrama belangeri,* which use to migrate earlier from the Chindwin-Irrawady River in Myanmar to the upstream areas of Manipur River, were not recorded. Construction of Ithai barrage, habitat destruction, water pollution, over fishing using sophisticated gears and methods were identified as important threats to these migratory fishes. Sources of pollution and conservation strategies of the fishes are discussed.

Keywords: Manipur river, Water quality, Migratory fishes.

---

* E-mail: laishramkosygin@yahoo.com

Introduction

Manipur River is one of the most important and largest rivers of Manipur. It drains the central valley and south-eastern parts of the state. Some of the important tributaries, which form the Manipur River, are Imphal, Iril, Thoubal and Sekmai rivers. The Manipur River originates in the north at Karong (Senapati district) then it flows southwards of Imphal valley and finally joins the Chindwin River in Myanmar. The river course along the valley, downstream of Imphal, the riverbed slopes very gently. The bed level is 773 m at Imphal and 768.6 m at Lilong (near its confluence with Iril River) and 762.2 m at Ithai, downstream of the confluence of the Manipur and Khuga rivers, about 65 km. south of Imphal. Notwithstanding the gentle slope, the river has been regulated by two barrages for irrigation and hydropower. Sand-Jensen (2001) remarked that major rivers in Europe and Russia are strongly or moderately affected by dams, barrages and regulations. These structures have broken the natural connection and dynamics of watercourses for the movement of water, solutes and living organisms.

The important feature of the Manipur River from the hydrological point of view is the natural blockage of its flow in its lower reach. About 27 km downstream of Ithai barrage, after sloping down to 756.7 m, the river bed suddenly rises by 8 m within a distance of 800 m and remains above 762.5 m for about 2.5 km. This rocky barrier to the flow is known as "Sugnu hump" (named after Sugnu village). It reduces the capacity of Manipur River to discharge its flow. At the same time, Chakpi River drains a large part of the southern hills and meets the Manipur River from opposite direction in its full capacity during the rainy seasons. At such times it acts as a barrier for the Manipur River. On the other hand sometimes the Chakpi River discharges into Manipur River earlier in the season filling the section between Ithai barrage and the Sugnu hump thereby create an obstruction for its flow in the Manipur River. The excessive flow of both Chakpi and Manipur rivers spread backwards flooding large areas of the southern parts of the valley.

In the past, various scientists and researchers have believed that many riverine species migrate from the Chindwin and Irrawady River in Myanmar to the upstream areas of Manipur River and bred in various shallow lakes in the valley (Singh, 1993). He further reported that these fishes disappeared from the lakes since the construction of Ithai barrage across the Manipur River at Ithai which seem to have blocked their migratory route. Thus, it was pertinent to understand the present migratory pattern of fishes in the downstream Ithai barrage. The study was undertaken to find out present distribution pattern of fishes and physico-chemical characteristics of the river water.

Materials and Methods

The survey was carried out between 18–26 October 2000 in the downstream part of the Manipur River following Ithai Barrage near the Chakpi Confluence (29°16′31.5″ and 93°52′35.9″) to the Khaukaul Lui (24°03′13.0″ and 93°45′42.9″) near the Indo-Myanmar border (Figure 2.1). The entire study area was divided into three sections each comprising three zones of about 5 kms. each. Zone one was further divided into two sub-zones [Z1(S1) and (S2)]. The survey was undertaken at a time when flood level was high, considering the flood cycle as an essential element in the life history of most fishes in the river. It was carried out using rafts on the downstream river course and trekking upstream along the riverbanks and hills.

Water Quality Analysis

Surface water samples were collected from all the 9 zones along the Manipur River and 2 feeder streams namely Kanalok and Khaukual Lui. Temperature (air and surface water) was measured in the field using a standard centigrade thermometer. Transparency was measured with the help of a Sacchi

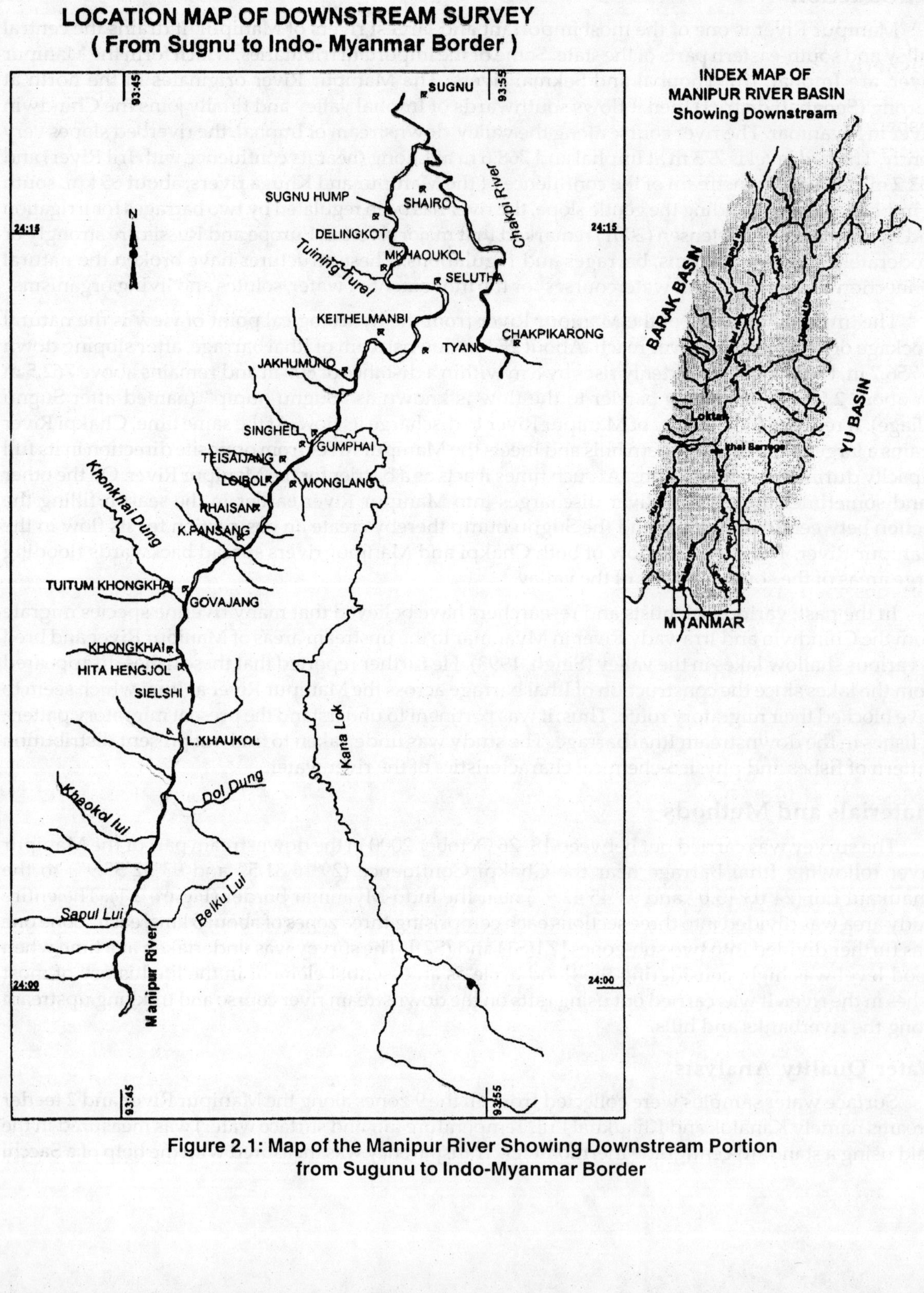

Figure 2.1: Map of the Manipur River Showing Downstream Portion from Sugunu to Indo-Myanmar Border

disc. Dissolved oxygen (DO) and pH were measured using DO meter and pH meter, respectively. Free CO_2 was determined by titremetric method (APHA, 1989).

Fisheries Survey

Experimental fishing was conducted at different zones along the river course. Three expert fishermen were engaged for conducting the day and night experimental fishing. Important fishing gears employed were gillnets (mesh size ranging from 25 mm to 150 mm), cast net, and hooks and lines. Fishes were identified following Jayaram (1999) Talwar and Jhingran (1991) and Vishwanath (2000). Specimens collected were preserved in 10 per cent formalin.

Observation and Discussion

Physico-chemical Characteristics of Water and Distribution of Fishes

The air and surface water temperature varied from 26.0 to 32.0°C and 24.9 to 27.2°C, respectively. Water temperatures of feeder streams from the nearby hills were much cooler than the River water (23.5, 18.8 vs 26.3). pH showed a narrow range of fluctuation between 6.8 and 7.3. It showed an increasing trend along the downstream. The average value of all the sites during the survey period remained almost neutral (7.02). However, pH values were alkaline in the feeder streams (7.7, 8.1).

Transparency value was low recording a minimum value at zone-1 before meeting Chakpi stream (0.13 m) and maximum value at zone-4 (0.2 m). The river water was very turbid due to heavy rainfall, and carried heavy load of silt and solid waste. On the contrary, the stream water was very clear and transparent to the bottom. There was a distinct difference in the clarity of the water at their confluence.

The dissolved oxygen (DO) content varied to a great extent at different zones. The highest value was observed at zone-9 (7.4 mg/L) and lowest at zone-1 before the confluence of Chakpi stream (1.6 mg/L). However, the value increased after meeting the Chakpi stream (4.1 mg/L). Level of DO was often hypoxic (<4 mg/L) at zones-1, 3 and 4 which is not favourable for the survival of fish and their existence in the area (ISI, 1982). DO show an increasing trend along the downstream (Figure 2.2). Higher value of DO at zone-6 to 9 may be due to the higher velocity of water and these sections may be considered as recovery zones. DO of Kanalok and Khaukaul Lui was very high. On the contrary, high value of free CO_2 was observed in the present survey. Its concentration ranges from 7.0 mg/L (at zone-8) to 11.4 mg/L (at zone-1 and 6). The values exceed 6 mg/L, which is the tolerance limit for fish existence (ISI, 1982). Low values of DO and high concentration of free CO_2 in section-1 may be due to:

1. Discharge of pollutants from the Sugnu and Serou markets.
2. Runoffs rich in nutrients from the surrounding hills.
3. Sewage carried by the tributaries of Manipur River through the Imphal valley.
4. Pesticides and insecticides from the agricultural lands in the surrounding hills.
5. Out flow of polluted water from Loktak Lake due to opening of Ithai barrage.

During the present survey, the diversity of fish was poor with very low catch. Vishwanath *et al.* (1998) reported 56 fish species (representing 35 genera and 14 families) from Manipur River. Of this, only twenty species could be collected during the survey (Table 2.1). This low catch and poor diversity may be due to low DO, high free CO_2 and turbidity of water. FIshes seem to migrate in the feeder hill streams of Manipur river, which have good transparency (which enables them to see their food), high DO and less concentration of free CO_2. Distribution pattern of fishes in the River stretch was also affected by water quality. Only highly tolerant species like *Mystus microphthalmus, Mystus bleekeri,*

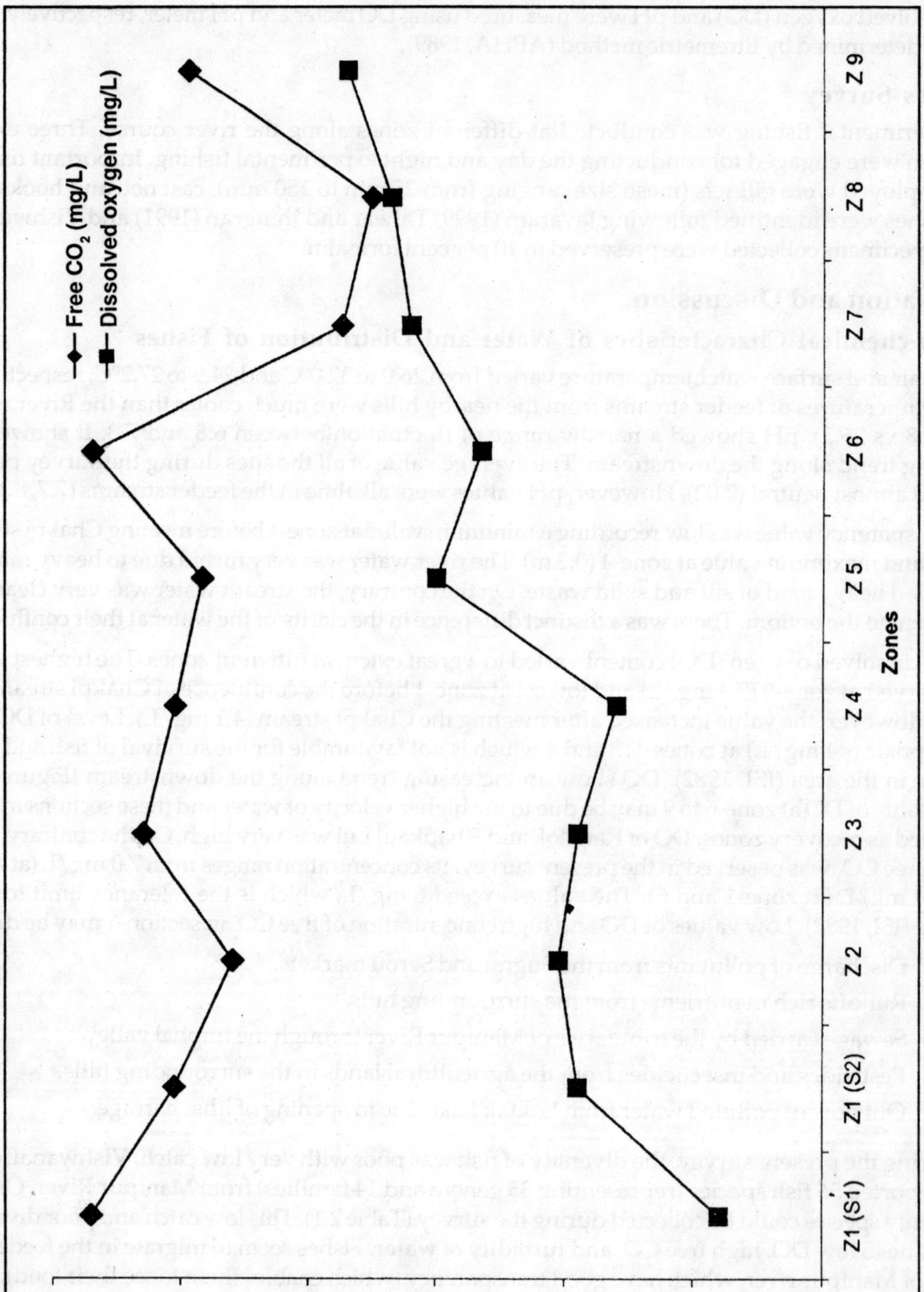

Figure 2.2: Variations in Dissolved Oxygen (DO) and Free CO_2 at Different Zones of the Manipur River

Puntius ticto and *Colisa fasciatus* were observed in section 1 in which water quality was poor. The distribution of fish was more concentrated at the confluence of streams with the Manipur River. Fish diversity was relatively high at section 2 and 3 where water quality is comparatively better.

Table 2.1: List of Fishes Collected from Downstream of Manipur River

Fishes	*Distribution*			*Status*
	Section 1	*Section 2*	*Section 3*	
Family: Anguillidae				
1. *Anguila bengalensis* (Gray)	–	–	+	Common
Family: Cyprinidae				
2. *Cirrhinus reba* (Ham.)	–	–	+	Rare
3. *Labeo pangusia* (Ham.)	–	+	–	Rare
4. *Labeo bata* (Ham.)	–	+	–	Rare
5. *Neolissochilus hexagonolepis* (McCl.)	–	–	+	Common
6. *Osteobrama cunma* (Day)	–	+	–	Common
7. *Puntius jayarami* (Vish. and Tombi)	–	+	+	Very common
8. *Puntius ticto* (Ham)	+	–	–	Common
9. *Barilius dogarsinghi* (Hora)	–	+	+	Common
10. *Barilius ngawa* (Vish. and Manoj)	–	+	+	Very common
11. *Raiamas guttatus* (Day)	–	+	–	Rare
12. *Crossocheilus burmanicus* Hora	+	+	–	Rare
Family: Balitoridae				
13. *Schistura manipurensis* (Chaudhuri)	–	–	+	Common
Family: Cobitidae				
14. *Lepidocephalus guntea* (Ham.)	–	+	+	Common
Family: Bagridae				
15. *Aorichthys aor* (Hami)	–	+	–	Rare
16. *Mystus bleekeri* (Day)	+	–	–	Common
17. *Mystus micropthalmus* (Day)	+	–	–	Rare
Family: Sisoridae				
18. *Bagarius* sp.	–	–	+	Common
Family: Belontiidae				
19. *Colisa faciatus* (Ham.)	+	–	–	Common
Family: Mastacembelidae				
20. *Mastacembelus armatus* (Lacèpéde)	–	+	+	Common

Fish Fauna and Status of Migration

Twenty fish species representing 8 families were caught during the survey. Fish catch was very poor at all the sections. The important species caught from the section-1 were *Colisa fasciatus, Puntius ticto, Mystus bleekeri and Mystus microphthalmus.* Eleven species *viz., Labeo bata, Puntius jayarami,*

Osteobrama cunma, Mastacembelus armatus, Aorichthys aor, Raiamus guttatus, Lepidocephalus guntea, Crossocheilus burmanicus, Barilius dogarsinghi, Barilius ngawa and *Labeo pungusia* were caught from section-2. Of the 11 species, only *Labeo bata* was a short distance migratory fish. Catches from section-3 comprised of *Anguilla bengalensis, N. hexagonolepis, Barilius dogarsinghi, Barilius ngawa, Bagarius yarelli, Lepidocephalus guntea, Puntius jayarani, Mastacembelus armatus, Cirrhinus reba* and *Schistura manipurensis.* Among the species caught from the section 3, *A. bengalensis* is the only catadromous fish exhibiting long distance migration. Jenkins (1999) pointed out that the most extreme example of fish migration between freshwater and marine ecosystem involves freshwater eels of the genus *Anguilla.* These eels migrates from freshwater out to sea upon reaching sexual maturity and spawn in a centralized location far out to sea. These eels have an amazing ability to reach remote headwater streams using their muscular bodies to slither up rapids and waterfalls. During the survey not a single fish of *O. belangeri* was caught from the sections surveyed. List of fishes collected during the survey is presented in Table 2.1.

Singh (1991, 1993) remarked that riverine fishes like *Labeo dero, L. angra, L. bata, Cirrhinus reba* and *Osteobrama belangeri* migrate from the Chindwin-Irrawady River system in Myanmar to the upstream areas of Manipur River and breed in various shallow lakes in the valley. He further reports that these fishes grew in the lakes into fingerlings, which were known locally as 'Tharak' (for *O. belangeri*) or 'Ngaton' (for other species).

The fingerlings started their downstream migration with the onset of monsoon. Some of the fishes migrated to the rivers later during the winters (November–December). These fishes grew slightly larger in size and were locally known as 'Ngaton tolaba'. They attained sexual maturity in the larger river system and were known as 'Pengba' (*O. belangeri*) or 'Khabak' (other species). The adult fishes migrated in huge shoals of several thousands in a group and constituted a major seasonal fishery commonly called as Pengba and Tharak fishery (*O. belangeri*) and Ngaton and Khabak Tharak fishry (other species) and these fishes accounted for about 40 per cent of the natural fishery resources of the Manipur state. However, these fishes disappeared from the valley lakes since the construction of Ithai barrage, which seems to have blocked their migratory route. Further, they are threatened by habitat degradation, water pollution and over fishing due to the increase in number of fishermen using more sophisticated fishing gears and methods. In order to conserve these migratory fishes, it is imperative that integrated and holistic approach for conservation and management of watercourses and environmental quality may be developed using scientific techniques. Further, alternative routes may be constructed connecting the river with various lakes of valley to ensure the survival of the fishes at different stages of their life cycles.

Acknowledgements

We are thankful to Th. Manihar, Former Project Director, Loktak Development Authority (LDA), Manipur, Dr. C.L. Trisal, Director, Wetlands International South Asia (WISA), New Delhi and Robson Ivan, former coordinator WISA, New Delhi. We gratefully acknowledge S. Santosh, Ch. Bidan, N. Sanajaoba and all other who participated in the survey for their field support and valuable helps. Financial support for the survey was provided by India Canada Environment Facility (ICEF, New Delhi).

References

APHA, 1989. *Standard Methods for the Examination of Waters and Wastewaters*, 17th Edition, American Publication Health Association, Inc., New York.

ISI, 1982. *Indian Standard, Tolerance Limit for Inland Surface Waters Subject to Pollution.* Indian Standards Institution Publication, New Delhi, pp. 18.

Jayaram, K.C., 1999. *The Freshwater Fishes of the Indian Region*. Narendra Publishing House, Delhi, pp. 551.

Jenkin, A., 1999. Inter-habitat migration of fish, Kamiali wetlands, Papua New Guinea. *Wetlands*, 8: 12.

Sand-Jensen, K., 2001. Freshwater ecosystems, human impact on. In: *Encyclopedia of Biodiversity*, (Ed.) Levin S. A. Academic Press, New York, 3: 89–108.

Singh, H.T., 1991. Study on the biology and culturable possibilities of the indigenous fishes of Manipur, *Osteobrama belangeri* and *Monopterus albus.* Technical Report, ICAR, New Delhi.

Singh, H.T., 1993. Impact of Loktak national hydro-electric project on the environment of Manipur. Technical Report, Department of Science, Technology and Environment, Government of Manipur.

Talwar, P.K. and Jhingran, A.G., 1991. *Inland Fishes of India and Adjacent Countries.* I. Oxford and IBH Publ. Co. Pvt. Ltd., New Delhi, pp. 541.

Vishwanarh, W., 2000. *Fish Fauna of Manipur.* Manipur Association for Science and Society (MASS), Imphal, pp. 143.

Vishwanath, W., Manojkumar, W., Kosygin, L., and Selim, K.S., 1998. Biodiversity of freshwater fishes of Manipur, India. *Ital. J. Zoo.*, 65: 321–324.

Chapter 3

Impact of Distillery Spentwash, Biocompost, Biosuper and Inorganic Fertilizer on Soil Microflora

V. Davamani and A. Christopher Lourduraj*

Department of Environmental Sciences,
Tamil Nadu Agricultural University, Coimbatore – 641 003, India

ABSTRACT

A field experiment conducted in turmeric (*Curcuma longa* Linn.) during 2001–2002 revealed that the combined application of distillery spentwash along with composts prepared from pressmud with spentwash (biocompost and biosuper) and various levels of inorganic fertilizers, significantly increased the soil microbial population and enzyme activity. Among the different combinations, the treated spentwash at 50KL ha^{-1} as basal + 2.5 tonnes ha^{-1} Biosuper + 75 per cent NP recorded maximum enzyme (catalase and amylase) activity and microbial load of bacteria and fungi. However in case of actinomycetes, the treated spentwash at 50KL ha^{-1} as basal + 2.5 tonnes ha^{-1} Biocompost + 75 per cent NP + micronutrient recorded the highest Actinomycetes population. The microbial population and enzyme activity recorded gradual increase up to 150th day, after which it got reduced.

Keywords: *Spentwash, Biocompost, Biosuper, Microbial population, Enzyme activity, Turmeric.*

---

* E-mail: christoens2000@yahoo.co.in.

Introduction

The sugar industry is amongst the largest agro processing industries in India with average sugar production of about 176.75 lakh tonnes and an annual turnover of Rs. 150 billion. Pressmud, molasses and spentwash are some of the by products from sugar and distillery industries. Pressmud is a solid waste by-product from sugar factory. About three per cent of pressmud is obtained for the total quantity of cane crushed. The annual production of pressmud is around 18,000 tonnes from the sugar factory of 2,500 tonnes crushing capacity per day. The number of distilleries producing alcohol in the country has been increased two folds during the last decade and at present there are 285 distilleries in India producing around 2.7 million litres of alcohol. The total amount of spentwash generated by the 285 distilleries, amounts to nearly 40 billion litres annually (Joshi, 2001). The treated spentwash contains large amounts of organic carbon, potassium, calcium, magnesium and sulphur and moderate levels of nitrogen and phosphorus and traces of micronutrients. The soil microflora play a major role in the growth and productivity of any crop. Hence with a view to study the impact of sugar and distillery wastes on soil microbial population and enzyme activity, the present study was undertaken.

Materials and Methods

A field experiment was carried out at the research centre, Bannari Amman Sugars Ltd., Modhur, Erode District during 2001–2002 to assess the effect of application of spentwash and composts prepared from pressmud with spentwash (biocompost and biosuper) on soil microflora and enzyme activity with turmeric as test crop. In the study, the secondary treated spentwash and composts were combined with various level of inorganic fertilizers, as indicated below

- T_1: 2.5 tonnes ha^{-1} biocompost + 75 per cent NPK + micronutrient
- T_2: 2.5 tonnes ha^{-1} biosuper + 75 per cent NPK + No micronutrient
- T_3: 2.5 tonnes ha^{-1} biocompost + 100 per cent NPK + micronutrient
- T_4: 2.5 tonnes ha^{-1} biosuper + 100 per cent NPK + No micronutrient
- T_5: Absolute control
- T_6: Treated spentwash at 50 KL ha^{-1} as basal + 2.5 tonnes ha^{-1} biocompost +75 per cent NP + micronutrient
- T_7: Treated spentwash at 50 KL ha^{-1} as basal + 2.5 tonnes ha^{-1} biosuper + 75 per cent NP + No micronutrient

Soil microbial population was found out by the standard plate count method (Waksman and Fred, 1922). The amylase and catalase enzyme activity was assessed by the methods suggested by Ross (1966) and Skujins (1976), respectively. Prior to initiation of the field experiment, the microbial analysis for distillery spentwash, biocompost and biosuper and microbial analysis and enzyme activity for soil were done. The microbial population of secondary treated spentwash was 29×10^6 CFU ml^{-1} for bacteria, 20×10^3 CFU ml^{-1} for fungi and 13×10^4 CFU ml^{-1} for actinomycetes. In case of biocompost, the microbial population was 29×10^6 CFU.g^{-1} for bacteria, 22×10^3 CFU g^{-1} for fungi and 12×10^4 CFU g^{-1} for actinomycetes. The microbial population of biosuper was recorded as 29×10^6 CFU g^{-1} for bacteria, 24×10^3 CFU g^{-1} for fungi and 8×10^4 CFU g^{-1} for actinomycetes. The microbial population of initial soil was recorded as 9.3×10^6 CFU g^{-1} for bacteria, 5.1×10^3 CFU g^{-1} for fungi and 4.6×10^4 CFU g^{-1} for actinomycetes. The amylase and catalase activity of initial soil were 774.3 mg of glucose g^{-1} and 63 μmol of H_2O_2 g^{-1} respectively.

The microbial analysis and enzyme activity was assessed in fresh soil samples collected from 30th day onwards at 30 days interval upto maturity, as well as in postharvest soil samples. The experimental data were statistically analysed and the critical difference was worked out at 5 per cent (0.05) probability.

Results and Discussion

Soil Microbial Population

The microbial population in different treatments showed significant differences. The bacteria and fungi were found to be significantly high in treated spentwash at 50 KL ha^{-1} as basal + 2.5 tonnes ha^{-1} Biosuper + 75 per cent NP (T_7) followed by treated spentwash at 50 KL ha^{-1} as basal + 2.5 tonnes ha^{-1} biocompost + 75 per cent NP + micronutrient (T_6). In case of actinomycetes the highest value was recorded in T_6 followed by T_7 (Tables 3.1–3.3). The microbial population recorded gradual increase upto 150th day. This is in line with Mattiazo and Ada Gloria (1985) who reported increase of soil microbial activity due to oxidation of organic matter in treated spentwash applied soil. Devarajan *et al.* (1993) studied the population dynamics of bacteria, fungi, actinomycetes, in field soil grown with turmeric, paddy, gingelly, cotton and groundnut and found that 50 and 40 times diluted spentwash irrigations enhanced or maintained the population dynamics of microorganisms in soil. Gopal *et al.* (2001) observed increased microbial population in spentwash applied soils. Humic substances in compost also would have contributed to the increase in the level of soil microflora. Raman *et al.* (1999) reported that the application of pressmud composted with spentwash increased the population of soil *Azotobacter*, fungi and actinomycetes. After 150th day the microbial population gradually decreased and recorded lower values in post harvest soil (37×10^6 CFU g^{-1} for bacteria, 10.3×10^3 CFU g^{-1} for fungi and 9.9×10^1 CFU g^{-1} for actinomycetes). Decrease in microbial population in post harvest stage might be due to decreased nutrient availability as a result of plant uptake and leaching losses. The rhizosphere region could be influenced by various environmental factors and physiological condition of plant which might have been responsible for variation in the population dynamics of different groups of microflora Goyal *et al.* (1995) reported increased soil microbial dynamics and dehydrogenase enzyme activity upto 60 days after spentwash application and reduction in later stages.

Table 3.1: Influence of Spentwash, Biocompost and Biosuper on Bacterial Population ($\times 10^6$ CFU g^{-1}) of Soil

Treatments	*Sampling Period (Days)*								
	30	*60*	*90*	*120*	*150*	*180*	*210*	*Postharvest*	*Mean*
T_1	26.3	28.7	31.0	35.0	38.7	34.0	31.3	30.0	31.9
T_2	30.3	31.0	33.3	37.3	42.7	38.7	35.0	30.7	34.9
T_3	34.0	35.0	38.0	42.0	46.0	43.7	41.3	39.7	40.0
T_4	37.7	39.7	40.3	43.3	47.0	45.7	43.0	41.0	42.2
T_5	9.3	12.3	14.0	16.7	24.7	22.0	19.7	18.0	17.1
T_6	41.0	43.7	45.0	49.3	55.0	52.0	50.0	49.0	48.1
T_7	46.0	49.3	50.0	55.0	59.3	55.3	53.0	5.1.0	52.4
Mean	32.1	34.2	36.0	39.8	44.8	41.6	39.0	37.0	38.1

	T	*D*	*T × D*
SEd	0.419	0.448	1.185
CD (0.05)	0.830	0.888	2.348

Table 3.2: Influence of Spentwash, Biocompost and Biosuper on Fungal Population (× 10³ CFU g⁻¹) of Soil

Treatments	Sampling Period (Days)								
	30	*60*	*90*	*120*	*150*	*180*	*210*	*Postharvest*	*Mean*
T_1	3.7	3.7	5.7	6.7	9.3	9.0	8.0	8.3	6.8
T_2	4.7	4.7	6.7	8.7	12.3	10.7	9.7	9.3	8.3
T_3	6.0	6.3	7.0	9.3	16.0	12.7	9.7	9.0	9.5
T_4	5.3	6.3	8.7	11.7	17.7	15.3	13.3	10.7	11.1
T_5	4.7	5.3	5.7	7.0	7.7	6.7	6.3	7.0	6.3
T_6	5.7	6.7	10.7	14.7	16.0	15.3	13.0	11.7	11.7
T_7	6.3	8.0	12.3	17.7	22.3	21.3	19.3	16.0	15.4
Mean	5.2	5.9	8.1	10.8	14.5	13.0	11.3	10.3	9.9

	T	*D*	*T × D*
SEd	0.172	0.184	0.486
CD (0.05)	0.340	0.364	0.962

Table 3.3: Influence of Spentwash, Biocompost and Biosuper on Actinomycetes Population (× 10⁴ CFU g⁻¹) of Soil

Treatments	Sampling Period (Days)								
	30	*60*	*90*	*120*	*150*	*180*	*210*	*Postharvest*	*Mean*
T_1	5.7	6.6	8.7	10.3	13.0	11.6	8.6	7.0	9.0
T_2	4.7	5.6	6.7	10.6	11.3	10.6	9.3	7.7	8.3
T_3	5.0	7.7	7.3	10.6	13.7	12.3	11.0	11.0	9.8
T_4	5.3	7.3	8.7	12.7	14.0	12.7	10.3	8.7	10. 0
T_5	5.0	7.0	8.3	12.7	12.0	10.6	7.3	7.0	8.8
T_6	5.6	8.3	10.3	16.0	21.7	19.33	19.3	14.6	14.4
T_7	6.6	7.3	8.3	14.3	18.7	17.0	15.6	13.0	12.6
Mean	5.4	7.1	8.3	12.4	14.9	13.15	11.7	9.9	10.4

	T	*D*	*T × D*
SEd	0.177	0.190	0.502
CD (0.05)	0.352	0.376	0.995

Soil Enzyme Activity

The catalase and amylase enzyme activity were found to be significantly high in treated spentwash at 50 KL ha⁻¹ as basal + 2.5 tonnes ha⁻¹ Biosuper + 75 per cent NP (T_7) (314.3 µmol of H_2O_2 g⁻¹ and 1258.3 mg of glucose g⁻¹, respectively) compared to absolute control (T_5) (109.3 µmol of H_2O_2 g⁻¹ and 833.7 mg of glucose g⁻¹, respectively) (Tables 3.4 and 3.5). The catalase and amylase enzyme activity in soil were significantly increased over period and the 150th day sample recorded maximum catalase, and amylase activities of 296.5 µmol of H_2O_2 g⁻¹ and 1145.3 mg of glucose g⁻¹, while initial soil recorded only 143 μmol of H_2O_2 g^{-1} and 956.9 mg of glucose g^{-1}, respectively. Increased enzyme activity over periods might be due to increased microbial activity, which would have enhanced the organic matter degradation and mineralization. Rajannan *et al.* (1998) stated that catalase, dehydrogenases are considered as important enzymes for oxidative process in the soil. The application of composts has led to significant increase in activity of soil enzymes as reported by Perucci (1990) and Garcia *et al.* (1993). After 150th day, the activity of enzyme was reduced significantly. Decrease in enzyme activities in post harvest stage could be due to the reduction of nutrient contents as a result of plant utilisation. This is in accordance with the findings of Pallab De *et al.* (1990) who reported, that enzyme activities depend on the type and physiological stage of standing crop.

Table 3.4: Influence of Spentwash, Biocompost and Biosuper on Amylase Enzyme Activity (mg of glucose g^{-1}) of Soil

Treatments	*Sampling Period (Days)*								
	30	*60*	*90*	*120*	*150*	*180*	*210*	*Postharvest*	*Mean*
T_1	881.1	9.12.0	959.3	979.0	1017.0	1000.0	904.3	868.3	940.2
T_2	904.7	938.3	988.3	1002.7	1079.3	1022.7	931.0	893.7	970.1
T_3	973.0	994.7	1008.3	1064.3	1112.0	1089.0	989.0	938.3	1021.1
T_4	1013.0	1052.7	1091.3	1126.0	1194.0	1119.7	1039.0	1003.7	1079.9
T_5	174.3	798.7	823.3	872.0	9.13.0	900.0	808.3	780.0	833.7
T_6	1026.3	1084.3	1152.7	1211.3	1285.7	1220.0	1134.3	1094.3	1151.1
T_7	1125.0	1197.0	1243.7	1320.0	1416.0	1316.3	1209.3	1179.3	1258.3
Mean	956.9	996.8	1038.1	1082.2	1145.3	1104.0	1002.2	965.4	1036.4

	T	*D*	*T × D*
SEd	0.472	0.505	1.335
CD (0.05)	0.935	0.999	2.645

Table 3.5: Influence of Spentwash, Biocompost and Biosuper on Catalase Enzyme Activity (μmol of H_2O_2 g^{-1} of Soil

Treatments	*Sampling Period (Days)*								
	30	*60*	*90*	*120*	*150*	*180*	*210*	*Postharvest*	*Mean*
T_1	141.0	150.7	174.3	200.3	229.0	210.7	186.0	145.7	179.7
T_2	141.7	152.0	188.7	204.3	251.3	226.0	193.7	151.7	188.7
T_3	147.0	169.7	192.0	217.0	268.0	240.3	200.0	173.7	200.9
T_4	153.3	3174.3	200.0	222.0	288.7	267.0	217.3	197.3	215.0
T_5	63.0	71.3	97.3	123.7	172.3	157.3	111.7	77.3	109.3
T_6	170.0	215.7	238.3	301.0	370.7	337.3	283.3	234.3	268.8
T_7	184.7	227.3	270.3	345.7	495.7	411.3	318.7	260.7	314.3
Mean	143.0	165.9	194.4	230.6	296.5	264.3	215.8	177.2	211.0

	T	*D*	*T × D*
SEd	0.616	0.658	1.742
CD (0.05)	1.220	1.305	3.452

References

Devarajan, L., Rajannan, G., Ramanathan, G. and Oblisami, G., 1993. Sugarcane cultivation with distillery effluent. *SISTA Sugar J.*, 20(3): 23–25.

Garcia, C., Costa, F., Hernandez, T.L. and Masciandaro, G., 1993. Kinetics of phosphatase activity in organic wastes. *Soil Biol. Biochem.*, 25(5): 561–565.

Gopal, H., Kayalvizhi, C., Baskar, M., Sheik Dawood, M., Subash Chandra Bose, M. and Rajukkannu, K., 2001. Effect of distillery effluent on soil properties, microbial population dynamics, yield and quality of sugarcane. In: *National Seminar on Use of Poor Quality Water and Sugar Industrial Effluents in Agriculture*, held at ADAC and RI, Tiruchirapalli, February, 5, p. 75.

Goyal, S., Chander, K. and Kapoor, K.K., 1995. Effect of distillery wastewater application on soil microbial properties and plant, growth. *Environ. and Eco.*, 13: 189–193.

Joshi, H.C., 2001. *Bio-energy Potential of Distillery Effluents*. URL: http://www.undp.org. in/Programme/GEF/june/page10–15.htm.

Mattiazo, M.E. and Ada Gloria, N., 1985. Effect of vinasses on soil activity. *STAB (Portuguese)*, 4(2): 38–40.

Pallab De, K., Arun, K. and De, S.K., 1990. Soil enzymes : Activities and agricultural Importance. *Indian. J. Agric. Chem.*, 22(2–3): 223–224.

Rajannan, G., Devarajan, L., Prabakaran, J. and Oblisami, G., 1998. Assessment of impact on the use of distillery effluent as irrigation water to banana crop. In: *National Seminar on Application of Treated Effluent for Irrigation*, held at REC, Tiruchirapalli, March 23, pp. 24–31.

Raman, S., Pati, R.G. and Zalawadi, N.M., 1999. Pressmud as a potential source of nutrients-management and uses: *Fert. News*, 44(11): 25–31.

Ross, D.J., 1966. A survey of activities of enzymes hydrolyzing sucrose and starch in soil under pasture. *J. Soil Sci.*, 11: 1–15.

Skujins, J., 1976. *Extracellular Enzyme in Soil*. CRC, Cri. Rev. Microbiol., 4: 383–421.

Waksman, S.A. and Fred, E.B., 1922. A tentative outline of the plate method for determining the number of microorganisms in soil. *Soil Sci.*, 14: 27–28.

Chapter 4

Investigation on Sub Surface Water Quality of Tarikere Taluk with Special Reference to Physico-Chemical Characteristics

K. Harish Babu and E.T. Puttaiah
Department of P.G. Studies and Research in Environmental Science, Kuvempu University, Shankaraghatta – 577 451, Shimoga Distt., Karnataka

ABSTRACT

The research study was carried out in Tarikere taluk, Karnataka (India). This article mainly address the physico-chemical concentration of 56 groundwater sample during September 2004. The results of all the findings are discussed in details which reflect the present status of the groundwater quality of the study area.

Keywords: *Groundwater quality analysis, Tarikere taluk.*

Introduction

Rural India relies a mainly on groundwater for drinking and agriculture practices. Villages once relied on sources like wells, lakes, ponds and streams for their water needs. Contamination of most surface water sources has rendered them unfit for consumption. Also increase in water demand by an increasing population has necessitated resource to tapping groundwater. Unsustainable withdrawal of groundwater has led to the spectra of depleting the problem of water scarcity.

Every human society, be it rural or urban, industrially or technologically advanced, disposal of waste exceeds the limit of natural scavenging or removal process, they are bound to effect the normal functioning of the ecosystems and consequently they bear an adverse effect on the water resources (Miller, 1984).

Supplying inhabitants with safe and clean drinking water is one of the most common problems in developing countries like India, especially in arid and semi-arid regions. Hazardous pathogenic germs and anthropogenic substances do not only contaminate the available water quality but also geogenic substances is adversely affect the water supply of many regions.

The groundwater of Tarikere taluk had many threats such as anthropogenic activities, quality deterioration by agricultural activities and over exploitation. Persistence of continuous drought condition during 2001–2004 had declined the amount of groundwater in the region. With this background an effort has been made to know the quality of groundwater in this region.

Methodology

Study area

Tarikere taluk is located between 75°45′00″–75°52′30″ E and 13°30′00″–13°47′30″ N and it has a geographical area of 1,222 sq. kms. It comprises 45 Gram Panchayats and 245 villages with a total population of 2,54,093 as per 2001 census. As a whole, the region is a flat plain land with an average elevation of about 590.58m msl (Gazetteer of India, 1981).

Meteorology

The climate of Tarikere taluk is semi arid and enjoys all the three seasons *viz.*, pre-monsoon (February to May), monsoon (June to September) and post-monsoon (October to January). The monthly mean temperature ranges from 8.9 to 38°C. The relative humidity varies between 50 to 84 per cent and it is highest during the month of August and September and lowest humidity observed during the months of January and February.

Precipitation

The precipitation and distribution of rainfall in Tarikere taluk is highly erratic. The annual average rainfall is 980 mm received over 55 rainy days. It varies from as low as 621 mm in the east and as high as 920 mm in west. About 2/3rd of the geological area of Tarikere taluk receives less than 750 mm of annual rainfall. The study area had been receiving rainfall mainly from southwest monsoon and slightly from northeast monsoon.

Geology

Stratigraphically the study area comes under Bababudan belt of Dharwar super group. Bababudan belt is well known for its iron ore (horse shoe shape). Peninsular gneiss, which forms the basement to the younger green stone belts like Shimoga, Chitradurga and Tarikere (Swaminath and Ramakrishnan, 1981). An important gneissic exposure is near Tarikere that transect the structural trend lines of the schist belts while the gneissic foliation is conformable to the schist belt boundary. The tonalitic gneisses occur in Tarikere taluk (Naqvi, 1983).

Pillow breccias and conglomerates are also seen. Metagabbro with titaniferous magnatite is intersheeted with in the metabasites at north of the Tarikere dome of basement gneisses. Serpentinites and talcose schist form a minor component of the Meta volcanic association in the Bababudan group (Fareeddudin *et al.*, 1988). Geological status of Tarikere taluk were represented in Figure 4.1.

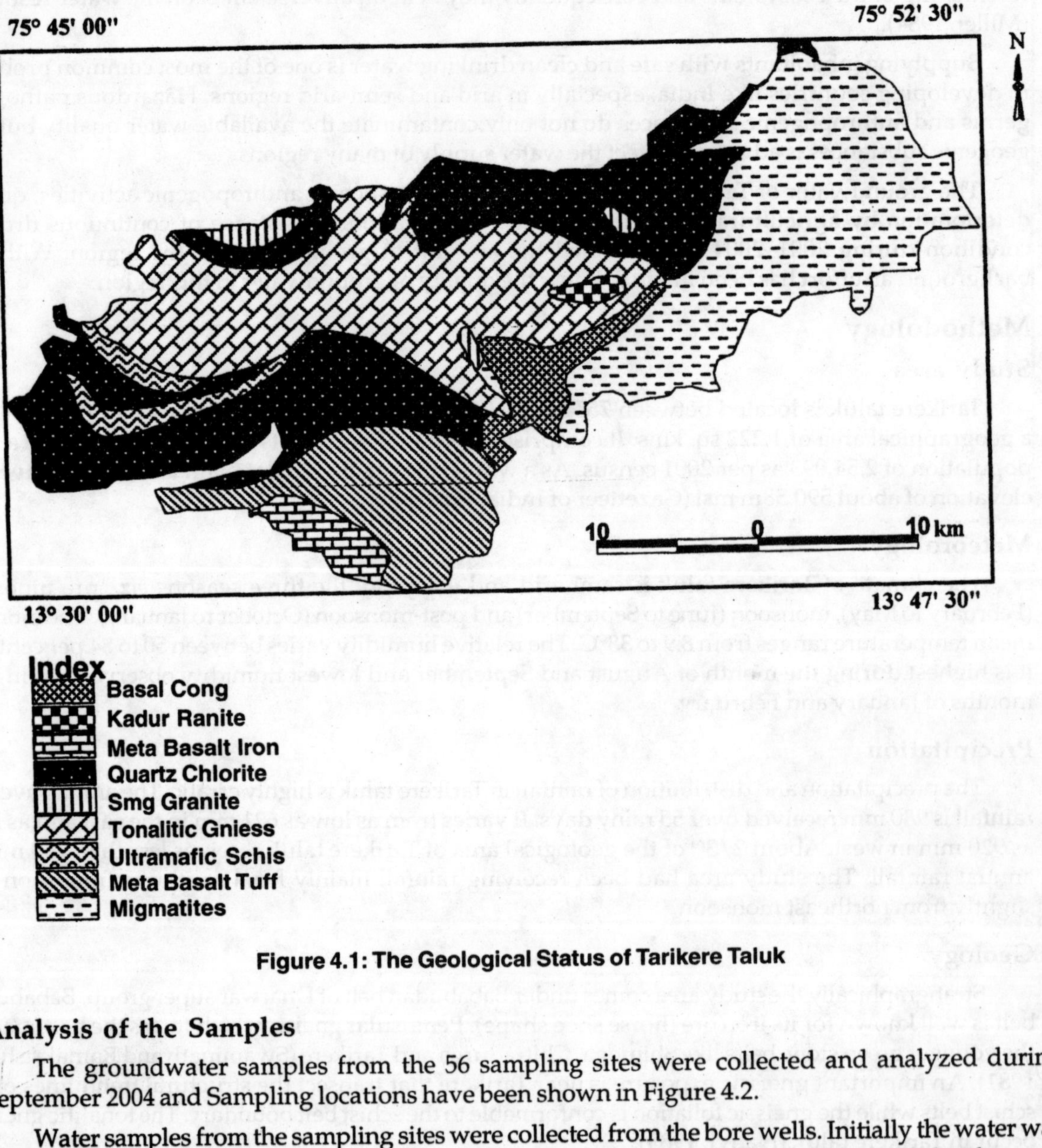

Figure 4.1: The Geological Status of Tarikere Taluk

Analysis of the Samples

The groundwater samples from the 56 sampling sites were collected and analyzed during September 2004 and Sampling locations have been shown in Figure 4.2.

Water samples from the sampling sites were collected from the bore wells. Initially the water was allowed to run for 15 minutes in order to flush out stationary water. Further, the sample bottles were also flushed with water before the samples were collected. The parameters of water such as dissolved oxygen; total dissolved solids, electrical conductivity and pH were analyzed on the spot with the help

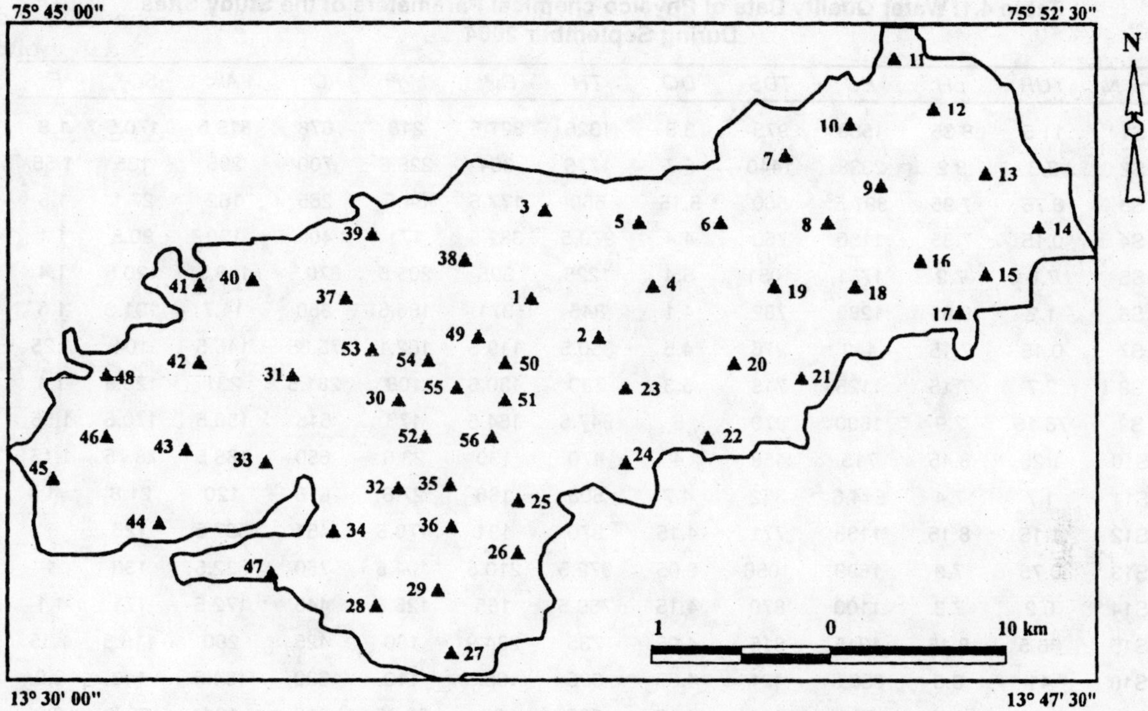

Figure 4.2: Map Showing the Groundwater Locations of the Tarikere Taluk

of water analysis kit (Elico). The remaining parameters were analyzed in the laboratory. Hence, the water was carried to the laboratory in suitable inert bottles. The samples were analyzed using analytical method of APHA (1995) and all analysis was done in triplicate.

Results and Discussion

Physico-chemical Parameters

Water analysis was carried out, by taking 12 parameters, which are very essential to know the water qualities for drinking purpose. The findings of the present investigation are summarized in Table 4.1, and it has been compared with BIS 1998 drinking water standards (Table 4.2) which provides the comprehensive picture of the physico-chemical characteristics of groundwater in the study area.

Turbidity (TUR)

In the present study the turbidity values ranged between 0.15 and 86.1 NTU with a mean value of 9.03. The BIS (1998) acceptable limit for turbidity is 25 NTU. It is proved from the present study, 12.5 per cent of total number of samples cross their permissible limit with reference to the BIS (1998) standards. It is observed from the results that these parameters which crossed their permissible limit are unfit for drinking purposes. It causes health problems like gastro-intestinal disorders, headache and also associated with respiratory diseases (Maiti, 1982; Peter, 1974).

Table 4.1: Water Quality Data of Physico-chemical Parameters of the Study Sites During September 2004

Sam. No.	*TUR*	*pH*	*EC*	*TDS*	*DO*	*TH*	*Ca^{2+}*	*Mg^{2+}*	*Cl^-*	*Alk*	*SO_4^{2-}*	*F^-*
S1	11.5	8.35	1558	975	3.8	1325	327.5	218	678	815.5	170.5	1.8
S2	3.3	8.2	2338	1440	5.7	1775	364	228.5	700	295	135	1.55
S3	6.75	7.95	897.5	560	5.15	560	177.5	34.5	265	162	27.1	1.5
S4	0.15	7.85	1150	750	4.4	970.5	332.5	171	409	190	90.5	1.1
S5	7.6	7.2	1751	1051	5.4	1225	395	205.5	670.5	188.5	120.5	1.4
S6	1.2	6.85	1289	782	4.1	845	371	186.5	350	14.7	101.3	1.6
S7	0.15	7.15	449	276	4.5	350.5	119.5	102.1	75.35	146.5	104	1.25
S8	2.7	7.15	1125	718	5.3	790	130.5	108	281.5	231	125.5	1.4
S9	76.15	7.9	1600	970	5	947.5	164.5	173	515	156.5	170.6	1.55
S10	1.25	8.45	715	458	4	470	130	23.6	650	188.5	78.75	1.15
S11	1.7	7.4	544.5	352	4.7	508	186	27.6	616	120	21.8	1
S12	1.15	8.15	1136	771	4.15	970	191	170.5	451	537.5	126	1
S13	0.75	7.9	1699	1056	6.05	972.5	210.5	104.6	750	532.5	133	1
S14	0.2	7.3	1100	670	4.15	759.5	165	125.5	445	172.5	171	1.1
S15	36.5	8.15	1315	815	4.05	735	234.9	160	425	260	118.5	1.15
S16	41	8.3	633.5	421	4.55	395	165.5	110	330	184.5	143	0.9
S17	69.5	6.7	595	370	4.15	505	68	39.95	565	194	22.7	0.9
S18	2.15	8.7	1125	705	5	800	275	9.45	426.5	615	215.2	0.6
S19	1.3	8.4	955	606	4	695	206	21	163.5	119.5	176	0.7
S20	1.6	7.65	1469	950	3.8	1175	417.5	25.35	650	913	117	0.7
S21	1.4	6.7	101.8	68	4.7	85	26.5	3.5	25.65	116	174.5	0.3
S22	1.85	7.25	210	121	6.04	177	24.5	16.3	33.1	130	115.5	0.55
S23	0.3	7.6	1047	636	4.4	594	124	63	241.5	110	98.5	0.45
S24	0.2	6.8	227	138	5.6	270	45	37.5	110.1	46	98.95	0.6
S25	1.3	7.3	1278	758	4.6	970	220.5	108	553	130	165.1	0.725
S26	10.9	8.05	1373	830	4.8	986	180.5	115.5	367	96	136.3	0.625
S27	0.6	7.3	242	147	5.15	385	103	57.05	346	87	128	0.7
S28	0.25	7.15	805	482	5.35	625	135.5	102	467	110	171.5	0.85
S29	0.65	8.1	1608	995	3.65	960	137	131	547	132	100	0.975
S30	0.15	7.1	603	362	5.2	456	122.5	66.2	183.1	144	140	1
S31	0.9	7.25	466.5	280	3.9	400.5	105.5	32.85	285.5	110	246.5	1.05
S32	2.4	6.75	486.5	286	4.8	460	132	92.35	69.35	168	149.1	0.95
S33	6	6.7	778	467	4.12	623	305.5	128.7	70.2	86	104.1	1.05
S34	1.15	7.75	923	554	4.3	644	47.72	110.5	251	68	231	0.85
S35	3.15	7.25	1037	622	4.4	610	127.5	100.5	206	91.8	130.1	0.95
S36	0.2	7.45	987.5	593	4.63	570	177	91.55	266.5	180	145.1	1.05

Contd...

Table 4.1–Contd...

Sam. No.	*TUR*	*pH*	*EC*	*TDS*	*DO*	*TH*	*Ca^{2+}*	*Mg^{2+}*	*Cl^-*	*Alk*	*SO_4^{2-}*	*F^-*
S37	0.2	8.4	840	520	3.75	695	139	84.45	177	82.5	249	1.1
S38	3.45	7.85	545	315	4.05	1700	395.5	229.5	670	145.5	231.5	0.8
S39	1.6	7.95	360	260	3.95	595	83.9	63.35	200.5	79	132	1.55
S40	0.25	8.25	252.5	150	5.1	1160	189	170	735	138.5	215	1.075
S41	0.15	7.7	324	195	4.5	370	76.55	23.05	124.5	44.95	112.1	0.975
S42	10.55	8.35	322.5	178	3.85	1065	181.5	142	535	765.5	171	0.85
S43	1.55	7	452	263	4.6	1125	299.5	119.5	406	74	151.5	0.6
S44	0.5	8.15	784.5	478	4.15	580	279	11.1	285	66.5	141.5	0.95
S45	0.5	7.1	475.5	294	5.1	283	54.45	35.5	156.5	172	161.2	0.8
S46	11.65	7.05	1193	705	4.9	672	234.5	101.5	93	81	139.5	0.9
S47	0.35	6.85	693	425	4	397	108.1	56.7	118	92.1	265.5	1
S48	10.7	7.5	1550	925	3.55	882	212.5	186.2	86.95	75	262	1.05
S49	0.2	7	957	577	5.6	625	78.85	92.5	142	149	152.8	1
S50	1.65	7.05	789	477	4.4	440	91.35	43	175	110	162	1.1
S51	0.6	6.95	900	553	4.4	800	237	45.1	453	126	210.5	1
S52	0.15	7	520.5	315	5.15	640	156.5	56.05	538.5	101	222.5	1.15
S53	86.1	7.05	784.5	478	3.8	1505	520	46.1	652.5	86.7	172	1.1
S54	69.2	7.55	705	416	4.35	219	36.5	25.5	120.5	89	156.4	1.05
S55	1.25	7.9	815	585	4.1	695	215	9.7	685	255	170.3	1.4
S56	7.05	7.8	1200	749	4	695	125	116.5	470	126	57.3	1.2
Mean	9.029	7.542	894.3	551.7	4.552	727.4	185	92.12	361.8	191.1	150.7	1.012
Min	0.15	6.7	101.8	68	3.55	85	24.5	3.5	25.65	14.7	21.8	0.3
Max	86.1	8.7	2338	1440	6.05	1775	520	229.5	750	913	249	1.8

pH

In the present investigation, pH values found to be 6.7 to 8.7 with a mean value of 7.54. The recommended value of pH for drinking purposes is between 6.5 to 8.5 (BIS, 1998). In the present study all the water samples analyzed are all well with in the safer limits except in few sampling stations. However, higher values of pH hasten the scale formation in water heaters and reduce the germicidal potential of chlorine (Mohapathra and Purohit, 2000).

Electrical Conductivity (EC)

The values of electrical conductivity ranged between 101.8 µmhos/cm to 2338 µmhos/cm. Mean values of EC showed 885.31 µmhos/cm. Higher the concentration of acid, base and salts in water, higher will be the EC. The variability of EC could be explained to the natural concentration of ionised substances present in water (Kataria and Jain, 1995). However the higher values of electrical conductivity (>2000 µmhos/cm), may be due to long residence time and lithology. Ballukraya and Ravi (1999) had proved the variation of the conductivity of the water due to the residential times and the geographical features of the sites.

Table 4.2: Comparison of Groundwater Quality Data with Drinking Water Standard (BIS, 1998)

Sl.No.	Parameters	BIS (1998)		Observed Values	
		P	E	Minimum	Maximum
1.	Turbidity	5	25	0.15	86.1
2.	pH	6.5	9.2	6.7	8.7
3.	EC	–	–	101.8	2338
4.	TDS	500	1000	68	1440
5	Dissolved oxygen	–	–	3.55	6.05
6.	Total hardness	300	600	85	775
7.	Calcium	75	200	24.5	520
8.	Magnesium	30	100	3.5	229.5
9.	Chloride	250	1000	25.65	750
10.	Alkalinity	200	600	14.7	913
11.	Sulfate	200	400	21.8	249
12.	Fluoride	0.6–1.2	1.5	0.3	1.8

Note: P: Permissible limit; E: Excessive limit.

All parameters are expressed in mg/L except pH, Turbidity (NTU) and conductivity (μmohs/cm).

Total Dissolved Solids (TDS)

It is justified from the analytical report TDS values ranges from 68 mg/L to 1440 mg/L with a mean value of 551.53 mg/L. However groundwater chemistry changes as the water flows through the subsurface and the increase in geological environment and dissolved solids and major ions. Chebotarev (1985), Ramababu and Somashekara Rao (1986) and Joseph (2001) expressed the dissolution of soil particles containing minerals under slightly alkaline condition; favour the TDS concentration in groundwater. However TDS concentration above the permissible limit (1500-ppm) causes gastrointestinal irritation (Shankar and Muttukrishnan, 1994).

Dissolved Oxygen (DO)

The present research findings revealed for the values of DO varies from 3.55 mg/L to 6.05 mg/L, and observed mean value is 4.55 mg/L respectively.

In the present study, all the samples showed for DO values were well with in the permissible limits for drinking and domestic purposes. However, the presence of oxygen demanding pollutants (like organic wastes) causes rapid depletion of dissolved oxygen from water. Oxidizable inorganic substances like hydrogen, sulphide, ammonia, nitrites, iron etc., decrease the dissolved oxygen concentration (Sawant *et al.*, 2000). And also due to physical, chemical and biological activities in water the level of DO may vary (Jameel, 2002).

Total Hardness (TH)

Total hardness levels varied from 85 mg/L to 775 mg/L, associated with a mean value of 727.36 mg/L. The BIS (1998) acceptable limit for total hardness is 600 mg/L. In the present study, revealed that 60.7 per cent of total number of water samples crosses the permissible limits of BIS (1998) drinking water standards. Owing to fact that higher amount of hardness in the study area comes mainly from the leaching of igneous rock and carbonate rocks (dolomite, calcite and limestone). Water containing the soluble salts of calcium and magnesium such as chlorides, sulphates and bicarbonates are also governs the quality of water (Ramaswamy and Rajaguru, 1991). The adverse effects of total hardness are formation of kidney stone and the heart diseases (Sastry and Rathee, 1998). Nevertheless, groundwater chemistry is controlled by the composition of its recharge components as well as by geological and hydrological variations (Narayana and Suresh, 1989).

Calcium (Ca^{2+})

Present investigation reports stated for calcium values ranged from 24.5 mg/L to 520 mg/L, with a mean value of 184.99 mg/L. BIS (1998) acceptable limit for calcium is 200 mg/L. However, in the present study 30.2 per cent of water samples crosses the permissible limit. Presence of higher amount of calcium in the study area may be due to groundwater receives the calcium minerals leached from the rocks and other deposits like limestone, dolomites, calcite, gypsum, amphiboles, feldspar, and clay minerals leaching or weathering of igneous rocks. Sewage and domestic waste are also important sources of calcium (Mishra and Saxena, 1989).

Magnesium (Mg^{2+})

Investigation report reveals for the magnesium values ranged from 3.5 mg/L to a 229.5 mg/L with a mean value of 92.11 mg/L.

BIS (1998) acceptable limit for magnesium is 100 mg/L and in the present study 48.2 per cent of the water samples crossed the permissible range. Magnesium arises principally from the weathering of rocks contain ferro-magnesium minerals and some carbonate rocks. High concentration of magnesium proves to be diuretic and laxative (Schroeder 1960).

Chloride (Cl^{-})

Chloride is also one of the important parameter to know the quality of water. Anthropogenic sources of chlorides include fertilizer, road salt, human and animal waste. Concentration of chlorides is considered to be an indicator of organic pollution of animal origin (Kumara, 2002). Here Chloride values ranged from 25.65 mg/L to 750 mg/L, however the mean value observed in the present study is 361.82 mg/L.

The BIS (1998) acceptable limit for chloride is 1000 mg/L. In the present investigation, the values of chloride for all the sampling sites are with in the permissible range as prescribed by BIS (1998) drinking water standards. However, dissolving of the soil constituents had contributed the chloride into the groundwater and also the soil characteristics play an important role in contributing the chloride content in the groundwater (Shivasankaran, 1997).

Total Alkalinity

In the present study total alkalinity values ranged from a 14.7 mg/L to 913 mg/L and a mean value of 189.31 mg/L. The BIS (1998) acceptable limit for total alkalinity is 600mg/L. In the present study, the data revealed that 10.7 per cent of water samples in the study area crossed the permissible limit. When alkalinity of water exceeds the permissible limits, it is likely to produce incrustation sediment deposits, difficulties in chlorination, certain physiological effects on human systems etc. (Raviprakash and Rao, 1989). The constituents of alkalinity result from dissolution of mineral substances in the soil and atmosphere contributes to alkalinity in groundwater (Mittal and Verma, 1997).

Sulphate (SO_4^-)

It is very interesting to note that, sulphate values ranged from 21.8 mg/L to 249 mg/L with a mean value of 150.2 mg/L. The BIS (1998) permissible range for sulphate is 400 mg/L. In the present investigation, the sulphate values for all the samples are within the prescribed limit of BIS (1998) drinking water standards. Sulphate in groundwater takes place the break down of organic substances in the soil. However, geological, hydrological and geomorphologic characteristics showed a remarkable variation in sulphur content (Alexander, 1961).

Fluoride (F^-)

In the present investigation, fluoride values varied from 0.3 mg/L to 1.8 mg/L with a mean value of 1.01 mg/L. The BIS (1998) acceptable limit for fluoride is 1.5 mg/L. In present study 7.14 per cent of total number of water samples in present investigations have crossed the permissible range as prescribed by BIS (1998) drinking water standards.

Degree of weathering and leachable fluoride in terrain is of great significance for the fluoride present in groundwater than the mere presence of fluoride bearing minerals in rocks (Kumar *et al.*, 2000).

Intake of excess fluoride causes dental, skeletal and non-skeletal fluorosis. The non-skeletal fluorosis can be observed such as gastrointestinal complaints, intermittent diarrhoea and flatulence in expectant and lactating mothers hard working young adults and children. Therefore, fluorosis has been considered as one of the incurable diseases. Hence, prevention is the only solution for the disease (Hem, 1985).

Conclusion

Currently carried research investigation should give more precise answer on influence of Geomorphological condition than anthropogenic activities in the examined groundwater samples of the study area. Local geological settings may supports the increasing concentration of physico-chemical characteristics in groundwater. The factors like slow circulation, longer period of contact between aquifer and water, dissolving of minerals at the time of weathering, residential time, drainage pattern and surface water link. Porosity of the soil and rock also alters the characteristics of the groundwater.

The high level contents of the parameters observed may be minimized if the groundwater is recharged with the available water in the rainy season. This not only dilutes the constituents of the groundwater but also raises the groundwater level that depletes due to large-scale exploitation.

Groundwater is extremely important to the future economy and growth of rural India. If the resource is to remain available as high quality water for future generation it is important to protect from possible contamination. Hence it is recommended that suitable water quality management is essential to avoid any further contamination.

References

Alexander. 1961. *Introduction to Soil Microbiology*. Wiley, New York, London, p. 171–172.

APHA. 1995. *Standard Methods for the Examination of Water and Wastewater*; 18th edition, AWWA, WPCF, New York, pp. 1120.

Ballukraya, P.N. and R. Ravi. 1999. Characterization of Groundwater the unconfined aquifers of Chennai city. *Indian Journal of Geological Society of India*, 54: 1–11.

BIS. 1998. *Specifications for Drinking Water*, New Delhi, p. 171–178.

Chebotarev. 1985. Metamorphism of natural waters in the crust of weathering. *Geochem. Cosmochim. Acta*, 8: 22–28.

Fareeddudin, A.S. Janardhan and B. Basavalingu. 1988. Sedimentology, minerology and geochemistry of the Kalasapura conglomerate. *Geol. Soc. India. Mem.*, 9: 65– 82.

Gazetteer of India, Karnataka State, Chikmagalure district. 1981. p. 8–15 and 630–633.

Hem, J.D. 1985. *Study and Interpretation of the Chemical Characteristics of Natural Water*, 3rd ed. U.S. Geological Survey, Water Supply Paper, 225: 263.

Jameel, A. 2002. Physico-chemical studies in Vyakondan channel water of river Cauvery. *Pollution Research*, 17(2): 111–114.

Joseph, W. 2001. Groundwater chemistry in the valley De Yabucoa alluvial aquifer, South-eastern Pnerto Rico. *AWRA 3rd International Symposium on Tropical Hydrology*, San Juan U.S.A.

Kataria, H.C. and O.P. Jain. 1995. Physico-chemical analysis of river Ajhar. *Indian Journal of Environmental Protection*, 5: 569–571.

Kumar, P. Kumari and L.K.R. Singh. 2000. *Environmental Biology*. P.G. Dept. of Zoology, S.K. University, Dumka, Jharkand, p. 180–182.

Kumar. 2002. *Ecology of Polluted Waters*. 1: 144–180.

Maiti, T.C. 1982. *Science Reporter*, 19: 360–361.

Miller, D.R. 1984. Chemicals in the environment. In: *Effects of Pollutant at the Ecosystem Level*. John Wiley and Sons, Chinchester, pp. 7.

Mishra, S.R. and Saxena. 1989. Industrial effluent pollution at Birla Nagar, Gwalior. *Pollution Research*, 8(2): 77–86.

Mittal, S.K and N. Verma. 1997. Critical analysis of groundwater quality parameters. *Indian Journal of Environmental Protection*, 17: 426–429.

Mohapatra, T.K. and K.M. Purohit. 2000. Qualitative aspects of surface and groundwater for drinking purpose in Paradeep area. *Ecology of Polluted Waters*, 1: 144.

Naqvi, S.M. 1983. Geochemistry of gneisses from Hassan district and adjoining areas, Karnataka, India. In: *Precambrian of South India*, (Eds.) Naqvi, S.M. and J.J.W. Rogers. Geological Society of India Mem., 4: 401–416.

Narayana, A.C. and G.C. Suresh. 1989. Chemical quality of groundwater of Mangalore city, Karnataka. *Indian Journai of Environmental Health*, 31: 228–236.

Ramaswamy, V. and P. Rajaguru. 1991. *Indian Journal of Environmental Health*, 33(2): 187–191.

Rambabu, C. and K. Somashekara Rao. 1986. Studies on the quality of bore well water by Nuzuid. *Indian Journal of Environmental Protection*, 16(7).

Raviprakash, S. and K.G. Rao. 1989. The chemistry of groundwater, Parvada area with regard to their suitability for domestic and irrigation purposes. *Indian Journal of Geochem.*, 4: 39–54.

Sastry, K.V. and P. Rathee. 1988. Physico-chemical and microbiological characteristics of water of village Kanneli, (Distt. Rohtak) Haryana. *Proc. Academic, Environmental Biology*, 7(1): 103–108.

Sawant, C.P., Saxena, G.C. and Shrivastava, V.S. 2000. Trace Metals in and around the Industrial belt. *Ecology Environment and Conservation*, 6(6): 135–137.

Schroeder, H.A. 1960. Relation between hardness of water and death rates from certain chronic and degeneration diseases in the US. *J. Chron. Dis.*, 12: 568–573.

Shankar and Muttukrishan. 1994. *In situ* bioremediation of contaminated groundwater. *Proceedings of National Seminar on EPCR–04*, UBDT Engineering College, Davangere, pp. 32.

Shivashankaran, M.A. 1997. Hydrogeochemical assessment and current status of pollutants in groundwater of Pondicherry region, South India. *Ph.D. Thesis*, Anna University, Chennai, p. 80–87.

Swaminath, J. and M. Ramkrishnan. 1981. In: *Early Precambrian Supracrustals of South India*, (Eds.) Swaminath, J. and M. Ramakrishnan. Geol. Surv. Ind. Mem., 3: 23–38.

Chapter 5

Growth Analysis, Nutrient Uptake (kg ha^{-1}) and Nitrogen Use Efficiency of Hybrid Rice CoRH2 as Influenced by Crop Establishment Methods and Split Applications of N and K

R. Gobi, S. Ramesh, B.J. Pandian and B. Chandrasekaran
Department of Agronomy, Tamil Nadu Agricultural University, Coimbatore – 641 003, Tamil Nadu, India

ABSTRACT

Field experiment was conducted at Agricultural College and Research Institute, Killikulam during late *Pishanam* season (November–April) to study the effect of crop density, establishment methods and split application of N and K on the growth rate, nutrient uptake and Nitrogen use efficiency of hybrid rice CoRH2. The results revealed that crop growth rate was significantly superior in seedling broadcasting with 40 hills m^{-2} (186.94 gm^{-2}d^{-1}) during tillering to flowering stages whereas line planting with 50 hills m^{-2} registered significantly higher crop growth rate during flowering to harvest. At all the crop growth stages, five splits of N and four splits of K produced significantly higher crop growth rate (CGR). Treatment combinations did not significantly influence the RGR at all the crop growth stages. Net assimilation rate was maximum in seedling broadcasting with 50 hills m^{-2} and four splits of N and three splits of K application (S_2). Line planting with 50 hills m^{-2} (P_2) registered significantly higher N uptake at tillering stage whereas broadcasting with 40 hills m^{-2} (P_3) recorded higher N uptake during flowering and harvest stages respectively. Seedling broadcasting with 40 hills m^{-2} (P_3) registered significantly higher P and K uptake. Three splits of N and two splits of K exhibited higher N and K uptake at tillering stages whereas five splits of N and four splits of K (S_3) recorded higher N and K uptake at flowering and harvest stages. Application of N as five splits and K as four splits (S_3) recorded higher P uptake which was on par with four splits of N (S_2) and three splits of K. There is significant interaction between the crop establishment method and split application of N, P and K uptake at all crop growth stages. Higher N use efficiency was observed with treatment combination of seedling broadcasting with 40 hills m^{-2} (P_3) and five split application of N and four splits of K (P_3S_3).

Keywords: *Growth rate, Nutrient uptake, Nitrogen use efficiency, Hybrid rice, Establishment method.*

Introduction

Rice is cultivated in more than 90 countries spread over six continents, more than 92 per cent of the rice is produced and consumed in Asian countries alone. In India rice is cultivated in an area of 42.2 million ha with a production of 110.9 million tones of rice, while china produces 187.45 million tones of rice in 33.1 million ha. The productivity of rice in India and China are 2.69 and 5.73 tha^{-1} respectively. To sustain the self-sufficiency in the coming decades the rice production need to be geared up every year by almost two million tones. This is a daunting task, in view of plateauing trend observed in yield potential of high yielding varieties, and decreasing and declining natural resource base. Among various possible biological approaches to break the yield barriers, hybrid rice technology is the most promising and gaining impetus in rice growing countries of the world. The increase in rice productivity of China is mainly due to cultivation of hybrids (Mohamed Ali, 2000). The average yield of hybrid rice is 6 to 7 tonnes ha^{-1} with a yield advantage of 30 per cent over conventional varieties (Pandian *et al.*, 2001). Growing hybrid rice is a complex process and agronomic management especially optimum plant density and fertilizer management are considered as challenges in hybrid rice cultivation. Method of stand establishment is a very vital cultural practices which influence the hybrid rice performance through its effect on growth and development. Agronomic technologies have to be standardized for achieving higher productivity for hybrid rice.

Adequate nitrogen supply is crucial in realizing the full potential of hybrid rice. Phosphorus is mobile within the plant and promote tillering, root development, early flowering and ripening. Potassium as a major nutrient, accounts for greater share of total nutrient removal by rice crop and it enhances the plants resistance against lodging. Split application of N and K also play vital role in enhancing their use efficiency. Delayed application of N and K, coinciding with flowering could help to realize the potential yield of hybrids. In this context, the present study was, undertaken with the objective of studying the growth rate, nutrient uptake and nitrogen use efficiency of CoRH2 hybrid as influenced by crop density and split application of N and K under the Tambiraparani command area.

Materials and Methods

Field experiment was conducted at Agricultural college and Research Institute Killikulam, during late *Pishanam* season (November–April) to study the effect of establishment methods and split application of N and K on the plant population, flowering and post harvest fertility status of hybrid rice CoRH2. The soil of the experimental field was sandy clay loam with a pH of 6.8. Regarding fertility status, the soil was classified as low in available nitrogen, high in available Phosphorus and medium in available potassium content. The detailed physico-chemical characteristics are given in Table 5.1. The hybrid CoRH2 is a medium duration hybrid rice (125 days) developed by three line breeding system using cytoplasmic male sterile line, maintainer line and restorer line. The gain is medium slender and white. The experiment was laid out in factorial randomized block design and replicated thrice.

Table 5.1: Physico-chemical Properties of the Experimental Field

Particulars	*Unit*	*Value*
Available N	kg ha^{-1}	173.00
Available P_2O_5	kg ha^{-1}	27.90
Available K_2O	kg ha^{-1}	214.50
Organic carbon	%	0.48
pH (1 : 2 soil water suspension)		6.80
EC (1 : 2 soil water suspension)	Dsm^{-1}	0.21

The details of treatments are given below.

Establishment Methods

P_1: Line planting at 25 × 10 cm (40 hill m^{-2})

P_2: Line planting at 20 × 10 cm (50 hill m^{-2})

P_3: Seedling broadcasting with 40 hills m^{-2}

P_4: Seedling broadcasting with 50 hills m^{-2}

Split Application of N and K

S_1: Three splits of N at 50 per cent B, 25 per cent AT, 25 per cent PI and two splits of K at 50 per cent B, 50 per cent PI

S_2: Four splits of N at 50 per cent B, 16.5 per cent AT, 16.5 per cent PI, 16.5 per cent H and three splits of K at 50 per cent B, 25 per cent PI, 25 per cent F

S_3: Five splits of N at 50 per cent B, 12.5 per cent AT, 12.5 per cent PI, 12.5 per cent H, 12.5 per cent F and four splits of K at 50 per cent B, 25 per cent PI, 12.5 per cent H, 12.5 per cent F

(B-basal, AT-active tillering, PI-panicle initiation, H-heading and F-flowering)

Under line planting seedlings of 25 days old were planted with single seedling hill^{-1} by adopting spacing as per the treatments. Seedlings were broadcasted as per the treatments over the field by maintaining thin film of water. The recommended dose of 200 : 50 : 50 kg of NPK ha^{-1} was applied to CoRH2 rice hybrid in the form of urea, superphosphate and potash respectively. The entire dose of Super phosphate was applied as basal whereas potash and nitrogenous fertilizers as muriate of potash and urea respectively were applied in different splits as per the treatments.

Crop Growth Rate (CGR)

The crop growth rate was estimated by using the formula suggested by Buttery (1970) and expressed in gm^{-2} d^{-1},

$$CGR = W_2 - W_1 / (t_2 - t_1)$$

Relative Growth Rate (RGR)

The relative growth rate was calculated by using the formula suggested by Williams (1946) and expressed in mg $g^{-1}d^{-1}$,

$$RGR = \text{Log } ew_2 - \text{Log } ew_1/(t_2 - t_1)$$

Net Assimilation Rate

The method proposed by Williams (1946) was employed for calculating the NAR and expressed in mg cm^{-2} day^{-1},

$$NAR = W_2 - W_1/t_2 - t_1 \times \ln L_2 - \ln L_1/(L_2 - L_1)$$

where,

W_1 and W_2 are whole plant dry weight at t_1 and t_2 stages respectively

L_1 and L_2 are leaf areas at t_1 and t_2 respectively, $(t_2 - t_1)$ is the time intervals in days.

Nutrient Uptake

Plant samples collected for recording dry matter production at different stages were ground into fine powder in Willey mill and dried samples were anlaysed for N, P and K content using standard procedure and the uptake of nutrients was calculated by multiplying the drymatter with respective nutrient content and expressed in kg ha^{-1}.

Nitrogen Use Efficiency (NUE)

The NUE was calculated by using the formula as given below.

$$NUE = \text{Grain yield (kg ha}^{-1}) / \text{Applied nitrogen (kg ha}^{-1})$$

Results and Discussion

Crop Growth Rate (CGR)

A significant difference existed in CGR, RGR and NAR of CoRH2 rice hybrid grown under different stand establishment methods. The results revealed that during tillering to flowering, CGR significantly varied from 171.67 to 190.55 $gm^{-2}d^{-1}$ and it declined from flowering to harvesting. Among the stand establishment method, CGR was maximum in seedling broadcasting with 40 hills m^{-2} (186.94 g $m^{-2}d^{-1}$), minimum in line planting of 50 hills m^{-2} (174.19 g $m^{-2}d^{-1}$). With regard to split application of N and K, five splits of N and four splits of K produced higher CGR of 184.03 g $m^{-2}d^{-1}$. The treatment combinations P_3S_3 produced significantly higher CGR of 190.55 g $m^{-2}d^{-1}$. Lowest CGR was registered in P_2S_1 (171.67 g $m^{-2}d^{-1}$) (Table 5.2). From flowering to harvesting line planting with 50 hills m^{-2} significantly produced significantly higher CGR of 66.39 g $m^{-3}d^{-1}$ as compared to other treatments. With regard to split application of N and K, five splits of N and four splits (S_3) recorded higher CGR of 64.82 g $m^{-2}d^{-1}$ followed by three splits of N and two splits of K (S_1) 63.77 g $m^{-2}d^{-1}$. Similar findings were observed by Biswas (1976) and Senthilvel and Palaniappan (1985). The increased CGR in seedling broadcasting might be due to zero depth of planting resulting in better inducement of root growth for anchorage. This results corroborates the findings of Matsushima (1979) and Sanbagavalli *et al.* (1999).

Relative Growth Rate (RGR)

During tillering to flowering and flowering to maturity, treatment combinations did not influence the RGR.

Table 5.2: Growth Rate of CoRH2 as Influenced by Crop Stand Establishment Methods and Split Application of N and K

Treatments	Crop Growth Rate ($gm^{-2}d^{-1}$)		Relative Growth Rate ($gm^{-2}d^{-1}$)		Net Assimilation Rate ($mgcm^{-2}d^{-1}$)	
	Tillering to Flowering	Flowering to Harvest	Tillering to Flowering	Flowering to Harvest	Tillering to Flowering	Flowering to Harvest
A. Establishment						
P_1	184.56	61.97	0.012	0.001	14.11	4.73
P_2	174.19	66.39	0.011	0.002	17.78	6.61
P_3	186.94	62.86	0.012	0.001	12.45	4.74
P_4	179.1	61.92	0.012	0.001	15.36	5.86
SEd	0.4045	0.4097	–	–	0.2492	0.8711
CD (0.05)	0.8389	0.8497	–	–	0.5169	1.8066
B. Split application						
S_1	176.94	63.77	0.011	0.002	15.11	5.57
S_2	182.62	61.26	0.012	0.001	15.23	5.44
S_3	184.03	64.82	0.012	0.002	14.44	5.49
SEd	0.3503	0.3548	–	–	0.2158	0.7544
CD (0.05)	0.7265	0.7358	–	–	0.4476	1.5646
PS						
SEd	0.7006	0.7096			0.4317	1.5088
CD (0.05)	1.4530	0.4717			0.8953	3.1292

Net Assimilation Rate (NAR)

During tillering to flowering stage, the NAR significantly varied from 11.75 to 18.39 mg $cm^{-2}d^{-1}$. Among the stand establishment methods, NAR was significantly higher (15.36 mg $cm^{-2}d^{-1}$) in seedling broadcasting with 50 hills m^{-2} (P_4), minimum (12.45 mg $cm^{-2}d^{-1}$) in seedling broadcasting with 40 hills m^{-2} (P_3) (Table 5.2). With regard to split application of N and K four splits of N and three splits of K (S_2) produced maximum NAR of 15.23 mg $cm^{-2}d^{-1}$. The treatment combination P_2S_1 produced maximum NAR of 18.39 mg $cm^{-2}d^{-1}$ followed by P_2S_2 18.26 mg $cm^{-2}d^{-1}$. NAR was not significantly influenced by the treatments and its combinations.

Nutrient Uptake

Nitrogen uptake was significantly influenced by stand establishment methods and split application of N and K. Line planting with 50 hills m^{-2} (P_2) registered the higher N uptake of 35.06 kg ha^{-1} at tillering stage. Seedling broadcasting with 40 hills m^{-2} (P_3) recorded significantly higher N uptake of 92.67 and 101.08 kg ha^{-1} at flowering and harvesting stages respectively. Significantly lower N uptake was observed in seedling broadcasting with 50 hills m^{-2} (P_4) at tillering, flowering and harvesting stages (Table 5.3). Three splits of N and two splits of K (S_1) registered higher N uptake of 35.25 kg ha^{-1} at tillering stage. Five splits of N and four splits of K (S_3) recorded significantly higher N uptake at flowering and harvest stages. Significant interaction effect was noticed between treatments for N uptake at all stages of crop growth, where P_3S_3 recorded higher N uptake of 92.67 and 101.93 kg ha^{-1} at flowering and harvesting stages respectively. Seedling broadcasting with 40 hills m^{-2} (P_3) registered significantly higher P and K uptake at tillering, flowering and harvest stages. Five splits of N and four splits of K (S_3) recorded higher P uptake of 25.89 kg ha^{-1} which was on par with (S_2) four splits of N and three splits of K (25.79 kg ha^{-1}). Application of N as three splits and two splits of K (S_1) registered higher K uptake at tillering stage whereas five splits of N and four splits of K (S_3) recorded significantly higher K uptake of 75.17 and 94.64 kg ha^{-1} respectively. Significant interaction was observed between treatments for N and K uptake at all the stages of crop growth. Increased uptake of N, P and K under five splits of N and foul splits of K at all the stages were due to increased amount of dry matter accumulation. Combined application of N and K at critical stages reduced the volatilization loss of ammonia and fixation of ions in the exchange complex and thus increased the N and K recovery by rice right from tillering to harvest stages. At the time of transplanting, seedlings were not able to absorb nutrients due to transplantation shock. The results are in accordance with Vaiyapuri *et al.* (1998), Dheebakaran and Ramasamy (1999) and Devasenamma *et al.* (1999).

Table 5.3: Effect of Crop Stand Establishment Methods and Split Application of N and K on Nutrient Uptake (kg ha^{-1})

Treatments	*N Uptake*			*P Uptake*	*K Uptake*		
	Tillering	*Flowering*	*Harvest*	*Harvest*	*Tillering*	*Flowering*	*Harvest*
A. Establishment							
P_1	34.84	92.05	100.37	26.23	30.92	75.67	95.61
P_2	35.06	86.98	94.81	24.53	31.23	71.80	91.15
P_3	34.31	92.67	101.08	26.49	31.39	76.45	95.99
P_4	34.08	89.68	98.08	25.44	30.95	74.26	94.16
SEd	0.1266	0.1472	0.1090	0.1166	0.0982	0.1024	0.1126
CD (0.05)	0.2626	0.3053	0.2262	0.2419	0.2024	0.2125	0.2335
B. Split application							
S_1	35.25	89.58	97.67	25.34	31.97	73.78	93.50
S_2	34.20	90.38	98.67	25.79	30.78	74.68	94.53
S_3	34.27	91.07	99.42	25.89	30.62	75.17	94.64
SEd	0.1096	0.1274	0.0944	0.1010	0.0850	0.0887	0.0975
CD (0.05)	0.2274	0.2644	0.1959	0.2095	0.1764	0.1840	0.2022
PS							
SEd	0.2193	0.1274	0.1889	0.2021	0.1701	0.1774	0.1950
CD (0.05)	0.4548	0.2644	0.3918	0.4191	0.3528	0.3680	0.4044

Nitrogen Use Efficiency

Seedling broadcasting with 40 hills m^{-2} (P_3) registered higher N use efficiency 36.03 kg ha^{-1} kg$^{-}{}_{1}$ N. Lower N use efficiency (33.15 kg ha^{-1} kg^{-1} N) registered in line planting with 50 hills m^{-1} (P_2). Application of N as five splits and four splits of K (P_3) recorded higher N use efficiency of 34.83 kg ha^{-1} kg^{-1} N. Lower N use efficiency 34.07 kg ha^{-1} kg^{-1} N was registered in three splits of N and two splits of K (S_1). Among the treatment combinations the higher N use efficiency was recorded in P_3S_3 (36.58 kg ha^{-1} kg^{-1} N) and lower N use efficiency was recorded in P_2S_1 (32.91 kg ha^{-1} kg^{-1} N) (Table 5.4).

Table 5.4: Effect of Stand Establishment Methods and Split Application of N and K on Nitrogen Use Efficiency (kg ha^{-1} kg^{-1} of N) of CoRH2 Rice Hybrid

Treatments	S_1	S_2	S_3	*Mean*
P_1	34.60	34.93	34.96	34.83
P_2	32.91	33.20	33.35	33.15
P_3	35.33	36.18	36.58	36.03
P_4	33.46	33.91	34.45	33.94
Mean	34.07	34.55	34.83	

* Data not statistically analysed

References

Biswas, C.R., 1976. Effect of split application of potash on the efficiency of nitrogen utilization by paddy in saline soils in West Bengal. *Indian Potash J.*, 1(2): 33–36.

Devasenamma, V., Reddeppa Reddy, M. and Soundarajan, M.S., 1999. Effect of varying levels of nitrogen on growth and nitrogen uptake of rice hybrids. *The Andhra Agric. J.*, 46(1 and 2): 124–125.

Dheebakaran, G. and Ramasamy, S., 1999. Source and times of splits of nitrogen on N uptake and its efficiency on irrigated lowland rice. *Madras Agric. J.*, 86(7–9): 421–431.

Matsuhima, S., 1979. Rice cultivation for the million: Diagnosis of rice cultivation and technique of yield increase. University of Tokyo Press, Tokyo, pp. 267.

Mohamed Ali, A., 2000. Current status of hybrid rice research and development in India. In: *Training on Hybrid Rice Seed Production Technology-Rabi* 1999–2000 held at Agril. College and Research Institute, Killikulam, March 9–10, pp. 1–3.

Pandian, B.J., Arumugaperumal, A. and S. Avudaithai, 2001. Medium duration hybrid rice. *The Hindu*, dated 15-03-01.

Sanbagavalli, S., Kandasamy, O.S. and Singaram, P., 1999. Seedling throwing method: A new concept for lowland rice production technology. *Madras Agric. J.*, 86(4–6): 191–195.

Vaiyapuri, V., Ravichandran, M. and Srirama Chandrasekaran, M.V., 1998. Effect of graded levels of *Sesbania speciosa* with time of N application on growth and yield in rice-rice *sequence. Ann. Agri. Res.*, 19(1): 1–3.

Williams, R.S., 1946. The physiology of plant growth with special reference to the concept of NAR. *Ann. Bot.*, 10: 41–74.

Chapter 6

Studies on Water Pollution at Gorakhpur: A Review

A.K. Srivastava[1] and Gopal Chandra[2]
[1]Department of Botany, M.G. Post Graduate College Gorakhpur – 273 001
[2]Department of Botany, S.H.K.P.G. College, Basti (U.P.)

Introduction

Land is prime natural resource and water resources are finite, which need to be conserved, developed and managed (Delhi declaration, 2000). *Water,* the most abundant and wonderful natural resources is extremely essential for survival of all living organism, but today clean water has become a precious commodity and its quantity is threatened by many sources of pollution. Thus freshwater is a precious which needs to be protected by all means by men. But unfortunately it has not been the case. The World Bank estimate that water pollution in India results in a very high health toll, with economic costs to the tune of Rs. 20,000 crores annually (Sharma, 1999).

Water pollution may be defined as *"any impairment of the suitability of water for any or its beneficial uses actual or potential, by man caused changes in the quality of the water."*

Basically, there are three main sources of water pollution–*domestic, agricultural* and *industrial*. The pollutants contributed to the aquatic environment by these sources are diverse. Rao (1983) categorized them as (1) disease causing organisms (via human wastes) (2) synthetic organic compounds (household and industrial chemicals, pesticides) (3) Inorganic compounds (acids, *fibers* and *heavy metal* mostly from industrial effluents. (4) Radio active substances from nuclear power plants. (5) Oxygen demanding wastes (Sewage and industrial effluents (6) Plants nutrients through sewage and agricultural runoff. (7) Sediments (through soil erosion) and (8) *Thermal discharge* (from power plants). Details of water borne diseases (*bacterial, viral* and *protozoa* are also available elsewhere.

Central and state boards for the prevention and control of water pollution observed that the major sources of pollution of our natural water sources including coastal waters are due to discharge of community wastes from human settlements. The community and industrial wastewaters go straight into the water sources making them unfit for drinking, bathing and agricultural purposes. Only 8 cities in India are provided with complete sewage treatment facilities *i.e.* Ahmedabad, Bangalore, Bizapur, Sangli, Nanded, Nashik, Thane and Durgapur. Referring to water quantity requirement Rao (1983) pointed out that parameters which are generally altered by pollution and which one wishes to specify quantitative standards may be–physico-chemical such as temperature, pH, turbidity, dissolved oxygen and inorganic nutrients or biological such as counts of potentially pathogenic bacterial population, densities of sensitive species, and species diversity of specified community. The presence of Coliform bacteia is regarded as a positive indicator of sewage pollution, and their number as direct–measure of the magnitude.

As per report of Dr. C.D. Thattey, Chairman, International Commission for Irrigation and Sewage only 30–40 per cent of sewage water discharges after treatment in rivers due the drinking of these polluted water the diseases spread are–Hepatitis–A-B, Cholera, Typhoid, Polio, Dysentry and Amoebic dysentry. Other water borne diseases are–Malaria, Filaria, Dengu, Meru and Hepatitis. According to recent study in August 2003 in the Laboratory of Centre for Science and Environment (C.S.E.) New Delhi. The branded Cold drink and mineral water contain poisonous pesticides. It is 36 times high than the standard of European countries and contain harmful pesticides like Linden, DDT, Melthian, Chloropyriphos. These pesticides may cause Cancer and weakened bones in human being. (Naryan, 2003).

Pollution Due Stagnant Water

The complete range of tropical communicable diseases such as Malaria, filariasis, dengue, guinea worms, Japanese encephalitis, etc. are becoming more widespread. The main reason for this condition for the causal organism and their vectors. The breeding places of these diseases vectors are stagnant water bodies such as reservoirs, irrigation channels, pond, burrow pits and ditches. Thus it is necessary to eliminate these water bodies and such conditions which are favourable for the breeding of vectors of human and animal diseases (Sangal 1983). It is very well recognised that unplanned urbanisation has changed the ecological conditions which now favour the spread of filariasis. Kalazar has also now come back in an unprecedential way.

Besides it has been found that creation of large water reservoirs can result in the elevation of subsoil water in the adjacent areas, and consequent changes in the levels of fluoride, calcium and trace metals etc. in soil sediments. This in turn can give rise to diseases such as flurosis in people who use the contaminated water. Of course the National Institute of Nutrition in Hyderabad has observed the seriousness of flurosis in neighbouring areas of the Nagarjun Sagar Dam.

The other diseases transmitted by water include skin infections, trachoma, guinea worms and schistosomiasis. The cost for the lack of recognition and control of these water related diseases, is paid not only in terms of human and animal health but also in turns of price of medical care and pest control (Sangal, 1983) besides the polluted and contaminated water due to biomedical wastes also causes cancer. Since bones, urinary bladder, kidney, liver and brain can easily affected by contaminated water and develop cancer diseases (Jaiswal, 2001).

Water is the only life giving components of "*Panchtatva*" (Sky, Water, Fire, Clouds and Air). According to great philosopher Goethe "Every life is generated and supported by water". Most of the water present on the earth is in the form of Salt water (Sea), it is about 97 per cent. The remaining 3 per cent are in the form of ice, underground water and water vapours. Only 0.6 per cent of water is available to us in the form of surface water, like rivers and lakes. Our body too has 2/3 part as water and the human blood contains 80 per cent water. Due to increasing population, industrization, urbanisation, the water resources becoming highly polluted. The World Health Organisation (WHO) says that out of total death 70 per cent is caused by the use of polluted water. As per record of National Environmental Engineering Institute (NEERI) Nagpur, 70 per cent or more available water is polluted in India and about two million people were affected every year with various water borne diseases like–Hepatitis, typhoid, diarrhoea and cholera. In India about 30,000 villages still has water which can produce cholera.

Investigation in the Gorakhpur Region

Gorakhpur is a fast Growing Industrial Centre of the eastern Uttar Pradesh, its geographical position is 25.5 to 26.2 North and 83.25–84.20 East. The total area of the city is about 147 sq km. The area falls within the subtropical zone of upper-gangetic plain with an annual rainfall about 1200 mm.

Rapti is the main and longest river of the area which flows from north of Nepal tarai and join *Ghagra* river in the south near Barhaj Bazar. The river *Aami* originate from the Sikhar lake of Basti, *Rohini* from the hills of Butwal, Nepal and *Kuano* from the well of Bahraich district also join the river *Rapti* at different places. Most of the major industrial units and township are established along these water bodies. Besides so many temporary ponds and lakes the main are *Chilwa, Ramgarh, Ashuran* and *Mansarovar.* The domestic wastes from the townships, washing ghats, organic matter and solid wastes too discharges into the adjacent water bodies causing deterioration of the water quality.

Not only rivers and Lakes our ground water upto 70 feet too become polluted as such various epidemic diseases were noticed among local peoples from time to time. The city generate besides 5 metric tonnes of biomedical wastes, about 700 metric tonns of silt garbage and wastes per day. Since there is no electric crimatoria and incinerators in the hospitals. The unburnt dead bodies, flashes, organs of men and dead animals also found here and there in the river and highway side, which greatly affects not only quality of water but also of air (Srivastava, 1991, 1995, 1996 and Jaiswal 2001).

The biomedical wastes, also spread the multiple drug resistant bacteria to the municipal sewage system of the area. The effluent sample from the hospital included most of the antibiotics used for treating human and animal infection. The worst fear seen is the transfer of such resistance bacterial pathogens causing infections. In such case most of the presently available antibiotic will betutile against the infectious organism. Thus hospital effluents can also be considered as potential health hazard by mixing multiple drug resistance bacteria to the city sewage system. (Srivastava and Srivastava, 2003). This need further extensive studies.

Physico-chemical and biological characteristics of several factory effluents of Gorakhpur and adjacent district have been, studied in detailed by Srivastava, 1988, 1988a, 1989, 1990, 1998 and 1999 which revealed high level of pollution and an adverse effects not only on plant life including Algae but also to the local people (Table 6.1).

Since, the suitability of water depends upon its chemical, physical biochemical and microbiological composition. Most important of all, it must be free from pathogenic microorganisms and harmful chemicals. Spectral and bacteriological analysis of drinking water form different sources *viz.*, water from filter using Dowex–50 candle, hand pipe, municipal supply tube well in the Gorakhpur region all showed the presence of *E. coli* and heavy metal ions (Lakshmi *et al.*, 1995) and it was found that only boiled water free from microorganisms and other pollutants. A method for purification of drinking water (Municipal water supply) has also been devised using activated charcoal, potassium

Table 6.1: Distributional pattern of Algal Flora

Sl.No.	Genera Recorded	Habitats
1.	**Cyanophyceae**	
	Anabaena	++
	Calothrix	++
	Microcystis	++++
	Nodularia	+
	Nostoc	++++
	Oscillatoria	++++
	Spirulina	+++
	Stechosiphon	++
	Synechococcus	++
2.	**Chlorophyceae**	
	Chara	+++
	Closterium	+++
	Hydrodictyon	+++
	Spriogyra	++
	Volvox	++
3.	**Xanthophyceae**	
	Vaucheria	++
4.	**Bacillariophyceae**	
	Novicula	++
	Nitzschia	++
	Pinnularia	+++
	Phormidium	++++

++++: Abundant; +++: Common; ++: Frequent; +: Rare

permanganate and silica gel beds Lakshmi *et al.* (1989). Use of bleaching powder was not found to be safe.

The excessive use of nitrogenous fertilizer and pesticides by the farmers to gain maximum crop also affect the quality of these water bodies, as the same runoff with rain water and ultimately contaminated the water bodies of the area making them unfit for drinking bathing and pisciculture purposes. The sewage water developed due to increasing township also reduces the quality of water as these flows untreated into the water bodies. The hand pumps located in the premises of the industrial area revealed that the ground water is becoming polluted. (Srivastava 1980, 1989 and 1997).

The physico-chemical characteristic of the factory effluents revealed that they are highly toxic and exhibited wide range of pH *i.e.*, 6.4–8.2. The water samples contain high concentration of ammonia, bicarbonate and free carbon dioxide. The BOD was found to be 2800 mg dissolved oxygen was found to be zero at some places. As such various epidemic diseases were reported among local people. The effluents discharges from the industrial units also decreases the productive potential of the cultivated field of the area. The main crops like *Brassica compestris, Cajanus cajan, Oryza sativa, Triticum aestivans* and *Zea mays* were found greatly affected (Srivastava, 1988, 1989, 1990 and 1991) (Tables 6.2–6.5).

Table 6.2: Amount of Effluent Discharges in Cubic Meters per Day from Various Courses of the Factory and their pH

Sl.No.	Sources	Amounts in Cubic Metres	pH
1.	Beaters	450	5.5
2.	Brown stock washes	1600	9.0
3.	Digester	250	13–14
4.	Floor cleaner	300	7.8
5.	Open decher and centri cleaner	800	8.5
6.	Paper M/c	750	5.5

Table 6.3: Physico-chemical Parameters of Paper Factory Effluent

Factors	*Contents*
Colour	Dark brown
Smell	Caustic
Suspended solid	2624 mg/l
Volatile solid	4252 mg/l
pH	9–14
BOD	1662 mg/l
COD	4575 mg/l
Dissolved oxygen	Nil
Phosphorus	25 mg/l
Calcium	45.0 mg/l
Chloride	29.50 mg/l

Phytotoxic effect of an insecticides can be judged by the damage such as chlorosis, discoloration, burns, necrosis and abscission of leaves, puckering and, distortion of lamina, stunting, malformation and finally death of plants. Various insecticides have been reported to interfere with the metabolic pathways of plants in general and photosynthesis in particulars. Since such studies are useful in deciding the suitability of any insecticides for a particulars crop plant and in establishing its proper dose and use. The phytotoxic effects of insecticides on plants were studied in details by Chandra *et al.* (2001).

Table 6.4: Effect of Effluent on Plant Distribution

Plant Species	*Distribution Density*
Chara sp.	–
Nostoc sp.	+++
Oscillatoria sp.	+++
Phormidium sp.	+++
Spirogyra sp.	–
Ulorthrix sp.	–
Zygnema sp.	–
Alternanthera sessilis	+++
Amaranthus spinosus	+++
Ceratophyllum demursum	–
Cyndon dactylon	+
Cyperus rotundus	+
Eichhornia crassipes	–
Hydrilla verticillate	–
Jussiea perennis	–
Lemna sp.	–
Marsilea sp.	–
Polygonum sp.	–
Potamogeton indicus	–
Ranunculus scleratus	–
Rumex dentatus	+
Setaria glauca	–
Vallisneria spirallis	–
Veronica cineria	+

+++: Better growth; +: Normal growth; –: Less or absent.

Table 6.5: Physico-chemical Parameters of Sugar Factory and Distillery Wastes

Factors	*Contents*	
	Sugar Factory (mg/lit)	*Distillery (mg/lit)*
Colour	Dark brownish	Dark brown
Smell	Rotten Egg	Decaying molasses
Temperature	16–40°C	42–70°C
pH	6.4–8.2	3.4–4.9
Total solid	730–3560	11830–82300
Dissolved oxygen	Nil	Nil
Free carbon dioxide	315.00	–
Bicarbonate	1435–2100	3168–6760

Contd...

Table 6.5–Contd...

Factors	*Contents*	
	Sugar Factory (mg/lit)	*Distillery (mg/lit)*
Ammonia	3.80	–
Calcium	210–280	2950–4180
Nitrate	1.82–10	30–40
Phosphate	7.3–16.0	21–45
Sugar contents	0.25–0.38%	–
BOD	1.2–28000	25000–49800

References

Chandra, G., A.K. Srivastava and A.K. Pandey, 2001. Phytotoxic effect of insecticides on plants. *J. Liv.*, 8(1): 33–37.

Jaiswal, T.B.L., 2001. "Jal pradushan aur cancer", (Water pollution and cancer). *Dainik Jagran*, Gorakhpur.

Lakshmi, Srivastava, A.K., Mishra, A.N. and Pandey, K.N., 1989. Purification of drinking water (municipal water supply) by activated charcoal potassium permanganate and silica gel beds vis-à-vis to bleaching powder and its effect on fish life. *Acta Hydrochem. Hydrobiol.*, 17(3): 335–339.

Lakshmi, Shukla, S. and Chandra, G., 1995. Spectral and bacterial analysis of drinking water from different sources. *J. Liv. World*, 2(1): 1–7.

Rao, T.R., 1983. *Freshwater Pollution in the Environment Concerned*, (Ed.) S.C. Bhatia. Centre for Adult, Continuing Education and Extension, University of Delhi.

Sangal, P.S., 1983. Environment and Law. *Ibid*.

Sharma, M., 1999. Presidential address, New bioscience: Opportunities and challenges as we move into the next millennium. In: *86th Session Ind. Sci. Cong. Asso.*, Chennai.

Srivastava, A.K., 1988. Studies on sugar factory effluents and their effects on plant distribution. *Ad. Plant Sci.*, 1(2 Suppl.): 290–294.

Srivastava, A.K., 1988a. Physico-chemical and biological characteristics of a sugar factory effluent. *Ind. J. Eco.*, 15(2): 192–193.

Srivastava, A.K., 1989. Pollution of faren stream by the effluents of sugar factory and distillery. *Ad. Plant Sci.*, 2(2): 308–310.

Srivastava, A.K., 1990. Study of desastrons effects of sugar factory effluent and their effect n Algal flora. *Env. Poll. and Health Hazard*, p. 183– 186.

Srivastava, A.K., 1991. Pollution study of the natural water bodies (river) at Gorakhpur. In: *Current Trends in Limnological Research*, 1: 153–158.

Srivastava, A.K., 1992. Studies of wetland flora Gorakhpur and their conservation. INTECOL'S IV. Int. Wetland Conf. Columbus, Ohio (USA).

Srivastava, A.K., 1996. Effect of factories effluents on crop plants. In: *83rd Session of Ind. Sci. Cong.*, Patiala.

Srivastava, A.K., 1996a. Influence of industrial effluents on biomass of crop plants. *Int. Conf. on Plant and Env. Pollution* (ICPEP) Lucknow, 4: 1, 134.

Srivastava, A.K., 1997. Studies on toxicity status of Rapti river at Gorakhpur. In: *XII Ann. Conf. of the Acad. on Women and Environment and Nat. Symp. on Env. and Toxicology*, New Delhi, ET/36.5.

Srivastava, A.K., 1998. Pollution status of water resources at Gorakhpur. In: *Sem. on Keep Our Water Resources Clean*, LWS, p. 8.

Srivastava, A.K., 1999. Studies on toxicity of the factories effluents on the vegetation and river ecosystem in Mahrajganj and Kushinagar. In: *Nat. Semi. on Water and Sanitation in Relation on to Public Health*, GSA-Basti, 10.

Srivastava, A.K. and Priyanka Srivastava, 2003. Biomedical wastes at Gorakhpur: A case study. *J. Liv. World*, 10(2): 4–6.

Thattey, C.D., 2002. Report of chairman, *International Commission for Irrigation and Sewage.*

Chapter 7

Water Quality Index for Ground Water Affected with Bicycle Manufacturing Industrial Wastes: An Environmental Quality Audit

Vineeta Shukla[1], Sharda Abusaria[2], Monika Dhankhar[1] and K.V. Sastry[2]

[1]Department of Biosciences, Maharshi Dayanand University, Rohtak – 124 001, Haryana

[2]Lecturer, Department of Zoology, A.I.J.H.M. College, Rohtak

ABSTRACT

The present study is intended to calculate water quality index for ground water at Sonepat in Haryana State of India from various parameters, *viz.*, pH, dissolved oxygen, total hardness, chlorides, total alkalinity, biological oxygen demand, sulphates, iron, chromium, cyanide and MPN coliforms to ascertain the quality of water for human, consumption and to assess the impact of bicycle manufacturing industrial wastes on ground water quality. The high values of water quality indices for ground water samples from various sources located in the industrial premises and vicinity of the industry indicate the influence of industrial wastes on ground water quality and make it unsuitable for drinking purposes. The trend of variation in water quality index values collected around and inside the industry was irregular while for control samples, water quality index values were well below the permissible limit throughout the study period and did not vary widely. The importance of water quality index and pollution control has also been discussed and emphasized.

Keywords: *Ground water, Bicycle manufacturing industrial wastes, Water quality index, Pollution control.*

Introduction

Environment is our biological capital. It makes up a huge, enormously complex living machine on which our total productivity depends. Water is one of the basic natural resources needed for survival of life. Nobel laureate A. Szent-Gyorgy, physician and biochemist, called water the "matrix of life" many organisms can live without air, but none without water. The use of this under ground fluid for agriculture and drinking purposes has a long history. But ground water is safe only as long as it is free from disease causing organisms and chemicals that are deleterious to human health. Unfortunately, in recent years industrial growth has created severe environmental pollution, and degradation has not spared even ground water aquifer system. A number of places in the country face serious ground water contamination problem due to indiscriminate dumping of toxic chemicals by polluting industries (Singh and Parwana, 1999). Water soluble chemicals are not retained by soil matrix, they percolate through porous medium and enter the ground water (Rao and Shantaram, 1995). Environmental protection and mankind's survival are closely linked. Ground water once polluted cannot be treated, therefore, the effect is long lasting and problem is of vital concern. Hence, there is need for protection and management of ground water quality. Accurate and timely information on water quality is essential to frame a sound public policy and to implement the water quality improvement programs.

Water quality index is the most efficient tool of environmental audit process to convey the overall water quality of a water resource. It is a rating of water quality parameters by a single numerical expression reflecting the complete influence of water quality parameters. Unfortunately in India water quality index determination has been much neglected. Few attempts have been made to calculate water quality for river waters (Tiwari *et al.*, 1986; Singh and Ghosh, 1999). Studies on ground water quality index are scanty (Venkata Mohan *et al.*, 1998; Mariappan *et al.*, 1999).

Sonepat in Haryana State is known for bicycle manufacturing industry whose product is available globally. But the industrial effluents and sludge generated as byproduct may cause great environmental stress locally. Spent waters of different operations such as electroplating and heat treatment are normally stored inside the industrial premises before discharge. For the past more than forty five years, surrounding land has been receiving untreated/partially treated effluents of the industry. Sludge is being dumped within the premises and in low lying areas located near the industry. The dumping and holding of such chemical wastes are likely to affect the ground water quality in these areas. So far, no detailed investigation on the effect of these industrial wastes on ground water quality has been carried out. The present study has, therefore, been undertaken to calculate the water quality index for ground water in and around the industry to measure the extent of damage to water quality.

Material and Methods

In the industry and surrounding regions ground water is the only source of drinking water and is being tapped by tube wells. Four sampling sites were selected for the present investigation. Details of these sites are given in Table 7.1. In the formulation of water quality index, the importance of various parameters depends upon the intended use of water. In this study eleven water quality parameters of ground water are studied from the point of view of suitability of water from different sources at Sonepat for human consumption. For this purpose monthly samples of ground water from selected sites were collected and analysed as per Standard Methods during March, 1999 to February, 2000.

Based on the ground water analysis data, water quality index was calculated for each sample by a simple method. The 'standards' (permissible values) of various parameters for drinking water, recommended by the ISI (ISI, 1991) are used for comparison. For dissolved oxygen and BOD, European Economic Community (EEC) Standard and United Public Health Service (UPHS) have been considered respectively.

Table 7.1: Particulars of Ground Water Sampling Sites

Reference Number	*Name and Location*	*Description*	*Age (Yrs)*	*Depth (M)*
GW1	Tubewell No. 3 in bicycle industry premises in contaminated area	20M from effluents bodies	40	115 up to May, 1999. June onwards 120. Closed in August
GW2	Tubewell No. 7 in bicycle industry premises in contaminated area	Earlier 30M, now 1000M from sludge dumping area and effluents bodies	8	105
GW3	Sujan Singh Tubewell in Agricultural field contaminated with industrial effluents	Near cyanide sludge dumping area of industry and about 1000M from effluents flow channel	8	75
GW4	Public Health Tube well in unpolluted area	Beyond the influence of industrial wastes and it served as control ground water	5	83

Calculation of Subindex (SI)

To calculate the Water Quality Index, first of all subindex (SI) i for the ith parameter was calculated which was the product of the quality rating (qi) and the unit weight (wi) of the ith parameter *i.e.*

$$(S)\, i = qi\, wi \qquad ....(1)$$

Calculation of Quality Rating (qi)

The quality rating, qi for the ith water quality parameter was obtained from the equation:

$$qi = 100\,(Vi/Si) \qquad ....(2)$$

where,

Vi: Value of the ith parameter at a given sampling station; and

Si: Recommended standard for the ith parameter.

Equation (2) ensures that qi = 0 when pollutant (the ith parameter) is absent in the water, while qi = 100 if the value of this parameter is just equal to its permissible value for drinking water.

Quality ratings for pH, dissolved oxygen and MPN coliforms required special handling.

The permissible range of pH for drinking water is 6.5 to 8.5. Therefore, the quality rating qpH was calculated from the relation:

$$qpH = 100[(V_{pH} \sim 6.5)/(8.5-6.5)] \qquad ....(3)$$

where,

V_{pH}: Value of pH; and

Symbol "~" means simply the numerical differences between V_{pH} and 6.5 ignoring the algebraic sign.

The situation is slightly complicated in the case of dissolved oxygen, since, in contrast to other pollutants, the quality of water is enhanced if it contains more dissolved oxygen. Therefore quality rating qDO was calculated from the relation.

$$qDO = 100\left[\frac{14.6}{14.6 j 5.0} j \frac{V_{DO}}{14.6 j 5.0}\right] \qquad \text{.........(4)}$$

where,

V_{DO}: Value of dissolved oxygen.

14.6 is the solubility of oxygen (mg/l) in distilled water at 0°C; and 5.0 is the standard (mg/l) for drinking water.

In case of MPN coliforms, the permissible value for the drinking water is 1/100 ml, while their actual number in untreated water may be several thousands/100 ml. Therefore to get a convenient number for quality rating qMPN was calculated from the relation

$$qMPN = 100\left[\frac{\text{Log}_{10} V_{MPN}{}^{+1}}{\text{Log}_{10} S_{MPN}{}^{+1}}\right] \qquad \text{.........(5)}$$

where,

V_{MPN}: MPN count/100 ml; and

S_{MPN}: MPN standard for drinking water.

Calculation of Unit Weight (wi)

$$wi = k/Si \qquad \text{.........(6)}$$

where,

wi: Unit weight;

Si: Standard for the ith parameter (i = 1, 2..............11); and

k: Constant of proportionality which was determined from the condition

$$\sum_{i=1}^{i=11} wi = k \sum_{i=1}^{i=11}\left(\frac{1}{si}\right) = 1 \qquad \text{.........(7)}$$

Final Value of Water Quality Index

Finally the water quality index was obtained by aggregating these eleven subindices (SI) linearly

$$W.Q.I. = \sum_{i=1}^{i=11}(qiwi) \qquad \text{.........(8)}$$

Results and Discussion

Unit weights assigned to water quality parameters and a sample calculation of overall water quality index from the subindices for ground water samples from different sources have been shown in Table 7.2. Water quality indices for one year study period are depicted in Table 7.3. On extensive comparison with control sample to document the impact of industrial pollution on groundwater quality present study reveals an alarming feature that in 100 per cent samples from affected area, quality of ground water is much altered with respected to natural and local background indicating pollution status of ground water and is not fit for human consumption at any time period of the year due to non-conforming water quality index with drinking water limit (100). A critical study of Table 7.2 indicates that for GW1, GW2 and DW3 samples the main dominating contribution to water quality index comes from the existence of un-natural toxic chemicals–chromium and cyanide–which are used in bicycle manufacturing industry. It is further evidenced by the fact that the more harmful a given pollutant is, the smaller is its permissible value for drinking water. So the, unit weights for various water quality parameters are assumed to be inversely proportional to the recommended standards for the corresponding parameters. Maximum value of the unit weight for water quality parameters in the present study was in the order of chromium and cyanide > iron > MPN coliforms > BOD and dissolved oxygen> pH > total alkalinity and sulphates > chlorides> total hardness, consequently due to high values of chromium and cyanide subindices WQI increased drastically. It may thus be inferred that higher water quality indices for ground water samples from different sources–in and around the bicycle manufacturing industry–may be due to leaching of chemicals from industrial wastes through soil into ground water. Further, WQI for GW1, GW2 and GW3 samples varied depending upon the location of sampling site and variation was sizeable. Very high value of WQI for GW1 indicates the influence of proximity of effluents tanks. There was a wide difference in the values of WQI between GW1 and other two samples coinciding with levels of pollutants. Amongst above three samples, GW1 water exhibited a maximum value of WQI in the month of April which is about 39 times greater than the limit for drinking water. However, GW2 and GW3 samples also exhibited deteriorated status of water quality. On the other hand chromium and cyanide were not the constituents of WQI for GW4 sample (control) coinciding with their nil quality rating values. WQI for GW4 was quite less than other three samples and remained below the limit for drinking water. The variability in WQI among different ground water samples was measured statistically by calculating the coefficient of variation. It confirmed well marked variation in WQI for ground water samples from different sources (Table 7.3).

Table 7.2: Water Quality Index of Ground Water Samples (March)

Water Quality Parameters	*Drinking Water Standard (Si)*	*Unit Weight (wt)*	*Parameter Sub-index (qiwi)*			
			GW1	*GW2*	*GW3*	*GW4*
pH	6.5–8.5	0.00262	0.1179	0.1048	0.1179	0.1048
Dissolved oxygen	5.0 mg/l	0.00446	0.3836	0.3791	0.4371	0.4371
Total hardness	300.0 mg/l	0.00007	0.0104	0.0238	0.0149	0.0141
Chlorides	250.0 mg/l	0.00008	0.0028	0.0214	0.0071	0.0053
Total alkalinity	200.0 mg/l	0.00011	0.0157	0.0092	0.0109	0.0144
BOD	5.0 mg/l	0.00446	1.6484	0.8028	0.0347	0.2382
Sulphates	200.0 mg/l	0.00011	0.0037	0.0049	0.0037	0.0028
Iron	0.3 mg/l	0.7430	0.2194	3.4178	3.2194	12.4764
Total chromium	0.05 mg/l	0.44580	3584.2320	53.4960	53.4960	0.0000
Cyanide	0.05 mg/l	0.44580	71.3280	53.4960	53.4960	0.0000
MPN coliforms	1/100 mg/l	0.02229	0.0000	0.0000	0.0000	0.0000
Water quality index			3660.9619	111.7528	111.8377	3.2931

Table 7.3: Monthly Variation in Water Quality Index of Ground Water Samples

Months	*Water Quality Index (WQI)*			
	GW1	*GW2*	*GW3*	*GW4*
Mar	3360.96	111.75	111.84	3.29
Apr	3911.12	111.02	120.73	3.43
May	3633.90	114.32	142.60	3.07
Jun	3635.88	110.46	123.77	4.28
Jul	3654.14*	110.69	148.53	9.48
Aug		119.45	113.44	7.43
Sep		132.22	131.19	3.18
Oct		141.64	139.60	3.19
Nov		129.55	122.18	3.74
Dec		119.80	130.35	3.42
Jan		129.87	121.90	3.73
Feb		138.14	130.91	3.42
Average	3639.20	122.41	128.09	4.30

* GWI (Tube well No.3) was closed by the industry from August, 1999 CV (%) = 182.64.

Another striking feature emerging from the present investigation is that, different sources do not synchronize in monthly WQI variation. This may be attributed to the fact that an index significantly increases or decreases when some of the critical parameters exceed or become less than the specified permissible value. Over a period of about half a century enormous quantity of pollutants has been added to the soil which may act as a bank of pollutants. Thus, chemical characteristics of percolating effluents, and nature and status of soil may be the possible reasons for irregular variations in WQI for affected sample. However, WQI for control samples indicates uniformly good water quality, reflecting pollution free status of ground water of area contaminated, without any irregular variation. But if the bicycle industry is allowed to dump its wastes without proper treatment the possibility of spread of pollution of ground water in this area also cannot be ruled out.

It is clearly brought out by the findings of this study that the ground water environment of bicycle manufacturing industrial area has already become grossly polluted. Operational phase of the industry established beyond doubt the potential threat to ground water. It is unfortunate that groundwater, a gift of nature, turns out to be a man-made curse. Good quality ground water, as of now is a rare and scarce commodity in the area. Therefore, there is an urgent need for strict enforcement of Environment (Protection) Act, 1986. Violation of these rules can be disastrous and can have far reaching consequences. Present study also clearly confirms the utility of WQI as a proper indicator for conveying the idea of the overall quality of water with a single number. This type of information is helpful for public to understand the quality of drinking water and may have an administrative use in the field of water quality management.

References

APHA 1985. *Standard Methods for the Examination of Water and Wastewater*. American Public Health Association, Washington, D.C.

Horton, R.K. 1965. An index number system for rating water quality. *J. Water Poll. Cant. Fed.*, 37(3): 300–306.

ISI 1991. *Indian Standard Specification for Drinking Water* (IS: 10500), Indian Standards Institution, New Delhi.

Jeevan Rao, K. and Shantaram, M.V. 1995. Ground water pollution from refuse dumps at Hyderabad. *Indian J. Environ. Hlth.*, 37(3): 197–204.

Mariappan, P., Vasudevan, T. and Yegnaraman, V. (1999). Surveillance of ground water quality in Thiruppathur Block of Sivagangai district. *Ind. J. Environ. Prot.*, 19(4): 250–254.

Singh, A.P. and Ghosh, S.K. 1999. Water quality index for river Yamuna. *Poll. Res.*, 18(4): 435–439.

Singh, K.P. and Parwana, B.K. 1999. Ground water pollution due to industrial wastewater in Punjab State and strategies for its control. *Ind. J. Environ. Prot.*, 19(4): 241–244.

Tiwari, T.N., Das, S.C. and Hose, P.K. 1986. Water quality index for the river Jhelum in Kashmir and its seasonal variation. *Poll. Res.*, 5(1): 1–5.

Venkata Mohan, S., Rama Mohan, S. and Jayaraman Reddy, S. 1998. Determination of water quality index (GWQI) of ground water: An application. *J. Environ. Poll.*, 5(3): 219–223.

Chapter 8

Isolation of Bacteriophages Specific to *Pseudomonas aeuroginosa*

P. Dhevagi[1], M.A. Shibu[2], M. Rabeeth[2] and V. Bharath Kumar[1]
[1]Department of Environmental Sciences, Tamil Nadu Agricultural University
[2]Kongunadu Arts and Science College, Coimbatore – 641 029

ABSTRACT

Bacteriophages specific to *Pseudomonas aeuroginosa* have been demonstrated in albino mice under *in vitro* condition. *Pseudomonas aeuroginosa* was isolated from infected person and characterised. After characterization bacteriophages specific to *Pseudomonas aeuroginosa* was isolated from phage enriched sewage sample. Minimal lethal dose of *Pseudomonas aeuroginosa* was found to be 3×10^3 CFU/ml. Phages specific to *Pseudomonas aeuroginosa* were injected to mice after injecting with the pathogen. Results showed that mice treated with lower concentration of lysate showed the symptoms of lethargy. As the phage dose increased from 200 to 800×10^{14} PFU/ml complete recovery of the animal same as control was observed.

Keywords: *Isolation, Bacteriophages, Pseudomonas aeuroginosa, Bacteriophage therapy, Animal study.*

Introduction

Over evolutionary time, bacteria becoming resistance to antibiotics by acquiring genes for proteins that either destroy the antibiotic molecule or expel it from the cell. In addition to it, increasing concern is about the impact of antibiotics on food products. Bacteria such as *Staphylococcus aureus, Mycobacterium tuberculosis, Klebsiella pneumoniae, Pseudomonas aeuroginosa* are some of the dreadful pathogens resistant to antibiotics. *Pseudomonas aeuroginosa* are gram negative, fermentative, motile and opportunistic human pathogen. It is responsible for ventilator-associated pneumonia in incubated patients reaching the death rate of about 38 per cent. It is also responsible for urinary tract infections, surgical wound infections, burns etc (Brewer *et al.*, 1996). This opportunitic pathogen can be eliminated by a pre historic era usage–Bacteriophages. Recent studies and decade long use by Georgians have brought renewed interest in the possibilities of bacteriophage therapy, the use of specific kind of viruses that attack only bacteria (Carlton 1999, Sulakvelidze *et al.*, 2001). Their specificity and harmless nature to eukaryotes makes them ideal as therapeutic agents (Merill *et al.*, 2003). In recent years, well controlled animal models have been demonstrated that phages can rescue animals from a variety of fatal infections, while non controlled clinical reports have shown that phages can be effective in treating multi drug resistant infections in human. The potential of bacteriophages as therapeutic agents against infection has yet to be fully explored. Hence the present investigation was planned on the possibility of using phages specific to *Pseudomonas aeuroginosa* as *in vitro*–therapy for infection

Materials and Methods

This study was carried out in four phases:

1. Isolation of *P. aeuroginosa* and phages specific to *P. aeuroginosa.*
2. Establishing the MLD in the animal model.
3. Test for phage toxicity.
4. Bacteriophage therapy.

Isolation of *P. aeuroginosa* and Phages Specific to *P. aeuroginosa*

Isolation of Pathogen

Pseudomonas aeuroginosa was obtained from the blood samples collected from the Hospitals located in and around Coimbatore. Collected samples were transported in sterile peptone water to laboratory and used for further studies. Isolation and characterization of the pathogen was done by the method suggested by Johnson and Case (1995) and identification was done by the method suggested by Holt *et al.* (1994).

Enrichment and Isolation of Bacteriophages

Bacteriophage was isolated from sewage samples at three different places by the method suggested by Nuttall *et al.* (1987). Sewage water was collected in sterile container. Five ml of Decca strength phage broth (DSPB) media and five ml of 24 hrs old broth culture of *Pseudomonas aeuroginosa* was added to forty five ml of sewage and incubated at 37°C for 24 hrs. After 24 hrs the suspension was centrifuged at 10,000 rpm for 10 min. Supernatant was filtered using Millipore filter (0.2 μ) and filtrate was collected and stored at 4°C. Phages in the filtrate were determined by seeding-agar overlay method (Sambrook and Russell, 2001). The plaques appeared on the *P. aeuroginosa* lawn was individually isolated, and used as a phage stock. The plaques appeared on the double layer plating method was individually confirmed for the specificity against *P. aeuroginosa.*

Establishing the MLD in the Animal Model

Germ free Swiss albino mice of the same race was used for the study. *P. aeuroginosa* was grown in Kings B Media at 37°C for 24 hrs. Log phase cultures of *P. aeuroginosa* were used at different concentrations in order to fix the minimal lethal dose (MLD). The treatment used was 0 (T_1), 0.1 (T_2) 0.25 (T_3), 0.50 (T_4), 1.00 (T_5) ml of 99 g × 10^9 CFU, where T_1 was used as control. All the treatments were replicated four times. The pathogen was injected intraperitoneally to mice. The animals were observed for symptoms within 24 hrs (Behavioral study of mice *i.e.* temperature and bodyweight was done to check the efficacy of the bacteria and bacteriophages). Mice inoculated with bacteria were scored for their state of health on a scale of 5 to 0, based on progressive disease states reflected by several clinical signs. A normal and unremarkable condition was scored as 5, slight illness defined as lethargy, as ruffled fur was scored as 4, moderate illness defined as severe lethargy, ruffled fur and hunched back was scored as 3, severe illness with above sign plus exudative accumulation around partially closed eyes was scored as 2, a moribund state was scored as 1, and death was scored as 0. To verify if the cause of death was really due to *P. aeuroginosa*, the serum was collected from the dead mice and streaked on *P. aeuroginosa* plate. Colonies growing on plates were then compared to the original *P. aeuroginosa* culture (Biswas *et al.*, 2002).

Test for Phage Toxicity

The efficacy of phages was evaluated in the animal model with four different treatments to find out whether phages were safe for injection and inoculation into the mice. Dilutions of 0 (T_1), 0.25 (T_2), 0.50 (T_3), 1.00 (T_4) ml of 800 × 10^{14} PFU of phages were prepared in PBS and injected intra-peritoneally, where T_1 was used as control. All the treatments were replicated four times.

Bacteriophage Therapy

The MLD for *P. aeuroginosa* was found to be 0.1 ml and was selected for inoculation. It was injected peritoneally followed by phages after 15 minutes were administrated with different concentrations 0 (T_1), 0.25 (T_2), 0.50 (T_3), 1.00 (T_4) ml, where T_1 was used as control mice. Behavioral changes and survival of mice were observed.

Results and Discussion

P. aeuroginosa were isolated from hospital sample and identified as *P. aeuroginosa* using, morphological and biochemical characterization (Tables 8.1a and 8.1b). The isolated *P. aeuroginosa* was designated as KDB-1. Isolated strain showed resistance to most of the antibiotics indicates the multi-drug resistance of the pathogens (Ahmad, 2002). Infection by the ubiquitous multi-drug-resistant *Pseudomonas* spp. creates no additional complexity to the infected patients. Levin and Bull (1996), Lederberg (1996), Barrow and Soothil (1997) also isolated the multi-drug resistant bacteria.

Table 8.1(a): Morphological, Biochemical Characterization of *P. aeuroginosa* KDB-1

Tests Performed	*Result*
Shape	
Gram reaction	Positive
Motility	Motile
Gelatin utilization	+
Citrate utilization	+
Methy red	+
Voges proskeur	–
Acid from glucose	+
Gas from glucose	+
Urease production	+
Indole production	+

+: Positive result; –: Negative result

Table 8.1(b): Antibiotic Sensitivity Test of *P. aeuroginosa* KDB-1

Antibiotics	*Result*
Gentamycin	R
Ciproflaxin	I
Chloramphenical	R
Tetracycline	R
Streptomycin	R
Kanamycin	R
Ampicillin	R
Erythromycin	R
Penicillin	R
Cephalosporin	R
Rifampicin	R

R: Resistance, I: Intermediate

Bacteriophages specific to *P. aeuroginosa* were isolated from sewage using chloroform technique. The isolated phage was quantified using double overlay technique for plaque assay and designated as PAD-1. The Result on plaque assay was given in Table 8.2. Isolation of bacteriophages from sewage has been demonstrated by Nutall *et al.* (1987) in which a total of 27-INCH plasmid dependent phages were isolated and all were shown to produce turbid plaques. Gowrishankar *et al.* (1998) also isolated phages from sewage.

The concentration 0.1, 0.25, 0.50, 1.00 ml of 99 × 10^9 CFU/ml of *P. aeuroginosa* (Table 8.3) was tested in mice to check the MLD. Mice were found to be dead at the minimum dosage of 0.1ml. The state of mice health was observed for 24hrs and the results given in Figure 8.1. No mortality was observed in the uninfected animals, on contrary the inoculated mice showed the symptoms of illness and had a very short survival period of half an hour. Difference in progression of disease can be ascribed to the innate variations in their immunity and defense in response to bacterial colonization, and for the varied transition of the disease from bacterial colonization to septic shock (Baron *et al.*, 1994).

Table 8.2: No of Plaque Forming Units (PFU/ml) of Lysate (PAD-1)

Sl.No.	*Dilution*	*Number of Plaques Formed per ml of Lysate*
1.	10^{-6}	800 × 10^{8}
2.	10^{-8}	737 × 10^{10}
3.	10^{-10}	529 × 10^{12}
4.	10^{-12}	412 × 10^{16}
5.	10^{-14}	200 × 10^{18}

Values represent mean of three replications.

Table 8.3: Cell Count of *P. aeuroginosa* (CFU/ml) in Different Dilution (KDB-1)

Sl.No.	*Dilution*	*CFU/ml of Sample*
1.	10^{-1}	300 × 10^{2}
2.	10^{-2}	203 × 10^{4}
3.	10^{-3}	99 × 10^{6}
4.	10^{-4}	30 × 10^{8}
5.	10^{-5}	9 × 10^{10}

Values represent mean of three replications.

Bacteriophages for *Pseudomonas aeruginosa* were tested in experimental infections of mice to investigate their potential for the treatment of infections of man. The phage lysates of 0.25, 0.50, 1.00 ml of 800 × 10^{14} PFU/ml were not toxic as they induced only slight ill condition in mice (Figure 8.2). Mice showed symptoms of slight illness at about half an hour after phage injection and this persisted until the 12th hour, after which all mice recovered their normal health. This decline in the health status of mice may be attributed to the endotoxins that are always present in crude phage lysates (Merril *et al.*, 2003). Studies by Soothill (1992) support the view that bacteriophages could be useful in the treatment of human infections caused by antibiotic-resistant strains of bacteria. Soothill (1994) reported that, bacteriophages prevents destruction of skin grafts by *Pseudomonas aeruginosa.*

Mice treated with lower concentration of lysate lead to expression of lethargy, reduced appetite, hunched back and severe illness. As the phage dose increases from 200 to 800 × 10^{14} PFU/ml, resulting in increase multiplicity of bacteriophage leads to survival of animal. After 24 hrs all the mice showed no sign of illness and were comparable with that of healthy mice. At lower concentration of lysate (0.25) the expression of symptoms were ruffled, severe lethargy and hunched back. Mice showed symptoms of slight illness at about 6 hrs after phage injection and after which all mice recovered their normal health (Figure 8.3). Further analysis of animal tissue and serum indicates the absence of bacterial load in response to phage administration, which proves the potential of phage for treatment of *P. aeuroginosa* infection.

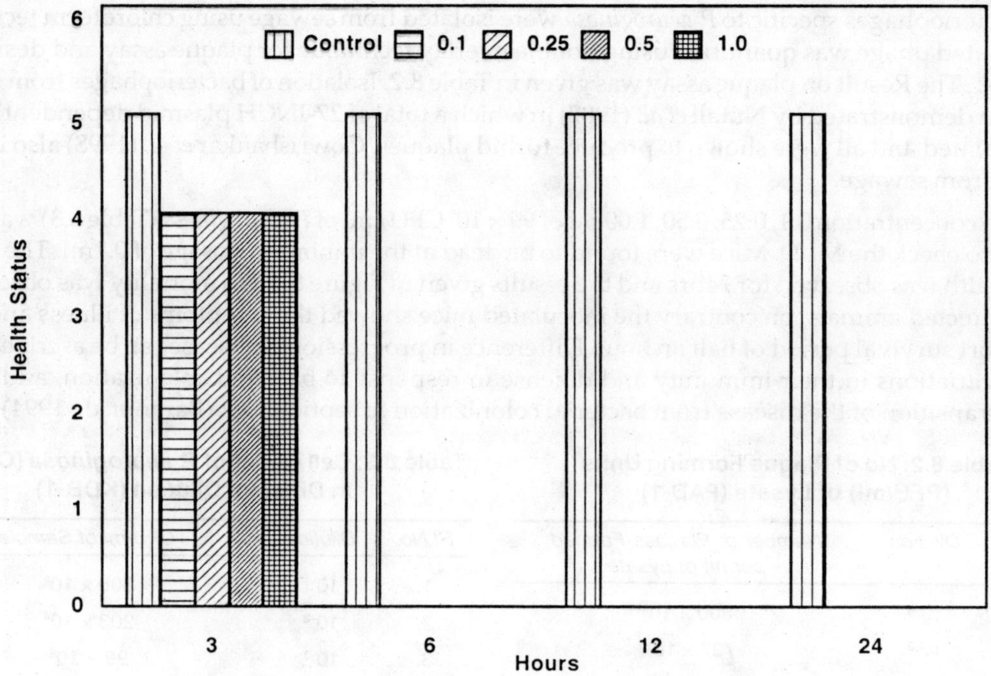

Figure 8.1: *Pseudomonas aeruginosa* Treated Mice

Figure 8.2: Lysate Treated Mice

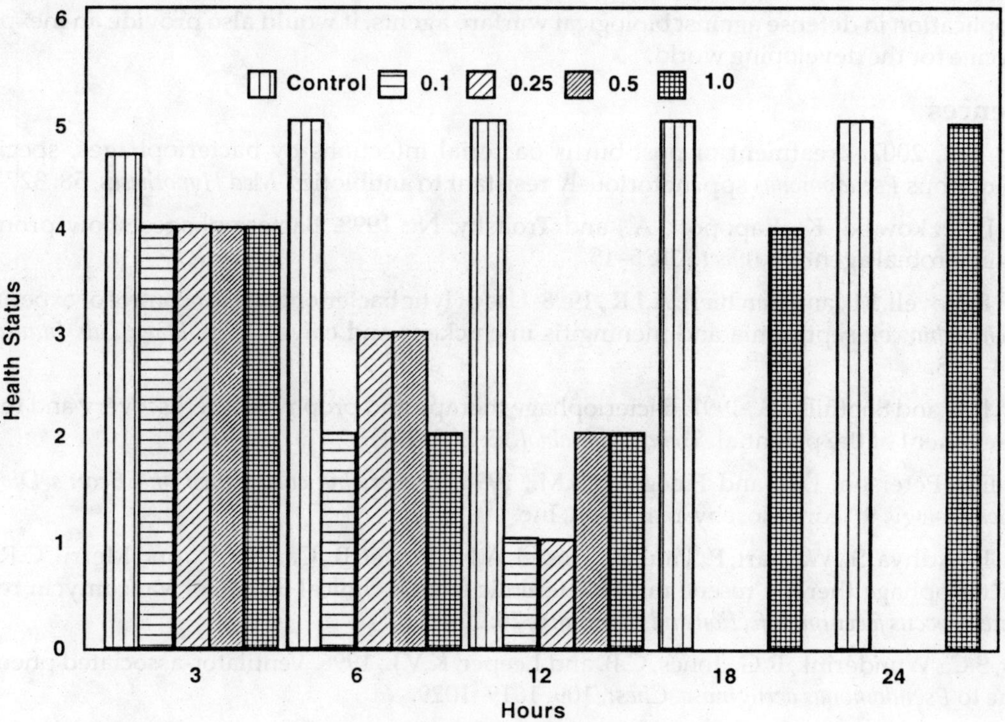

Figure 8.3: Phage Theraphy

Dubos *et al.* (1943) inoculated mice intracerebrally with 10^6 colony-forming units (cfu) of *Shigella dysenteriae,* which leads to death. Simultaneous intraperitoneal inoculation with 10^9 plaque-forming units (Cfu) of phage suspension produced 6/8 animals, beside he inoculated with a lower dose of 10^5 pfu, protected 2/8 mice. Geier *et al.* (1973) and Barrow *et al.* (1998) also isolated phages specific to bacterial infections.

Gowrishankar *et al.* (1998) showed clearly that animal protected with phage ingestion exhibit no sign of illness when compared to unprotected ones. Bacteriophage for *Acinetobacter baumanii, Pseudomonas aeuroginosa* and *Staphylococcus aureus* were tested in experimental infections of mice to investigate their potential for the treatment of infections of man (Soothill, 1992). Kozminska *et al.* (1997) examined bacteriophages for the treatment of bacteriology infection by influencing bacteriophages in producing cytokines by leukocytes of human peripheral blood.

In case of acute human infection, phage therapy have been considered valuable adjunct when compared with antibiotics. This present experiment reveals that lytic phages are effective therapy for *P. aeuroginosa* infection and indicates clearly the relevance of phage therapy, to rescue animals infected with *P. aeuroginosa* with no sign of illness. The ability of phage to rescue animals infected with these high titers of bacteria is especially putting forth the use of bacteriophages as therapeutic agent. This study further shows that a single IP dose of phage lysates was enough to bring about full recovery of mice infected by lethal concentrations of *P. aeuroginosa* (3×10^3 CFU/ml). The potential of phage therapy has been given by several persons (Alisky *et al.*, 1998; Carlton 1999; Pirisi 2000; Soothill and Barrow, 1997) and the present study reinforces the view that this study is worth exploring. In addition to its application in defense against biological warfare agents, it would also provide an inexpensive oral vaccine for the developing world.

References

Ahmad, S.I., 2002. Treatment of post-burns bacterial infections by bacteriophages, specifically ubiquitous *Pseudomonas* spp. notoriously resistant to antibiotics. *Med Hypotheses*, 58: 327–331.

Alisky, J., Iczkowski, K., Rapoport, A., and Troitsky, N., 1998. Bacteriophages show promise as antimicrobial agents. *J. Infect.*, 36: 5–15.

Barrow, P., Lovell, M., and Berchieri, A.J.R., 1998. Use of lytic bacteriophage for control of experimental *Escherichia coli* septicemia and meningitis in chickens and calves. *Clin. Diag. Lab. Immunol.*, 5: 294–298.

Barrow, P.A. and Soothill, J.S., 1997. Bacteriophage therapy and prophylaxis rediscovery and renewed assessment of the potential. *Trends Microbiol.*, 5: 268–271.

Baron, E.J., Peterson, L.R. and Finegold, S.M., 1994. *Escherichia coli. Bailey and Scott's Diagnostic Microbiology*, 9th edn. Moseby Year Book, Inc.

Biswas, B., Adhya, S., Washart, P., Paul, B., Trostel, A.N., Powel, B., Carlton, R., and Merril, C.R., 2002. Bacteriophage therapy rescue mice bacteremic from a clinical isolate of Vancomycin resistant *Enterococcus faecium. Infection and Immunity*, 70: 204–210.

Brewer, S.C., Wunderink, R.G., Jones, C.B. and Leeper, K.V.J., 1996. Ventilator-associated pneumonia due to *Pseudomonas aeruginosa. Chest*, 109: 1019–1029.

Carlton, R.M., 1999. Phage therapy: past history and future prospectous. *Archivum Immunologii et therapiae Experiements*, 47: 267–274.

Dubos, R.J., Straus, J.H. and Pierce, C., 1943. The multiplication of bacteriophage in vivo and its protective effects against an experimental infection with *Shigella dysenteriae. J. Exp. Med.*, 78(20): 161–168.

Geier, M.R., Trigg, M.E. and Mereil, C.R., 1973. The fate of Bacteriophage lambda in non-immune germ free mice. *Nature*, 246: 221–223.

Gowri Sankar, R., Madhusudhari, V. and Palaniappan, R., 1998. Evaluation of phage therapy to treat experimental infections in mice. *Indian Journal of Microbiology*, 38: 101–103.

Holt, G.J., Keieg, R.N., Sneath, A.H.P., Statley, T.J. and Williams, T.S., 1994. *Bergey's Manual Determinative Bacteriology*, 9th edn. Lippincott Williams and Wilkins Company.

Johnson, L.E. and Case, C.L., 1992. *Laboratory Experiments in Microbiology*. The Benjamin/Cummings Publishing Company, Inc., p. 125–129.

Kluytmans, J., 1997. Surgical infections including burns. In: *Prevention and Control of Nosocomial Infections*, 3rd edn., (Ed.) R.P. Wenzel. Williams and Wilkins, Baltimore, p. 841–65.

Kozminska, J., Weber-Dabrowska, B. and Mulczyk, M., 1997. The study on biology of bacteriophages and their usage in the treatment of bacterial diseases and on the influence of different bacteriophages on cytokine production by leukocytes in human peripheral blood. *OTOLARYNGOLOGIA POLSKA*, 51(Suppl 25): 195–198.

Lederberg, 1996. Samller fleas ad infinitum. *Therapeutic Bacteriophage, PNAS*, 93: 3167–3168.

Levin, B. and Bull, J.J., 1996. Phage therapy revisited: The population Biology of a Bacterial infection and its treatment with Bacteriophage and antiobiotics. *The American Naturalist*, 147: 881–898.

Merril, C.R., Scholl, D. and Adhya, S.L., 2003. The prospect for bacteriophage therapy in Western Medicine. *Nat. Rev.*, 2: 489–497.

Nuttall, D., Maher, D and Collerean, E., 1987. A method for direct isolation of INcH plasmid dependent bacteriophage. *Applied Microbiology*, 5: 37–40.

Pirisi, A., 2000. Phage therapy: Advantages over antiobiotics? *Lancet*, 356: 1418.

Sambrook, J. and Russell, R.W.D., 2001. *Molecular Cloning: A Laboratory Manual*, 2nd ed., 1: 2.25 to 2.33. Cold Spring Harbor Laboratory Press, Cold Spring Harbor, NY.

Soothill, J.S., 1992. Treatment of experimental infections of mice with bacteriophages. *J. Med. Microbiol.*, 37: 258–262.

Soothill, J.S., 1994. Bacteriophage prevents destruction of skin grafts by *Pseudomonas aeruginosa*. *Burns*, 20: 209–211.

Soothill, J.S. and Barrow, P.A., 1997. Bacteriophage therapy and prophylaxis: Rediscovery and renewed assessment of potential. *Trends Microbial.*, 5: 268–271.

Sulakvelidze, A., Alavidze, A.Z., and Morris, J.G., 2001. Bacteriophage therapy: Mini review. *Antimicrobial Agents and Chemotherapy*, 45(3): 649–659.

Chapter 9

Occupational Health Problems Observed Among the Saltpan Workers of Thoothukudi District, South East Coast of India

R. Sornaraj, A. Amutha Jaisheeba and A. Pushparaj
Research Department of Zoology, Kamaraj College, Thoothukudi – 628 003

ABSTRACT

A survey was made extensively to study the saltpan environmental impact on the workers. The information were collected from 490 respondents working in different areas of saltpan located in and around Thoothukudi. The various physical and physiological disorders of the workers were screened in cooperation with physician. Among the various disorders observed, the eye disorder ranks first and its percentage about 45.91. Next to eye disorder, the other common disorders observed are respiratory disorder (23.46 per cent), Circulatory disorder (20.40 per cent), skin disorder (18.36 per cent) and renal disorder (9.18 per cent). The number of workers get affected in relation to the increase of age and also the year of experience of the workers. Female workers are more affected than the male workers.

Keywords: *Saltpan environment, Physiological disorder, Eye disorder, Respiratory disorder, Skin disorder, Circulatory disorder.*

Introduction

To obtain the daily bread and to lead a healthy comfortable life, every individual in the globe engages himself in any one of the available jobs. Most of the jobs, the people are doing are not providing sufficient money and satisfaction. Most of the jobs are not even providing a healthy condition to the workers. Knowingly or unknowingly the workers are suffering a lot due to the bad environmental conditions in theirs working place. Occupational health is an important one in the working spot. The prevention of disease and maintenance of the highest degree of physical, mental and social well-being of workers in all occupations is essential (Park 1997).

India with about 61,000 km long coastal line stands fourth in production of salt in the world with an average annual production of 13.5 million tones. In India Gujarat, Tamil Nadu and Rajasthan contribute 95 per cent of the total salt production. Thoothukudi district constitutes 70 per cent of the total salt production of Tamil Nadu and meets 30 per cent of salt requirement of our country.

Thoothukudi center produces an average of 25 lakh tones per year. It stands next to Gujarat in salt production in the country. This production comes from about 3100 manufactures working in an area of 4100 acres including one public sector Unit (Tamil Nadu Salt Corporation) and one joint sector unit. Both sea and the subsoil brine are the chief sources in Thoothukudi. Being an intensive labour oriented industry; the salt industry has vast employment potential. In India, it provides employment to the tune of 20 million people.

Salt industry in Thoothukudi district provides large-scale employment to labourers. Every year during January and February the saltpans are prepared for the salt production *i.e.* mending of bunds, desilting of channels, preparation of crystallizer bed. All the salt manufacture related works are attended only manually. As the manual works are tough, no children is employed as child labourer in the salt field.

To excavate the various types of occupational hazards found among the saltpan workers of Thoothukudi, an observation was made and valuable finding were obtained.

Materials and Methods

General physiological conditions of the saltpan workers were collected by randomly selecting 490 workers and interviewing them with a prepared standardized questionnaire. The collected information was counter checked and confirmed with the help of a medical practitioner. The major disorders observed among the workers are skin disease, eye disorder, respiratory disorder, renal disease and circulatory disorders. The information were collected from 5 major salt producing spots namely Thoothukudi, Mullakkadu, Muthaiahpuram, Pullavezhi, Mathagiri and Mudukkukadu situated within the 25 km radius of Thoothukudi. Since the workers are illiterates, the information were collected by interviewing them and the questionnaire was filled up by the researcher in the interview spot itself. Every day around 20 respondents were met and the information were collected. Eye disorders are identified using the ophthalmic kit and the blood pressure of the workers are identified using sphygmomanometer. The statistical tool chi-square was adopted to test the validity of the results.

Results

In the saltpan Industry, workers of different age groups are involved. Their work is very hard and unsafe. They are working for about 10–12 hours in a scorching sun. The environmental condition in the working place is very poor. The job causes lot of stress and unhappiness to the workers.

Physical disorder is a notorious feature among saltpan workers. Cent per cent of the workers are affected by anyone type of pain. Apart from general body pain the workers are also affected by anyone of the other types of physical disorders. The various disorders observed among the saltpan workers are given in Table 9.1 and Figure 9.1 to 9.4. The general disorders observed was general body pain, eye disorder, skin disorder, respiratory disorder, renal disorder and circulatory disorder. Regarding eye disorder the long sight, short sight and reddening of the eye was observed. Regarding skin disorder most of them are affected by dry and scaly skin (keratosin). As far as circulatory disorder is concerned the BP of the patients alone identified and it was found that most of them are having low blood pressure. Regarding the respiratory disorder, most of the workers are frequently affected by cold and some of them are having breathing problems.

Table 9.1: Showing the Physically Affected Saltpan Workers in Relation to their Age. The Chi-Square values of various diseases in relation to age of the workers are also represented in the table.

Age	*Eye Disease*	*Percent-age*	*Skin Disease*	*Percent-age*	*Circu-latory Disorder*	*Percent-age*	*Renal Disorder*	*Percent-age*	*Respi-ratory Disorder*	*Percent-age*
15–25	20	4.08	–	–	–	–	–	–	10	2.04
26–35	35	7.14	5	1.02	5	1.02	5	1.02	25	5.10
36–45	65	13.27	55	11.22	10	2.04	10	2.04	45	9.18
46–55	50	10.20	10	2.04	35	7.14	5	1.02	20	4.08
56–65	55	11.22	20	4.08	50	10.20	25	5.10	15	3.06
Total	225	45.91	90	18.36	100	20.40	45	9.18	115	23.46
Chi-square value	9.27 $P > 0.05$ S		37.37 $P > 0.05$ S		5.84 $P > 0.05$ S		17.37 $P > 0.05$ S		20.56 $P > 0.05$ S	

Figure 9.1 clearly tells of the physical disorders that commonly occur among the saltpan workers. The common physical disorders observed among the saltpan workers are eye disease, skin disease, circulatory disorder, renal disorders and respiratory disorders in the percentage of 45.95, 18.36, 20.40, 9.18 and 23.00 respectively. Among the various disorders observed, the eye disease is in the apex position. The second position is occupied by the respiratory disorders and the third position is occupied by the circulatory disorder. The lowest position is occupied by the renal disorders.

Figure 9.2 clearly states the relationship between the disorders and years of experience of the workers. The most experienced workers are more affected by eye and circulatory disorder in the percentage of 24.49 and 13.26 respectively. The workers who are having the experience of 10 to 15 years occupy the second position. Among them nearly 10.2 per cent is affected by eye disease.

Figure 9.3 shows the sex-wise distribution of physically affected persons. The workers in salt industry are affected by more than one disease. The Figure 9.2 clearly shows the relationship between the year of experience and the disease appeared among the workers. Age old and experienced workers are affected by more than one disease. By visualizing the Figure 9.3 we can clearly recognize the different types of diseases and the number of individuals affected and so on. Majority of the workers (45.91 per cent) are having eye disorders. The eye disease equally affects both the sex. The percentage of eye defect among male is 20.4 per cent and among female is 25.51 per cent. The respiratory diseases take the second position among saltpan workers. The percentage is about 23.46 per cent. The other diseases such as skin, heart and kidney represents 18.36, 20.40 and 9.18 percentage respectively.

Among the total observed population 54.08 percentage of workers are having good vision. The remaining 45.9 per cent is affected by anyone of the eye disorder. Most of the affected workers are

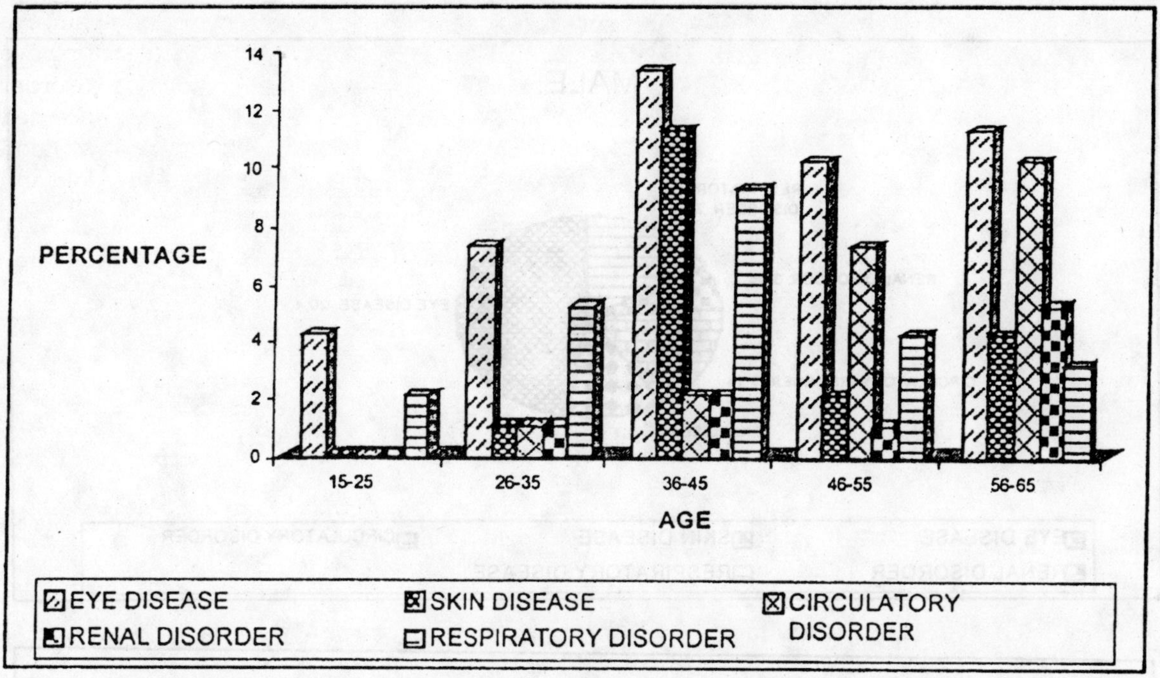

Figure 9.1: Showing the Physically Affected Saltpan Workers in Relation to their Age in the Thoothukudi Area

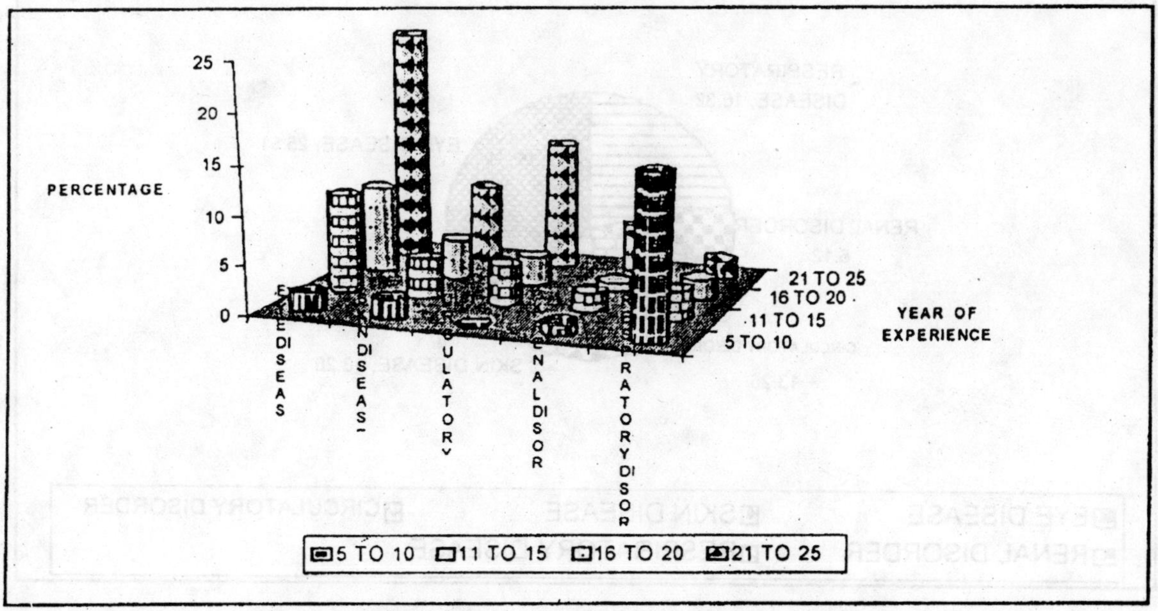

Figure 9.2: Showing the Relationship Between the Experience of the Saltpan Workers and their Nature of Physical Illness in the Thoothukudi Area

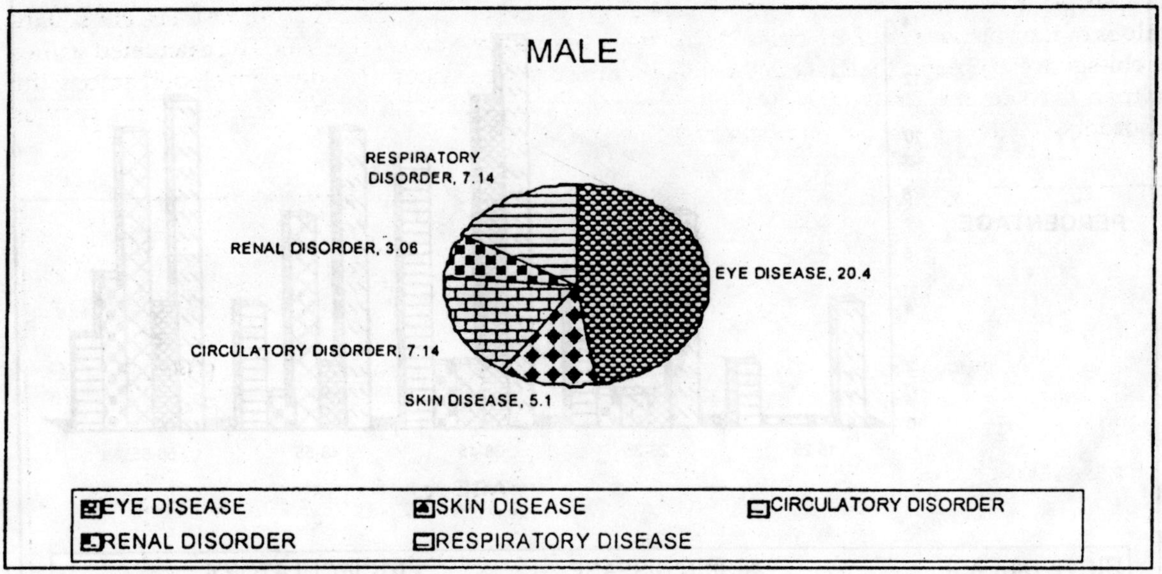

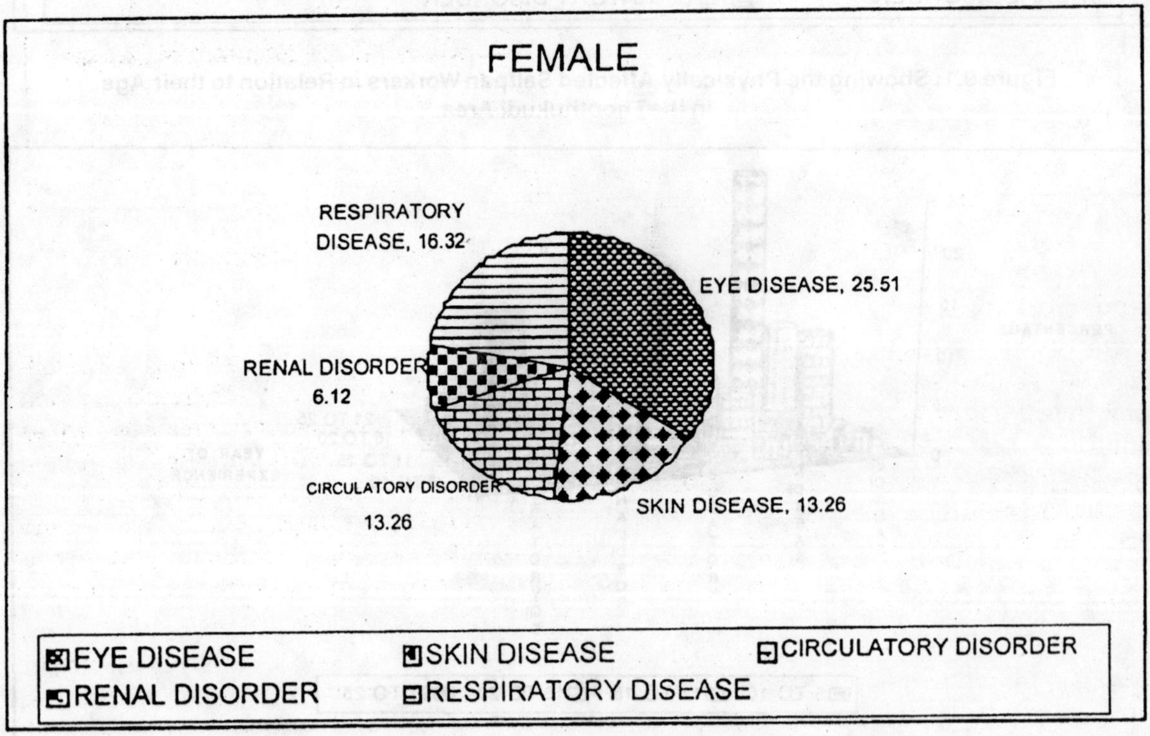

Figure 9.3: Showing the Sex-wise Distribution of Physically Affected Saltpan Workers in Thoothukudi Area

having short sight (41.83 per cent) only 4 per cent alone are affected by long sight. Eye defect occupies its apex position among the workers whose age group is between 35–45 (Figure 9.4). The chi-square values of the various results calculated showed a significant trend (Table 9.1). The calculated values of chi-square is greater than its table value, that means the various disorders developed among the saltpan workers have relationship with age of the individuals. As the age increased, the various disorders developed among the workers are also increased.

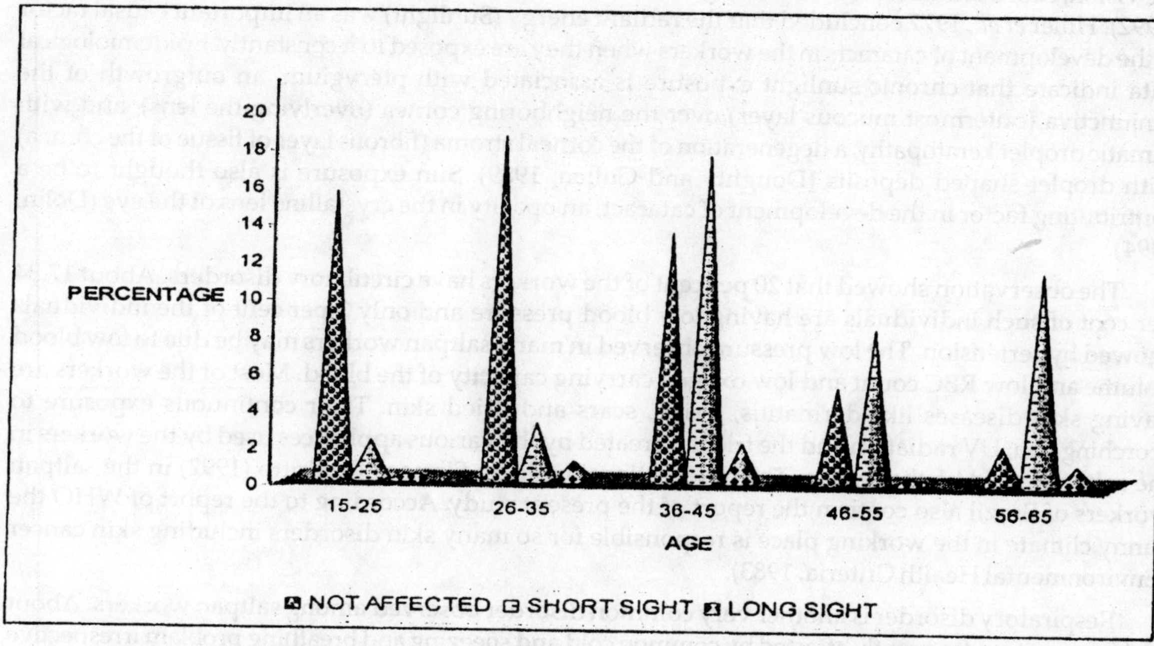

Figure 9.4: Showing the Saltpan Workers who are Affected by Various Eye Disorders

Discussion

The World Health Organization (WHO) has defined health as "physical, mental and social well-being". Well-built physical condition is essential for every individual in a society. There is a common saying. "Health is Wealth". The health of an individual is affected due to so many reasons such as malnutrition, environment in which he is living or working, nature of work and the like. As the saltpan workers are doing a hard and much painful work and are constantly exposed to the salt and saltpan environment, defects in their physical and mental health is inevitable. Generally speaking the physical disorder is a common occurrence in all kind of workers including the women who engage themselves in kitchen and other household works. In case of saltpan workers also the physical disorders are common. But the degree of disorder is entirely different from other kinds of workers.

Irrespective of the age, sex and year of experience the saltpan workers are affected by certain ailment. But the degree of damage corresponds to the year of experience in saltpan environment. The common physical disorders are the eye disease, skin disease, circulatory disorder, renal disorder and respiratory disorders. Most of the workers are very severely affected by the eye, heart and respiratory disorders.

Nearly 41.83 per cent of workers are having short sight and 4 per cent of the workers are having long sight the high percentage of short sight observed among the saltpan workers may be due to the excessive radiation from the saltpan and the heaps of salt crystals. The opinion of the author is supported by other researchers. Finch and Moriyama, (1980) has the opinion that due to severe radiation in the environment the nature of the eye lens and other accessory structures are affected resulted in eye defect. The workers exposed to excessive light rays comprised of ultra violet and infrared rays affect the vision, cause retinal injury, lens changes and acute or chronic conjunctiva as reported by Firmeza (1992). Hiller *et al.*, 1977 concluded that the radiant energy (Sunlight) was an important causal factor in the development of cataracts in the workers when they are exposed to it constantly. Epidemiological data indicate that chronic sunlight exposure is associated with pterygium, an outgrowth of the conjunctiva (outermost mucous layer) over the neighboring cornea (overlying the lens), and with climatic droplet keratopathy, a degeneration of the corneal stroma (fibrous layer of tissue of the cornea) with droplet-shaped deposits (Doughty and Cullen, 1989). Sun exposure is also thought to be a contributing factor in the development of cataract, an opacity in the crystalline lens of the eye (Dolin, 1994).

The observation showed that 20 per cent of the workers have circulatory disorders. About 17.34 per cent of such individuals are having low blood pressure and only 3 per cent of the individuals showed hypertension. The low pressure observed in many saltpan workers may be due to low blood volume and low RBC count and low oxygen carrying capacity of the blood. Most of the workers are having skin diseases like dermatitis, lesion, scars and dried skin. Their continuous exposure to scorching sun, UV radiation and the friction created by the various appliances used by the workers in the saltpan could be the reason. The observations made by Firmeza De Britto (1992) in the saltpan workers of Brazil also confirm the report of the present study. According to the report of WHO the sunny climate in the working place is responsible for so many skin disorders including skin cancer (Environmental Health Criteria, 1983).

Respiratory disorder is another very common disorder observed among saltpan workers. About 23.46 per cent are frequently affected by common cold and sneezing and breathing problem irrespective of age and sex. The respiratory disorders among the saltpan workers may be due to the deposition of suspended micro particles and salt dust in the nasal cavity and upper respiratory tract during their normal work. During windy days, the inhalation of the salt dust may be very high and that may result in serious respiratory disorders and the deposits may affect the epithelial layers of the respiratory passage and it also affects the easy inhalation resulting in wheezing and other respiratory disorders. Chatterjee (1998) reported that an excessive deposition of dust particles in the nasal passage provokes the fibrous tissue causing permanent disability.

The various physical disorders observed among the saltpan workers in the study area could be due to the continuous exposure of the workers to the saltpan environment, their hard and restless work, under nutrition and absence of awareness about their environment and work. Sidorenko, (1978) has the opinion that the population's health could be an integral index of environmental quality and the effects of various environmental agents. According to Bustueva and Sluchanko, (1979) the effect of some environmental agents is specific, but most are non-specific in their action.

Acknowledgement

The financial support provided by UGC, Government of India, New Delhi is gratefully acknowledged by the authors.

References

Bustueva, K.A. and I.S. Slucanko, 1979. *Methods and Criteria of the Evaluation of the Health of the Population in Relation to Environmental Pollution*. Meditsyna, Moscow.

Chatterjee, C.C., 1998. *Human Physiology*, 11th edn. Medical Allied Agency, Kolkata, Special Reprint, pp. 366–367.

Dolin, P.J., 1994. Ultraviolet radiation and cataract: A review of the epidemiological evidence. *Brit. J. Ophthalmol.*, 78: 4788–482.

Doughty, M.J. and A.P. Cullen, 1989. Long-term effects of a single dose of ultraviolet-B on albino rabbit cornea–I *in vivo* analysis, *Photochem. Photobiol.*, 49: 185– 96.

Environmental Health Criteria, 1983. *Guidelines on Studies in Environmental Epidemiology*. World Health Organization, Geneva.

Finch, S.C. and I.M. Moriyama, 1980. The delayed effects of radiation exposure among atomic bomb survivors, Hiroshima and Nagasaki, 1945–79: A brief summary. Radiation effects research foundation technical report, pp. 16–78

Firmeza De Brito, H., 1992. Pathogenic activity of Sodium chloride dust in eye. In: *Encyclopedia of Occupational Health and Safety*, 3rd edition. International Labour Office, Geneva.

Hiller, R., Giacometti, L., and K. Yuen, 1977. Sun light and cataract, an epidemiological investigation *Am. J. Epidemiol.*, 105: 450–459.

Sidorenko, G.I., 1978. Modern problems of environmental health science. *Gigi. Sanit.*, 10: 9–15.

Chapter 10

Effect of Nitrogen Management on Growth and Micro-climatic Parameters of Bed Planted *durum* Wheat (*Triticum durum* Desf.)

Navneet Kaur Brar and D.S. Kler
Department of Agronomy and Agrometeorology, Punjab Agricultural University, Ludhiana – 141 004, Punjab, India

ABSTRACT

Field trial on *durum* wheat (*Triticum durum* Desf.) was conducted during Rabi 2002–03 at Punjab Agricultural University, Ludhiana. The experiment replicated thrice consisted of 16 treatment combinations arranged in Randomized Block Design with three nitrogen management levels (120 kg N; 120 kg N + 15 t FYM and 150 kg N ha^{-1}) and five nitrogen application times [1/2 at sowing (S) + 1/2 at crown root initiation (CRI); 1/3 at S + 1/3 at CRI + 1/3 at tillering (T); 1/4 at S + 1/2 at CRI + 1/4 at T; 1/3 at S + 1/3 at CRI + 1/3 at T + 3 per cent urea spray (US) at flowering (F) and 1/4 at S + 1/2 at CRI + 1/4 at T + 3 per cent US at F] on beds and one flat sown treatment (unidirection 22.5 cm). Mean grain yield of 40.02 and 39.53 q ha^{-1} of bed planted and flat sown crop were statistically at par with each other. One extra treatment where whole N was applied at sowing to bed planted wheat, yielded 38.31 q ha^{-1} which was comparable to 2 split application of N (Rec.) either to flat sown (39.53 q ha^{-1} or to bed planted (38.33 q ha^{-1}) crop indicating that whole N could be applied at sowing to bed planted *durum* wheat. 120 kg N + 15 t FYM ha^{-1} gave 9.9 per cent and 7.9 per cent higher grain yield than 120 kg N ha^{-1} and 150 kg N ha^{-1} respectively. 120 kg N + 15 t FYM ha^{-1} on an average tended to record higher values of growth parameters and favorable values of microclimatic parameters.

Keywords: *Durum wheat, Bed planting, Split application, Growth parameters, Micro-climatic parameters.*

Introduction

Durum wheat occupies 8 to 10 per cent area in Punjab covering about 250 to 300 thousand hectares with a production of about 800 to 1000 thousand mt annually (Anon, 1995). Introduction of *durum* wheat has helped create varietal diversity for a balanced and stable agro ecosystem to sustain high production, caring the industry needs and reduce the risk of crop failures resulting from pest and disease outbreaks. In view of export for foreign exchange of *durum* wheat it becomes imperative to improve its productivity from the present almost a static level being obtained at recommended levels of inputs. To make *durum* wheat competitive in international market it becomes imperative to improve its productivity and quality by manipulating agronomy of the crop through variation in planting methods and nitrogen management with the emphasis on use of organic manures. Hence, present investigations on *durum* wheat were conducted to analyze crop growth and yield under bed planting and different nitrogen management combinations.

Materials and Methods

The experiment was conducted at the Punjab Agricultural University, Ludhiana during 2002–03 on a loamy sand irrigated soil. The soil of the experimental field tested low in available nitrogen and medium in available phosphorous and potash. The 16 treatment combinations (15 bed planted + 1 flat sown) consisting of 3 nitrogen management levels [120 kg N (Rec.); 120 kg N + 15 t FYM and 150 kg N ha^{-1}] with 5 times of N application [1/2 at sowing (S) + 1/2 at crown root initiation (CRI) (Rec.); 1/3 at S + 1/3 at CRI + 1/3 at tillering (T); 1/4 at S + 1/2 at CRI + 1/4 at T; 1/3 at S + 1/3 at CRI + 1/3 at T + 3 per cent urea spray (US) at flowering (F) and 1/4 at S + 1/2 at CRI + 1/4 at T + 3 per cent US at F to bed sown wheat including 1 recommended treatment of flat sown *durum* wheat (Rec.) and an extra treatment where whole N (120 kg ha^{-1}) was applied at sowing to bed planted wheat for comparison were replicated thrice in a randomized block design having gross plot size of 6.00 m × 4.05 m while net plot size measured 5.00 m × 2.70 m. A seed rate of 100 kg ha^{-1} of *durum* wheat (cv. PDW 274) was used. Two rows per bed (67.5 cm) were sown and row to row spacing of 22.5 cm was kept for flat sowing. All other agronomic management practices of the crop were uniformly adopted. The crop was sown on November 9, 2002 and harvested on April 24, 2003. The data on yield attributes and yield were recorded at harvest. Data on various growth parameters and microclimatic parameters such as tiller count, plant height, dry matter accumulation, canopy temperature, soil temperature, relative humidity and Photosynthetically active radiation interception were recorded at periodic intervals of 45, 75, 105 and 135 days after sowing (DAS).

Assman Psychrometer was used for recording data on canopy temperature and relative humidity between 10.00 am to 11.00 am and 2.00 pm to 3.00 pm. Soil temperature (0°C) was recorded at 5 cm soil depth with a Bimetal soil thermometer between 2.00 pm to 3.00 pm. Penetration of photosynthetically active radiation (PAR) in the range of 0.4 to 0.7 microns was measured between 10.00 am to 11.00 am, 12.00 noon to 1.00 pm and 2.00 pm to 3.00 ppm and mean values for the day were calculated for further use. Line quantum sensor (LI-COR Photometer model LI-191-SA) which has quantum (photon) response through wavelength range of 400–700 nm for photosynthetic photon flux density (PPFD) was used for PAR measurements in µmoles m^{-2} sec^{-1}. The percentage photosynthetically active radiation intercepted (PARI) by the crop was computed for reporting.

$$\text{PARI (\%)} = \frac{\text{PAR above the crop canopy} - \text{PAR at the soil surface}}{\text{PAR above the crop canopy}} \times 100$$

All interaction effects were found to be non significant therefore the mean effects are discussed in results.

Results and Discussion

Tiller Count

Tillering is one of the most important determinant of grain yield in wheat. Good tillering leads to production of more number of ears per unit area leading to higher grain yield. The data on tiller count are presented in Table 10.1. The differences between the bed planted and flat sown crop were non-significant.

Table 10.1: Effect of Nitrogen Management and Time of Nitrogen Application on Number of Tillers in Bed Planted *durum* Wheat

Treatments	*Total Tillers m⁻²*			*Effective Tillers m⁻²*
	Days After Sowing			
	45	*75*	*105*	*135*
Nitrogen management (ha⁻¹)				
120 kg N	325	364	285	239
120 kg N +15 t FYM	380	397	294	245
150 kg N	340	369	287	243
CD (P = 0.05)	NS	NS	NS	NS
Time of N application				
1/2S + 1/2CRI	344	354	273	231
1/3S + 1/3CRI + 1/3T	352	379	282	240
1/4S + 1/2CRI + 1/4T	348	389	284	242
1/3S + 1/3CRI +1/3T + 3 per cent US	354	380	303	247
1/4S + 1/2CRI +1/4T + 3 per cent US	345	380	303	248
CD (P = 0.05)	NS	NS	NS	NS
Interaction	NS	NS	NS	NS
Flat	535.6	494.6	258.5	250.9
CD (P = 0.05) Bed vs Flat	NS	NS	NS	NS
Bed (S)	340	357	270	234

S; Sowing; CRI: Crown root initiation; T: Tillering; US: Urea spray at flowering stage.

Nitrogen management did not influence the tiller count significantly throughout the crop growth as the results are not significant statistically. Similar results were reported by Dhiman *et al.* (1982) and Kaur (2000). Application of FYM along with 120 kg N ha^{-1} tended to produce higher tiller count throughout the crop growth over 120 kg N ha^{-1} or 150 kg N ha^{-1} when applied alone. Similar results have been reported by Negi and Mahajan (2000) and Sushila and Giri (2000).

Time of application of nitrogen also resulted in non-significant differences in tiller count throughout the crop growth. The lowest tiller count was recorded in two splits followed by three splits and the highest with three splits combined with urea spray. Tiller count values with whole N applied at sowing were comparable with the various split applications on the beds. Similar results have been reported by Uppal (1999) and Dhuka *et al.* (1991).

Plant Height

Plant height is a useful growth index of the developmental phase of plant to give an idea about the dry matter production leading to ultimate yield. The data on plant height are presented in Table 10.2. The plant height was almost similar between the bed planted and flat sown crop throughout the crop growth as they were statistically at par.

Table 10.2: Effect of Nitrogen Management and Time of Nitrogen Application on Periodic Plant Height in Bed Planted *durum* Wheat

Treatments	*Plant Height (cm)*			
	Days After Sowing			
	45	*75*	*105*	*135*
Nitrogen management (ha^{-1})				
120 kg N	28.8	41.9	64.2	77.8
120 kg N + 15 t FYM	34.2	44.5	70.2	85.0
150 kg N	30.6	42.7	65.2	79.3
CD (P = 0.05)	4.1	2.1	3.2	3.3
Time of N application				
1/2S + 1/2CRI	30.9	41.8	65.6	79.2
1/3S + 1/3CRI + 1/3T	31.3	43.0	66.2	79.5
1/4S + 1/2CRI + 1/4T	31.6	43.0	67.7	79.6
1/3S + 1/3CRI + 1/3T + 3 per cent US	31.1	43.7	67.0	82.4
1/4S + 1/2CRI + 1/4T + 3 per cent US	31.0	43.7	67.0	82.1
CD (P = 0.05)	NS	NS	NS	NS
Interaction	NS	NS	NS	NS
Flat	32.1	40.7	62.0	77.1
CD (P = 0.05) Bed vs Flat	NS	NS	NS	7.6
Bed (S)	30.5	40.5	63.7	78.0

The perusal of data indicated an increase in height of the plants with advancement in age and maximum height was obtained at 135 DAS. The increase in height was much rapid during the period between 75 to 105 DAS.

Nitrogen management significantly influenced plant height throughout crop growth. At 45 DAS minimum height was recorded at 120 kg N ha^{-1} which was significantly lower than 120 kg N + FYM and 150 kg N ha^{-1} both being at par between themselves. Similar trend was followed at 75 DAS. However at 105 and 135 DAS 120 kg N + 15 t FYM ha^{-1} recorded maximum plant height which was significantly higher than other two treatments which were at par statistically. The increase in plant height is attributed to the role of nitrogen in plant metabolism and growth (Tisdale, 1985). Similar results have been reported earlier also (Singh and Agarwal, 2001). Significant increase in plant height with application of FYM at 10 t + 120 kg N ha^{-1} over 120/t kg N ha^{-1} was reported by Negi and Mahajan, 2000).

The split application of nitrogen did not affect plant height significantly. The earlier work done by Dhuka *et al.* (1991), Singh (1996) and Uppal (1999) also revealed that plant height did not differ significantly with different times of nitrogen application.

Dry Matter Accumulation

Accumulation of dry matter is good growth index to express the photosynthetic efficiency of plants. The dry matter accumulation (DMA) was not influenced significantly by sowing methods *i.e.* bed planting and flat sowing of wheat (Table 10.3) throughout crop growth. Similar results were reported by Kaur (2000). The maximum DMA was produced during 75–105 DAS (Table 10.3).

Table 10.3: Effect of Nitrogen Management and Time of Nitrogen Application on Periodic Dry Matter Accumulation in Bed Planted *durum* Wheat

Treatments	*Dry Matter (q ha^{-1})*			
	Days After Sowing			
	45	*75*	*105*	*135*
Nitrogen management (ha^{-1})				
120 kg N	6.01	31.44	75.29	90.95
120 kg N + 15 t FYM	6.85	35.12	79.68	97.66
150 kg N	6.21	34.41	77.83	94.99
CD (P = 0.05)	NS	NS	NS	NS
Time of N application				
1/2S + 1/2CRI	6.25	30.85	72.44	84.71
1/3S + 1/3CRI + 1/3T	6.34	34.16	76.29	93.81
1/4S + 1/2CRI + 1/4T	6.47	35.25	75.77	93.97
1/3S + 1/3CRI + 1/3 T + 3 per cent US	6.36	34.01	82.75	99.23
1/4S + 1/2CRI + 1/4 T + 3 per cent US	6.47	34.01	80.76	100.94
CD (P = 0.05)	NS	NS	NS	NS
Interaction	NS	NS	NS	NS
Flat	5.48	31.90	81.98	99.98
CD (P = 0.05) Bed vs Flat	2.28	NS	NS	NS
Bed (S)	6.00	30.20	72.00	82.50

Nitrogen management did not affect the dry matter accumulation by the crop significantly throughout the growth. Application of FYM along with chemical fertilizers (120 kg N + 15 t FYM ha^{-1}) tended to result in higher dry matter accumulation than 120 kg N and 150 kg N ha^{-1} when applied only through chemical fertilizers. This effect of FYM may be due to increased availability of nitrogen to crop for longer period. However, significantly higher dry matter production by application of FYM at 15 t ha^{-1} along with recommended dose of fertilizer (120 kg N ha^{-1}) have been reported by Nehra *et al.* (2001).

The effect of split application of nitrogen was also found non-significant on dry matter accumulation throughout the crop growth. Application of nitrogen through three splits and three splits + urea spray tended to accumulate more dry matter than two splits and whole N applied at the sowing time due to the reason that the application of nitrogen in three splits in light textured soils made the nitrogen available to plants for longer period and synthesis of photosynthates continued and thus increased the dry matter accumulation CGR followed the dry matter accumulation pattern of the crop. Similar results have been reported by Uppal *et al.* (2000) and Singh (1996).

Canopy Temperature

The data on canopy temperature are given in Table 10.4. In general the canopy temperature increased with increase in age of crop. At 45 and 75 DAS the canopy temperature on an average was lower in flat sown as compared to bed sown crop except at 135 DAS. At 135 DAS LAI was less due to more senescence of leaves. Similar results have also been reported by Kaur *et al.* (2001). More LAI captured higher radiation and resulted in lower temperature.

Table 10.4: Effect of Nitrogen Management and Time of Nitrogen Application on Periodic Canopy Air Temperature in Bed Planted *durum* Wheat

Treatments	*Canopy Air Temperature (°C)*			
	Days After Sowing			
	45	*75*	*105*	*135*
Nitrogen management (ha^{-1})				
120 kg N	16.5	20.1	24.1	28.6
120 kg N + 15 t FYM	15.8	19.4	22.1	27.4
150 kg N	15.6	19.8	22.8	28.0
Time of N application				
1/2S + 1/2CRI	17.2	20.6	23.3	28.6
1/3S + 1/3CRI + 1/3T	15.5	20.3	23.1	28.0
1/4S + 1/2CRI + 1/4T	15.9	19.6	23.0	28.4
1/3S + 1/3CRI +1/3T + 3 per cent US	15.6	19.0	22.8	27.6
1/4S + 1/2CRI + 1/4T + 3 per cent US	15.6	19.3	22.9	27.3
Flat	14.5	18.6	22.0	28.9
Bed (S)	15.0	18.9	22.9	28.0

Among nitrogen management, with increase in nitrogen level from 120 kg N to 150 kg N ha^{-1} the canopy air temperature decreased and the lowest was recorded in 120 kg N + 15 t FYM ha^{-1} throughout the crop growth. It may be due to more nitrogen availability which helped in sustaining higher LAI and interception of high solar radiation resulting in lower temperature. Grewal (1987) has reported decreased canopy temperature with increased nitrogen level from 40 to 200 kg N ha^{-1}.

With the increase in the number of splits the canopy air temperature decreased. The maximum value throughout the crop growth was recorded with two split application with a general order of two splits > three splits. Almost same canopy temperature values were obtained with whole N dose applied at sowing on beds. While after 3 per cent urea spray at flowering stage, the air temperature recorded at 105 and 135 DAS showed the trend as two splits > three splits > three splits + urea spray.

Soil Temperature

The data pertaining to soil temperature are given in Table 10.5. In general, the soil temperature remained lower in flat sown as compared to bed planted wheat. Light was falling due to border effect on all rows through vacant space in furrows and keeping the soil warm in bed sown crop. Similar results have been reported by Kaur (2000).

Table 10.5: Effect of Nitrogen Management and Time of Nitrogen Application on Periodic Soil Temperature in Bed Planted *durum* Wheat

Treatments	*Soil Temperature (°C)*			
	Days After Sowing			
	45	*75*	*105*	*135*
Nitrogen management (ha^{-1})				
120 kg N	15.0	17.7	24.1	28.4
120 kg N + 15 t FYM	14.8	17.2	23.2	27.7
150 kg N	14.9	17.4	23.5	28.2
Time of N application				
1/2S + 1/2CRI	15.0	18.6	25.0	29.0
1/3S + 1/3CRI + 1/3T	14.9	17.2	24.0	28.7
1/4S + 1/2CRI + 1/4T	14.9	17.2	23.8	28.6
1/3S + 1/3CRI +1/3T + 3 per cent US	14.9	17.1	22.8	27.5
1/4S + 1/2CRI + 1/4T + 3 per cent US	14.9	17.1	22.7	27.4
Flat	14.2	16.8	23.0	28.0
Bed (S)	14.0	16.1	23.5	29.0

With increase in nitrogen level from 120 to 150 kg ha^{-1} the soil temperature tended to decrease and it further declined under 120 kg N + 15 t FYM ha^{-1}. The highest temperature was recorded under 120 kg N ha^{-1} and the lowest with 120 kg N + 15t FYM ha^{-1}. The general order was 120 kg N ha^{-1} > 150 kg N ha^{-1} > 120 kg N + 15 t FYM ha^{-1}. This is due to better crop growth and interception of more solar radiation by the crop canopy. Grewal (1987) has reported decrease in soil temperature with increase in nitrogen level from 40 to 200 kg N ha^{-1}. The soil temperature recorded was lower where light interception was more because incoming radiation was intercepted by crop and lesser of it reached the ground level. Such results have been reported by Kler and Bains (1985), Kler *et al.* (1987), Dhaliwal and Kler (1995), where light interception showed inverse relation with soil temperature and direct correlation with yield of various crops.

Among the time of N application the soil temperature decreased with increase in number of splits. The value of soil temperature tended to be higher in two split than three split application. After 3 per cent urea spray the trend was two splits > three splits > three splits + 3 per cent urea spray. The application of urea as spray further tended to increase the canopy growth and PAR interception resulting in lower temperature. However, soil temperature on the beds was slightly lower than two and three splits at 45 and 75 DAS and almost same at 75 and 135 DAS where whole N was applied at sowing, there were no differences statistically.

Relative Humidity

The periodic data on relative humidity was taken at regular intervals and are presented in Table 10.6. The relative humidity tended to be higher in case of flat sown compared to bed planted crop. Among the nitrogen management treatments, with increase in nitrogen level from 120 to 150 kg ha^{-1} there was increase in relative humidity and it further increased under 120 kg N + 15 t FYM ha^{-1} through out the crop growth. Increase in relative humidity with increase in nitrogen level has been reported by Kaur *et al.* (2001). As PARI (per cent) increased by the crop, relative humidity also increased.

Table 10.6. Effect of Nitrogen Management and Time of Nitrogen Application on Periodic Relative Humidity in Bed Planted *durum* Wheat

Treatments	*Relative Humidity (%)*			
	Days After Sowing			
	45	*75*	*105*	*135*
Nitrogen management (ha^{-1})				
120 kg N	50.4	79.8	77.0	54.5
120 kg N + 15 t FYM	54.2	86.4	79.2	61.0
150 kg N	51.2	83.0	77.8	57.0
Time of N application				
1/2S + 1/2CRI	48.0	81.0	76.3	53.0
1/3S + 1/3CRI + 1/3T	52.7	83.7	78.1	56.8
1/4S + 1/2CRI + 1/4T	53.0	83.3	78.2	57.0
1/3S + 1/3CRI + 1/3T + 3 per cent US	52.8	83.6	78.7	57.1
1/4S + 1/2CRI + 1/4T + 3 per cent US	53	83.3	78.7	57.2
Flat	53	83.5	79.5	54.5
Bed (S)	48.5	81.0	77.5	53.2

With increase in the number of N splits there was increase in the relative humidity. Minimum values of relative humidity were recorded under two split application of nitrogen followed by three split doses at 45 and 75 DAS where as at 105 DAS and 135 DAS after 3 per cent urea spray it was further reduced in the plots receiving urea spray. Whole N on beds applied at sowing recorded almost similar values of relative humidity as two splits.

Photosynthetically Active Radiation Interception

The visible light interception (PARI per cent) increased with increase in crop age (Table 10.7) upto 105 days and found slightly less at 135 days of crop age. The effect of sowing methods on PARI (per cent) was found significant. The PARI (per cent) was recorded significantly higher in flat sown crop than bed planted crop at 45 and 75 DAS and tended to be higher at 105 DAS whereas at 135 DAS PARI (per cent) was slightly less in flat sown crop. Significantly higher PARI (per cent) by flat sown over bed planted crop throughout the crop growth has been reported by Kaur (2000). Lower interception of visible radiation by the bed planted crop supports the fact that the bed planting system of sowing has border effect on all the rows of the crop.

Table 10.7: Effect of Nitrogen Management and Time of Nitrogen Application on Photosynthetically Active Radiation Interception in Bed Planted *durum* Wheat

Treatments	*Photosynthetically Active Radiation Interception (PARI per cent)*			
	Days After Sowing			
	45	*75*	*105*	*135*
Nitrogen management (ha^{-1})				
120 kg N	40.2	58.7	74.2	56.2
120 kg N + 15 t FYM	48.4	66.3	75.6	62.2
150 kg N	41.1	62.1	74.6	58.5
CD (P = 0.05)	5.4	6.75	NS	NS
Time of N application				
1/2S + 1/2CRI	40.4	56.5	72.0	55.2
1/3S + 1/3CRI + 1/3T	43.6	63.2	73.3	57.1
1/4S + 1/2CRI + 1/4T	44.7	64.2	75.2	60.0
1/3S + 1/3CRI + 1/3T + 3 per cent US	44.0	63.5	75.5	60.9
1/4S +1/2CRI + 1/4T + 3 per cent US	44.1	64.3	77.0	61.6
CD (P = 0.05)	NS	NS	NS	NS
Interaction	NS	NS	NS	NS
Flat	74.4	98.0	98.4	88.4
CD (P = 0.05) Bed vs Flat	11.9	1.46	NS	13.3
Bed (S)	49.9	60.0	71.0	54.9

The nitrogen management had a significant effect on PARI (per cent) by the crop at initial stages of crop growth at 45 and 75 DAS. The highest values for PARI (per cent) were recorded at 120 kg N + 15 t FYM ha^{-1} throughout the crop growth. At 45 DAS the PARI (per cent) by the crop with 120 kg N + 15 t FYM ha^{-1} (48.4 per cent) was significantly higher than both 120 (40.2 per cent) and 150 kg N ha^{-1} (41.1 per cent). However at 75 DAS the 120 kg N + 15 t FYM intercepted significantly higher PAR (66.3 per cent) than 120 kg N ha^{-1} (58.7 per cent) but was statistically at par with 150 kg N ha^{-1} (62.1 per cent). While at 105 and 135 DAS the PARI (per cent) was not affected significantly by nitrogen management and the observed order of values was 120 kg N + 15 t FYM > 150 kg N ha^{-1} > 120 kg N ha^{-1}. Addition of FYM continued the available supply of nutrients to plant photosynthetic apparatus showing more LAI and evidently more capture of radiant energy.

The time of nitrogen application did not influence the PARI (per cent). However with increase in number of splits from two to three the PARI (per cent) tended to increase. After the 3 per cent urea spray the observed order was two splits < three splits < three splits + 3 per cent urea spray. More LAI captured higher light and hence higher PARI (per cent) values. Thus showing that urea spray helped in maintaining the leaf area for longer period preventing senescence. Application of whole N at sowing on beds also resulted in comparable value of PARI (per cent). With increase in number of splits, nitrogen was available to plants for longer period helping delayed senescence of leaves for trapping more PARI (per cent) at later stages of crop growth.

Grain Yield

The ultimate effect of experimental variables is reflected in the final yield of the crop and thus it is a major criterion to identify the efficiency of various treatments in a given situation. The efficiency of different factors can be judged mainly by the addition they make to yield which is the ultimate aim of the field research investigations. The data on the grain yield as affected by various treatments are given in Table 10.8. The grain yield of bed planted *durum* wheat was statistically on par with the flat sown crop.

Table 10.8: Effect of Nitrogen Management and Time of Nitrogen Application on Grain Yield and Harvest Index in Bed Planted *durum* Wheat

Treatments	*Grain Yield (q ha^{-1})*	*Harvest Index*
Nitrogen management (ha^{-1})		
120 kg N	38.33	0.379
170 kg N +15 t FYM	42.54	0.380
150 kg N	39.19	0.372
CD (P = 0.05)	NS	–
Time of N application		
1/2S + 1/2CRI	38.33	0.373
1/3S + 1/3CRI + 1/3T	39.00	0.372
1/4S + 1/2CRI + 1/4T	39.78	0.379
1/3S + 1/3CRI + 1/3T + 3 per cent US	41.12	0.378
1/4S + 1/2CRI + 1/4T + 3 per cent US	41.88	0.383
CD (P = 0.05)	NS	–
Interaction	NS	–
Flat (Rec.)	39.53	0.377
CD (P = 0.05) Bed vs Flat	NS	–
Bed (S)	38.31	0.373

Nitrogen management levels did not influence the grain yield significantly. However, the application of combined use of 15 t FYM with 120 kg N ha^{-1} as fertilizer in bed planted wheat tended to increase the grain yield over both 120 kg N ha^{-1} (4.21 q ha^{-1} or 9.9 per cent) and 150 kg N ha^{-1} (3.35 q ha^{-1} or 7.9 per cent) as fertilizers. This increase in the yield was due to cumulative increased effect of various yield attributing characters in 120 kg N + 15 t FYM ha^{-1} treatment (Negi and Gulshan, 2000) also reported 10.5 per cent increase in wheat yield with FYM application over no FYM application. Similar results were reported by (Nehra *et al.*, 2001; Thakur *et al.*, 1999) and (Singh and Agarwal, 2001). Time of nitrogen application did not influence the grain yield significantly and all treatments were at par statistically. Similar results were reported by (Robinson *et al.*, 1979; Dhuka *et al.*, 1991; Mairona *et al.*, 1992; Singh, 1996; Uppal, 1999).

Harvest Index

The data regarding harvest index were presented in Table 10.8. The harvest index values for bed planted and flat sown crop were almost same. Among nitrogen management treatments the highest harvest index value was obtained with 120 kg N + 15t FYM ha^{-1} (0.380) followed by 120 kg Nha^{-1} (0.379) and lowest was recorded in 150 kg N ha^{-1} (0.372). Thus showing that increasing the level of nitrogen through FYM manure increased or at least maintained harvest index when compared with 120 kg N ha^{-1} whereas when nitrogen level was increased only through fertilizer at 150 kg N ha^{-1} the harvest index decreased because there was more increase in straw yield than grain yield. Slow availability of nitrogen from FYM improved the yield attributes and thus grain yield.

Time of N application did not influence the harvest index value and 3 per cent urea spray tended to increase the harvest index marginally. When whole dose of N was applied at sowing on beds almost similar harvest index value was obtained as compared with two split application on beds and flat.

Thus it may be concluded that the performance of *durum* wheat with respect of level of nitrogen management and time of nitrogen application under bed and flat sowing was statistically on par between each other. Under bed sown, management of nitrogen through integration of chemical fertilizers with FYM (120 kg N + 15t FYM ha^{-1}) tended to improve yield attributes and yield over alone chemical fertilizers in *durum* wheat. With increase in times of nitrogen application from whole dose of N applied at sowing to two splits, three splits and three splits + 3 per cent urea spray, there was tendency of beneficial effect on growth and microclimatic parameters tending to improve grain yield in bed sown *durum* wheat.

References

Dhaliwal, G. S. and Kler, D. S., 1995. *Principles of Agricultural Ecology*. Himalaya Publishing House, Bombay.

Dhiman, S.D., Mohan, R.D.S. and Sharma, H.C., 1982. Evaluation of nitrogen requirement for triple gene dwarf wheat. *Indian J. Agron.*, 27: 383–393.

Dhuka, A.K., Sadaria, G.S., Patel, J.C. and Vyas, M.N., 1991. Response of irrigated late sown wheat to nitrogen application. *Indian J. Agron.*, 36: 276–277.

Grewal, D.S., 1987. Microenvironmental parameters in wheat field as influenced by geometric pattern and nitrogen fertility. *Indian J.Ecol.*, 14: 72–77.

Kaur, G., 2000. Studies on planting techniques at higher nitrogen nutrition for microclimate, lodging, growth and yield of wheat (*Triticum aestivum* L.). *M.Sc. Thesis*, Punjab Agricultural University, Ludhiana, India.

Kaur, G., Kler, D.S. and Singh, S., 2001. Influence of planting techniques and nitrogen levels on leaf area index, dry matter, canopy air temperature and relative humidity of wheat (*Triticum aestivum* L.). *Indian J. Environ. Ecoplan.*, 5(2): 265–269.

Kler, D.S. and Bains, D.S., 1985. Relationship between soil temperature and light interception in *durum* wheat (*Triticum durum* Desf.) under different sowing pattern. *Haryana J. Agron.*, 1: 107–112.

Kler, D.S., Dhaliwal, J.S. and Atwal, A.S., 1987. Effect of ploughing frequency, stand geometry and nitrogen levels on the performance of *durum* wheat in Bajra-wheat rotation. *Transactions of Indian Society of Desert Technology*, 12: 21–26.

Kler, D.S., Walia, A.S. and Singh, S., 1987. Utility of light interception and soil temperature relations for some crops as influenced by directional orientations of sowing. *Environ. Ecol.*, 5: 422–429.

Negi, S.C. and Mahajan, G., 2000. Effect of FYM, planting methods and fertilizer levels on rainfed wheat. *Crop Res.*, Hisar, 20: 534–36.

Nehra, A.S., Hooda, I.S. and Singh, K.P., 2001. Effect of integrated nutrient management on growth and yield of wheat (*Triticum aestivum*). *Indian J. Agron.*, 46(1): 112–117.

Singh, K., 1996. Effect of irrigation regime and nitrogen levels on growth, yield and quality of *durum* wheat (*Triticum durum* desf.). *M.Sc. Thesis*, Punjab Agricultural University, Ludhiana, India.

Singh, R. and Agarwal, S.K., 2001. Growth and yield of wheat (*Triticum aestivum*) as influenced by levels of farmyard manure and nitrogen. *Indian J. Agron.*, 46(3): 462–467.

Sushila, R. and Giri, G., 2000. Influence of farmyard manure, nitrogen and biofertilizers on growth, yield attributes and yield of wheat (*Triticum aestivum*) under limited water supply. *Indian J. Agron.*, 45(3): 590–595.

Tisdale, S.L., Nelson, W.L. and Beaton, J.D., 1985. *Soil Fertility and Fertilizer.* MacMillan Publishing Company, New York.

Uppal, R.S., 1999. Effect of levels of NPK and time of nitrogen application on yield and quality of *durum* wheat (*Triticum durum* Desf.). *Ph.D. Dissertation*, Punjab Agricultural University, Ludhiana, India.

Uppal, R.S., Singh, J. and Kler, D.S., 2000. Effect of split application of nitrogen on growth and yield of *Durum* wheat. *Indian J. Environ. Ecoplan.*, 3: 527–534.

Chapter 11

Impact of Organic Mulberry Cultivation Practices on Soil Microbial and Fertility of M_5 Mulberry Garden

K. Rashmi[1], M.A. Shankar[2], T.K. Narayanaswamy[1], K.R. Sreeramulu[3] and Rajegowda[1]

[1]Department of Sericulture, [2]Dryland Agriculture and [3]Agricultural Microbiology University of Agricultural Sciences, GKVK, Bangalore – 560 065

ABSTRACT

The effect of organic mulberry cultivation practices on the microbial development and fertility status of soil was studied in M_5 mulberry garden during 2004–05 at Main Research Station, Hebbal, University of Agricultural Sciences, Bangalore. The results revealed that, application of biofertilizers 10 kg each of *Azospirillum brasilense* + *Aspergillus awamori* + 20 per cent each of recommended N through compost, green manure (*Glyricidia maculata*), castor cake, vermicompost, fertilizer + remaining P and K through fertilizers (T_{11}) significantly increased the available macro and micronutrient contents in the soil. Whereas, the increased number of beneficial microbes *viz., A. awamori* and *A. brasilense* population (15.00×10^3 and 15.00×10^6 cfu/g of soil, respectively) were found at 10 kg each of *A. brasilense* + *A. awamori* + 25 per cent each of recommended N through compost, green manure (*Glyricidia maculata*), castor cake, vermicompost (T_{10}) compared to control.

Keywords: *Mulberry, Organic manures, Soil microbial, Soil fertility.*

Introduction

Quality of mulberry leaf plays an important role in the nutrition of silkworm. Production of quality leaf depends on several factors including the native soil fertility. The fertility of soil not only depends on its chemical composition but also on the nature of microorganisms living in it. However, organic manures are known to improve soil fertility, physical, chemical and biological properties of soil. In addition, the inorganic fertilizers are also recognized as a key component for improved crop productivity and sustainable fertility management. However excess use of chemical fertilizers in an unbalanced manner has a deleterious effect on soil structure and soil properties such as alkalinity, acidity and nutrient deficiencies which leads to total loss of soil health. In order to balance this situation, the different organic manures *viz.*, farm yard manure (FYM), compost, green manure, vermicompost, oil cakes *etc.* acts as a better source of nutrients either alone or in combination with fertilizers. Hence, the present investigation was undertaken to study the impact of organic farming practices for mulberry cultivation and their influence on microbial and fertility status of soil.

Materials and Method

A field experiment was carried out during 2004–2005 in established irrigated M_5 mulberry garden planted at a spacing of 60 × 60 cm. The experiment was laid out in randomized complete block design (RCBD) with 13 treatments of 3 replications each. Before the commencement of experiment the soil analysis report revealed the presence of carbon content (10.6 per cent), nitrogen (200 kg/ha), phosphorous (50 kg/ha), potassium (86 kg/ha), soil pH 6.8 and the microbial population such as bacteria (7×10^6 cfu/g soil), fungi (2×10^3 cfu/g of soil), actinomycetes (1×10^4 cfu/g of soil), *Azospirillum brasilense* (4×10^6 cfu/g of soil) and PSB (1×10^3 cfu/g of soil). The cultivation practices for mulberry were followed as suggested by Dandin *et al.* (2003). The treatments were as follows.

T_1: 100 per cent recommended N through compost; T_2: 50 per cent recommended 'N' through compost + 50 per cent recommended 'N' and remaining P, K through fertilizer; T_3: 100 per cent recommended N through green manure (*Glyricidia maculata*); T_4: 50 per cent recommended N through green manure + 50 per cent recommended N and remaining P and K through fertilizer; T_5: 100 per cent recommended N through castor cake; T_6: 50 per cent recommended N-castor cake + 50 per cent recommended N and remaining P and K through fertilizer; T_7: 35 per cent recommended N through compost + 30 per cent recommended N through castor cake + 35 per cent recommended N through green Manure; T_8: 100 per cent recommended N through vermicompost; T_9: 50 per cent recommended N through vermicompost + 50 per cent recommended N and remaining P and K through fertilizer; T_{10}: Biofertilizer 10 kg each of *Azospirillum brasilense* + *Aspergillus awamori* + 25 per cent recommended N through compost, green manure, castor cake, vermicompost; T_{11}: Biofertilizer 10 kg each of *Azospirillum brasilense* + *Aspergillus awamori* + 20 per cent each recommended N through compost green manure, castor cake, vermicompost + 20 per cent recommended N and remaining P and K through fertilizers; T_{12}: Recommended 20 tonnes FYM + 300 : 120 : 120 NPK kg/ha/year through fertilizer (control); T_{13}: Only fertilizer 300 : 120 : 120 NPK kg/ha/year (control).

The soil fertility and microbial population was estimated after the harvest of third crop. The available nitrogen in soil was determined by alkaline potassium permanganate method. Available phosphorous was estimated by Brays No. 1 method, available potassium by Flame photometer method. Calcium and available magnesium were determined by Versanate titration method and available sulphur by Turbidometric method as per the procedure outlined by Jackson, 1973. The available copper, manganese and zinc were estimated by Atomic absorption spectrophotometer method. The data were analysed statistically by using simple RCBD as outlined by Cochran and Cox (2000).

Results and Discussion

The fertility status and microbial population of M_5 mulberry garden was significantly superior when treated with biofertilizers at 10 kg each of *A. brasilense* and *A. awamori* + 20 per cent each of recommended N through compost, green manure, castor cake, vermicompost, fertilizers + remaining P and K through fertilizers (T_{11}).

After harvest of third crop the soil analysis indicated that, among the different treatments, T_{11} has recorded significantly higher available nitrogen (274 kg/ha), phosphorous (80.60 kg/ha), potassium (259 kg/ha), calcium (4.85 c.mol/ kg), magnesium (4.10 c.mol/kg), sulphur (38 ppm), copper (1.40 ppm), manganese (22.00 ppm) and zinc (2.00 ppm) contents (Table 11.1). Availability of nutrients from both organic and inorganic sources was greatly influenced by nitrogen fixation by *A. brasilense* and solubalization of P by *A. awamori*.

Table 11.1: Macro, Secondary and Micro Nutrient Contents in Soil Grown with M_5 Mulberry as Influenced by Different Organic Manures and Inorganic Fertilizers

Treatments	*Available N (kg/ha)*	*Available P (kg/ha)*	*Available K (kg/ha)*	*Available Ca (C.mol\Kg)*	*Available Mg (C.mol\Kg)*	*Available S (ppm)*	*Available Cu (ppm)*	*Available Mn (ppm)*	*Available Zn (ppm)*
T_1	232	243	64.30	4.39	3.45	26.66	1.00	18.00	1.30
T_2	201	216	60.80,	4.20	3.00	30.00	0.90	17.00	1.30
T_3	185	224	62.20	4.12	2.88	37.00	1.00	16.00	1.20
T_4	239	232	70.90	4.35	3.33	29.00	1.10	17.00	1.50
T_5	195	239	69.80	4.34	3.30	27.00	0.80	19.00	1.40
T_6	245	240	75.40	4.52	3.70	33.00	1.30	20.00	1.70
T_7	191	236	71.60	4.00	2.22	27.00	1.00	17.00	1.60
T_8	183	226	72.90	3.90	2.98	31.00	0.90	18.00	1.40
T_9	242	249	74.80	4.50	3.60	33.00	1.30	17.00	1.70
T_{10}	212	240	73.56	4.38	3.45	32.00	1.20	18.00	1.50
T_{11}	274	259	80.60	4.85	4.10	38.00	1.40	22.00	2.00
T_{12}	268	255	78.70	4.69	3.90	35.00	1.30	20.00	1.90
T_{13}	188	214	60.40	3.70	2.26	20.00	0.80	14.00	1.20
F test	*	*	NS	*	*	*	*	*	*
SEm ±	5.94	5.88	–	0.06	0.05	3.08	0.13	1.25	0.15
CD at 5%	17.35	17.18	–	0.17	0.15	9.01	0.39	3.67	0.46

Significantly higher bacterial (65,00 × 10^6 cfu/g of soil), fungal (30.66 × 10^3 cfu/g of soil) and actinomycetes population (7.00 × 10^4 cfu/g of soil) were recorded in T_{11}. Whereas, maximum *A. awamori* (15.00 × 10^3 cfu/g of soil) and *A. brasilense* population (15.00 × 10^6 cfu/g of soil) were recorded in the plots manured with bio fertilizers 10 kg each of *A. brasilense* and *A. awamori* + 25 per cent each of recommended N through compost, green manure, castor cake, vermicompost and fertilizers (T_{10}) (Table 11.2). Prabhuraj *et al.* (2005) reported that, combination of FYM, sericulture wastes compost, green manure and biofertilizers with NPK recorded significantly higher population of inoculated Phosphate solubalizing microorganisms (6.1 × 10^4 cfu/g of soil) and Nitrogen fixing bacteria (5.1 × 10^4 cfu/g of soil).

Table 11.2: Microbial Population (cfu/g) in Soil of M_5 Mulberry Garden as Influenced by Different Organic Manures and Inorganic Fertilizers

Treatments	*Bacteria (No. × 10^6)*	*Fungi (No. × 10^3)*	*Actimycetes (No. × 10^4)*	*Aspergillus awamori (No. × 10^3)*	*Azospirillum brasilense (No. × 10^6)*
T_1	33.00	10.00	03.00	10.00	08.00
T_2	32.00	10.33	03.00	06.00	07.00
T_3	31.33	18.33	03.00	09.00	06.00
T_4	33.00	21.66	04.00	10.00	08.00
T_5	25.66	21.66	04.33	09.00	07.00
T_6	42.66	30.33	06.00	10.00	07.00
T_7	34.66	12.66	05.00	11.00	10.00
T_8	33.66	19.00	05.33	08.00	08.00
T_9	38.00	27.00	06.00	11.00	11.00
T_{10}	32.00	24.00	05.66	20.00	16.00
T_{11}	65.00	30.66	07.00	15.00	15.00
T_{12}	51.00	30.66	06.66	13.00	12.00
T_{13}	24.33	09.00	02.33	06.00	07.00
F test	*	*	*	*	*
SEm ±	2.15	1.39	0.89	1.32	1.19
CD at 5%	6.28	4.05	2.61	3.86	3.48

From the above results it is concluded that, the integrated use of organic manures, biofertilizers with lower dose of inorganic fertilizers not only economized the N requirement of crop but also helped in sustained availability of nutrients for crop production. The application of organic manures in turn provides adequate biomass as a feed for the microbes and helps in increasing microbial population in soil.

Acknowledgement

The research investigation is part of the Department of Science and Technology (DST) New Delhi funded project entitled "Significance of organically grown mulberry on bio-chemical and reproductive biology of silkworm seed cocoons (*Bombyx mori* L.). The financial help received from the DST is greatly acknowledged.

References

Cochran and Cox, 2000. *Experimental Design*, Second edition. Wiley Classics Publishers, p. 95–145.

Dandin, S.B., Jayanth Jayaswal and Giridhar, K., 2003. *Handbook of Sericulture Technologies*. CSB, Bangalore, pp. 259

Jackson, M.L., 1973. *Soil Chemical Analysis*. Prentice Hall of India (Pvt.) Ltd., New Delhi, p. 259–260.

Prabhuraj, D.K., Bongale, U.D., Sukumar, J., Sanaulla, H. and Thimma Reddy, H., 2005. Comparative study on the organic and integrated nutrient management in mulberry. In: *Progress of Research in Organic Sericulture and Seri-Byproducts Utilization*, 2005, pp. 146–148.

Chapter 12

Himalayas: Seat of Pristine Biodiversity

Rajan Kumar Gupta and Shiwani Kaushik
Department of Botany, Government Post Graduate College, Rishikesh –249 201, Dehradun, U.A.

ABSTRACT

India is a country of vast biodiversity in the world, 45,000 wild species of plants and over 77,000 wild species of animals have so far been recorded (Botanical Survey of India undated; Zoological survey of India 1991), together comprising about 6.5 per cent of the world's known wildlife. India also has a very rich element of "endemics" in its flora. The richest area from the biodiversity point of view lies in the Himalayas, in the Western Ghats, in the two 1ndian islands (Andamans and Lakshadweep) and the North-eastern hills of India. The great mountain chain of the Himalayas with its associated hill tracts in the north-eastern India is the biological treasure trove of the country. The tremendous altitudinal range of these hills coupled with the large variation in rainfall from dry conditions at the western edge in Jammu and Kashmir to the wettest place on earth–the Meghalaya have created an unparalleled diversity of plants and animals in this region. Due to highly dissected mountain topography and intense geographical isolation, there is also high degree of endemism in this region. It is estimated that the Himalayas and the north-east India possesses over 3,000 endemic species of dicotyledons alone.

Biological diversity is, in fact the very basis of human existence on earth its conservation and wise management is crucial to the very survival of humanity.

Keywords: *Biological, Diversity, Treasure, Endemics, Himalayas, Altitude.*

Introduction

Human civilization depends on the survival of the biological world. Perhaps in short term a species lost here and there may be of little consequences for overall ecosystem stability but in the long term the cumulative effect of such losses may same day threaten our existence. Some 100 species are lost every day. Most of them vanish unknown (UNEP Report, Anonymous 1992f). This loss will hurt mankind. As the global biological heritage erodes, the potential for new and useful products disappears. Even the most insignificant species may play a crucial role in the ecosystem to which we all belong. We should realize how important are the forest and wildlife for our very survival on earth and how vulnerable we can be without for our very survival on earth and how vulnerable we can be without it.

Half of all medicines are derived from plants; there are even more countless curable herbal drugs still to be discovered. Food, medicine, clothing, housing, energy, other material, spiritual, intellectual inspirational needs also comes from wild and domesticated biological resources. In India majority of the people depend on traditional medicine, which is primarily based on plant and animal products. With each species that disappears, mankind moves closer to his doomsday and loses potential for sustainable development. Though the biological resources constitutes the basis of life on earth their destruction by us all continues in complete silence far from the televised 'hot spot'.

The conservation of genetic diversity, both as a matter of insurance and as an investment necessary to sustain and improve agriculture, forestry and fisheries production; to keep future option, as a buffer against harmful environmental changes, and as the new materials for scientific and industrial innovation and a matter of moral principle.

Materials and Methods

Because of great climatic and physiographic diversity the Himalayan region has a different and unique wealth of flora and fauna than the peninsular tropical-subtropical parts of the country. The vegetation of this area is very interesting as it has flora and fauna of higher as well as lower altitude being a transition zone between hills and plains. The survey of this area was made with help of relevant literature, monograph, flora and record books (Jain, 1981; Singh *et al.*, 1983). ICBN has been used for generic and specie names. Some of the important flora and faunal vegetation are described here.

Biogeographical Regions with Respect to Floral Diversity

The forest vegetation in the Himalaya shows altitudinal zonation. The flora is very rich in endemics, there is convincing evidence that hundreds of species in all plant groups remain to be discovered and described. In the foothills of the northwestern Himalaya below 1000 meter, such as in the Siwaliks, the forests are of "sub-tropical dry evergreen" type having species or *Olea; Punica, Acacia, Pistacia*, etc. the lower altitudes have "sub-tropical pine forests", chiefly composed of *Pinus roxburghii*, and species of *Quercus* or *Syzygium* or *Rhododendron* are occasionally associated. Between 1500 and 3300 meter: "moist temperate zone forests" or *Abies, Quercus, Cedrus, Picea* and *Acer* are found. The predominantly coniferous forests in temperate zone have species of *Pinus, Cedrus, Pices*, and *Tsuga*. At places, *Taxus* and *Betula* are also found. The inner ranges of mountains have "dry temperate forests" of *Pinus gerardiana* and species of *Cedrus, Quercus, Acer, Fraxinus, Celtis* and *Olea*. In sub-alpine zone the forests are mainly of *Pinus wallichiana* and species of *Abies, Picea, Taxus, Betula*, Rhododendrons and some rosaceous plants. Above 300 meter the vegetation is of "moist alpine scrub" comprising species of *Betula, Rhododendron, Lonicera* and *Berberis*. The uppermost limit of vegetation around 400 meters, has the "dry alpine scrub" formed by species or *Junipers, Lonicera, Artemisia*, etc.

Compared with Western Himalaya, the eastern Himalaya, are wetter and warmer, and hence more rich in biodiversity. The forests in foothills are known for species of *Dipterocarpus, Artocarpus* and *Shorea assamica.* The lower slopes in the eastern Himalaya have "tropical semi-evergreen forests" of *Terminalia myriocarpa, Michelia, Champaca, Phoebe, Dysoxylum, Canarium, Litsea, Mangifera* and *Castanopsis.* Patches of sub-tropical broad-leaved hill forest having species of *Quercus, Castanopsis, Schima* and *Terminalia* are also seen. Between the altitudes of 1800 and 3000 meter, the forests are of "montane werr or moist temperate" type having species chiefly or *Machilus, Cinnamomum, Litsea, Magnolia* and *Schina. Cedrus deodara* is absent in Eastern Himalaya. Dry temperate forests chiefly of coniferous trees are found in Sikkim in the inner ranges: they comprise species of *Picea, Larix, Abies* and *Tsuga.* Above 300 meter, the "moist alpine scrub" are found, as in the western Himalaya. The forests of eastern Himalayas are rich in "Orchids." About 600 species of orchids are reported from Sikkim and eastern Himalaya.

Uttaranchal lies between Himalayas and upper Gangetic plain, and have a great climatic and physiographic diversity. This state has about 17,910 sq.km dense forest areas and the 53,000 sq.km total forest area. The vegetation of this region is very rich in medicinal plants. Uttaranchal, a part of Himalaya has a huge natural wealth. The rich and rare biodiversity of Himalaya has to be conserved through strong, honest, dedicated, diligent knowledgeable planners and laborious workers through coordinated programme of local, tribal people, forest and government department. There is need to evaluate their therapeutic potential through scientific and clinical research. So, rich and rare biodiversity or Uttaranchal, which lies between Himalayas and upper Gangetic plain, should be protected through coordinated programme. The local and tribal communities would have greater responsibility in forest protection.

The important objective should be protection of the ecosystem of the Shivalik (Uttaranchal), biodiversity conservation and protection as well as support to endangered and threatened species through community management. If we love, respect, give proper and utilization, then these so-called natural doctors will certainly take care of our health.

Therefore a great concern should be given upon the threat to survival of natural plant diversity. The main objective should be protections as well as support to endangered and threatened species through community management.

The vegetation of this area is very interesting as it has plants of higher as well as lower altitude being a transition zone between hills and plains. The forests are of tropical deciduous type. Extensive survey of this area was made for wild medicinal plants. Identification work was made with the help of relevant literature, monographs, flora and record books (Singh and Arora, 1978; Jain, 1981; Singh *et al.*, 1983). ICBN has been used for the generic and species names. The plants are arranged alphabetically on the basis of native information, their local names. Some of the important wild medicinal plants are described here along with their therapeutic uses.

Description of Plants

Botanical Name: *Adhotada vasica* Nees.

Family

Acanthaceae

Vern. Names

Hindi: Bansa, *English*: Adulsa

Description

A gregarious shrub leaves white with red spots. Part used: Leaves, flowers and roots.

Therapeutic Uses

It is well known drug of Ayurveda and Unani system of medicine. Leaves and roots are useful in cough, asthma and phitisis, leaves are antirheumatism and insecticidal, possesses anthelmintic and medicinal properties. Flowers and roots are antispasmodic. Oil from leaves and roots are active against Tubercle bacilli

Chemical Constituent

Alkaloids like vasicine and small amount of essential oils are reported. Seeds yield a fatty oil (25.6 per cent), consisting of arachidic, behenic, lignoceric, cerotic, oleic and linoleic acid and sitosterol.

Botanical Name: *Azadirachta indica* Juss.

Family

Meliceae

Vern. Names

Hindi: Neem; *English*: Margosa tree.

Description

A big tree. Flowers in short axillary branches. Fruits one seeded.

Part Used

Fruits, seeds, leaves, flowers, twigs, stem and root bark.

Therapeutic Uses

Plant is regard as "village dispensary" in India. Fruits are antihelmintic, astringent, purgative, tonic, useful in piles and urinary troubles. Seed oil is antiseptic and antirheumatic. Seeds used in snakebite and scorpion sting. Kernel is antimicrobial and antiseptic. Leaves are antimicrobial, useful for ulcers. Twig are used as toothbrush, poultice is applied to boils, eczema, ulcers and are antiseptic. Twigs used as toothbrush (antipyorrohoel). Flowers and gums demulcent, stimulant and tonic. Bark is used as bitter tonic, antispasmodic, useful in fevers, in skin diseases. Berries purgative, emolient or anthelmic.

Chemical Constituent

Some of the compounds reported were Azadirachtin, Nimbin, and Margosic acid. Bitter constituents from root and bark. The bitter constituents nimbin molecules contains an acetoxy, a lactone, an ester, a mithoxy and an aldehyde group.

Botanical Name: *Cassia fistula* L.

Family

Caesalpeniaceae

Vern. Names

Hindi: Amaltas; *English*: Indian laburnum.

Description

Moderate sized trees, wholly glabrous, bark greenish gray.

Part Used

Pods, flowers, leaves and roots.

Therapeutic Uses

Plant parts are laxative tonics, leaf juice is useful on skin diseases, in leprosy, and fruit applies in rheumatism and snakebite. Pulp of fruit is used as a purgative.

Chemical Constituent

Fistulic acid, fistulin and tannins.

Botanical Name: *Eucalyptus citridora* Hook.

Family

Myrtaceae

Vern. Names

English: Lemon scented, blue gum.

Description

Tall, slender perennial tree, trunk shiny with thin peppery light gray barks, leaves aromatic, seeds round tough.

Part Used

Leaves.

Therapeutics Uses

Oil extracted from leaves is antiseptic and used in cough, cold and respiratory troubles, skin diseases; oil ointment is beneficial for burns and as mosquito repellent, in chronic bronchitis an in asthma.

Chemical Constituent

Oil contains pinene, cineole, and sesquiterpene.

Botanical Name: *Ipomea fistulosa* L.

Family

Convolvulaceae

Vern. Name

Hindi: Besharam

Description

Erect perennial under shrub, flower with purple colour.

Part Used

Fruit, seeds, leaves, flowers, and twigs, stem and root bark.

Therapeutic Uses

Leaves slightly roasted with a little of mustard oil is applied in inflamed swollen areas externally. Seeds are purgative, laxative but over dosing cause irritation.

Chemical Constituent

Glycosides and alkaloids, Beta-sitosterol, resins.

Botanical Name: *Lantana camara* L.

Family

Verbenaceae

Vern. Names

Hindi: Aripple; *English*: Lantana.

Description

An erect perennial under shrub.

Part Used

Leaves and root.

Therapeutic Uses

Decoction is given in malaria, rheumatism, tetanus, used as tonic and toxy of abdominal viscera, plant is carminative antiseptic, diaphoretic, applied in treatment of asthma, bronchitis, arterial hypertension.

Chemical Constituent

Alkaloidism, leaves contain landene A, Landene B and D Lancamarone, the bark and root contains alkaloid Lantanine.

Botanical Name: *Ocimum sanctum* L.

Family

Labiatae

Vern. Name

Hindi: Tulsi; *English*: Holy basil.

Description

A small aromatic shrub.

Part Used

Seeds, leaves, stem.

Therapeutic Uses

The leaves are used in tea in the treatment of malarias fever, the seeds are used in chronic diarrhea and dysentery, decoction of the leaves is said to be useful in cold.

Chemical Constituent

Plant contains volatile oil, alkaloids, glycosides, saponins, and tannins.

Botanical Name: *Ricinus cummunis* L.

Family

Euphorbiaceae

Vern. Names

Hindi: Arandi; *English*: Castor oil plant.

Description

An evergreen shrub.

Part Used

Seeds, leaves, root bark.

Therapeutic Uses

Seeds used as purgative, used in scorpion sting and fish poison, oil from seed is a purgative, leaves applies to the head to relieve headache and as poultice for boils, the fruit is edible and are refrigerant and laxative.

Chemical Constituent

Alkaloids and ricinine from seeds and leaves.

Botanical Name: *Vinca rosea* L.

Family

Apocyanaceae

Vern. Name

Hindi: Sadabahar; *English*: Periwinkle.

Description

A suffruticose plant

Part Used

Fruit, seeds, leaves, flowers, twigs, stem, bark.

Therapeutic Uses

Plant used as remedy for diabetes, juice of leaves applied on wasp sting, used to cure Leukemia.

Chemical Constituent

Vincamine, carapanaubine, pubescine alkaloid are reported and non-alkaloid constituents include dembonitol, robinin, and ursolic acid.

Botanical Name: *Opuntia dilenii* Haw.

Family

Cactaceae.

Vern. Name

Hindi: Nagphani; *English*: Prickly pear.

Description

Erect shrub, bearing spines, flowers solitary, fruit pyriform.

Part Used

Fruit, leaves, and latex.

Therapeutic Uses

Fruit useful in gonorrhoea, whooping cough, spasmodic, cough expectorant. Milky juice is purgative. Leaves applied as poultice in inflamed areas.

Chemical Constituent

Isorhamnetin, quercetin and betanin.

Biogeographical Region with Respect to Faunal Diversity

Trans-Himalayan: Ladakh (Jammu and Kashmir) and Lahul Spiti (Himachal Pradesh)

It covers an area of 186,200 sq.km, in the cold and arid regions (4500–6000 mts above mls) with sparse alpine steppe vegetation having several endemic species. It harbour a good community of wild sheep and goats, urial, ibex, wild yak and tibetan ass, gazelle and antelope. Among the carnivores are the snow leopard, Tibetan wolf and the endemic pallas cat and smaller animals like marbled cat, pika and marmot. The brackish water lakes and marshes have a good variety of avifauna of which the most spectacular is the black-necked crane. It has two protected areas.

Himalaya (North-West, West, Central, Eastern)

Covering an area of about 236,300 sq.km in the Himalayas, it displays a wide altitudinal range and is among the richest zones in term of species and habitat diversity. Sambar, muntjac wild boar are found in the sub-tropical foothill; musk deer, serow, goral, tahr, koklas and pheasants in the temperate and sub-alpine regions; while bharal, snow leopard, brown bear, and snow cock inhabit in the alpine region. It has 56 protected areas. There are more endangered species in the Himalaya than anywhere else in the country. "Sikkim stag" may already have become extinct. The "Thar," "markhor" and "western tragopan" are the facing extinction.

North east (Brahmaputra Valley; Assam Hills)

"The biogeographic gateway" for much of India's flora and fauna. This zone is one of the richest in biological resources, both endemic and others. Brahmaputra valley contains extensive areas of natural vegetation-swamps, grasslands and fringe forests. Elephant, rhino, buffalo, swamp deer, hog deer, pygmy hog and hispid hare are wild lives or the zone. The diversity in plant communities and species is extremely high. The animals communities too are diverse. In fact smaller carnivores exhibit a richness not seen anywhere else in the world, the country's highest populations of elephants are here. The region also forms and important flyway on the route of migratory birds to and from Siberia and China. There are 17 protected areas in 1880 sq. km.

The Tropical Moist Deciduous and Evergreen Forests

The region is restricted to the lower altitude of the north-east spread over potential area of 113,000 sq.km with af1nual rainfall more than 2,500 mm. The important wildlife reserves located in the region are Kaziranga National Park (436 sq.km) in Assam, Intangki (202 sq.km) in the Nagaland, Balaphakram in Meghlaya and the unique wetland ecosystem the Keibul Lamjao National Park of Mainpur (40 sq.km), the only home of the last surviving population of the brow-antelered deer. There are also scattered several small "sacred groves", (the forest of God), with pristine flora and fauna protected by the tribal faiths and beliefs.

The Subtropical Forest of the Himalayas

It ranges from the dry evergreen "sclerophyllous" forest of the north-west Himalayas in the altitude of 300 to 1500 metres to the subtropical "chir-pine" forest in the altitude of 1000 to 1800 metres of the potential area of 65,000 sq.km 60 per cent is under forest with rich biodiversity. There are some good natural reserves in this region protecting the biota, the largest being Govind Pashu Vihar (320 sq.km) in the Uttar Kashi region of U.P.

The Temperate of the Himalayas

At the altitudes of 1800 to 3000 metres and annual rainfall regime of 1300 to 2000 mm there is only "Oak" and "Coniferous" forest, while at the higher altitudes of 2000 to 2,800 metres and annual rainfall regime of 1000 to 1800 mm there is only "Coniferous" forest with little biodiversity. Of the potential areas of 61,000 sq.km 60 per cent forest still persists with major conservation areas like Kistwar (200 sq.km), Overa-Aran (212 sq.km)" Gulmarg (180 sq.km) and Machigam (141 sq.km), in Kashmir, Shikari Devi (213 sq.km) in Himachal Pradesh; Kedarnath (478 sq.km) and part of Govind Pashu Vihar in U.P. Among the rich wet temperate forest of the north-east the most important is "Namdapha" in Arunachal Pradesh (2334 sq.km) which has been declared as the Biosphere Reserve of India, and the Nagaland Central Forest (104 sq.km).

The Sub-Alpine Altimontane Himalayan Forest

It lies below the snow slides from Kashmir to the eastern Himalayan at altitudes from 2900 to 3500 metres. Of the potential 50,000 sq.km only 15 per cent is under forest cover. There are some natural reserves like Henus High Altitude National Park (150 sq.km) in Ladakh and Kedernath Sanctuary (478 sq.km) in U.P. Nanda Devi Biosphere have been created in the U.P. region.

Alpine Scrub and Meadows of the Himalayas

It covers the higher reaches of Himalaya in the altitude range 3,500 to 4,900 metres. This biome suffers greatly from overgrazing by animals and invasion of tourists. The "Valley of Flowers" (90 sq.km) and the Govind Pashu Vihar national park of U.P. are important natural reserves protecting wild flora and fauna.

The Alpine Steppes of the Himalayas

Over altitude of more than 4,500 metres, the Himalayan ecosystem is highly fragile here are however several natural reserves such as "Sang Gauri", "Henus", " Aovera-Aran", and "Kistwar" in Aladakh Kashmir and Singalila National Park in West Bengal hills.

Conclusion

Biodiversity Erosion in India

At least 10 per cent of India's recorded wild flora, and possibility a larger fraction of its fauna, are on the threatened list, many of them are on the verge of extinction. The hunting leopard or the Asiatic Cheetah (*Acinonyx jubatus*) and the pink headed duck (*Rhodonessa caryophyllacea*) are already extinct.

To this habitat destruction hunting, over-exploitation, botanical and zoological collections, replacement by exotics, and a host of other activities have taken a heavy toll of biodiversity.

The Himalayan ecosystem once rich in biodiversity of elusive plants and animal species is under threat of destruction under population pressure and due to the ill conceived developmental projects particularly hydel power projects. Road buildings to promote tourism and for strategic reasons (military preparedness) has taken heavy toll of biodiversity. The Himalayan ecosystem is highly fragile because the mountain is of relatively recent geological origin and therefore much unstable.

Hundreds of species of plants and animals endemic to the Himalayas have been wiped out due to over grazing, faulty agricultural practices, and the rapacity of timber gangs. Commercial exploitation of the Himalayan forest began in the middle of the 19th century with the demand of railway sleepers by the then British government. In the upper regions of Himalayas, which were demarcated as reserve forest, commercially valuable conifers (*Pinus* and *Cedrus*) were known for timber, resins, and broad-leaved oak trees were eliminated. Pine forest has led to loss of soil fertility drying up of water sources and scarcity of fodder and fuel wood for the local people who are forced to exploit other species thus further aggravation deforestation and biodiversity destruction. Horticultural practices of apple production in the Himalayas and stress on growing species like "poplars" for apple packing cases and other commercial purposes have been on important factor in biodiversity erosion of the Himalayas.

Himalayas, particularly the eastern parts are considered as one of the 10 global biodiversity "hot-spots", along with the Amazonian rain forest and island of Madagascar. A similar fate has hit India's domesticated flora and fauna. Perhaps the greatest factor in the disappearance of thousands of varieties of crops has been the spread of modern chemical based agriculture. A handful of HYV seeds generated in the laboratory have replaced countless indigenous varieties. This is most evident in crop like rice, of which at one time about 50,000 varieties are said to have been grown in the country, which have now been reduced to few hundreds. This broad genetic base provided insurance against pests and diseases, and a bank for desired characteristics like productivity, nutritional value, and resistance to drought.

Table 12.1: Ecological Status of Some Selected Plant Species of Himalayas

Name of Species	*Ecological Status*	*Habitat*
Abies pindrow	Endangered	Himalayas
Aconitum kerox	Rare	Himalayas
Astragalus strobiliferus	Vulnerable	Himalayas
Atropa acuminata	Endangered	Himalayas
Balanophora dioica	Endangered	Himalayas
Colchicum luteum	Vulnerable	Himalayas
Cyathea gigantea (Giant tree fern)	Indeterminate	Himalayas
Osmunda regalis (Royal Fern)	Indeterminate	Himalayas
Dianthus coschemiricus	Rare	Himalayas
Drosera burmannivhi	Endangered	Khasi Hills (Eastern Himalaya)
Insectivorious Plants		
Nepenthes khasiana	Critical	Khasi Hills (Eastern Himalaya)
Magnolia griffithii	Vulnerable	Himalayas
Orchis latifolia	Endangered	Western Himalaya

Contd...

Table 12.1–Condt...

Name of Species	Ecological Status	Habitat
Angelica glauca	Endangered	Western Himalaya
Polygonum alpinum	Endangered	Western Himalaya
Polygonum verticillatum	Endangered	Western Himalaya
Ginkgo biloba	Critical	Himalayas
Dioscorea deltoidea	Rare	Himalayas
Delphinium denudatum	Rare	Himalayas
Acorcus calamus	Vulnerable	Himalayas
Saussurea lappa	Indeterminate	Himalayas
Podophyllum hexandrum	Rare	Himalayas
Diplomeria hirsute	Rare	Himalayas
Picrorhiza kurroa	Endangered	Himalayas
Swertia chirayita	Endangered	Himalayas
Juniperus communis	Endangered	Western Himalaya
Citrus assanensis	Endangered	Western Himalaya

Source: Various Reports of Botanical Survey of India and NBRI.

Table 12.2: Ecological Status of Some Selected Animal Species of Himalayas

Common Name	Scientific Name	Ecological Status
Peacock pheasant	*Polyplectron bicalcaratum*	Critical
Snow leopard	*Panthera leo persica*	Critical
Giant flying squirrel	*Pataurista candidulus*	Endangered
Kashmir stag	*Cerevus elaphus hanglu*	Endangered
Musk deer	*Moschus moschus*	Critical
Himalayan ibex	*Capra ibex*	Rare
Red panda	*Ailurus fulgens*	Critical
Himalayan brown bear	*Ures arctos*	Rare
Blue sheep	Pseudovis nahoor	Vulnerable
Western tragopan	*Tragopan melanocephalus*	Rare

Source: Wildlife Institute of India and Various Reports of Zoological Survey of India.

Other wildlife species of India which are at risk are *"Himalayan Quair"* and *"Himalayan Duck"* and *"Himalayan Newt"* from the Western Himalayas; the *"While Winged Wood Duck"*, *"White Garo Hills Tree Toad"* from the North-East India *"Himalayan Dragonfly"* from the Himalayas.

The wildlife society of India recently reported sighting of the rare trans-Himalayan bird the "mountain quail" in the Uttarakhand region. The mountain quail is regarded as one of the three rarest Indian birds along with the "Pink headed duck" (now extinct) and the Jurdon's courser which are listed in Schedule I, Section 3, of the Indian Wildlife Preservation Act, 1972. Prof. Salim Ali's book on Indian birds records that the mountain quail was last sighted in India in 1876.

Acknowledgement

The authors wish to thank Principal, Government Post-Graduate College, Rishikesh, for providing necessary facilities for the above work and the UGC for financial help.

References

Annonymous, 1990. Endangered species of India. *Monographs of Indian Institute of Ecology and Environment*, New Delhi.

Annonymous, 1992f. *Saving our Planet*. UNEP Report on the state of the environment (1972–92), Nairobi, Kenya.

Annonymous, 1992. *The Wealth of India, Raw Material*. CSIR, New Delhi.

Annonymous, 1994. *IUCN Red list of Threatened Animals*. World Conservation Monitoring Center, Cambridge, U.K.

Jain, S.K., 1981. *Glimpses of Indian Ethnobotany*. Oxford and IBH Publishing Co., New Delhi.

Meyers Norman, 1986. Biological resources of the tropics. In: *Conservation for Productive Agriculture*, ICAR, New Delhi.

Singh, H.B. and Arora, R.K., 1978. *Wild Edible Plants of India*. ICAR, New Delhi.

Singh U., A.M. Wadhwani and B.M. Johri, 1983. *Dictionary of Economic Plants in India*.

Chapter 13

Evaluation of Antifungal Property of *Catharanthus roseus* in Relation to Amino Acid Content of the Seeds of *Pisum sativum*

Sucharita Das Gupta and Padma Chatterjee*
Plant Physiology and Biochemistry Laboratory, Department of Botany, University of Kalyani, Kalyani – 741 235, West Bengal

ABSTRACT

The antifungal property of *Catharanthus roseus* was examined in terms of amino acid content of healthy and fungal infected seeds of *Pisum sativum*. The crude 50 per cent aqueous ethanolic extract of leaves of *Catharanthus roseus* and its different fractions were tested for its antifungal effect by agar cup method. It was revealed that the crude, Di ethyl ether and Chloroform fractions were active against *Fusarium oxysporum* and *Curvularia* sp. The most active fraction was chloroform fraction. Fungal infected seeds were treated with different doses of the fraction (2, 2.4, 2.8, 3.2 mg/ml) and in each case a control set was maintained. The effect of the fraction was compared with chemical fungicide Griseofulvin. It was observed that fungal infection caused a serious threat to the amino acid content of the seeds and a sharp reduction in the level of amino acid was noticed in the infected ones. Treatment of the fungal infected seeds with different doses of the compound caused increase of the amino acid levels almost reaching the same as those in the healthy ones. Furthermore the bioactive fraction had a similar effect like that of Griseofulvin and it was non-toxic towards healthy individuals.

Keywords: *Catharanthus roseus, Griseofulvin.*

* E-mail: schatterjeecal@yahoo.co.in

Introduction

Since long, phytochemicals, economically important to mankind with no toxic effect have been isolated, purified and identified. Different antimicrobial, antifungal and anticancerous properties of *Catharanthus roseus* were shown by different workers. Such as Rozas Hernandez NM showed antimicrobial activity of indole alkaloid obtained from *C. roseus* against some bacterial and fungal strains (1979). Nwafor, S.V. *et al.*, reported anticancer property of *C. roseus* (2001). Landini *et al.* (2001) reported an antitumor alkaloid from *Vinca rosea*. At the beginning of the 1960's the discovery of the anticancer drugs, Vinblastine and Vincristine, from *C. roseus* has highlighted the importance of plants as the raw materials in the search for natural drugs.

Materials and Methods

Matured plants of *Catharanthus roseus* were collected from Kalyani during June–July. Seeds of *Pisum sativum* were collected from Bidhan Chandra Krishi Vishyavidyalaya, Mohanpur.

50 per cent aqueous ethanolic extract of leaves of *Catharanthus roseus* was concentrated under reduced pressure and fractionated over petroleum ether, Di ethyl ether and Chloroform. The crude extract and the three fractions were tested for its antifungal effects against *Curvularia sp* and *Fusarium oxysporum* by agar cup method of Purohit *et al.*, 1995. The most active fraction was Chloroform fraction. Seeds of *Pisum sativum* were soaked in distilled water for 24 hours. Then the seeds were divided into twelve experimental sets. Fungal infected seeds were treated by presoaking method with different

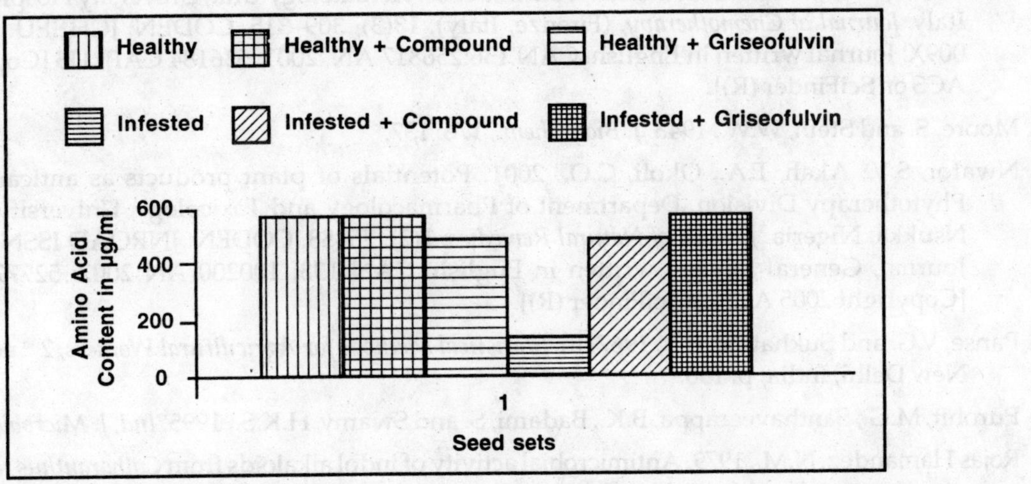

Figure 13.1: Effect of the Fraction on Amino Acid Content of Seeds *Pisum sativum*

***N.B.*: For convenience of presentation, treatment with the highest dose (3.2 mg/ml) has been shown in the figure.**

	LSD AT 5 per cent	*SE*
Treatment	2.0422	0.6996
Type	1.1790	0.4039
Tr × Ty	2.8881	0.9894

concentrations (2, 2.4, 2.8 and 3.2 mg/ml) of the compound 0.5 gm of each of the seeds were subjected to the treatments. A set was also treated with chemical fungicide Griseofulvin (500 μg/ml). In each case a control set was maintained. Quantitative estimation of amino acid content was done following the method of Moore and Stein (1948) as modified by Bhattacharya (1984).

Results and Discussion

The effect of Chloroform fraction on amino acid content of the seeds is presented in Figure 13.1. Two-factor factorial analysis of variance is done as per standard technique (Panse and Sukhatme, 1967). It is evident from the results that fungal infection causes a serious threat to level of amino acid content and a marked reduction in its level was noted in fungal infected seeds. However the presoaking of the seeds by different concentrations of the compound reduced the fungal incidence and the amino acid content was increased along with increase in the doses of the compound, almost reaching towards normal like that of Griseofulvin. The herbal compound was also non-toxic to healthy individuals.

Acknowledgement

I am indebted to Department of Environment, Govt. of West Bengal, for providing research grant and fellowship.

References

Landini, I., Bartolozzi, B., Banchelli, I., Innocenti,A. Degli, Nocentini, O., Bernabei, P.A., 2001. *In vitro* activity or Vinorelbine on human leukemia cells. Hematology Unit, University Hospital, Florence, Italy. *Journal of Chemotherapy,* (Firenze, Italy), 13(3), 309–315. CODEN: JCHEEU. ISSN: 1120-009X. Journal written in English. CAN 136:256817 AN. 2001: 536184 CAPLUS [Copyright 2005 ACS on SciFinder (R)].

Moore, S. and Stein, W.W., 1948. *J. Biol. Chem.*, 176: 137.

Nwafor, S.V., Akah, P.A., Okoli, C.O., 2001. Potentials of plant products as anticancer agents. Phytotherapy Division, Department of Pharmacology and Toxicology, University of Nigeria, Nsukka, Nigeria. *Journal of Natural Remedies,* 1(2): 75–88. CODEN: JNROAD ISSN: 0972–5547. Journal, General Review written in English. CAN 138: 100200 AN 2002: 527776 CAPLUS [Copyright 2005 ACS on SciFinder (R)].

Panse, V.G. and Sukhatme, P.T., 1967. In: *Statistical Methods for Agricultural Workers,* 2nd edn. I.C.A.R., New Delhi, India, p. 150.

Purohit, M.G., Santhaveerappa, B.K., Badami, S. and Swamy, H.K.S., 1995. *Ind, J. Microbiol.*, 35(1): 77.

Rojas Hamandez, N.M., 1979. Antimicrobial activity of indol alkaloids from *Catharanthus roseus* against some bacterial and fungal strains. *Rev. Cubana Med. Trop.*, 31(3): 199–204.

Chapter 14

Rapid Composting of Irrigated Pearl Millet Straw

J. Kannan[1] and P. Singaram[2]
[1]Krishi Vigyan Kendra, Pechiparai
[2]Department of Environmental Sciences, TNAU, Coimbatore – 3

ABSTRACT

One experiment was conducted to fasten the composting of pearl millet straw with different combination of microorganisms. The microorganisms used were *Bacillus* sp., *Pseudomonas* sp., *Trichoderma* sp., *Pleurotus* sp. and *Streptomyces* sp. Among the different combination of microorganisms utilized, the combination of microbes *Trichoderma* sp. and *Pleurotus* sp. decomposed the straw rapidly and fully matured compost was obtained within sixty (60) days. The nitrogen and potassium content of the matured compost was highest in T_2, T_4 and T_5, all the fungi inoculated treatments. In the fungi inoculated treatments, the total nitrogen content was 15 per cent more than control and the total phosphorus and total potassium contents were 13 per cent and 33 per cent more than control.

Keywords: Compost, Microorganisms and Matured compost.

Introduction

The disposal of organic materials through burning can vanish valuable resources and contribute to environmental problems. When applied properly and adequately, these products can positively impact our environment by improving soil and plant health, conserving water, reducing erosion and minimising the use of fertilizers and pesticides. Composting is a biological process in which biological wastes are stabilized and converted into a product to be used as a soil conditioner and organic manure. Composting is considered to be a part of recycling in which the nutrients are put back into the soil and plant life.

Pearl millet is the fourth important food crop in India next to rice, wheat and sorghum and also the second major coarse cereal grain after sorghum. Pearl millet is extensively cultivated in poor fertile and water-deficit soils, mostly with low and erratic rainfall. India is the largest producers of pearl millet in the world and in Tamil Nadu, the pearl millet is cultivated in about 215 thousand hectares (Balasubramaniyan and Palaniappan, 2003). The pearl millet straw contains lot of nutrients and these nutrients can be recycled, through proper composting. Generally, it takes six to sevn months to obtain good, finished compost from agricultural wastes and one of the important fields of research is to reduce the period of composting of agricultural residues through rapid composting. The quick composting process may be a great boon to the farmers for sustaining the soil fertility.

Materials and Methods

A heap method of composting experiment was carried out in order to quicken the composting of irrigated pearl millet straw at the department of Environmental Sciences, TNAU, Coimbatore, Tamil Nadu. Treatments fixed were six replicated four times. The composting was carried out for a period of 105 days, during February–May, 2004. The size of the heap uniformly was 250 × 95 × 55 cm as length, width and height. The cultures used for composting are as follows:

Cultures used: Bacteria–*Bacillus* sp., *Pseudomonas* sp.

Fungi–*Trichoderma* sp., *Pleurotus* sp.

Actinomycetes–*Streptomyces* sp.

Treatments

T_1: Pearl millet straw + cowdung + *Bacillus* sp. + *Pseudomonas* sp.*

T_2: Pearl millet straw + cowdung + *Trichoderma* sp. + *Pleurotus* sp.*

T_3: Pearl millet straw + cowdung + *Streptomyces* sp.*

T_4: Pearl millet straw + cowdung + Existing microbial inoculum**

T_5: Pearl millet straw + cowdung + mixed microbial inoculum*

T_6: Pearl millet straw + cowdung (control)

Table 14.1

Sl.No.	*Cultures*	*Broth Used*
1.	*Bacillus* sp. and *Pseudomonas* sp.	Nutrient broth, Kings broth
2.	*Trichoderma* sp. and *Pleurotus* sp.	Potato dextrose broth
3.	*Streptomyces* sp.	Kusters broth

Preparation of Inoculants

The cultures were isolated from the compost pits by the standard serial dilution plate technique and by streak plate method. The isolated cultures were purified by single hypal tip method and by streak plate method.

* Cultures were isolated during mesophilic and thermophilic stages of composting from the compost pits of Regional Research Station, Aruppukottai. Mixed microbial inoculum consists of *Bacillus* sp., *Pseudomonas* sp., *Trichoderma* sp., *Pleurotus* sp. and *Streptomyces* sp.

** Existing microbial inoculum contains *Bacillus* sp., *Pseudomonas* sp. and *Pleurotus* sp.

The purified cultures were inoculated in the conical flask containing concerned sterilized broth. Then the bacterial cultures were kept in the electrical shaker for two days, whereas the fungal and actinomycetes cultures were kept for eight days for multiplication.

Composting Methodology

The pearl millet straw was chopped into bits of 10 to 15 cm length. One hundred kilograms of straw on wet weight basis was heaped for each treatment which was then sprayed with water to maintain the moisture content of 75 to 80 per cent. Cultures of different organisms were inoculated at 1 per cent level, as per the treatments. Accurately 10 kg of fresh cowdung was prepared as slurry by mixing with 10 litres of water and sprinkled over each heap. The pearl millet straw was allowed to decompose by maintaining the moisture content of 75 to 80 per cent during the period of composting. The second time of inoculation of cultures at 1 per cent level was done on 15th day of composting. Every fortnight, all the compost heaps were turned upside down. The samples were collected from five different spots of each heap and composite samples were obtained for the initial, 7th day of composting and at fortnight interval starting from 15th day of composting. The samples were analysed for their physico-chemical and biochemical properties by following standard procedure. The statistical analysis of data was carried out by completely randomised block design.

Results and Discussion

Table 14.2: Characteristics of the Fresh Irrigated Pearl Millet Straw

Sl.No.	Characters	Pearl Millet Straw
1.	pH	6.23
2.	Electrical Conductivity (dSm^{-1})	0.98
3.	Organic carbon (per cent)	38.18
4.	Total nitrogen (per cent)	0.51
5.	C : N ratio	74.86
6.	Total Ca (per cent)	0.49
7.	Total Mg (per cent)	0.20
8.	Cellulose (per cent)	33.80
9.	Hemicellulose (per cent)	25.60
10.	Lignin (per cent)	12.40
11.	Silica (per cent)	8.20
12.	Total phosphorus (per cent)	0.018
13.	Total potassium (per cent)	1.14
14.	Manganese (ppm)	58.96
15.	Zinc (ppm)	25.42
16.	Iron (ppm)	18.74
17.	Copper (ppm)	23.80
18.	Total solids (per cent)	81.75
19.	Ash content (per cent)	9.20
20.	Moisture (per cent)	18.25

Table 14.3: Characteristics of the Cowdung

Sl.No.	Characters	Cowdung
1.	pH	8.18
2.	Electrical Conductivity (dSm^{-1})	1.94
3.	Total nitrogen (per cent)	0.83
4.	Total phosphorus (per cent)	0.09
5.	Total potassium (per cent)	0.24
6.	Total calcium (per cent)	3.12
7.	Total magnesium (per cent)	0.48
8.	Organic carbon (per cent)	17.10

Application of immatured compost with high C : N ratio, can result in immobilization of available nitrogen, causing an nitrogen deficiency in plants (Bannick and Joergensen, 1993). So, testing the maturity and quality of compost is essential. C : N ratio of the compost should be less than 20. Final C : N/Initial C : N ratio should be less than 0.6.

Wide C : N ratio and higher amounts of resistant constituents like lignin are responsible for the slow degradation of cereal residues (Bharadwaj and Gaur, 1985). The fresh pearl millet straw had a C : N ratio of 74.86 and lignin content of 12.4 per cent. In irrigated pearl millet straw composting, the treatments T_2, T_4 and T_5 matured on 60th day. The treatment control T_6 (with no microbial inoculation) had not matured completely, even after 90 days of composting [Table 14.5 showing CEC of 59.3 C mol (P+) kg^{-1} which is less than 60]. The treatments T_2 and T_5 matured earlier as they were inoculated with *Pleurotus* sp. and *Trichoderma* sp. This is in accordance with the results of Nandi *et al.*, 2000 in which they reported that the inoculation of lignin degrading fungi along with cellulose decomposing fungi deserves its application in disposing of cereal residues by rapid composting.

Table 14.4: Compost Maturity Test Conducted

Treatments	Tests Conducted (60th Day of Composting)					Tests Conducted (75th Day of Composting)				
	pH		CEC	C : N Ratio	Final CN/ Initial CN Ratio	pH		CEC	C : N Ratio	Final CN/ Initial CN Ratio
	Before Incubation	After Incubation	Cmol $(p+)kg^{-1}$			Before Incubation	After Incubation	Cmol $(p+)kg^{-1}$		
T_1	7.40	6.43	51.2	21.78	0.29	7.39	7.14	59.4	20.48	0.27
T_2	7.33	7.08	60.1	19.94	0.27	7.32	7.20	60.4	18.15	0.24
T_3	7.54	6.53	52.4	22.65	0.30	7.48	7.27	59.5	20.63	0.28
T_4	7.41	7.11	60.3	19.94	0.27	7.40	7.21	61.0	18.35	0.25
T_5	7.45	7.09	60.5	19.81	0.26	7.44	7.18	60.8	17.61	0.24
T_6	7.21	6.30	50.7	34.42	0.46	7.18	6.61	58.2	25.60	0.34

According to Kalaiselvi and Ramasamy (1996), none of the parameters are 90 per cent sure of deciding the quality. Combination of methods is required to test the quality standards. Hence, four parameters *viz.*, C : N ratio, CEC, final C : N/initial C : N ratio and pH of the water extract before and after incubation at 55°C, under anaerobic condition for 24 hours were tested to confirm the maturity of the compost. During composting of irrigated pearl millet straw the C : N ratio was reduced to 14.08 after 105 days. In contrast, Talashilkar *et al.*, (2002) reported that the C : N ratio of farm wastes reduced to 10 after 105 days of composting with *Trichoderma* sp. The microbial inoculants containing *Pleurotus* sp. and *Trichoderma* sp. were applied twice at an interval of 15 days. The treatments that received the above mentioned cultures T_2, T_4 and T_5 were matured during 60th day itself. This is in accordance with the study conducted by Kumaresan *et al.* (2003), in which they revealed that sugarcane trash, coirpith and pressmud can be composted with microorganisms *Pleurotus sajor-caju* and *Trichoderma viride* along with 1 per cent urea and 10 per cent cowdung slurry within 60 days. In contrast, Singh and Amberger (1998) reported that wheat straw compost enriched with nitrogen, molasses and rock phosphate should be used only after 120 days of decomposition for sustainable crop production. During composting of irrigated pearl millet straw, the treatments (T_2, T_4 and T_5) with fungal inoculum registered significantly higher nitrogen and potassium contents than the treatments not applied with fungal inoculants (Table 14.7). This was in confirmation with the reports of Nakas and Klein (1980) and they indicated that mineralization was high in fungal treatments because of their capacity to produce enzymes capable of polymeric cleavage.

Table 14.5: Compost Maturity Tests Conducted During 90th Day of Composting

Treatments	*Test Conducted*				
	pH		*CEC*	*C : N Ratio*	*Final C : N/Initial C : N*
	Before Incubation	*After Incubation*	*Cmol (p+) kg⁻¹*		
T_1	7.39	7.31	62.7	16.52	0.22
T_2	7.25	7.14	64.3	15.23	0.20
T_3	7.44	7.36	63.8	17.27	0.23
T_4	7.38	7.29	64.5	15.50	0.21
T_5	7.42	7.34	63.9	15.62	0.21
T_6	7.15	6.91	59.3	19.80	0.26

T_1: T_6 + Bacterial culture; T_2: T_6 + Fungal culture; T_3: T_6 + Actinomycetes culture; T_4: T_6 + Existing microbial inoculum; T_5: T_6 + Mixed microbial inoculum; T_6: Pearl millet straw + cowdung (control).

Table 14.6: Chemical Characteristics of the Matured Pearl Millet Straw Compost

Treatments	*pH*	*EC (dSm⁻¹)*	*Organic Caron (%)*	*Total N (%)*	*C : N Ratio*	*CEC Cmol (p+) kg⁻¹*	*Heavy Metals (ppm)*		
							Lead	*Cadmium*	*Chromium*
T_1	7.36	0.78	19.04	1.19	16.0	70.9	5.00	0.98	17.80
T_2	7.24	0.88	17.73	1.24	14.3	73.6	5.25	1.26	23.08
T_3	7.43	0.85	18.54	1.13	16.41	70.2	6.75	1.15	16.90
T_4	7.37	0.87	17.61	1.23	14.32	72.9	5.75	1.23	24.56
T_5	7.40	0.85	17.32	1.23	14.08	73.1	6.25	1.29	25.78
T_6	7.11	0.82	19.16	1.05	18.25	67.2	5.00	0.95	15.50
SED	0.14	0.02	0.35	0.02	0.29	1.23	0.02	0.02	0.02
CD (P = 0.05)	0.31	0.03	0.75	0.05	0.63	2.64	0.04	0.05	0.04

T_1: T_6 + Bacterial culture; T_2: T_6 + Fungal culture; T_3: T_6 + Actinomycetes culture; T_4: T_6 + Existing microbial inoculum; T_5: T_6 + Mixed microbial inoculum; T_6: Pearl millet straw + cowdung (control).

Table 14.7: Nutrients Content of the Matured Pearl Millet Straw Compost

Treatments	*Total N (%)*	*Total P (%)*	*Total K (%)*	*Total Ca (%)*	*Total Mg (%)*
T_1	1.19	0.35	2.11	0.60	0.29
T_2	1.24	0.37	2.51	0.69	0.32
T_3	1.13	0.35	1.96	0.61	0.29
T_4	1.23	0.36	2.51	0.69	0.30
T_5	1.23	0.39	2.49	0.67	0.32
T_6	1.05	0.32	1.67	0.55	0.26
SED	0.02	0.02	0.05	0.01	0.006
CD (P = 0.05)	0.05	0.04	0.10	0.03	0.012

T_1: T_6 + Bacterial culture; T_2: T_6 + Fungal culture; T_3: T_6 + Actinomycetes culture; T_4: T_6 + Existing microbial inoculum; T_5: T_6 + Mixed microbial inoculum; T_6: Pearl millet straw + cowdung (control).

References

Balasubramaniyan, P. and Palaniappan, S.P., 2003. *Principles and Practices of Agronomy*, pp: 29–30.

Bannick, C.G. and Joergensen, R.G., 1993. Changes in N fractions during composting of wheat straw. *Biol. Fert. Soils*, 16: 269–274.

Bharadwaj, K.K.R. and Gaur, A.C., 1985. *Recycling of Organic Wastes*, ICAR, New Delhi, pp. 104.

Kalaiselvi, T. and Ramasamy, K., 1996. Compost maturity: Can it be evaluated? *Madras Agric. J.*, 83: 609–618.

Kurnaresan, M., Shanmugasundaram, V.S. and Balasubramanian, T.N., 2003. Biocomposting, of organic wastes. *Agric. Sci. Digest*, 23: 67–68.

Nakas, J.P. and Klein, D.A., 1980. Mineralisation capacity of bacteria and fungi from the rhizosphere-rhizoplane of a semiarid grassland. *Appl. Environ. Microbiol.*, 39: 113–117.

Nandi, N., Rahman, F.R., Sinha, N.B. and Hajra, J.N., 2000. Compatibility of lignin degrading and cellulose decomposing fungi during decomposition of Rice straw. *J. Indian Soc. Soil Sci.*, 48: 387–389.

Singh, C.P. and Amberger, A., 1998. Organic acids and phosphorus solubilization in straw composted with rock phosphate. *Siores. Technol.*, 63: 13–16.

Talashilkar, S.C., Dosani, A.A.K. and Chatterjee, A., 2002. Soil fertility and organic recycling. *Indian J. Agric. Chem.*, 35: 1–31.

Chapter 15

Influence of Pollutants of Soil Bacteria

D. Vijaya Lakshmi[1], C. Krishna Kanth[2] and M.A. Singara Charya[3]

[1]Department of Microbiology, [2]Department of Physics
Kavitha Memorial P.G. College, Khammam – 507 001
[3]Department of Microbiology, Kakatiya University, Warangal – 506 009

ABSTRACT

The impact of pollutant generated from Dairy, Sewage, hospital and Automobile site on the soil bacteria was studied during 1996–1999. The minimum bacterial colonies (230×10^6) were recorded during rainy and maximum (606×10^6) during winter months. The dumps of automobiles caused for drastic reduction of bacteria in the soils (59×10^6). No significant relations were drawn between physico-chemical characteristics and bacterial densities in these polluted sites. pH correlated positively with bacterial densities of dairy soils ($r = 0.621$) at 5 per cent significant level. The date revealed the impact of toxic pollutants and the densities and diversities of bacteria and in turn the management of polluted soil systems.

Keywords: Pollutants, Bacterial population, Degradative enzymes.

Introduction

Soil is undoubtedly one of the most important need of the environment. Various physical, chemical and biological process are constantly changing soils over geologic time. Soil is a natural, three-dimensional array of vertically differentiated material at the surface of earths crust.

Pollution is often thought of as resulting from chemical contaminant generating from dairy industries, drug factories, sewage, automobiles etc. It is well known that microorganisms play the significant role in dynamics of organic matter by production of enzymes. (Hattori 1976).

Soil microorganisms in general bacteria in particular reproduce rapidly and adapt to new environmental situations is important to the decomposition and transformation of both natural and anthropogenic products. Some of the functions performed either entirely or in part by bacteria include nutrient cycling, decomposition of organic materials, nitrogen fixation, pesticide, detoxification and oxidation-reduction reactions. Without bacteria to mediate these process is impossible. Soil conditions that affect the growth of bacteria are many. However, the most important factors are the O_2, temperature, moisture, acidity, inorganic and organic nutrient status of the soil environment. Microbes also play an important role in mineralization process. The enzymatic activity of accumulated enzymes and from those in proliferating microorganisms. Accumulated enzymes in soils are regarded as enzymes present and active in a soil in which no microbial proliferation occurs (Kiss *et al.*, 1975). Sources of enzymes present and active in a soil in which no microbial proliferation occurs. Source of enzymes in soil are primarily the microbial biomass, although they can also originate from plant and animal residues. Enzymes can be immobilized on organic and inorganic constituents of soils. It is well known that microorganisms, the most significant sources of soil enzymes, are absorbed by soil organic and inorganic constituents. (Trivedi *et al.*, 2003), Arnaud *et al.* (2004). In view of this to understand this rapid depilation of bacteria at pollution sites. A study was undertaken for three year in different seasons and the facts were established and presented in Tables 15.1–15.3. A correlation was drawn between the physical, chemical parameters and bacterial densities in the sites are increased pollution status were analyzed and presented in Tables 15.4–15.6.

Table 15.1: Bacterial Densities (1 × 10^6)/gm of Four Polluted Sites in Khammam During 1996–1999

Polluted Sites	*Jun*	*Jul*	*Aug*	*Sept*	*Oct*	*Nov*	*Dec*	*Jan*	*Feb*	*Mar*	*Apr*	*May*
Dairy	402	500	320	230	570	606	272	398	406	270	368	255
Sewage	270	472	310	310	450	242	292	142	136	142	329	242
Hospital	392	266	276	175	272	190	169	252	190	284	284	271
Automobile	256	370	390	180	182	139	144	59	76	85	163	148

Table 15.2: Bacterial Densities (1 × 10^6)/gm of Four Polluted Sites in Khammam During 1997–1998

Polluted Sites	*Jun*	*Jul*	*Aug*	*Sept*	*Oct*	*Nov*	*Dec*	*Jan*	*Feb*	*Mar*	*Apr*	*May*
Dairy	260	300	100	232	317	250	180	62	105	185	272	200
Sewage	205	180	166	300	472	362	309	180	286	192	250	360
Hospital	300	325	380	262	165	326	204	189	300	110	170	220
Automobile	145	172	180	175	170	168	108	72	142	140	148	189

Table 15.3: Bacterial Densities (1 × 10^6)/gm of Four Polluted Sites in Khammam During 1998–1999

Polluted Sites	*Jun*	*Jul*	*Aug*	*Sept*	*Oct*	*Nov*	*Dec*	*Jan*	*Feb*	*Mar*	*Apr*	*May*
Dairy	150	320	108	125	310	216	172	187	165	207	216	117
Sewage	230	175	250	256	200	207	162	160	120	272	200	276
Hospital	85	100	95	162	161	150	171	106	162	106	101	106
Automobile	98	176	186	100	121	82	142	197	143	100	65	75

Table 15.4: Correlation Coefficient (r) Between Bacterial Densities and Physico-chemical Characteristics in Four Polluted Sites at Khammam During 1996–1997

Parameters	*Polluted Sites*			
	Dairy	*Sewage*	*Hospital*	*Automobile*
pH	–0.039	0.257	0.455	–0.143
Temperature	–0.069	0.218	0.781***	0.393
Conductivity	0.029	0.443	0.342	0.183
Turbidity	0.009	–0.176	–	–
Dissolved Oxygen	–0.156	0.410	–	–
Biological Oxygen Demand	–0.112	0.295	–	–
Chemical Oxygen Demand	0.284	0.230	–	–
Organic matter	0.135	0.162	–0.024	0.570
Phosphates	0.580	0.408	–0.111	0.118
Sulphates	–0.227	0.465	–0.272	0.423
Nitrates	0.223	0.307	–0.136	0.072

***$P < 0.01$; –: Not detected.

Table 15.5: Correlation Coefficient (r) Between Bacterial Densities and Physico-chemical Characteristics in Four Polluted Sites at Khammam During 1997–1998

Parameters	*Polluted Sites*			
	Dairy	*Sewage*	*Hospital*	*Automobile*
pH	–0.153	0.010	–0.097	–0.145
Temperature	–0.327	–0.307	–0.098	–0.102
Conductivity	–0.391	–0.164	–0.310	–0.757**
Turbidity	0.022	–0.309	–	–
Dissolved Oxygen	0.042	–0.332	–	–
Biological Oxygen Demand	0.386	0.044	–	–
Chemical Oxygen Demand	0.205	–0.006	–	–
Organic matter	0.224	0.227	–0.413	–0.331
Phosphates	0.005	–0.073	–0.153	–0.576
Sulphates	0.435	0.113	–0.068	0.227
Nitrates	–0.453	–0.496	–0.217	–0.619*

*$P < 0.05$; **$P < 0.01$; –: Not detected.

Materials and Methods

The physical Parameters (pH, temperature, electrical conductivity and turbidity) and chemical characteristics (Dissolved Oxygen, BOD, COD, Organic matter, Phosphates, Sulphates and Nitrates) of the soils collected from four polluted sites in Khammam town were analyzed with monthly intervals for a period of three years. The analysis of these parameters were followed as suggested by Bradley

(1999). The serial dilution technique was used to isolate the bacteria from the polluted sites. From different habitats, each 10 gm of soil was suspended or agitated in a known volume of sterile water (90 ml) to make microbial suspension.

Table 15.6: Correlation Coefficient (r) Between Bacterial Densities and Physico-chemical Characteristics in Four Polluted Sites at Khammam During 1998–1999

Parameters	*Polluted Sites*			
	Dairy	*Sewage*	*Hospital*	*Automobile*
pH	0.621*	–0.384	–0.323	–0.041
Temperature	0.102	0.251	–0.589	–0.194
Conductivity	–0.299	0.131	0.393	–0.304
Turbidity	0.432	0.621*	–	–
Dissolved Oxygen	0.256	0.122	–	–
Biological Oxygen Demand	–0.207	–0.058	–	–
Chemical Oxygen Demand	–0.372	0.290	–	–
Organic matter	–0.413	0.056	0.055	0.197 –
Phosphates	0.298	–0.334	–0.474	0.069
Sulphates	0.062	–0.056	–0.291	–0.049
Nitrates	0.437	–0204	–0.179	–0.020

*$P < 0.05$; –: Not detected.

Serial dilutions were made and desired amount of suspension transferred to culture plates and incubated at 37°C for 24–48 hrs. and analyzed the plates for bacterial count. Standard method used for the analysis of desired samples and two way analysis of variance and correlation coefficient values between physico-chemical and bacterial densities were calculated and recorded in Tables (15.1–15.6).

Results and Discussion

From the Tables, it was clear that the bacterial densities varied significantly with seasons and with the influence of effluents released from different industries. The maximum bacterial colonies were recorded in the month of November, 1996 (606×10^6) while minimum in September, 1996 (230×10^6) in the soils amended with dairy wastes. Remarkable reduction in the bacterial colonies were noticed in the same site during 1997–98, where the maximum bacterial colonies (370×10^6) were observed in the month of October 1997 and the minimum (62×10^6) in the month of January 1998. Subsequently, the bacterial colonies were recovered and stabilized during 1998–99 with maximum and minimum colony count as 320×10^6 and 125×10^6 respectively.

The enumeration of bacterial colonies in the sewage sites of Khammam town revealed that the constituents of the sewage promoted the growth of bacteria where, seasons were not showed their influence on the proliferation of the colonies.

The maximum colonies were 450×10^6 in the month of October while minimum 142×10^6 in the month of March 1997. The fluctuation of bacterial colonies in (1997–98) the site ranged from 166×10^6 (August 1997) to 472×10^6 (October 1997). The moderate densities of bacteria were noticed in all other months. The stability of the bacterial colonies were witnessed during 1998–99 with the range of densities from 120×10^6 (February 1999) to 272×10^6 (March 1999).

The surroundings of the hospital with higher possibilities of contamination with diversified bacterial genera were observed. The soil samples were collected around hospital site (where the waste dumped areas were selected) waste material showed high incidence of bacterial colonies. However, the densities varied with seasonal influence. The maximum colony count (392×10^6) was noticed in the month of June, 1996, while minimum (9×10^6) in December 1996. (The fluctuations in all other months of the year were in between these two values). The bacterial counts were varied from 110×10^6 (March 1998) to 380×10^6 (August 1997) significant decrease was observed in the colony counts of bacteria during 1998–99 (85×10^6 to 162×10^6). It was very difficult to identity the exact reasons and influence of different factors responsible for these bacterial decreases.

The fourth site is amended with used oils, burned or partially burned engine oils and varieties of hydrocarbon related compounds. These soils were shown a variety of bacterial genera with varied densities. The maximum (390×10^6) bacterial colonies were recorded in the month of August, while minimum 59×10^6 in January. In general the months June to October showed moderate to high densities of bacterial population. Similar trend was continued in the enumeration of bacterial colonies. The highest densities of bacteria in the soils contaminated with oils and its related compounds was from May to October (1997–98). The minimum (72×10^6) number of colonies was noticed in the January 1998. The bacterial densities varied from 65×10^6 (April 1999) to 197×10^6, (January 1999). The gradual percolation and attachment of oil globules on the soil moieties could not show any remarkable fluctuations in densities of bacteria.

Similar to our observations many microbiologists (Border *et al.*, 1985), Varada Raj and Ayyappan (1989), Douglas *et al.* (1990), Lindsay and Riley (1994), Jon *et al.* (1999) from different parts of world tried to isolate and identify the different bacterial genera in various polluted environments. They further signified the relationships between the bacteria and environmental conditions, Phyllis (1989).

Akira (1989) characterized the bacterial populations and structure in different degraded environments on the basis of respiratory quinine profiles (Mehta *et al.*, 2002; Datke *et al.*, 2003; Pawar *et al.*, 2004; Amaud *et al.*, 2004; Medina *et al.*, 2005). The effect of pollutant stress on the bacteria flora and relationships between nutritional requirements and presence of available bacteria aerobic and anaerobic systems.

Lalitha Kumari and Singara Charya (1996), they studied on the assessment of industrial and organic pollution and biological diversity of the soils in Warangal-India, observed the high incidence of bacteria in the soils amended by sewage waters. Recently Baranyi and Carmen (1999), isolated variety of bacterial generation in UK with the help of numerical method. Greg *et al.* (1999), observed the dominance of *Pseudomonus aeruginosa* in different decayed environments and elaborated its physical characterization.

The correlation coefficient was determined between various physico-chemical parameters and bacterial densities of four polluted sites were presented (Tables 15.3–15.5). In general no major significant correlation was recorded between physico-chemical and bacterial densities. In most case the significance was positive or negative above 5 per cent level. From the Table 15.4 it was observed that the temperature showed a positive correlation with bacterial densities in soils near to the hospital site ($r = 0.781$) ($P < 0.01$). The correlation during (1997–98) were not in significant level in polluted sites of dairy, sewage and hospital but negative correlation was observed between conductivity and nitrates ($r = -0.757$) ($P < 0.01$) and ($r = -0.619$) ($P < 0.05$) with bacterial densities in the soils of automobile site. The significant correlations were observed in dairy and sewage sites (1998–99). pH correlated positively with bacterial densities of dairy soils ($r = 0.621$) at 5 per cent significant level. In the sewage soils the correlation was significant (r = 0.621) at 5 per cent significant level, and also turbidity and bacterial densities showed a negative correlation at sewage site (r = – 0.621) (P < 0.05). Recently, Goovaerts (1999) Mehta *et al.* (2002), Datke *et al.* (2003), Pawar *et al.* (2004), Arnaud *et al.* (2004), Medina *et al.* (2005), reviewed the statistics of soil science, its state of the art and perspective and concluded the close relations between various abiotic factor and their positive role in the bio-productivity of the soils.

Acknowledgement

The authors are thankful to Dr. S. Girisham, Head, Department of Microbiology, Kakatiya University, Warangal for his encouragement. Two of us (DVL, CKK) are grateful to management, Sri Kavitha Educational Society for providing Lab facilities.

References

Akira, H., Kiyoshi, M. and Kitamura, H., 1989. Characterisation of the bacteria Population structure in an anaerobic-aerobic activated sludge system on the basis of respiratory quinone profiles. *Appl. Enviom. Microbiol.*, 55: 897–901.

Arnaud, D., Mathews, R. and Jones, J.E., 2004. Perceptions and assessment of soil fertility by farmer in the mid hills of Nepal. *Agriculture Ecosystem and Environment*, 103: 191–206.

Border, M.M., Firehammer, B.D., Shoop, D.S. and Myer, 1985. Isolation of *Bacteriods fragilis* from the *feces diarrheric* calves and lambs. *J. Clin. Microbiol.*, 21: 472–473.

Baranyi, J. and Carmen, P., 1999. Estimating Bacterial growth parameters by means of detection time. *Appl. Environ. Microbiol.*, 65: 732–736.

Bradley, P.D., 1999. Microbial functional diversity can be influenced by the addition of simple organic substrates to soil. *Soil Biol. Biochem.*, 30: 1981–1988.

Datke, S.B., Minimol, J.S. and Ujjainkar, V.V., 2003. Effect of season on yield and yield components in groundnut. *PKV Res. J.*, 27: 57–60.

Douglas, S., Myers, L. and Julie, B.L., 1990. Enumeration of Enbtero toxigenic *Bactriods fragilis* in Municipal sewage. *Appl. Environ. Microbiol.*, 56: 2243–2244.

Goovaerts, P., 1999. Geostastics in soil science: State of the art and perspectives. *Goederema*, 89: 1–45.

Jon, H., Gatlin, L.E., K. Charles, G.H. Choi and M.S. Hanson, 1999. Identification and characterization of Sir A: An iron regulated protein from *Staphylococcus aureus*. *J. Bacteriol.*, 181: 1436–1443.

Kiss, S., M.D. Bularda and D. Radulescn, 1975. Nutrient cycling in soil. *Adv. Agron.*, 27: 25–87.

Lalitha Kumari, B., 1996. Assessment of industrial an pollution biological diversity and enzyme productions of the soils in warangal, A.P. *Ph.D. Thesis*, Kakatiya University.

Lindsay, J.A. and T.V. Riley, 1994. Staphylococcus iron requirements siderophore production and iron regulated protein expression. *In fact Immun.*, 62: 2309–2314.

Kumar, A.S., 2005. Effect of salt stress on germination of three varieties of *Vigna radiata*. *Ind. J. Environ. and Ecoplan.*, 8: 201–203.

Madhuya, D.A., S. Adak and M.K. Purohit, 2003. Assessment of effluents from cement industries: A pollution problem. *Indian J. Environ. and Ecoplan.*, 7: 497–500.

Medina, A., A. Probanza and R. Azcon, 2005. Interactions of arbuscular mycorrhizal fungi and bacilli strains and their effects on plant growth microbial rhizosphere activity. *Applied Soil Ecology*, 22:

15–28.

Pawar, S.N., S.P. Divekar, S.B. Glule and A.S. Kadale, 2004. Effect of mulching on moisture conservation and yield of summer groundnut crop. *J. Sci. Crops*, 14: 410–413.

Phyllis, A., W. Martin and S.T. Rusel, 1989. World wide abundance and distribution of *Bacillus thuringenues* isolates. *Appl. Environ. Microbiol.*, 55: 2437–2442.

Hattori, J., 1976. The physical environment in soil microbiology: An attempt to extend principles of microbiology to soil microorganisms. *CRC Crit. Rev. Microbiol.*, 4: 423–461.

Trivedi, B.S., 2003. Use of farm residues for integrated nutrient management. *Indian J. Agric. Environ. and Biotech*, 1: 75–93.

Varadaraj, P. and Ayyappan, S., 1989. Population and distribution of heterotrophic bacterial flora in freshwater bodies of Bangalore. *Ind. J. Environ. Health*, 31: 162–170.

Chapter 16

Protective Effect of *Mucuna pruriens* Seed on Ethanol Treated Rat Brain ATPases

G. Krishnamoorthy and A. Sivamady
Department of Zoology, K.M. Centre for Post Graduate Studies, Pondicherry – 605 008, India

ABSTRACT

The impact of *Mucuna pruriens* seed on ethanol treated rat brain ATPases was assessed in the present study. A group of rats were given ethanol orally (25 per cent 1 ml) + *Mucuna pruriens* seed extract (15 mg/kilogram body weight) for 30 days (twice daily). The ATPase activities of these rat brain were compared with ethanol treated and control group of rats. Suppressive action of ethanol on these enzymes correlated to the disturbed nerve physiology and functional behaviour. Mucuna seed extract stimulate the brain performance by activating the energy cleaving enzyme ATPase in alcoholic rats. The present study suggest that the *Mucuna pruriens* seed treatment may be beneficial to prevent ethanol induced alterations on ATPase mediated and normal transport functions of brain.

Keywords: *Mucuna pruriens, Rat, Brain, Ethanol, ATPase.*

Introduction

Ethanol consumption disturbs a variety of physiological and behavioural aspects in the central nervous system. Changes in memory, uncontrolled behaviour and inappropriate confidence are usually associated with chronic alcoholism. Intake of ethanol for a longer period interferes in the electrical and communication properties of nerve cells (Rodrigo *et al.*, 2002). Earlier studies reveals the disturbed neural transport functions and membrane bound enzyme activities of brain after alcoholization (Oner *et al.*, 2002; Mischuk *et al.*, 2003). Further it is stated that ethanol induce injury on astroglial cell by generating free radical species and change its viability (Muscoli *et al.*, 2002).

The poor response of allopathic drugs on alcoholics instigated the researchers to work on some traditional medicines. Moreover, a series of principles isolated from Indian medicinal plants act as a antioxidant and scavengers of free radicals (Sanz *et al.*, 1994; Gyanji *et al.*, 1999). *Mucuna pruriens* a potent herb was used to treat the Parkinson's disease in ancient time. In addition it is the best source of many principles which regulates the neurotransmitters of brain cells (Manyam, 1995) and also known to enhance mental alertness and coordination (Manyam, 2004). The present study was designed to delineate the protective effect of *Mucuna pruriens* seed on rat brain in relation to ethanol toxicity.

Materials and Methods

Healthy adult male albino rats of Wistar strain (*Rattus norvegicus*) weighing 230 to 300 g of body weight were used. The rats were kept in clean cages in a temperature controlled room with 12 hours light/dark scheduled. They were fed with balanced diet with free access to water. Fifteen rats were selected for this study and randomly divided into three groups of five each. Group I includes control, (Rats received an isocaloric quantity of sucrose in the same volume as experimental rats that received ethanol). Group II includes Ethanol treated (Ethanol was administered twice daily at regular intervals by gastric incubation at a dose of as 25 per cent (1 ml) aqueous solution for 30 days. Group III includes Ethanol (1 ml of 25 per cent) + *Mucuna pruriens* seed extract (15 mg/kilogram body weight) treated for 30 days, twice daily at regular intervals by gastric incubation. At the end of the experimental period (30 days) the animals were sacrificed by decapitation. The brain, was removed carefully and cleaned from adjoining tissues. The sodium-potassium (Na^+ K^+ ATPase), magnesium (Mg^{++} ATPase) and calcium (Ca^{++} ATPases) dependent adenosine triphosphatases were estimated according to the method of Takeo and Sakanashi (1985).

Results and Discussion

The activities of Na^+ K^+ ATPase, Mg^{++} ATPase and Ca^{++} ATPases generally shows decreased trend in ethanol treated group than the control rats. The reduction in the activities of ATPases correlated to altered ionic transport and decreased ATP breakdown which drastically impairs the metabolic and vital physiological activities. Previous studies also reported the altered neural membrane function associated with decreased ATPase activities in ethanol treated animals (Kanbak *et al.*, 2001; Babich *et al.*, 2002; Oner *et al.*, 2002). Further ethanol treatment is known to suppress the ATPase activities through activation of lipid peroxidation in the neural membranes (Zamai *et al.*, 2002). The free radicals generated during the catalytic cycle of ethanol would have induced the peroxidation process in membrane lipids (Warner and Gustafsson, 1994).

The combined treatment of *Mucuna pruriens* and ethanol prevents the adverse effect of alcohol on brain ATPases. The activities of different ATPases in brain was found increased in these group compared to ethanol treated animals. *Mucuna pruriens* seed is the best natural source of L-Dopa and easily cross the blood brain barrier (Manyam, 1995). L-Dopa is the precursor for neurotransmitter synthesis and enhance the neuronal activity (Manyam *et al.*, 2004). Besides the L-dopa, the other bioactive agents of Mucuna seed would have acted on brain cells as free radical scavenger/inhibitors of lipid peroxidation and exert stimulatory effect on brain (Sinha,1992). The present study suggest that the bioactive principles of Mucuna seed would have counteracted the ethanol toxicity at molecular level. In addition the increased activities of brain ATPases after Mucuna seed treatment may be the

Table 16.1: Effect of Ethanol and *Mucuna pruriens* Seed on ATPases in Rat Brain

Group	*Mg^{++} ATPase*	*Ca^{++} ATPase*	*Na^{+} K^{+} ATPase*
Control	170.77±8.35	199.27±5.20	124.42±5.13
Ethanol treated (E)	104.44±6.5***	109.35± 8.49***	46.17± 4.80***
Ethanol + *Mucuna pruriens* seed treated (EM)	159.22^{C}±3.53	169.14^{C}±4.82**	101.47^{C}±4.13

Enzyme activities are expressed as μ moles of Pi formed/hr/mg protein.

Each value is Mean±SEM of five animals.

***: P < 0.001 = Control Vs Ethanol (E) treated animals.

**: P < 0.01 = Control Vs Ethanol + *Mucuna pruriens* seed (EM) treated animals.

C: P < 0.001 = Ethanol treated (E) Vs Ethanol + *Mucuna pruriens* seed (EM) treated animals.

basic mechanism for the improvement of mental alertness and coordination in ethanol toxicated animals.

References

Babich, L.G., Shlykov, S.G. and Borisova, L.A. 2002. Effect of ethanol on intracellular Ca^{++} metabolism. *Ukr. Biokhim. Zh.*, 74: 19–26.

Gyanji, M.A., Yonamine, M. and Aniya, Y., 1999. Free-radical scavenging action of medicinal herbs from Ghana: *Thonningia sanguinea* on experimentally-induced injuries. *Gen. Pharmacol.*, 32: 661–667.

Kanbak, G., Akyuz, F. and Inal, M., 2001. Preventive effect of betaine on ethanol induced membrane lipid composition and membranes ATPases. *Arch. Toxicol.* 75: 59–61.

Manyam, B.V., 1995. An herbal alternative. *J. of Alternative and Complement Med.*, 1: 249–255.

Manyam, B.V., Dhanasekaran, M. and Hare, T.A., 2004. Effect of antiparkinson drug HP-200 (*Mucuna pruriens*) on the central mono aminergic neurotransmitters. *Phytother. J. Res.*, 18: 97–101.

Mishchuk, D.O. and Kaplia, A.A., 2003. Structural and functional state of microsomal membranes in the brain cortex during long-term ethanol. *Ukr. Biokhim. Zh.*, 75: 97–100.

Muscoli, C., Fresta, M., Cardile, V., Palumbio, M., Remis, M., Puglisi, G., Paoline, D., Nistoco, S., Rotiroti, D. and Mollace, V., 2002. Ethanol induced injury in rat primary cortical astrocytes involves oxidative stress: Effect of indebenone. *Neurosci. Lett.*, 23: 21–24.

Oner, P., Cinar, F., Kocak, H. and Gurdol, F., 2002. Effect of exogenous melatonin on ethanol induced changes in Na(+), K(+) and Ca^{++} ATPase activities in rat synaptosomes. *Neurochem. Res.*, 27: 1619–1623.

Rodrigo, R., Trujilla, S., Bosco, C., Orellana, M. and Araya, L., 2002. Changes (Na^{+} K^{+}) ATPase activity and ultrastructure associated with oxidative stress: Induced by ethanol intoxication. *Chest*, 121: 589–596.

Sanz, M.J., Ferrandiz, M.L., Ejudo, M., Terencio, M.C., Gill, B., Bustos, G., Ubeda, A., Gunasegaran, R. and Alcaraz, M.J., 1994. Influence of a series of natural flavonoids on free radical generating systems and oxidative stress. *Xenobiotoca*, 24: 689–699.

Sinha, R., 1992. Principle constituents of *Mucuna pruriens. J. Res. Educ. Indian Med.*, 11: 15–18.

Takeo, S. and Sakanashi, M., 1985. Characterization of membrane bound adenosine triphosphatase activity of enriched fraction from vascular smooth muscle. *Enzyme*, 34: 152–165.

Warner, S. and Gustafsson, K., 1994. Effect of ethanol on cytochrome P450 in the rat brain. *Prac. Nat. Acad. Sci.*, USA, 91: 1019–1023.

Zamai, T.N., Titova, N.M., Zamai, A.S., Usol'tseva, O.S., Yulenkova, O.V. and Shumkova, D.A., 2002. Effect of alcoholic intoxication on water content and activity of Na, K-ATPase and Ca-ATPase in rat brain. *Bull. Exp. Biol. Med.*, 134: 541–543.

Chapter 17

The Allergenicity of Pollen on Environmental Influences

M. Venkateshwarlu, Ch. Sridhar Rao and B. Digamber Rao
Department of Botany, Kakatiya University, Warangal – 9

ABSTRACT

The raising importance of wild relations of crop makes it obligatory for the cryopreservation of pollen of wild species related to cereals, millets, weed plants and other anemophilus agro-crops. The importance parameters of pollen systems in the biomonitoring of phytotoxicants in problems related to environmental health, as enumerated by Rosen (1981). The discussion about the influence of environmental factors, in particular air pollution has not ended yet. If air pollution would have influence on the allergencity of pollen we would have to assume that it influences both *Hypties* and *Eukalyptus* pollen. The increase of sensitizations against *Parthinium* pollen correlates with the increasing number of pollen in the air.

Keywords: *Allergenicity, Allergenic potency, Air mass, Sensitiation.*

Introduction

The biological materials contained in the air are not only relevant to the allergy problem, but are also of application as bioindicators of air pollution, Air mass is also a repository of organic materials like plant and animal parts. Algal filaments in association with pollen and spores and a medium of life and transport of many organisms, seem or un seen. Another area in which pollen and spores found early application was in coal prospecting consequent to the discovery of sporae dispersae "so named by Van post in (1916). The important parameters of pollen systems in the biomonitoring of phytotoxicants in problems related to environmental health, as enumerated by Rosen (1981). The increase of sensitizations against *Parthinium* pollen correlates with the increasing number of pollen in the air. If air pollution would have influence on the allergenicity of pollen, we would have to assume that it influences both *Hypties* and *Eukalyptus pollen.* The general discussion about the increase of allergic diseases includes the question of environmental influences on the allergenic capacity of pollen.

Materials and Methods

Aerobiology deals not only with different types of organisms but also with their products such as spores of bacteria, fungi, viruses, associated metabolites, allergic pollen of higher plants, fragments and faecal pellets, plant and animal cells during industrial processing; and products of biotechnological processes, mainly enzymes. The impact on other organisms includes infection, allergy and toxicosis in man and animals and infection of plants.

In such cases the pollen is generally coated with nutrients and pigments, which allow the *Parthinium, Hyptis* and *Eucalyptus* pollen grains to adhere to the insect body and serve as an insect-attractant for eventual dispersal. Wind pollinated flowers show adaptations towards reduction of the floral parts, increased pollen production and size of stigma together with a more effective presentation of both organs. In addition, the entomophilous pollen are generally larger, heavier with various types of exine ornamentations. Birds, bats and some small mammals eat pollen and nectar as a major part of their diet, and may act as pollinators. Early speculation as to the origins of disease by Hippocrates and Lucretins and later, the discovery of fungal spores by Micheli, had lid which is aerobiology was first concerned with airborne bacteria and disputes over spontaneous generation extending from the time of Antohine van Leen wenhoek. It is only during the last 40–50 years that more efficient sampling methods have been adopted. New sampling instruments have often originated as instruments used in industrial hygiene, which have been applied first in plant pathology and then in the study of allergens.

Results and Discussion

The data sets for our comparison were annual totals of airborne pollen of *Parthinium, Hypties* and *Eukalyptus.* The RAST data have been performed by use of phadabas RAST in the period of July to December, we decided to consider positive results to be those of RAST class 2 and higher or >0.34 PRO, respectively only specimens related to inhalation allergies have been taken into account pollen spores, house dust mite, and animal dander. The total number of included subjects was 2179. For the statistical evaluation, we determined the significance to be on the 5 per cent error probability level (Table 17.1). For the annual total of *Parthinium Hypties* and *Eukalyptus* trends have been calculated by correlation of the totals against the consecutive months. Whilst there was almost no trend obvious for *Parthinium,* a weak trend to ward higher values was observed in *Hypties* pollen, but a significant trend to higher totals was detected in the annual totals of *Eukalyptus* pollen (Figure 17.2) correspondingly, we investigated the trends in the relative amount of sensitizations against those pollen for the same period. The correspondingly, we investigated the trends in the relative amount of sensitisations against those pollen for the same period. The correlation between annual totals of the pollen and RAST sensitivity was not significant, but near to significance for *Grass* correlation in order to prove the response of sensitization rates to increasing pollen numbers taking into consideration a reaction delay of one or two months respectively, also failed to become significant (Figure 17.3).

Table 17.1: Pollen and Percentage of Rast Positive Results Out of All Positive Rast Test on Inhalation Allergens

Month	*Hyptis Pollen (1000)*	*Pos. Rast*	*Eucalyptus Pollens (1000)*	*Pos. Rast*	*Grass Pollen (1000)*	*Pos. Rast*
July	558	40.9%	231	74.4%	182	17.8%
August	120	27.5%	411	66.8%	409	17.2%
September	552	43.8%	298	55.1%	193	23.1%
October	592	31.8%	325	53.6%	801	14.7%
November	554	40.5%	250	64.9%	130	16.5%
December	295	51.1%	262	59.1%	594	15.9%
January	432	46.1%	302	67.1%	389	15.8%
February	591	53.7%	431	66.1%	114	26.9%
March	234	48.7%	338	66.2%	186	24.9%
April	255	33.7%	502	63.9%	131	16.6%

Investigation of Jilek *et al.* (1993) have proved that stressed individuals of birch produce more Bet VI and Bet V II, known as profiline related to the human stress hormone Valenta *et. at.* (1991). Suggest that a number of environmental influences cannot only have influence on human health by itself, but would also influence the pollen grain and its allergenic potency. Schata *et al.* (1993) found no differences in the aggressively of pure pollen and those exposed to high concentrations of diesel exhaust. In their study, there is no need to investigate the influence of air pollution factors, as they should have the same influence on birch pollen as well as on and *Hyptis* pollen (Figure 17.1). A possible explanation

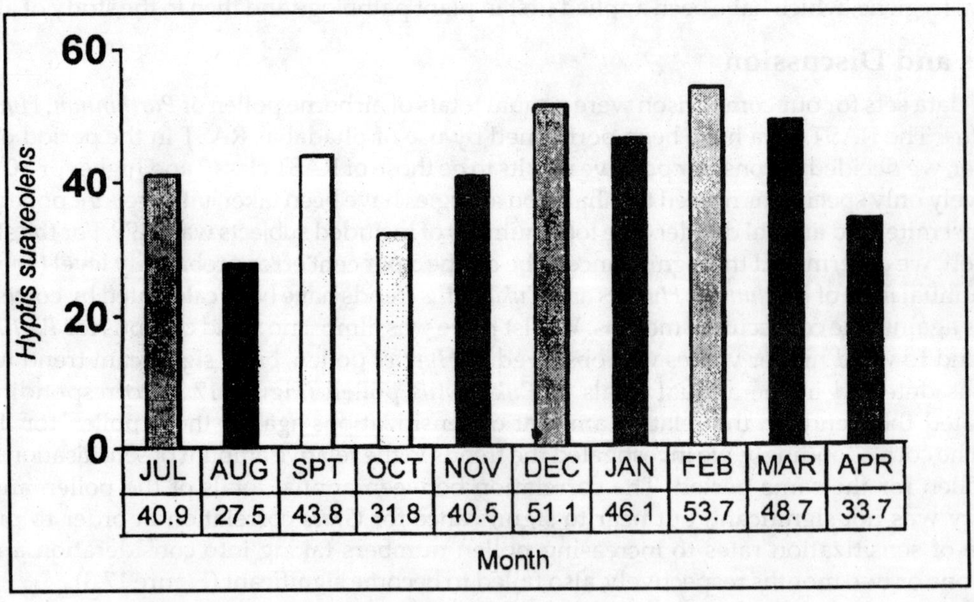

Figure 17.1: Incidence of *Hyptis suaveolens* (July to April)

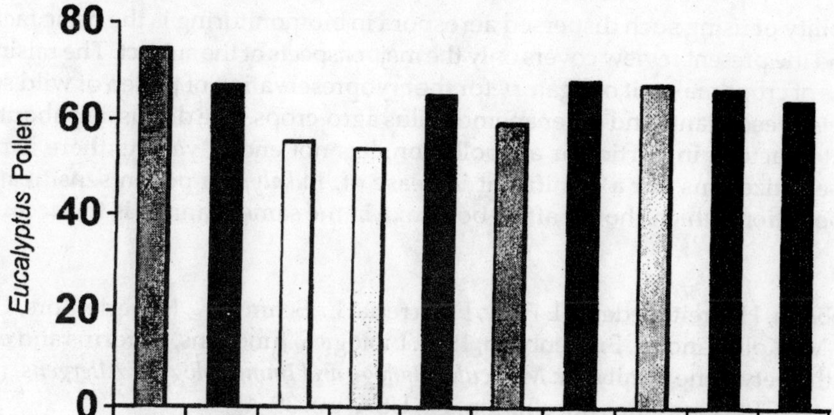

JUL	AUG	SPT	OCT	NOV	DEC	JAN	FEB	MAR	APR
74.4	66.8	55.1	53.6	64.9	59.1	67.9	67.2	66.2	63.9

Month

Figure 17.2: Incidence of *Eucalyptus* Pollen (July to April)

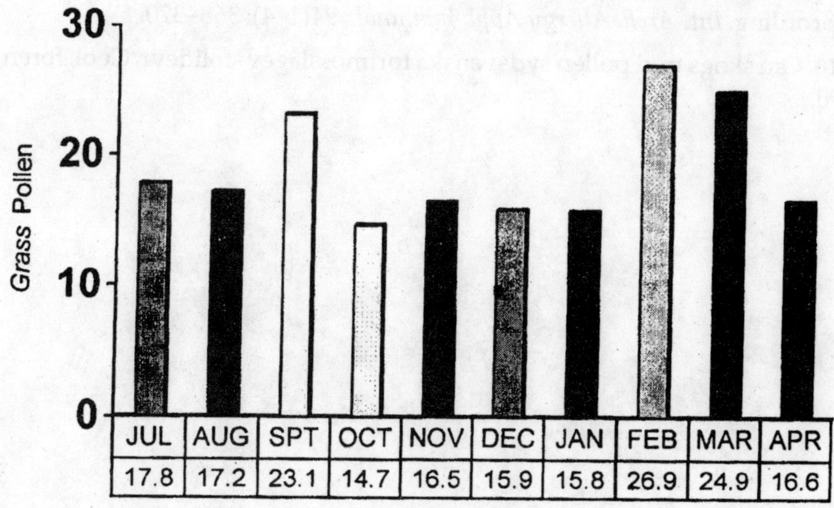

JUL	AUG	SPT	OCT	NOV	DEC	JAN	FEB	MAR	APR
17.8	17.2	23.1	14.7	16.5	15.9	15.8	26.9	24.9	16.6

Month

Figure 17.3: Incidence of *Grass* Pollen (July to April)

is, that in the region of the birch, being an element of the cold and moderate climates, is should to the margins of its natural distribution area because of a permanent warning of the atmosphere.

Conclusion

The possibility of using such dispersed aerospora in biomonitoring is the basic factor in applied Aerobiology, and the present review covers only the major aspects of the subject. The raising importance of wild relations of crops makes it obligatory for the cryopreservation of pollen of wild species related to cereals, millets, weed plants and other anemophilus agro-crops. The discussion about the influence of environmental factors, in particular air pollution, has not ended yet. As there is no increase of *Hyptis* pollen sensitizations but a significant increase of, *Eukalyptus* pollen sensitizations, it is not likely that air pollution, which should affect both taxa in the same manner, is the accountable factor.

References

Jilek, A., I. Swoboda, H. Breitenedeve, I. Fogy, F. Ferreria, E., Schmid E., Heberle-Bors, O. Scheiner, H. Rumpold, M., Koller and M. Breitenbach, 1993. Biological functions, isoforms and environmental control in the betvigene family. In: *Molecular Biology and Immunology of Allergens*, (Eds.) D. Kraff and A. Sehon. CRC Press Inc., Boca Raton, Fl., USA, pp. 39–45.

Rosen, G.W., 1981. Pollen systems to detect phytotoxicants in the environment an introduction. *Environ. Hlth. Persp.*, 17: 105.

Schata, M.R., Wahl, H.D., Baner, D., Dahmann and H.W. Stockmann, 1993. Immunological comparison on pure pollen and those exposed to high concentrations of diesel exhaust. *Allergy* 48(Suppl. 16): 184.

Valenta, R., M. Duchece, M. Breitenbach, K. Petteuburgev, I., Koller, H., Rumpold O. Scheiner and D. Krait, 1991. A low molecular weight allergen of white birch (*Betula verrucosa*) is highly homologous to human profiling. *Int. Arch. Allergy Appl. Immunol.*, 94(1–4): 368–370.

Vonpost, L., 1916. Om skogs trad pollen sydsvenska torfmosslagev–folidevr, Geol, foren, stockh, forh., 38: 384–390.

Chapter 18

Neem Derivatives as a Larvicide for the Management of Country Bean Pest *Pericallia ricini* F. (Lepidoptera : Arctidae)

N. Revathi[1] and S. Kingsley[2]*

[1]Department of Advanced Zoology and Biotechnology, Loyola College, Chennai – 600 034, India

[2]Entomology Research Institute, Loyola College, Chennai – 600 034, India

ABSTRACT

Larvicidal property of neem oil, neem cake extract and monocrotophos at different concentrations was evaluated on fourth, fifth and sixth instar larvae of *Pericallia ricini.* It was found that neem oil was highly effective. Larval mortality was found to be lesser on the larvae of older age. An increase in the concentration of the test solutions resulted with an increase in the rate of larvicidal effect.

Keywords: Pericallia ricini, Azadirachta indica, Bioefficacy, Larvicidal effect, Country bean.

Introduction

Due to the increased awareness on indiscriminate use of synthetic pesticides and their persistent harmful effects, much interest has been evinced in the use of botanical products to control the

* Corresponding Author.

agricultural pests. Of the numerous plants investigated, the neem tree *Azadirachta indica* A. Juss, has shown promising results for the control of agricultural pests. More than 400 species of insects, including key pests of agriculture are susceptible to various behavioural and physiological effects of neem (Schmutterer and Singh, 1995). The practical utility of neem as a pest control agent was first demonstrated by Pradhan *et al.* (1962). Patnaik *et al.* (1987) observed larvicidal effect of *Crocidolomia binotolis* with neem oil. Larvicidal effect of neem oil was also observed earlier by Azam (1991) on *Liriyomyza trifolii*, Jhansi and Singh (1993) on *Helicoverpa armigera*, Brar *et al.* (1994) on *Earias vittella*, Jeyakumar and Uthamasamy (1994) on *Liriyomyza trifolii.* Chari *et al.* (1999) on *Spodoptera litura*, Ramamurthy and Rajaram (2000) on *Amrasca biguttula* and Vastrad *et al.* (2002) on *Plutella xylostella.* Pest control methods are aimed to combat the different stages of insects of which the larval stage is vulnerable. Controlling the pest population in the larval stage itself decreases the next generation development. This factor is one of the key aspects of the present investigation for choosing the larval stage of *P. ricini* for the evaluation of bioefficacy of neem derivatives.

The larva of lepidopteron moth *Pericallia ricini* is a serious polyphagous pest of agricultural crops. The black hairy caterpillar is a voracious leaf feeder of country bean *Dolichos lablab*, which is one of the main vegetable crops of Tamil Nadu. During the initial stages, the larvae feed on the lower side of the leaves, and when they reach third instar stage they begin to feed voraciously and defoliate the entire plant. When the larval population was uncontrolled, the crop yield loss may extend up to 70 per cent to 80 per cent. Botanical pesticides offer a viable alternative to synthetic insecticides as they are comparatively less toxic, biodegradable and are safe to non-target organisms (Verma and Pandey, 1981). This consideration was behind this study in evaluating the larvicidal effect of neem derivatives on the country bean pest *P. ricini.*

Materials and Methods

The larvicidal effect of neem oil 1.0 per cent, 1.5 per cent and 2.0 per cent, neem cake extract 1.0 per cent, 3.0 per cent and 5.0 per cent along with monocrotophos 0.05 per cent concentrations were evaluated against fourth, fifth and sixth instar larvae of *P. ricini.* The eggs of *P. ricini* were collected along with the leaves from the field during morning hours from 6.00 to 8.30 am and evening from 5.00 to 6.30 pm. The eggs were placed in a well ventilated plastic container under room temperature 28±2°C. After hatching, the larvae were transferred to plastic basin covered with nylon net. Larvae were provided with fresh leaves of country bean. The larvae were released in a container and neem derivatives were sprayed with a sprayer. The topically treated larvae were transferred into well ventilated plastic boxes, with filter paper covered bottom. Fresh leaves were supplied for treated larvae. Three replications were maintained for each treatment including control. Observations on the lalval mortality were recorded after 24 hours of treatment and percentages were calculated.

Results and Discussion

Larvicidal activity of neem oil, neem cake extract and monocrotophos at different concentrations tested against the fourth, fifth arid sixth instar larvae of *P. ricini* were presented in the Table 18.1. In the present investigation neem oil had higher larvicidal effect and it was maximum with 79.99 per cent, 71.10 per cent and 66.66 per cent when fourth, fifth and sixth instar larvae were treated with neem oil 2.0 per cent concentration, respectively. An increase in the concentration of the test solutions resulted in an increase in the rate of larval mortality of *P. ricini.* Similar observation was also reported earlier on *Helicoverpa armigera* (Behera and Satapathy, 1996); *Liriomyza trifoli* (Jayakumar and Uthamasamy,

1997); *Epilachna vigintioctopunctata* (Shanmugapriyan and Kingsley, 2001); *Helicoverpa armigera* (Sahayaraj and Paulraj, 2001); *Plutella xylostella* (Vastrad *et al.*, 2002); *Spodoptera litura* (Leatemia and Isman, 2004); *Helicoverpa armigera* (Pushpa *et al.*, 2004) and *Tribolium castaneum* (Gupta, 2006).

Table 18.1: Efficacy of Neem Oil, Neem Cake Extract and Monocrotophos on IV, V and VI Instar Larvae of *Pericallia ricini*

Treatment	*Concentration %*	*Larval Mortality %*		
		IV Instar	*V Instar*	*VI Instar*
Neem Oil	1.0	52.22 (46.29) bc	48.88 (44.35) bc	44.44 (41.71) bc
Neem Oil	1.5	61.10 (51.44) b	59.99 (50.87) abc	52.22 (46.29) ab
Neem Oil	2.0	79.99 (63.64) a	71.10 (58.02) a	66.66 (54.80) a
Neem Cake extract	1.0	39.99 (39.12) cd	32.21 (34.54) d	31.11 (33.77) cd
Neem Cake extract	3.0	55.55 (48.19) b	45.55 (42.42) cd	41.10 (39.84) bc
Neem Cake extract	5.0	75.55	64.44	63.33
Neem Cake extract		(60.58) a	(53.43) ab	(52.93) a
Monocrotophos	0.05	34.44 (35.86) d	29.99 (33.11) d	21.10 (27.25) d
Control	0.00	0.00 (0.19) e	0.00 (0.19) e	(0.19) e

Values mean of three replications.

Mean followed by a common letter are not significantly different at the 5 per cent level by DMRT.

Larvicidal effect of neem derivatives on the sixth instar larvae was lower than fifth and fourth instar larvae. It was observed that with an increase in the age of the larvae, there was a corresponding decrease in the larvicidal effect of the test solutions. Shanmugapriyan and Kingsley (2001) also recorded a similar decrease of the larvicidal effect, when they treated second, third and fourth larval instars of *Epilachna vigintioctopunctata* with neem oil 1.5 per cent concentration.

In the present evaluation it was observed that the neem oil had higher mortality rate followed by neem cake extract and monocrotophos. These results suggest that neem oil could be used as a larvicide for the management of *P. ricini*. However, bioefficacy ofneem derivatives will be confirmed only after field study.

References

Azam, K.M., 1991. Toxicity of neem oil against leafminer, *Liriyomyza trifolii*, Burgess on cucumber. *Pl. Port. Qrtly.*, 6: 196–197.

Behera, U.K. and Satapathy, C.R., 1996. Screening indigenous plants for their insecticidal properties against *Spodoptera litura* Fb. *Insect Envi.*, 2(2): 43–44.

Brar, K.S., Arora, S.K. and Ghai, T.R., 1994. Efficacy of neem oil, dipel and insecticides for the control of cotton jassid and fruit borer on Okra. *Pestology*, 18(2): 14–16.

Chari, M.S., Ramaprasad, G., Sitaramaiah, S. and Murthy, P.S.N., 1999. Laboratory and field evaluation of neem (*Azadirachta indica* A. Juss), Pongamia (*Pongamia pinnata* L.) Chinaberry (*Melia azedarach* L.) extracts and commercial neem formulations against tobacco caterpillar *Spodoptera litura* F. In: *Azadirachta indica A. Juss*, (Eds.) Singh, R.P. and Saxena, R.C., pp. 111–129.

Gupta, S., 2006. Performance of *Mentha piperita* against *T. castaneum* Herbst (Coleoptera : Tenebrionidae). *Curr. Sci.*, 9(1): 393–398.

Jeyakumar, P. and Uthamasamy, S., 1997. Bioefficacy of some synthetic insecticides and botanicals against *Liriyomyza trifolii. Indian J. Ent.*, 59(4): 347–350.

Jhansi, B.R. and Singh, R.P., 1993. Identification of effective and inexpensive neem (*Azadirachta indica* A. Juss) seed kernel extract for management of *Helicoverpa armigera* Hubner. Abstract. *World Neem conference,* February, Bangalore, India, pp. 7.

Leatemia, J.A. and Isman, M.Bv2004. Insecticidal activity of crude seed extracts of *Annona* spp., *Lanium domesticum* and *Sandoricum koetjape* against Lepidopteran larvae. *Phytoparasitica*, pp. 30–37.

Murugan, K., Sivaramakrishnan, S., Senthilkumar, N., Jeyabalan, D. and Senthilnathan, S., 1998. Synergistic interaction of botanicals and biocides Nuclear Polyhedrosis Virus on pest control. *J. Scientific and Industrial Research*, 57: 732–739.

Patnaik, N.C., Panda, N., Patro, E.R and Misra, B.K., 1987. Effect of neem oil (*Azadirachta indica* A. Juss) on mustard web-worm, *Crocidolomia binotalis* zell. (Lepidoptera : Pyralidae). *Neem News Letter*, 4(2): 15–17.

Pradhan, S., Jotwani, M.G. and Rai, B.K., 1962. The neem seed deterrent to locusts. *Indian Fmg.*, 12: 7–11.

Pushpa, P., Mary, J., Vijay, T. and Swati, J., 2004. Bioactivity of non-edible oil seed extracts and purified extracts against *Helicoverpa armigera* (Hubner). *Indian J. Expt. Biol.*, 42: 91–95.

Ramamurthy, R. and Rajaram, V., 2000. Effect of botanicals and insecticides against rice leaf folder complex. *Bio. Sci. Res. Bull.*, 16(1): 43–44.

Sahayaraj, K. and Paulraj, M.G., 1998. Relative toxicity of some plant extracts to groundnut leaf miner, *Aproaerema modicella* Dev. *Int. Arach. Newslett.*, 18: 27–29.

Sahayaraj, K. and Gabriel Paulraj, M., 2001. Efficacy of chosen plants against gram pod borer *Helicoverpa armigera* Hubner (Lepidoptera : Noctuidae). *J. Adv. Zool.*, 22(1): 8–14.

Schmutterer, H. and Singh, R.P., 1995. List of insect pests susceptible to neem products. In: *The Neem Tree Azadirachta indica A. Juss and Other Meliaceous Plants,* (Ed.) H. Schmutterer. VCH Publications, Weinheim, Germany, pp. 326–365.

Shanmugapriyan, R. and Kingsley, S., 2001. Bioefficacy of neem oil on larvae of bitter guard beetle *Epilachna vigintioctopunctata* (Coccinellidae : Coleoptera). *J. Ecotoxicol. Environ. Monit.*, 11(3): 215–219.

Vastrad, A.S., Lingappa, S. and Basavangoud, K., 2002, Multi tool approach for the management of Diamond backmoth, *Plutella xylostella* Linn. Biopesticides and neem products. *Pestology,* 24(8): 47–49.

Verma, G.S. and Pandey, U.K., 1981. Studies on the effect of *Acorus calamus, Cimicifuga foetida* and *Glynandropsis gynandra* extracts against insect pests of cruciferous vegetables painted bug *Bagrada cruciferum* Kirk (Hemiptera : Pentatomidae). *Zeitschriftfur Angew Zool.*, 68: 109–113.

Chapter 19

Efficacy of Five Plant Extracts as Ovicide Against Pulse Beetle, *Callosobruchus chinensis* (L.)

S.C. Dwivedi and Padam Chand Bhati
Ecotoxicology Laboratory, Department of Zoology, University of Rajasthan, Jaipur – 302004, India

ABSTRACT

Potentiality of five plant extracts in acetone was assessed for their ovicidal properties in a laboratory trial, against *Callosobruchus chinensis.* Out of the plant extracts evaluated, seed extract of *Momordica charantia* was found to be most effective, depicting 100 per cent corrected mortality at 100 per cent dose level. However, other plant extracts also exhibited considerable effect on egg hatchability.

Keywords: Plant extract, Ovicides, Pulse beetle.

Introduction

Pulses, stored for all the year round availability as food article and seeds are invariably attacked by pulse beetles (Sharma, 1984; Giga *et al.*, 1990). *Callosobruchus chinensis* (L.) is one of the major destructive pest of pulses, belonging to order Coleoptera and family Bruchidae. Infestation of pulse beetle is manifested by presence of its eggs laid on surface of pulse grains. The larval stages feed on seed cotyledons, reducing it to mere seed coat or testa, leaving unfit for human consumption and cultivation purposes. The middle decades of twentieth century have been the years of revolution in the field of pest control. Ever since the discovery of insecticidal properties of DDT, there has been a constant efforts by entomologists to empower synthetic pesticides with capabilities to eliminate various pests but, at the same time, several potential hazards associated with the excessive use of synthetic pesticides were not taken into account. Development of resistance in target animals and high degree of residual toxicity coupled with health hazards to non-target animals including human beings are few of these. Further, high infant morality has also been reported from the places where high residues of DDT were found in human milk.

The insecticides of plant origin, on the other hand, have the additional advantage on synthetic insecticides in having biodegradability and being safe to non-target animals and are neither persistent nor phytotoxic. Many plants have been explored for their insecticidal properties. Sharma *et al.* (1999), Kamakshi *et al.* (2000), Dwivedi and Venugopalan (2001) have all studied the effects of various botanicals on the pulse beetle. This piece of work is also aimed to assess aboriginal plants for their ovicidal properties against eggs of *Callosobruchus chinensis* so that the extent of damage and loss is checked at very first stage itself.

Materials and Methods

Extraction of Plant Materials

Plant materials (Seeds: *Momordica charantia, Nyctanthes arbortristis*, Leaves: *Zinnia angustifolia, Prosopis spicegera* and Flower: *Tagetes erecta*) were collected from the botanical garden of University of Rajasthan, Jaipur. They were thoroughly washed to remove all the dust and dirt particles. These were dried in shade and further pulverized with the help of an electric grinder. The extract of these plant materials were prepared separately by Soxhlet Extraction Method (Deshmukh and Borle, 1975) using acetone as solvent. 30g of powder of plant material and 300 ml. of solvent were taken for extraction keeping the ratio 1 : 10. After filtration and vacuum evaporation it was kept in refrigerator as stock solution (100 per cent concentration). Further dilution was done with the solvent to get desired doses for experiments.

Rearing of Experimental Insect

Rearing of *Callosobruchus chinensis* was done in laboratory in presterlised jars containing disinfested cowpea seeds (*Vigna anguiculata*). An optimum condition of 27±2°C temperatures and 60±10 per cent relative humidity was maintained. Freshly laid eggs were taken for experiments.

Experimental Design

Freshly emerged adults were sorted out from the main culture jar and released on disinfested cowpea seeds for egg laying. After oviposition seeds containing freshly laid eggs on its surface were taken for the experiments. 20 eggs were taken in a patriplate and treated with 1 ml. of each desired dose level. After complete evaporation of solvent, eggs were transferred into plastic vials, which were later covered by muslin cloth for proper supply of Oxygen. The doses taken were 25 per cent, 50 per cent, 75 per cent, and 100 per cent of stock solution prepared by dilution with acetone. For each concentration level three replicates were run along with a control having eggs treated with solvent only. The vials were observed until the last egg hatched. The number of unhatched eggs in each vial was calculated and per cent corrected egg mortality was calculated by following formula (Abbot, 1925).

$$\%\ \text{Corrected Mortality} = \frac{\%\ \text{Kill in Treated} - \%\ \text{Kill in Control}}{100 - \%\ \text{Kill in Control}} \times 100$$

Results and Discussion

Results of ovicidal actions of plant extracts in acetone documented in Table 19.1 indicate significant reduction in egg hatching on seeds of cowpea (*Vigna anguiculata*). Acetone extract of seeds of *Momordica charantia* has been registered as most potent ovicide among five plant extracts evaluated, depicting 100 per cent corrected egg mortality at 100 per cent dose level. The same extract exhibited 89.74 per cent, 67.48 per cent and 40.15 per cent corrected egg mortality at 75 per cent, 50 per cent and 25 per cent dose levels, respectively. These results get support from the observations made by Garcia (1990), who demonstrated the pet ether extract of *Momordca charantia* as an effective ovicide against been weevil.

Table 19.1: Ovicidal Actions of Plant Extracts Against the Eggs of *Callosobruchus chinensis* (L.)

Sl.No.	*Name of Plant*	*Family*	*Dose Level*	*Average No. of Hatched Eggs (A)*	*Average No of Un-hatched Eggs (B)*	*% of (A)*	*% of (B)*	*% Corrected Mortality*
1.	*Nyctanthes arbortristis*	Oleaceae	25%	12.67	7.33	63.34	36.66	35.03
			50%	10.67	9.33	53.34	46.66	45.29
			75%	9.00	11.00	45.00	55.00	53.84
			100%	6.00	14.00	30.00	70.00	69.23
			Control	19.50	0.50	97.50	2.50	–
2.	*Momordica charantia*	Cucurditaceae	25%	11.67	8.33	58.35	41.65	40.15
			50%	6.34	13.66	31.70	68.30	67.48
			75%	2.00	18.00	10.00	90.00	89.74
			100%	0.00	20.00	0.00	100.00	100.00
			Control	19.50	0.50	97.50	2.50	–
3.	*Tagetes erecta*	Asteraceae	25%	15.67	4.33	78.35	21.65	21.65
			50%	13.67	6.33	68.35	31.65	31.65
			75%	12.34	7.66	61.70	38.30	38.30
			100%	8.34	11.66	41.67	58.33	58.33
			Control	20.00	0.00	100.00	0.00	–
4.	*Prosopis spicegera*	Leguminaceae	25%	17.34	2.66	86.70	13.30	13.30
			50%	13.00	7.00	65.00	35.00	35.00
			75%	12.34	7.66	61.70	38.30	38.30
			100%	10.67	9.33	53.35	46.65	46.65
			Control	20.00	0.00	100.00	0.00	–
5.	*Zinnia angustifolia*	Asteraceae	25%	17.34	2.66	86.70	13.30	8.73
			50%	14.34	5.66	71.70	28.30	24.52
			75%	12.00	8.00	60.00	40.00	36.84
			100%	10.00	10.00	50.00	50.00	47.36
			Control	19.00	1.00	95.00	5.00	–

Seed extract of *Nyctanthes arbortristis* showed 69.23 per cent corrected egg mortality at 100 per cent concentration. The ovicidal action of extracts tends to decrease with the decrease in concentration level. Hence the registered per cent corrected egg mortality at 75 per cent, 50 per cent and 25 per cent dose levels was 53.84 per cent, 45.29 per cent and 35.03 per cent, respectively.

Floral extract of *Tagetes erecta* displayed 58.33 per cent, 38.3 per cent, 31.65 per cent and 21.65 per cent corrected egg mortality at 100 per cent, 75 per cent, 50 per cent and 25 per cent dose levels, respectively. Weaver *et al.* (1994) also reported floral and foliar extract of *Tagetes minuta* as useful insecticides for management of stored product pests. Dwivedi and Garg (2001) explored toxicological properties of *Tagetes* against rice moth, *Corcyra cephalonica.*

Foliar extract of *Prosopis spicegera* and *Zinnia angustifolia* in acetone depicted moderate ovicidal properties. At 100 per cent dose level *Zinnia* showed superiority over *Prosopis* recording 47.36 per cent corrected egg mortality because *Prosopis* depicted 46.65 per cent corrected egg mortality at same dose level. However, at 75 per cent, 50 per cent and 25 per cent dose levels *Prosopis spicegera* depicted 38.3 per cent, 35 per cent and 13.3 per cent corrected egg mortality,which is more than that exhibited by *Zinnia angustifolia i.e.* 36.84 per cent, 24.52 per cent and 8.73 per cent at 75 per cent, 50 per cent and 25 per cent dose levels, respectively.

These results get support from the observations recorded by Naik and Dumbre (1987), Ofuya (1990), Ahmed and Ahmed (1992), Pathak *et al.* (1997) and Giga (1990) who explored toxicological properties of plant materials. The ovicidal properties of botanicals may be attributed to the active ingredients present in them, which may arrest the embryonic development. The work of Chander and Ahmed (1986) also provides support to this assumption by observations on the eggs of *Callosobruchus chinensis.* However, the study needs further investigations to find out active principle ingredients responsible for ovicidal properties of plant extracts.

Acknowledgement

The authors are thankful and express their deep gratitude to Professor P.P. Bakre, Head, Department of Zoology, University of Rajasthan for providing necessary facilities.

References

Abbot, W.S., 1925. A method of computing the effectiveness of insecticides. *J. Econ. Ent.*, 18(2): 265–267.

Ahmed, S.M. and Ahmed, A., 1992. Efficacy of some indigenous plants as pulse protectants against *Callosobruchus chinensis* (L.) infestation. *International Pest Control*, 34(2): 54–56.

Chander, H. and Ahmed, S.M., 1986. Efficacy of oils from medicinal plants as protectants of green gram against the pulse beetle, *Callosobruchus chinensis. Entomon.*, 11(1): 21–28.

Deshmukh, S.D. and Borle, M.N., 1975. Studies on the insecticidal properties on indigenous plant products. *Indian J. Ent.*, 37(1): 111–118.

Dwivedi, S.C. and Garg, S., 2001. Comparative efficacy of various botanical extracts as larvicide against rice moth *Corcyra cephalonica* (Lepidoptera : Pyralidae). *Indian J. Environ. and Ecoplan.*, 5(1): 185–188.

Dwivedi, S.C. and Venugopalan, S., 2001. Evaluation of leaf extracts for their ovicidal action against *Callosobruchus chinensis. Asian J. Exp. Sci.*, 16(1 and 2): 29–34.

Garcia, J.R. Jr., 1990. Bioassay of five botanical materials against been weevil, *Callosobruchus chinensis* (L.) on mungbean [*Vigna radiata* (L.) Wilczek]. College, Laguna (Phillipines) 67 leaves.

Giga, D.P. and Munetsi, M.M., 1990. The effectiveness of vegetable and citrus oils as protectants of cowpea against infestation by *Callosbruchus rhodesianus* (Pic.). *Plant Protction Quarterly*, 5(4): 148–151.

Kamakshi, B.S., Rabiah Ibrahim, S., Raja, N. and Ignacimuthu, S., 2000. Control of pulse beetle *Callosobruchus maculates* using edible plant leaf extracts. *Uttar Pradesh J. Zool.*, 20(2): 143–146.

Naik, R.L. and Dumbre, R.B., 1987. Surface coating of cowpea seeds with certain vegetable oils for protection from pulse beetles. *J. Maharashtra Agric. Univ.*, 12(1): 76–78.

Ofuya, T.I., 1990. Oviposition deterrence and ovicidal properties of some plant powders against *Callosobruchus maculatus* in stored cowpea (*Vigna unguiculata*) seeds. *J. of Agricultural Science*, 115: 343–345.

Pathak, Namrata, Yadav, T.D. and Vasudevan, P., 1997. Contact and fumigant action of volatile essential oil of *Murraya koenigii* against *Callosobruchus chinensis. Indian J. Ent.*, 59(2): 198–202.

Sharma, S., Sochan, G.C. and Bhattacharya, A.K., 1999. Effects of crucifer oils on the growth and development of *Callosobruchus chinensis* infesting pigeon pea *Cajanus cajan. Indian J. Ent.*, 61(1): 42–47.

Sharma, S.S., 1984. Review of literature on the losses caused by *Callosobruchus* species (Bruchidae : Coleoptera) during storage of pulses. *Bull. Grain Technol.*, 22(1): 62–71.

Weaver, David, K., Carl, D. Wells, Florence, V. Dunkel, Wolfgang, Bertsch, Sharlene, E. Singh and Shobha, Sriharn, 1994. Insecticidal activity of floral, foliar and root extracts of *Tagetes minuta* (Asterales : Asteraceae) against adult mexican bean weevils (Coleoptera : Bruchidae). *J. Econ. Entomol.*, 87(16) : 1718–1725.

Chapter 20

Aquatic Macrophyte Species with Potential to Remove Heavy Metals from Sewage Water of Faizabad (U.P.)

Arun K. Pandey and G. C. Pandey*

Department of Environmental Sciences,

Dr. R.M.L. Avadh University, Faizabad – 224 001

ABSTRACT

Several aquatic macrophytes [*Lemna minor* (A), *Phragmites karka* (B), *Potamogeton punctatus* (C), *Trapa natus* (D) and *Cyperus ambilis* (E)] were evaluated for their metal (Cu, Fe and Zn) accumulation potentials from municipal sewage under laboratory conditions. All the macrophytes have showed significant reduction of metals from city sewage at Faizabad (U.P.). The order of potential of aquatic macrophytes for the removal of these metals was found to be as, B > C > E > A > D.

Keywords*: Metals, Bioaccumulation, Aquatic macrophytes.*

Introduction

Heavy metals, a well-known contaminant of sewage fed aquatic systems, pose multiple hazards (Berti and Cunningham, 2000) to aquatic life forms. However, their low concentrations are required for growth and development of plants and are essential for several metabolic processes (Prasad *et al.*,.

* Corresponding Author.

1999). In urban sewage, copper, iron and zinc are potentially hazardous (Boller, 1997) so that, it is required to remove the toxic metals from wastewater before its disposal.

Conventional metal-removal technologies from aqueous system, which are based on physico-chemical treatments, are expensive and only partially effective. Aquatic macrophytes based *Phytoremediation* is an alternative approach, which is technologically feasible and economically viable (Pandey *et al.*, 1999; Sadowasky, 1999). Metals-bioaccumulation by aquatic macrophytes has been reported by several authors (Kara *et al.*, 2003; Tripathi *et al.*, 2003; Garcia *et al.*, 2004; Lu *et al.*, 2004; Singh *et al.*, 2005). Attempts have been made for the potential use of aquatic macrophytes in remediation of municipal sewage and industrial runoff (Srivastava and Pandey, 1998; Srivastava and Pandey, 1999a; Srivastava and Pandey, 1999b; Srivastava *et al.*, 2000; Srivastava *et al.*, 2004; Singh *et al.*, 2005). Though the municipal sewage of Faizabad was reported to be polluted due to high amount of BOD, COD, TDS and low DO (Pandey and Pandey, 2003; Srivastava *et al.*, 2004).

It is envisaged to assess the potential use of aquatic macrophytes (*Lemna minor, Phragmites karka, Potamogeton crispus, Trapa natas* and *Cyperus ambilis*) in accumulation of Cu, Fe and Zn from municipal sewage of Faizabad (UP) under laboratory conditions.

Materials and Methods

The sewage sample was collected from Sahebganj Drain at Faizabad (U.P.) municipal area in the month of June 2005 and was analyzed for determination of concentration of different metals (Cu, Fe and Zn) with the help of the flame atomic absorption spectrophotometer using Perkin-Elmer 5000 model with automatic burner control.

Young and healthy specimens of same sized aquatic macrophytes; *Lemna minor* (A), *Phragmites karka* (B), *Potamogeton punctatus* (C), *Trapa natus* (D), *Cyperus ambilis* (E) were collected from different ponds near Faizabad in July 2005. These plants were thoroughly washed, dried and digested in HNO_3 : $HCLO_4$ (5 : 1, v/v) at 80°C. These macrophytes were added to each set of aquarium with or without (control) sewage wastewater.

Subsequently, after periodic exposures of 7, 14 and 21 days 50 ml samples were taken from each set to analyze again the metal concentration in the samples. Data were subjected to test of significance (t-test) and values were expressed as Mean±SEM of three replicates.

Results and Discussion

The concentrations of Cu, Fe and Zn were found to be 3.76 μg/ml, 4.75 μg/ml and 5.90 μg/ml of sewage, respectively which are beyond permissible level (Table 20.1).

Table 20.1 summarized the reduction of metals from sewage after exposure of five macrophytes (*L. minor, P. karka, P. crispus, T. natas* and *C. ambilis*). All macrophytes were characterized by intense accumulation of heavy metals in their body parts. Table-l shows the reduction of metals (Cu, Fe and Zn) from urban sewage after treatment with aquatic macrophytes at 7, 14 and 21 days. The maximum per cent reduction of Cu, Fe and Zn from sewage were observed after 21days of exposure, as 39.13 per cent, 44.25 per cent, 54.30 per cent for A, 47.83 per cent, 53.71 per cent, 63.26 per cent for B, 45.29 per cent, 51.72 per cent, 60.43 per cent for C, 37.32 per cent, 43.14 per cent, 52.16 per cent for D, 44.93 per cent, 49.53 per cent, 58.21 per cent for E respectively. The order of reduction is found to be B > C > E > A > D for all the metals. Of the five aquatic macrophytes, *P. karka* showed better reduction potential than others to all these metals from the sewage of Faizabad.

Table 20.1: Reduction in Metal Contents (Cu, Fe and Zn) at Various Exposure Durations After Harvesting Aquatic Macrophytes from Sewage

Aquatic Plants	Exposure Durations (Days)								
	Cu			Fe			Zn		
	7	14	21	7	14	21	7	14	21
Lemna minor	*2.42±0.19	*2.34±0.21	*2.29±0.23	*2.84±0.30	*2.73±0.30	*2.64±0.30	*3.10±0.07	*3.05±0.32	*2.55±0.27
	(35.51)	(37.68)	(39.13)	(40.04)	(42.36)	(44.25)	(47.31)	(48.23)	(54.30)
Phragmites karka	*2.14±0.12	*2.10±0.22	*1.96±0.22	*2.49±0.29	*2.36±0.27	*2.20±0.20	*2.69±0.31	*2.56±0.27	*2.04±0.22
	(43.12)	(44.20)	(47.83)	(47.56)	(50.32)	(53.71)	(54.28)	(56.60)	(63.26)
Potamogeton crispus	*2.23±0.12	*2.15±0.23	*2.06±0.19	*2.60±0.27	*2.47±0.26	*2.29±0.21	*2.82±0.23	*2.69±0.28	*2.20±0.25
	(40.58)	(42.75)	(45.29)	(45.18)	(48.04)	(51.72)	(52.16)	(54.39)	(60.43)
Trapa natas	*2.51±0.13	*2.42±0.22	*2.36±0.25	*2.92±0.34	*2.89±0.32	*2.70±0.28	*3.25±0.35	*3.14±0.34	*2.66±0.28
	(33.33)	(35.51)	(37.32)	(37.75)	(39.16)	(43.14)	(44.91)	(46.73)	(52.16)
Cyperus ambilis	*2.30±0.18	*2.25±0.19	*2.07±0.18	*2.66±0.28	*2.56±0.28	*2.40±0.25	*2.77±0.295	*2.66±0.28	*2.32±0.25
	(38.77)	(40.22)	(44.93)	(44.10)	(46.07)	(49.53)	(50.20)	(52.18)	(58.21)

The concentration of metals are represented as µg/ml.
Values in parentheses represent per cent reducation of metal contents.
The concentrations of Cu, Fe and Zn in sewage were 3.76, 3.75 and 5.90 µg/ml respectively.
The permissible level of Cu, Fe and Zn in sewage were reported to be as 3, 3 and 5 µg/ml respectively (MoEF, 1993).
*$P < 0.02$ (t: Test, one sided).

It is evident from Table 20.1, that the maximum degree of reduction was achieved in 7 days exposure period in all cases of exposure of aquatic macrophytes. Further, increasing the treatment duration up to 21 days did not comparatively result higher level of reduction potential by all aquatic macrophytes. This is possibly because of metals binding sites both on the plant surface (absorbed) as well as inside the cell (absorbed metals) acts saturated. Similar results were obtained during the Cu, Fe and Zn accumulation after 7 days and exposure up to 21 days has resulted comparatively lower level of accumulation (Srivastava *et al.*, 2000).

Indeed, it has been reported by various authors that metals bioaccumulation from sewage and industrial wastewaters are the essential feature of aquatic macrophytes (Srivastava *et al.*, 2000; Fritioff and Greger, 2003; Hasan *et al.*, 2003; Kara *et al.*, 2003, Garcia *et al.*, 2004; Lu *et al.*, 2004; Osmolovskaya and Kurilenko, 2005). It is evident that the profuse root system in aquatic macrophytes; their higher surface/ biomass ratio; presence of metal chelators and ability to attain equilibrium with surrounding medium within a short period which are essential features for metal hyper-accumulation by plants (Srivastava *et al.*, 2000).

At the same time, the harvesting and excavation of macrophytes can be the right way to prevent a secondary water contamination. The harvested metal-rich macrophyte-mass can be subjected to composting to produce valuable and pollution free farm-manure.

References

Berti, W. and Cunningham, S.D., 2000. Phytostabilization of metals. In: *Phytoremediation of Toxic Metals: Using Plants to Clean up the Environment*, (Eds.) Ruskin and Ensley. Willy and Sons, Inc., New York.

Boller, M., 1997. Tracking heavy metals reveals sustainability deficits of urban drainage systems. *Water Science and Technology*, 35:77.

Fritioff, A. and Greger, M., 2003. Aquatic and terrestrial plant species with potential to remove heavy metals from storm water. *International J. Phytorem.*, 5(3): 211.

Garcia, G., Faz, A. and Cunhaa, M., 2004. Performance of *Piptatherum miliaceum* (Smilo grass) in edaphic Pb and Zn phytoremediation over a short growth period. *Internatl. Biorem. Biodegrad.*, 54: 245.

Hasan, S.H., Rai, S. and Rupainwar, D.C., 2003. Removal of zinc from wastewater by water hyacinth. *Indian J. Chem. Technol.*, May: 274.

Kara, V.I., Aran, D.B., Kara, V., Zeytunluodlu, A. and Genc, H., 2003. Bioaccumulation of nickel, by aquatic macrophyte *Lemna minor* (duckweed). *Internatl. J. Agri. and Biol.*, 6(3): 28

Lu, V., Kruatrachue, M., Pokethitiyook, P. and Homyok, K., 2004. Removal of cadmium and zinc by water hyacinth, *Eichhornia crassipes. Sci. Asia,* 30: 93.

MOEF, 1993. *General Standards for Discharge of Environmental Pollutants: Part A: Effluent* (Schedule–VI, issued in GSR, 422, Ministry of Environment and Forest, dated 19th May).

Osmolovskaya, N. and Kurilenko, V., 2005. Macrophytes in phytoremediation of heavy metal contaminated water and sediments in urban inland ponds. *Geophysical Res.*, Abstracts. 7(10510): 1607.

Pandey, A.K. and Pandey, G.C., 2003. Physico-chemical characterization of city sewage discharged into river Saryu at Faizabad, Ayodhya, India. *Him. J. Env. Zoo.*, 17: 85.

Pandey, G.C., Neralia, S., Srivastava, P.K. and Tripathi, G., 1999. Bioremediation: A biological tool for pollution abatement. In: *Current Topics in Environmental Sciences*, (Eds.) Tripathi and Pandey. ABD Publishers, Jaipur, p. 347.

Prasad, K.V.S.K., Pardha Sardhi, P. and Sharmila, P., 1999. Concerted action of antioxidant enzymes and curtailed growth under zinc toxicity in *Brassica juncea*. Environ. Exp. Bot., 42: 1.

Sadowsky, M.I., 1999. Phytoremediation: Past and future practices. In: *Proc. of 18th Internatl. Symp. on Microbial Ecology*. Atlantic Canada Society for Microbial Ecology, Halifax, Canada.

Singh, N.K., Pandey, G.C., Rai, U.N., Singh, H.B. and Gupta, D.K., 2005. Metal accumulation and ecophysiological effects of distillery effluent on *Potamogeton pectinatus* L. *Bull. Env. Contam. and Toxicol.*, 74: 857.

Srivastava, P.K. and Pandey, G.C., 1998. Bioremediation of distillery effluent using selected aquatic macrophytes. *Res. J. Chem. Env.*, 2: 43.

Srivastava, P.K. and Pandey, G.C., 1999a. Paper mill effluent induced toxicity in *Ecchornia crassipes* and *Spirodela polyrhiza*. *J. Env. Biol.*, 20: 315.

Srivastava, P.K. and Pandey, G.C., 1999b. Effect of fertilizer effluent on total chlorophyll content and biomass of some aquatic macrophytes. *J. Ecotoxicol. Env. Monit.*, 11: 1237

Srivastava, P.K., Pandey, G.C. and Neralia, S., 2000. Bioaccumulation of distillery effluent metals by some aquatic macrophytes. *Proc. Natl. Acad. Sci. India*, p. 311.

Srivastava, S.K., Pandey, G.C. and Chaudhary, V.K., 2004. Phytoremediation of sewage effluent by a common aquatic macrophyte, *Spirodella polyrrhiza*. *Nature Env. and Poll. Tech.*, 3(4): 545.

Tripathi, R.D., Rai, U.N., Vajpayee, P., Ali, M.B., Khan, E., Gupta, D.K., Mishra, S., Shukla, M.K. and Singh, S.N., 2003. Biochemical responses of *Potamogeton pectinatus* L. exposed to higher concentrations of Zinc. *Bull. Environ. Contam. Toxicol.*, 71: 255.

Chapter 21

Content and Meaning of Ecosystem Management

I. Sundar[1] and K. Sivakumar[2]
[1]Reader in Economics, Directorate of Distance Education, Annamalai University
[2]Research Scholar in Economics, Annamalai University

ABSTRACT

The study and management of biodiversity and of nature's services are the key research foci of the discipline of Ecosystem Management. Ecosystem Management simply means the manipulation of land and waterscapes for human ends. Accordingly, academic staff works in a broad range of environments with varied resource uses including nature conservation, mining, forestry and agriculture. The biodiversity that is the focus of management is composed of both indigenous and exotic biota. Most importantly, as a 'management' discipline, staff recognizes the importance and legitimacy of social, economic and political concerns as well as ecological considerations, and the need to encompass a broad range of spatial and temporal scales in devising socio-economic and biophysical solutions to natural resource management problems. This article deals with Ecosystem Management and its related components. The content themes discussed in multiple area and principle and constituents of ecosystem management have been discussed in this article. It outlines the possible ways of managing biodiversity focusing on holistic resources management principles.

Keywords: *Ecosystem management*, *Principle of ecosystem management*, *Biodiversity management*.

Introduction

"Ecosystem management focuses on the conditions of the ecosystem, with goals of maintaining soil productivity, gene conservation, biodiversity, landscape patterns, and the array of ecological processes." It is a process that integrates scientific knowledge of ecological relationships within a complex socio-political and values framework toward the general goal of protecting native ecosystem integrity over the long run." Further it is said to be resource management system designed to maintain or enhance ecosystem health and productivity while producing essential commodities and other values to meet human needs and desires within the limits of socially, biologically and economically acceptable risk."

Concept Theme

"Ecosystem management is the integration of ecological, economic, and social principles to manage biological and physical systems in a manner safeguarding the long-term ecological sustainability, natural diversity, and productivity of the landscape. The primary goal of ecosystem management is to develop and implement management that conserves, restores, and maintains the ecological integrity, productivity, and biological diversity of public lands. Sustainable ecosystems provide, many benefits for wildlife and humans such as habitat for fish and wildlife, clean drinking water for communities, wood, fiber, forage, recreational, and economic opportunities," (US Bureau of Land Management 1994).

Ecosystem management is the integration of ecological, economic, and social factors in order to maintain and enhance the quality of the environment to meet current and future needs. It is a holistic approach to natural resource management. Applying ecosystem management involves collaboration of partners, land managers and scientists, with inclusion and consent of the public, particularly on public land. "Ecosystem management, then, is a concept of management that is more inclusive of the variables that impinge on management, occurring at larger scales, and over more extended time horizons than has been common in the past." (Thomas, 1996).

Common Elements within Ecosystem Management Definitions

Merely discussing ecosystem management themes and definitions is unsatisfactory for practical use. Incorporating ecosystem management principles into land management is a process that must evolve over time. Part of the complexity is due to the scientific uncertainty about what comprises an ecosystem management plan. Even at the small individual forest unit we are still learning about the effect of different management practices on forest ecological processes and species. Given our limited knowledge at the stand level, how can we plan for cumulative effects at the larger landscape level? A famous ecologist once said that ecosystems are not only more complex than we think–they are more complex than we can think. Therefore, perhaps the most important aspect of Ecosystem Management is to view it as a learning process, where landowners and forest managers adapt their plans to changing information. Landowners should set clear goals about what they would like to achieve, and be flexible in addressing changing information and needs.

Five Main Ecosystem Management Goals

Maintain viable populations of all native species *in situ*. Represent, within protected areas, all native ecosystem types across their natural range of variation. Maintain evolutionary and ecological processes (*i.e.*, disturbance regimes, hydrological processes, nutrient cycles, etc.) Manage over period of time long enough to maintain the evolutionary potential of species and ecosystems. Accommodate human use and occupancy within these constraints (Grumbine, 1994). Further ecosystem Management includes the following elements:

Sustainability

Ecosystem Management does not focus primarily on "deliverables" but rather regards intergenerational sustainability as a precondition. Goals. Ecosystem Management establishes measurable goals that specify future processes and outcomes necessary for sustainability.

Sound Ecological Models and Understanding

Ecosystem Management relies on research performed at all levels of ecological organization.

Complexity and Connectedness

Ecosystem Management recognizes that biological diversity and structural complexity strengthen ecosystems against disturbance and supply the genetic resources necessary to adapt to long-term change.

The Dynamic Character of Ecosystems

Recognizing that change and evolution are inherent in ecosystem sustainability.

Context and Scale

Ecosystem processes operate over a wide range of spatial and temporal scales, and their behaviour at any given location is greatly affected by surrounding systems. Thus, there is no single appropriate scale or time frame for management.

Humans as Ecosystem Components

Ecosystem Management values the active role of humans in achieving sustainable management goals.

Adaptability and Accountability

Ecosystem Management acknowledges that current knowledge and paradigms of ecosystem function are provisional, incomplete, and subject to change. Management approaches must be viewed as hypotheses to be tested by research and monitoring programs.

Principles of Ecosystem Management

The ecosystem concept can be applied to all lands, not just those managed for natural values. Four biological and social system properties of ecosystems underlie successful environmental assessments and have implications for management:

1. Ecological systems are continually changing.
2. There may be substantial heterogeneity in impacts from a particular action.
3. Systems may exhibit several levels of stable behaviour; and
4. There is an organized connection between parts, but everything is not connected to everything else. Given the political boundaries of park and wilderness ecosystems and unavoidable and uncertain human influence, change will occur in these systems.

Given this reality, a four-step approach can be used as a process to meet the ecosystem goals:

1. Define goals and measurable targets for ecosystem condition;
2. Define the ecosystem boundaries for the primary components;
3. Develop management strategies to achieve goals that transcend political boundaries; and

4. Establish a program to assess the effectiveness of the management strategies in achieving the identified goals (Agee and Johnson, 1988).

Ecosystem management is management driven by explicit goals, executed by policies, protocols, and practices, and made adaptable by monitoring and research based on our best understanding of the ecological interactions and processes necessary to sustain ecosystem composition, structure, and function. The following are fundamental scientific precepts for ecosystem management: (1) Spatial and temporal scale are critical; (2) Ecosystem function depends on its structure, diversity, and integrity; (3) Ecosystems are dynamic in space and time; and (4) Uncertainty, surprise, and limits to knowledge. Ecosystem management is not a rejection of an anthropogenic for a totally biocentric worldview. Rather, it is management that acknowledges the importance of human needs while at the same time confronting the reality that the capacity of our world to meet those needs in perpetuity has limits and depends on the functioning of ecosystems (Christensen and Bartuska, 1996).

The transition to ecosystem management in parks and wilderness will involve a gradual shift in management thinking and behaviour. Top priority issues in ecosystem management are:

1. What is the important research needs associated with ecosystem management?
2. What is the important general management, planning, and communications issues associated with ecosystem management?
3. What are the important challenges to ecosystem management in the areas of conflict resolution and cooperation?
4. What are the important limits and constraints to ecosystem management?

Management can be defined as human intervention in the dynamic processes, which determine the composition of plant and animal communities, so as to maintain a particular desired pattern or series of processes. Conservation. Is here taken to be a wider process of sustainable management of renewable natural resources, as an essential foundation for the human future on this planet. Although the temperate communities are not great reservoirs of biological diversity, the scientific management of temperate communities for conservation is critical since the sustainable development of the temperate zones is essential for the human future, and the preservation of such diversity as they exhibit is important. Moreover, because these ecosystems are relatively well known, and their management receives greater resources than in most tropical areas, work here can be a valuable source of knowledge and methodology. In particular, such work can emphasize:

The value of detailed scientific insights obtained through the study of key elements in temperate ecosystems and species. The value of temperate communities in their own right as components of global diversity and as valued features of the countries in which they occur. The value of demonstrating in the temperate regions how human populations and demands can be brought into balance with ecosystems. The use of temperate examples to demonstrate the management techniques available to implement strategies for conservation and sustainable development and the utility of ecology as a basis for both conservation for human use and conservation of the diversity of nature in its own right.

However, it is crucial that temperate zone ecologists are outward looking and involve themselves in the daunting problems of the tropical zones (Holdgate, 1989).

Management of Biodiversity

Tropical forests are very rich in biological diversity and form an important economic and ecological resource. This biodiversity is of great value for communities living in or near these forests as a ready source of subsistence and cash income, and for the world at large as a source of tropical timber and non-timber products and a repository of genetic and chemical information. However, this biological complexity is diminishing rapidly. The main human (anthropogenic) actions causing loss of tropical forest biodiversity along with the strategies of management for forest biodiversity are analyzed. Not only is the biophysical component important in management for biodiversity, but the active participation and support of local people, the national government, and international cooperation as a whole, are essential for an effective and sustainable development of tropical forests (Khasa and Dancik, 1997).

The perspectives and constraints that need to be addressed to establish biodiversity as a new scientific field. The perspectives include biodiversity as:

1. The unifying factor within the science of biology;
2. The cornerstone for agriculture, animal husbandry, forestry and aquaculture utilization;
3. The source of information for understanding the diversity of landscapes and the diversity of populations necessary for land use and regional development;
4. Fundamental in the new era of industrial applications;
5. The tool to establish a much-needed bridge within the social and cultural role; and
6. A pillar of human development where a new synthesis between globalization and diversity is achieved (di Castri and Younes, 1996).

The biodiversity management requires holistic management Holistic Resource Management (HRM) is a process of goal setting, decision making and monitoring which integrates social, ecological, and economic factors. Biodiversity enhancement is a fundamental principle in HRM and students are taught that biodiversity is the foundation of sustainable profit. In the HRM process, practitioners develop a holistic goal which includes:

1. Quality of life values;
2. Forms of production to support those values; and
3. Landscape planning, which should protect and enhance biodiversity and support ecosystem processes of succession, energy flow, hydrological and nutrient cycling (Stinner, D., B. Stinner and E. Martslof, 1997). In this context the following points can be considered.

Use a regional perspective when considering biological diversity. Think beyond the boundaries of specific ownership when planning and Managing. Plan and manage over large areas rather than using a stand-by-stand approach. Consider the cumulative impact of individual projects on regional population and resources. Emphasize multi species and ecosystem management instead of single species and tree management. Provide habitat sufficient to maintain species of concern (*e.g.* large wide ranging mammals), not just sufficient habitat to attract immigrants from more productive sources. Maintain or create spatial patterns. Conduct ecological surveys and inventories. Monitor problem species and inventories (Probst and Crow, 1991).

Conclusion

It could be seen clearly from the above discussion that ecosystem and biodiversity management involves various issues and components. The issues and approaches discussed in the paper bring to insight in the area of ecosystem management principle.

References

Agee, J.K. and D.R. Johnson, 1988. Introduction to ecosystem management. In: *Ecosystem Management for Parks and Wilderness*, (Eds.) I.K. Agee and D.R. Johnson. University of Washington Press, Seattle and London, p. 3–14.

Christensen, N., A. Bartuska, J. Brown, S. Carpenter, C.D. Antonio, R. Francis, J. Franklin, J. MacMahon, R. Noss, D. Parsons, C. Peterson, M. Turner, and R. Woodmansee, 1996. The report of the ecological society of America committee on the scientific basis for ecosystem management. *Ecological Applications*, 6(3): 665–691.

di Castri, F. and T. Younes, 1996. Introduction: Biodiversity: The emergence of a new scientific field, its perspectives and constraints. In: *Biodiversity, Science and Development, Towards a New Partnership*, (Eds.) F. di Castri and T. Younes. CAB International, Wallingford, UK.

Grumbine, R.E., 1994. What is ecosystem management? *Conservation Biology*, 8: 27–39.

Holdgate, M., 1989. Conservation in a world context. In: The scientific management of temperate communities for conservation. *The 31st symposium of the British Ecological Society*, South Hampton, (Eds.) I. Spellerberg, F. Goldsmith and M. Morris. Blackwell Scientific Publications, Oxford, UK.

Khasa, P. and B. Dancik, 1997. Sustaining tropical forest biodiversity. In: *Sustainable Forests: Global Challenges and Local Solutions*, (Eds.) O. Bourman and D. Brand. The Haworth Press.

Probst, J. and T. Crow, 1991. Integrating biological diversity and resource management: An essential approach to productive, sustainable ecosystems. *Journal of Forestry*, 89(2): 12–17.

Stinner, D., B. Stinner, and E. Martslof, 1997. Biodiversity as an organizing principle in agroecosystem management: Case studies of holistic resource management practitioners in the USA. *Agriculture, Ecosystems and Environment*, 62: 199–213.

Thomas, J.W., 1993. Forest service perspective on ecosystem management. *Ecological Applications*, 6(3): 703–705.

U.S. Department of the Interior, 1994. Bureau of land management. Ecosystem management in the BLM. From concept to commitment. Goy. Pub. BLM/SC/Gi-94/005+1736.

Chapter 22

Economics, Soil Health and Productivity of Different Cropping Systems Under Various Levels of Irrigation in Maharashtra

H.M. Patil[1], A.G. Wani[2], C.B. Gaikwad[3] and B.T. Shete[1]

[1]Technical Assistants, [2]Chief Agronomist, [3]Head, Department of Agronomy
All India Coordinated Research Project on Cropping Systems,
Mahatma Phule Krishi Vidyapeeth, Rahuri, Maharashtra

ABSTRACT

The field experiment was conducted during 2001–2002 to 2004–05 at main Centre for Cropping Systems Research Project MPKV Rahuri (Maharashtra), under irrigated eco system. The experiment was conducted in split plot design with four replications, three main plot treatments and eight sub-plot treatments. Total four crop cycles have been completed. Three treatments combinations including optimal (irrigation at all critical growth stages). Irrigation skipped at one critical growth stage and irrigation skipped at two critical growth stages were studied for eight cropping systems. Out of eight crop sequences, in Soybean-Chickpea crop cycle, soil fertility status including garlic carbon, available nitrogen, phosphorus and potassium has been improved with significant increase in yield.

Soybean-Chickpea cropping system was significantly superior to other cropping systems as it gave significantly net monetary returns (Rs. 38,971/-). The B : C ratio registered by Soybean-chickpea (2.88) was also significantly higher than those recorded by the other cropping systems.

Irrigation applied at all critical growth stages registered significantly higher gross (Rs. 52,013/-), net monetary returns (Rs. 31,534/-) and B : C ratio (2.54) than those recorded by one and two irrigations skipped. Two irrigation skipped registered significantly lowest gross and net returns as well as B: C ratio.

After harvest of *rabi* crops, the values of available N, P, and K in soil increased over their initial values. During *rabi* season irrigation at all the critical growth stages of *rabi* crops showed higher N, P, K uptake.

Introduction

Water plays vital role in crop production. It is the most critical basic resource, which must be conserved as efficiently as possible. Crops like sorghum, pearl millet, groundnut, soybean and pigeon pea grown in *kharif* season while wheat and chickpea on receding moisture along with protective irrigations at critical growth stages. There are substantial evidences that water deficit during critical growth stages can significantly reduce crop yield.

As, critical growth stage of crop is sensitive to moisture stress, it affects the crop growth and yield. Timing and duration of drought stress in soybean during a growing season can limit and alter dry matter accumulation within plant. Irrigated plants generally had fewer, shallower and more evenly distributed roots than those of non-irrigated plant (Chuck *et al.*, 1986). In summer groundnut treatment, which received 15 irrigations, was found superior to other treatments with regard to all yield attributes (Venkateshwarlu *et al.*, 1990).

The present study entitled Economics, soil health and productivity of different cropping systems under various levels of irrigation in Maharashtra was undertaken to evaluate profitability and productivity of different cropping systems under various levels of irrigations at critical growth stages of crops with the following objectives:

1. To study the effect of irrigation on yield of different cropping systems.
2. To study the changes in soil fertility and nutrient uptake due to different cropping systems.
3. To workout the economics of different cropping systems.

Materials and Methods

An experiment was conducted during 2001 to 2005 in semi-arid climatic conditions under irrigated ecosystem at Cropping Systems Research Project, Mahatma Phule Krishi Vidyapeeth, Rahuri. The experimental design was split plot with four replications, three irrigation levels and eight cropping systems. The main plot treatments (Irrigation levels) were I_1: Optimal (All critical growth stages), I_2: Irrigation skipped at one critical growth stage and I_3: Irrigation skipped at two critical growth stages. Sub-plot treatment (Cropping Systems) were CS_1: Pearl millet–wheat, CS_2: Sorghum–wheat, CS_3: Pearl millet–*Rabi* Sorghum, CS_4: Pearl millet–chickpea, CS_5: Sorghum–chickpea, CS_6: Pearl millet + Pigeon pea (2 : 2), CS_7: Groundnut + Pigeon pea (2 : 2) and CS_8: Soybean–chickpea. The gross and net plot size was 6.00×3.00 m^2 and 5.10×2.70 m^2, respectively.

Results and Discussion

Effect of Irrigation Levels

The yields of most of the *kharif* crops (Table 22.1) were higher due to optimum irrigations to all the crops. However, there was no much difference between the irrigation levels for the *kharif* crops because of the intermittent occurrence of rain showers during *kharif* season. The groundnut yields were drastically reduced in groundnut + pigeon pea intercropping system. This is because of the fact that luxuriant growth of pigeon pea (BSMR–736) affected the growth of groundnut adversely and hence, the yields of groundnut were reduced drastically. The yield of *kharif* sorghum was highest (27.70 q/ha) where the irrigation applied at all critical growth stages.

Table 22.1: Yields (Q/ha) of *Kharif* and *Rabi* Crops as Influenced by Irrigation Levels and Cropping Systems

Sl.No.	*Cropping System*	*Irrigation Levels (Kharif)*				*Irrigation Levels (Rabi)*			
		I_1	I_2	I_3	*Mean*	I_1	I_2	I_3	*Mean*
1.	Pearlmillet–Wheat	24.91 (68.99)	20.85 (65.36)	17.98 (58.99)	21.25 (64.45)	39.41 (58.47)	36.32 (53.74)	29.62 (42.13)	35.12 (51.45)
2.	Sorghum–Wheat	26.60 (85.24)	22.33 (71.45)	19.15 (61.32)	22.69 (72.67)	37.29 (53.56)	34.97 (50.74)	27.37 (39.02)	33.21 (47.77)
3.	Pearl millet–Sorghum	25.16 (68.35)	19.79 (65.64)	17.62 (52.99)	20.86 (62.33)	27.86 (81.70)	25.33 (74.80)	20.61 (61.76)	24.60 (72.75)
4.	Pearl millet–Chickpea	25.24 (68.99)	21.31 (58.09)	19.75 (59.00)	22.10 (62.03)	24.98 (33.56)	21.17 (30.20)	16.35 (23.03)	20.83 (29.60)
5.	Sorghum–Chickpea	27.70 (86.28)	23.43 (72.45)	19.79 (62.86)	23.64 (73.86)	23.12 (31.10)	20.39 (27.84)	15.06 (21.67)	19.52 (27.20)
6.	Pearl millet + Pigeon pea	17.95 (55.35)	17.40 (55.36)	12.56 (43.54)	15.97 (51.42)	18.96 (44.28)	16.69 (38.83)	11.08 (26.57)	15.58 (36.56)
7.	Groundnut + Pigeon pea	1.25 (2.23)	1.08 (1.80)	0.94 (1.30)	1.09 (1.78)	23.13 (52.45)	20.52 (47.00)	15.31 (34.82)	19.65 (44.76)
8.	Soybean–Chickpea	26.15 (35.60)	23.22 (29.97)	18.03 (24.60)	22.47 (30.06)	26.25 (36.14)	21.21 (32.69)	17.62 (24.15)	21.03 (30.99)
	Mean	21.86 (58.88)	18.67 (52.51)	15.72 (45.57)	18.75 (52.32)	27.62 (48.91)	24.57 (44.48)	19.13 (34.14)	23.69 (42.63)

N.B.: Figures in parentheses indicate the fodder/straw yield.

During *rabi* season, the highest yields of all the crops recorded due to optimal level of irrigation. There was significant reduction in yield due to skipping of one irrigation and two irrigations at critical growth stages of *rabi* crops.

Effect of Cropping Systems

During *kharif* season productivity of sorghum–chickpea cropping system was found to be maximum followed by sorghum–wheat cropping system. However, during *rabi* season pearl millet–wheat and sorghum–wheat cropping systems registered the highest yields.

Economics

Gross and net monetary returns as well as B : C ratio, as influenced by different cropping and irrigation levels is presented in Table 22.2. Gross monetary returns obtained from soybean–chickpea (Rs. 59,676/ha) was significantly higher over other treatments and it was followed by pearl millet–wheat (Rs. 46,620/ha) and sorghum-wheat (Rs. 45,722/ha) cropping systems.

Table 22.2: Gross, Net Monetary Returns (Rs./ha) and B : C Ratio as Influenced by Different Cropping Systems and Irrigation Levels

Particulars	*Gross Monetary Returns (Rs/ha)*	*Cost Cultivation (Rs/ha)*	*Net Monetary Returns (Rs/ha)*	*B : C Ratio*
Irrigation Levels				
I_1: Optimal	52013	20479	31534	2.54
I_2: Irrigation skipped at one critical growth stage	45437	19827	25610	2.29
I_3: Irrigation skipped at two critical growth stages	35255	19165	16090	1.84
S.E.±	570.42	–	570.49	0.041
C.D. at 5 per cent	1974.34	–	1974.59	0.116
Cropping System				
Pearl millet–Wheat	46620	22727	23893	2.05
Sorghum–Wheat	45722	25040	20682	1.82
Pearl millet–Rabi Sorghum	44821	21187	23634	2.11
Pearl millet–Chickpea	43498	19178	24320	2.27
Sorghum–Chickpea	41226	21490	19736	1.92
Pearl millet + Pigeon pea	40234	14272	25962	2.82
Groundnut + Pigeon pea	32078	15023	17055	2.13
Soybean–Chickpea	59676	20705	38971	2.88
S.E.±	874.83	–	874.80	0.05
C.D. at 5 per cent	2474.39	–	2474.31	0.160
General Mean	44235.08	–	24282.35	2.23

Selling price:

Crop	*Kharif Crops*		*Crop*	*Rabi Crops*	
	Grain Price (Rs/q)	*Fodder Price (Rs/q)*		*Grain Price (Rs/q)*	*Fodder Price (Rs/q)*
Pearl millet	650	30	Wheat	900	20
Sorghum	450	40	Sorghum	825	125
Pigeon pea	1700	50	Chickpea	1550	50
Groundnut	1600	50			
Soybean	1250	25			

Interaction effect (Table 22.3) shows significantly higher gross monetary returns in soybean–chickpea cropping system at optimum level of irrigation (Rs. 71,225/ha) over rest of the treatments. Second highest gross monetary returns (Rs. 60,977/ha) were also received in soybean-chickpea cropping system at irrigation skipped at one critical growth stage.

Table 22.3. Interaction Effect on Gross Monetary Returns, Net Monetary Returns (Rs./ha) and B : C Ratio

Sl.No.	Treatment	Gross Monetary Returns (Rs/ha)			Net Monetary Returns (Rs/ha)			B : C Ratio		
		I_1	I_2	I_3	I_1	I_2	I_3	I_1	I_2	I_3
1.	Pearl millet–Wheat	52989	48003	38869	29610	25276	16794	2.27	2.11	1.76
2.	Sorghum–Wheat	51924	46668	38575	26232	21628	14187	2.02	1.86	1.58
3.	Pearl millet–Sorghum	51611	45084	37770	29772	23897	17235	2.36	2.13	1.84
4.	Pearl millet–Chickpea	52480	43417	34598	32650	24239	16072	2.65	2.26	1.86
5.	Sorghum–Chickpea	49444	42419	31817	27302	20930	10979	2.23	1.97	1.53
6.	Pearl millet + Pigeon pea	47782	43279	29643	32858	29007	16023	3.20	3.03	2.17
7.	Groundnut + Pigeon pea	38647	33648	23942	22972	18625	9571	2.46	2.24	1.66
8.	Soybean–Chickpea	71225	60977	46828	49868	40272	26775	3.33	2.70	2.33
	S.E.±	1515	–	–	1515	–	–	0.098	–	–
	C.D. at 5 per cent	NS	–	–	NS	–	–NS	–	–	

Net monetary returns by soybean–chickpea cropping system (Rs. 38,971/ha) were significantly higher than those observed in other cropping systems and it was followed by pearl millet + pigeon pea cropping system. Groundnut + pigeon pea cropping system received least net monetary returns (Rs. 17,055/ha). This is because of the total failure of groundnut crop. Interaction effect also showed significantly higher net monetary returns in soybean–chickpea cropping system at optimum level of irrigation (Rs. 49,868/ha) over rest of the treatments.

The highest B : C ratio was observed in soybean–chickpea cropping system (2.88) followed by pearl millet + pigeon pea (2.82) intercropping system. At optimum level of irrigation B : C ratio was significantly higher over the treatments at irrigation skipped at one critical growth stage and the irrigation skipped at two critical growth stages.

Soil Fertility Status

After completion of fourth crop cycle soil samples were analyzed for its pH, EC, Organic carbon and available NPK and plant samples for its nutrient uptake study. The data presented in (Table 22.4) revealed that, the pH and EC were increased over its initial status. The values of organic carbon also slightly increased. Slightly higher organic carbon observed in the cropping system due to effect of pulse crops in the system than those compared with the cereal–cereal cropping systems. Available N decreased after *kharif* crops, while increased in *rabi* season over its initial values in all the treatments. However, available P and K showed higher values over their initial values during both the seasons.

Uptake Studies

NPK uptake by *kharif* crops as influenced by irrigation levels and cropping systems are presented in Table 22.5. Uptake of NPK during *kharifwas* higher due to irrigation skipped at one critical growth stages than irrigation skipped at two critical growth stages. NPK uptake was highest at optimal irrigation level. In the cropping systems, Soybean–chickpea showed highest uptake (167.03 kg/ha) of nitrogen followed by sorghum–wheat cropping system. Groundnut + pigeon pea intercropping shows

Table 22.4: Soil Fertility Status

Treatment	*Kharif (2004–2005)*						*Rabi (2004–2005)*					
	pH (1 : 2.5)	*E.C. (dSm^{-1})*	*Org. C (%)*	*Avail. N (kg/ha)*	*Avail. P (kg/ha)*	*Avail. K (kg/ha)*	*pH (1 : 2.5)*	*E.C. (dSm^{-1})*	*Org. C (%)*	*Avail. N (kg/ha)*	*Avail. P (kg/ha)*	*Avail. K (kg/ha)*
Irrigation Levels												
I_1: Optimal	8.31	0.28	0.62	192.70	15.68	520.35	8.34	0.28	0.62	242.45	15.82	522.88
I_2: Irrigation skipped at one critical growth stage	8.31	0.27	0.60	190.23	16.78	540.45	8.31	0.28	0.59	240.35	15.92	520.16
I_3: Irrigation skipped at two critical growth stages	8.31	0.29	0.59	185.14	15.85	549.76	8.31	0.27	0.60	249.54	16.27	525.22
Cropping Systems												
Pearl millet–Wheat	8.31	0.29	0.59	196.87	16.45	551.83	8.30	0.28	0.60	235.15	15.32	586.15
Sorghum–Wheat	8.30	0.28	0.61	185.13	16.18	563.43	8.31	0.28	0.60	233.90	14.95	558.75
Pearl millet–Sorghum	8.33	0.28	0.60	178.11	16.22	560.08	8.31	0.28	0.61	231.15	14.89	505.35
Pearl millet–Chickpea	8.31	0.28	0.61	192.01	16.16	557.58	8.32	0.28	0.60	238.70	16.41	508.15
Sorghum–Chickpea	8.30	0.28	0.62	185.78	16.26	529.58	8.32	0.27	0.61	234.45	15.92	470.40
Pearl millet + Pigeon pea	8.31	0.29	0.62	178.51	16.91	551.58	8.34	0.28	0.61	256.90	16.85	490.05
Groundnut + Pigeon pea	8.30	0.29	0.59	184.92	16.04	527.08	8.33	0.29	0.59	252.85	17.30	510.83
Soybean–Chickpea	8.31	0.29	0.62	194.86	16.25	537.17	8.34	0.28	0.61	244.69	16.05	551.75
Initial status	8.20	0.27	0.58	232.00	13.00	430.00	8.20	0.27	0.58	232.00	13.00	430

least NPK uptake by groundnut in *kharif* season. Higher P and K uptake was observed by sorghum and pearl millet respectively.

Table 22.5: NPK Uptake (kg/ha) by *Kharif* and *Rabi* Crops (2004–2005)

Treatments	*Kharif*			*Rabi*		
	N	*P*	*K*	*N*	*P*	*K*
Irrigation Levels						
I_1: Optimal	83.14	16.67	72.35	115.25	15.58	95.85
I_2: Irrigation skipped at one critical growth stage	72.33	14.75	65.85	145.85	13.88	89.75
I_3: Irrigation skipped at two critical growth stages	64.53	12.60	59.25	115.95	10.43	82.32
Cropping Systems						
Pearl millet–Wheat	66.46	15.02	95.15	92.45	15.69	94.25
Sorghum–Wheat	106.25	17.59	62.12	95.25	16.07	90.49
Pearl millet–*Rabi* Sorghum	54.20	12.50	82.45	75.82	21.13	92.36
Pearl millet–Chickpea	56.12	12.90	88.75	122.35	7.85	72.15
Sorghum–Chickpea	97.85	17.90	61.30	126.55	7.96	75.32
Pearl millet + Pigeon pea	45.77	15.25	98.75	201.32	13.62	96.15
Groundnut + Pigeon pea	9.50	1.38	5.35	225.85	14.46	62.38
Soybean–Chickpea	167.03	17.67	65.30	115.26	9.61	79.45

During rabi season, irrigation at all critical growth stages shows higher P and K uptake followed by irrigation skipped at one critical growth stage. But nitrogen uptake is more (145.85 kg/ha) in irrigation skipped at one critical growth stage. Highest N and K uptake observed by pigeon pea crop, while P uptake was found more (21.13 Kg/ha) by *rabi* sorghum.

Conclusion

Soybean–Chickpea cropping system was significantly superior to other cropping systems as it gave significantly higher net monetary returns (Rs. 38,971/-). The B : C ratio registered by Soybean–chickpea (2.88) was also significantly higher than those recorded by the other cropping systems.

Irrigation applied at all critical growth stages registered significantly higher gross (Rs. 52,013/-), net monetary returns (Rs. 31,534/-) and B : C ratio (2.54) than those recorded by one and two irrigations skipped. Two irrigation skipped registered significantly lowest gross and net returns as well as B : C ratio.

After harvest of *rabi* crops, the status of available N, P, and K in soil increased over their initial status. During *rabi* season, irrigation at all the critical growth stages of *rabi* crops showed higher N, P, K uptake. Based on gross and net monetary returns and B : C ratio soybean–chickpea cropping system found more remunerative as compared to other systems tried.

References

Jadhav, A.S., Kamble, A.B., Jagtap, S.E. and Mane, V.S., 1994. Effects of irrigation practices and fertilizer levels on the nutrient uptake, consumptive use of water, and seed yield. *Journal of Maharashtra Agril. Universities*, 19(3): 335–338.

Ruck, M.G., Peterson Curt, M., Gerrit Roogenboom and Busch, C.D., 1986. Distribution of dry matter between shoots and roots of irrigated and non-irrigated determinate soybeans. *Agronomy Journal*, 78: 807–813.

Saran Ganga and Mahendra Pal, 1981. Effect of soil moisture stress at different stages of plant development on the yield of grain sorghum. *Indian Journal of Agronomy*, 26(4): 463–464.

Singh Ved, Ram Deo, Singh, V. and Deo, R., 1998. Productivity and economics of different cropping systems under various levels of irrigation. *Indian Journal of Agronomy*, 43(3): 419–425.

Venkateswarlu, B., Prasad, P.R.K., Hanumantha, Y., Rao and Reddy, T. Ramakrishna, 1990. Effect of irrigation frequencies on the growth and yield of summer groundnut (*Arachis hypogea* Linn.). *The Andhra Agril. Journal*, 37(3): 315–317.

Chapter 23

To Study Irrigation Methods, Planting Layouts and Integrated Nutrient Management on Yield Attributes of Kabuli Chickpea

V.J. Dhage[1], B.T. Shete[2], H.M. Patil[2] and S.S. Ihle[3]
[1]M.Sc. (Agri.) Student; [2]Junior Research Assistant; [3]Assistant Professor
Department of Agronomy, Mahatma Phule Krishi Vidyapeeth, Rahuri – 413 722, M.S., India

ABSTRACT

The field experiment was carried out during *rabi* season of 2003–04 at All India Co-ordinated Research Project on Water Management, Mahatma Phule Krishi Vidyapeeth, Rahuri in Split plot design with three replications. The twelve treatment combinations, consisting two irrigation methods, *viz.*, sprinkler and surface irrigation, two planting layouts, *viz.*, ridges and furrow and flat bed layouts to the main plot and three treatments of bio-fertilizers and inorganic fertilizers *viz.*, RDF, RDF + Rhizobium + PSB and RDF + Rhizobium + PSB + foliar spray were to sub plots. The results revealed that the RDF of fertilizer to chickpea + inoculation of Rhizobium and PSB + 2 per cent foliar spray of DAP at pre flowering and Urea at pod filling recorded significantly higher pods per plant, seed per plant, seed weight per plant (27.46 g), 100 seed weight (42.42 g) seed yield 18.84 q ha^{-1} and straw yield (28.75 q ha^{-1}) over application of fertilizer with and without biofertilizers.

Keywords: Integrated nutrient management, Yield chickpea kabuli.

Introduction

Pulses play significant role in sustainable agriculture and provide nutritional security to predominant vegetarian population of the country. Chickpea is the second most important pulse crop cultivated in *rabi* season in Maharashtra, covering 22.47 per cent of the total area (0.79 million ha) and (0.44 million tonnes) with 560 kg ha^{-1} productivity (Jamdagni, 2003). The low fertility status of the soil with little or no fertilizers application is the major constraint in increasing the productivity of the pulses. The dose of 25 kg N + 50 kg P_2O_5 ha^{-1} is proved to be the best as a recommended dose for chickpea (Sarawgi and Singh, 1989). Application of phosphatic fertilizers in addition to inoculation of seeds with Phosphorus Solubilizing Bacteria (PSB) was found advantageous in many crops. Similarly, Rhizobium is one of the important components of nutrient management which fix atmospheric free nitrogen and maintain biological productivity of soil. Therefore, it is essential to adopt a strategy of organic and inorganic fertilizer management. Better yield potential and good market prices of kabuli chickpea is one of the demanded pulse in market. Hence, there is imperative need to screen out irrigation methods, planting layouts and integrated nutrient management in kabuli chickpea. In view of these considerations the present investigation was carried out to evaluate effect of irrigation methods, planting layouts, fertilizers and biofertilizers on yield of kabuli chickpea.

Materials and Methods

The present field investigation was carried out during *rabi* season of 2003–2004 at All India Co-ordinated Research Project on Water Management, M.P.K.V., Rahuri. Dist. Ahmednagar (M.S.). The soil of the experimental field was clayey loam in texture, low in nitrogen (171.62 kg ha^{-1}) medium in available phosphorus (17.92 kg ha^{-1}) and high in available potassium (251.61 kg/ha). The experiment was laid out in split plot design with twelve treatments. Main treatments comprised two irrigation methods *viz.* (1) Sprinkler irrigation at 50 mm CPU with 5 cm depth. (2) Surface irrigation at critical growth stages of chickpea and planting layouts *viz.* (*i*) Ridges and furrows (dibbling on both sides of ridges). (*ii*) Flat bed. The sub-treatments of biofertilzers and inorganic fertilizer, (1) Recommended dose of fertilizer 25 : 50 kg N : P_2O_5 ha^{-1}) (2) RDF + Rhizobium + PSB, (3) RDF + Rhizobium + PSB with DAP spray (2 per cent) at pre flowering + Urea spray (2 per cent) at pod filling. The gross plot size of each sub plot was 10.8 × 5.4 m^2 and net plot size was 8.4 × 4.5 m^2. The seeds were inoculated with Rhizobium and PSB @ 25 g kg^{-1} seed of kabuli chickpea variety Virat (PG-95418) by using jaggary and water 125g/lit of water) solution.

Results and Discussion

Irrigation Methods

Application of irrigation through sprinkler irrigation at 50 mm CPE produced significantly more number of pods per plant (69.34), number of seeds (70.26) and their weights per plant (28.16 g), 100 seed weight (42.50 g), seed yield (17.15 q ha^{-1}) and straw yield (26.47 q ha^{-1}) than the application of irrigation at critical growth stages of chickpea through surface irrigation. Dahiwalkar *et al.* (1998) noted increase in yield parameters by application of sprinkler irrigation in chickpea crop.

Planting Layouts

Sowing of chickpea on both sides of ridges and furrows (45 × 10 cm) layout registered significantly more number of pods per plant (68.70), number of seed per plant (26.85 g), 100 seed weight (41.78 g), seed yield (16.55 q ha^{-1}) and straw yield (25.42 q ha^{-1}) than sowing of chickpea at spacing of (30 × 15 cm) on flat bud. Similar results were recorded by Jadhav and Pawar (1999) and Shinde *et al.* (2000).

Integrated Nutrient Management

The important yield contributing characters were influenced, significantly by integrated nutrient management. These were increased significantly by use of biofertilizers and application of foliar spray in addition of RDF than application of RDF alone. However, number pods (69.99), number seeds (71.77) per plant and their weight (27.46 g), 100 seed weight (42.42 g), seed yield (18.84 q ha^{-1}) and straw yield (28.75 q ha^{-1}). The beneficial effect of RDF + Rhizobium + PSB + foliar spray of DAP at reproductive phase, is in conformity with the findings of Vadvaia *et al.* (1994).

Table 23.1: Yield Contributing Characters as Influenced by Different Treatments

Sl.No.	*Treatments*	*Number of Pods per Plant*	*Number of Seeds per Plant*	*Weight of Seeds per Plant (g)*	*100 Seed Weight (g)*	*Seed Yield (q)*	*Straw Yield (q)*
A.	**Irrigation Methods**						
1.	Sprinkler	69.34	70.26	28.61	42.50	17.15	26.47
2.	Surface	63.30	67.90	24.25	37.11	14.91	21.84
	S.E. ±	0.71	0.19	0.29	0.73	0.19	0.70
	C.D. at 5 per cent	4.30	1.18	1.76	4.43	1.18	4.25
B.	**Planting layouts**						
1.	Ridges and furrows	68.70	69.94	26.85	41.78	16.55	25.42
2.	Flat bed	63.93	68.21	26.01	37.83	15.52	22.88
	S.E. ±	0.58	0.36	0.18	0.68	0.10	0.61
	C.D. at 5 per cent	2.26	1.40	0.69	2.65	0.39	2.39
C.	**Integrated nutrient management**						
1.	RDF	61.61	65.07	24.64	37.50	13.91	19.94
2.	RDF + Rhizobium + PSB	67.35	70.40	27.19	39.50	15.35	23.76
3.	RDF + Rhizobium + PSB + foliar spray	69.99	71.77	27.46	42.42	18.84	28.75
	S.E. ±	0.61	0.54	0.63	0.60	0.26	1.16
	C.D. at 5 per cent	1.84	1.60	1.89	1.81	0.76	3.47
D.	**Interaction effect**						
	A × B	N.S.	N.S.	N.S.	N.S.	N.S.	N.S.
	A × C	N.S.	N.S.	N.S.	N.S.	N.S.	N.S.
	B × C	N.S.	N.S.	N.S.	N.S.	N.S.	N.S.
	A × B × C	N.S.	N.S.	N.S.	N.S.	N.S.	N.S.

Interaction

The interaction effects between irrigation methods, planting layouts and integrated nutrient management on yield contributing characters were found to be non significant.

References

Dahiwalkar, S.D., Aher, M.B., Bhosale, R.S. and Berad, S.M., 1998. Chickpea yield response to moisture status and irrigation methods. *J. Maharashtra Agric. Univ.*, 23(2): 169–197.

Jadhav, K.J., and Pawar, V.S., 1999. Effect of spacing, layouts and varieties on yield of chickpea (*Cicer arietimum*). *Indian J. Agric. Sci.*, 69(7): 516–518.

Jamadagni, B.M., 2003. *Research Review Committee on Pulses*, pp. III.

Sarawgi, S.K., Tiwari, P.K. and Tripathi, R.S., 1999. Uptake and balance sheet of nitrogen and phosphours in chickpea (*Cicer arietinum* L.) as influenced by phosphorus, biofertilizer and micronutrients under rainfed condition. *Indian J. Agron.*, 44(4): 768–772.

Shinde, S.W., Thakur, N.T. and Bhilare, R.L., 2000. Effect of field layouts and fertilizer levels on productivity of chickpea. *J. Maharashtra Agric. Univ.*, 25(1): 76–77.

Vadavia, A.T., Kalaria, K.K., Patel, J.C. and Baldha, N.M., 1994. Influence of organic, inorganic and biofertilizers on growth, yield of chickpea. *Indian J. Agron.*, 36(3): 263–264.

Chapter 24

Status of Black Pepper in High Altitude and Tribal Areas of Andhra Pradesh

M.M. Naidu and V. Prasanna Kumari
Regional Agricultural Research Station, Chintapalli – 531 111, Andhra Pradesh

Introduction

Black paper of Commerce, the whole dried fruits obtained from the perennial climbing vine, *Piper nigrum* which is indigenous to the tropical forests of Western Ghats of South India is one of the important and earliest known spices produced and exported from India. India dominated the world markets in pepper from time immemorial till the 1950s, Indonesia, Malaysia and Brazil took up commercial cultivation of pepper mostly as monocrop with scientific managements in the beginning of the nineteenth century and the production of pepper in these countries increased very fast. Black Pepper occupies a supreme position in the world trade of spices, Black pepper alone accounts for about 40 per cent in volume of the total world trade. But in Andhra Pradesh the production of Black Pepper is very meagre and the area is also negligible. This article outlines the area, production, productivity and major constraints in cultivation of Black Pepper.

Area and Production

Area and Production in high altitude and tribal area zone (Visakhapatnam and East Godavari Districts) shown in Table 24.1

Black Pepper was introduced in 1963 by Andhra Pradesh Forest Development Corporation as intercrop in coffee. In high altitude and tribal area zone 56 per cent of the Black Pepper area is under public sector agencies like Andhra Pradesh Forest Development Corporation and Coffee Board. The Andhra Pradesh Government established Girijan Coffee Plantation Development Corporation in 1975 and supplied coffee seedlings as well as 2 lakh Black Pepper rooted cuttings to farmers which account for 400 ha area. The total area under black pepper as mixed cropping is 920 ha. The area and production can't be correlated as all the vines are not in bearing stage. Similarly the area and no. of vines also cannot be correlated as the standards are very much scattered. 73 per cent of the production is from plantations of Andhra Pradesh Forest Development Corporation and the production from farmer's plantation is very meagre as most of the farmers are having 2–3 years old 7 plantations. The standard plants mostly used are Silver oak, Jack, Wild mango, Ficus, Terminalia and Berserra serrata. The natural standards are very much scattered *i.e.,* without proper spacing and some of the area is also not listed in revenue records. The Bureau of statistics is also not collecting the data on area and production. Because of these complications, the area estimation is not realistic.

Table 24.1: Area and Production in High Altitude and Tribal Area Zone (Visakhapatnam and East Godavari Districts)

		Area (ha)	*Total No. of Vines*	*Bearing Vines*	*Production (Tons)*
A.	**Public Sector**				
(*i*)	Coffee Board	20	20,000	10,000	2.5
(*ii*)	A.P. Forest Development Corporation	500	1,00,000	55,000	30
B.	**Private Sector**				
(*i*)	Farmer's plantation	400	1,00,000	30,000	15
	Total	**920**	**2,20,000**	**95,000**	**47.5**

*Ninety per cent of above area is in Visakhapatnam District.

Varieties

Most of the area is under Karrimunda, Kottanadan and Panniyur-1 varieties.

Productivity

The average productivity is 0.4 kg to 0.8 kg per vine. In general the production of Black Pepper is very much fluctuating *i.e.,* one year the yield is high and in the succeeding year there will be 30–40 per cent reduction in yield. This is due to variation in climatic conditions and also due to alternate bearing habit in some of the varieties.

Pests and Diseases

Phytophthora foot-rot disease is the major production constraint in the area followed by stunt disease caused by virus. The other pests and diseases noticed are fungal pollu disease; scales, stem borer and pollu beetle.

Quality Parameters of Trader

Mostly traders are concentrating on Boldness, density and colour of berries.

Quality Parameters of Farmer

As per the farmer's operation, berries harvested at full maturity are getting good price in the market.

Scope for Increasing Area

In view of favourable climatic conditions and availability of fertile virgin lands, there is good scope for increasing area under Black Pepper. The farmers will be getting additional income by growing Black Pepper along with coffee.

Major Constraints

1. Absence of effective extension strategy.
2. Inadequate availability of quality planting material.
3. Long duration of rain less period.
4. Fluctuating prices in the market.

References

Anonymous, 1990. *Status Paper on Spices*. Spices Board, Ministry of Commerce, Government of India, Cochin.

Anonymous, 1991. Spices in Different States (Facts and Figures). Trade information Services, Spices Board, Ministry of Commerce, Government of India, Cochin.

Chapter 25

A Case Study of Lake Jaisamand with Special Reference to Algal Flora, India

Premlata Vikal* and Sandhya Tyagi
Department of Botany, Mohanlal Sukhadia University, Udaipur – 313 002, Rajasthan, India

ABSTRACT

The present article deals with the study of algal flora in Lake Jaisamand during the year, 2005. A total of 29 genera and 43 species belonging to Cyanophyceae, Chlorophyceae, Euglenophyceae, Bacillariophyceae and Xanthophyceae have been recorded. During the study period blooming conditions developed twice. The bloom forming algae were *Gloeothece palaea* and *Microcystis aeruginosa*, respectively. Various physico-chemical parameters, *e.g.*, surface water temperature, pH, dissolved oxygen, bio-chemical oxygen demand, alkalinity, hardness, sulphate, nitrate, phosphate, total dissolved solids and conductivity were also conducted as it affects aquatic flora and fauna.

Keywords: *Gloeothece palaea, Microcystis aeruginosa, Aquatic ecosystem, Physico-chemical parameters.*

Introduction

Lake Jaisamand, the chief source of water supplies to fulfil various requirements like domestic and irrigational, etc. It covers an area of 36 square kilometers and its capacity is 27.5 ft. It is 51 kilometers far from Udaipur City. It was constructed by Maharaja Jai Singh during 1685–1691 in the

* E-mail: vikal16@yahoo.co.in

form of dam at river Gomati and later converted to lake. It is also known for fish cultivation. The man-made aquatic ecosystems play a vital role in inland fishery development in India due to their vast area and huge production potential (Fasihuddin and Puttaiah, 2004).

Algae form the base of food chain in an aquatic ecosystem. Therefore, it plays a major role in maintaining aquatic life and water quality. As phytoplanktons are the basis of aquatic production, any serious alteration in their constitution may have detrimental consequences on the food chain and concurrently entire community structure (Adesalu and Nwankwo, 2005). Today algal assays are very commonly used in assessing water quality (Wong, 1989 and Whitton *et al.*, 1991). Micro- as well as, macroalgae are known to be most sensitive organisms to pollutants (Kapoor and Solanki, 1992). Type of algae growing in a water body and its abundance or any changes in their population may helps in assessing the quality of water and its nutrient status, respectively as they are sensitive indicators of water quality. Therefore, present study was designed to study the algal flora of the lake in relation to its water characteristics.

Materials and Methods

At Lake Jaisamand two sampling stations were selected, *i.e.*, Boating site (Station I) and Shiva temple (Station II). Algal and water samples were collected monthly in the morning between 9.00 to 10.00 A.M. About 1.5–2.0 litres of water sample was collected in polythene cans and brought to laboratory for analysis as per standards (Trivedi and Goel, 1986; Manivasakam, 1996 and APHA, 1998).

Physico-chemical Analysis

Water was analyzed for various physico-chemical parameters, *e.g.*, air and water temperature, pH, dissolved oxygen (DO), biochemical oxygen demand (BOD), total alkalinity, hardness, sulphate, nitrate, phosphate, total dissolved solids (TDS) and conductivity. Air and water temperature were measured in the field using a mercury thermometer of 110°C and DO was also fixed on the spot with Winkler A and Winkler B. TDS and conductivity were determined using Systronics TDS-Conductivity meter-308 and pH with Systronics Digital pH meter-335.

Identification of Algae

Algal samples were identified using Desikachary (1959), Palmer (1980) and Bold and Wynne (1978).

Results and Discussion

Algae represent an essential and basic part of the cycle of living organisms as, small aquatic organisms feed on algae. Algae may prove to be nuisance, if it starts forming bloom or mats. Blooms may cause problems of tastes and odours, filter and screen clogging and slime accumulation in pipes. In the year 2005, a variety of algal species have been recorded which belong to Cyanophyceae, Chlorophyceae, Euglenophyceae, Xanthophyceae and Bacillariophyceae. Water temperature, pH and available nutrients play an important role in the growth of particular alga: The capacity of synthesizing organic matter by any aqualic ecosystem depends largely on physico-chemical parameters like temperature, nutrients and solar radiations besides producer populations (Fasihuddin and Puttaiah, 2004).

In January, 2005 different members of cyanophyceae and diatoms were recorded. *Gloeothece palaea* were abundant in January, 05 which formed blooming conditions in February, 05 at Station I. During that period, the water temperature was between 19.0–19.5°C, which favoured its growth. It vanished completely in March, 05 may be due to increased temperature. Liken (1975) observed a close relationship between temperature and productivity. The pH range 8.5 to 9.0 favours the growth of blue-green algae. Blooming has a direct correlation with pH, temperature, DO and available phosphates contributed by sewage and sedimentary cycle (Kodarkar *et al.*, 1991). The vigorous growth of algae has the ability to reduce hardness, as much as, one-third (Palmer, 1980) and it can be observed very clearly after February, 2005 (Table 25.2).

In February and March 05, abundance of diatoms was recorded. *Fragilaria capucina* were found attached to *Cladophora aegagropila.* Later, these diatoms were found free floating. The presence of *Synedra ulna, Cymbella ventricosa* and *Achnanthes* sp. indicated polluted water condition (Table 25.1). During this period TDS were high ranging between 517 to 766 ppm. High dissolved solids may be attributed for higher electrical conductivity values. The growth of algae depends on the physico-chemical characteristics of water. In turn, the abundance of algae is also capable of altering these water characteristics (Palmer, 1980). Total alkalinity was highest in April, 05 at both these stations I and II *i.e.*, 210 and 204 ppm, respectively. Alkalinity below 500 ppm indicates low photosynthetic rate (Das, 2002). Decreased photosynthetic activity lowers dissolved oxygen.

Table 25.1: Algal Flora in Lake Jaisamand during the Year 2005

Algal Species	*Jan.*	*Feb.*	*Mar.*	*Apr.*	*May*	*June*	*July*	*Aug.*	*Sept.*	*Oct.*	*Nov.*	*Dec.*
Cyanophyceae												
Gloeothece palaea	++	++	–	–	–	–	–	–	–	–	–	–
Microcystis aeruginosa	–	–	–	–	–	–	–	–	+	++	+	+
M. scripta	–	–	–	–	–	–	–	–	–	++	+	+
Oscillatoria anguina	+	+	–	–	–	–	–	–	–	–	–	–
O. agardhii	–	–	–	–	+	–	–	–	–	–	–	+
O. chlorina	–	–	–	–	+	–	–	–	–	–	–	–
O. lauterborni	–	–	–	–	+	–	–	–	–	–	–	–
O. limosa	–	–	–	–	–	–	–	–	–	–	–	+
Phormidium anomala	–	–	+	–	–	–	–	–	–	–	–	–
P. cevennense	+	–	–	–	–	–	–	–	–	–	–	–
Lyngbya digueti	–	–	–	–	+	–	–	–	–	–	–	–
L. gracilis	–	–	+	+	–	–	–	–	–	–	–	–
L. polysiphonia	–	–	+	–	–	–	–	–	–	–	–	–
L. corbierei	–	–	–	+	–	–	–	–	–	–	–	–
Anabaena planktonica	–	–	–	–	–	–	–	–	–	–	–	+
Stigeoclonium lubricum	+	+	+	+	+	–	–	–	–	–	–	–
Chlorophyceae												
Chlorococcum multinucleatum	–	+	–	–	–	–	–	–	–	–	–	–
Pediastrum tetras	–	+	+	–	–	–	–	–	–	–	–	–
P. simplex	–	–	+	+	–	–	–	–	–	–	–	–
Chlorela vulgaris	–	–	–	–	–	–	–	–	–	–	–	+

Contd...

Table 25.1–Contd...

Algal Species	Jan.	Feb.	Mar.	Apr.	May	June	July	Aug.	Sept.	Oct.	Nov.	Dec.
Tetraedron muticum	–	–	–	–	+	–	–	–	–	–	–	–
Chlamydomonas reinhardi	–	–	–	–	–	–	–	–	–	–	–	+
Tetraspora gelatinosa	–	–	–	–	–	–	–	–	+	+	+	–
Planktosphaeria gelatinosa	–	–	–	–	–	–	–	–	–	–	+	–
Scenedesmus quadricauda	–	–	+	+	–	–	–	–	–	–	–	–
Cladophora aegagropila	–	+	+	+	+	–	–	–	–	–	–	–
C. glomerata	–	–	–	–	–	–	–	–	–	–	–	+
C. crispata	–	–	–	–	–	–	–	–	–	–	+	–
Oedogonium suecium	–	–	–	+	–	–	–	–	–	–	–	–
O. boscii	–	–	–	+	–	–	–	–	–	–	–	–
Spirogyra fluviatilis	–	–	–	–	+	+	–	–	–	–	+	+
Microspora wittrockii	–	–	–	–	–	–	–	–	–	–	+	–
Euglenophyceae												
Euglena viridis	–	–	–	–	–	–	–	–	–	–	–	+
E. sangvlnea	–	–	–	–	–	–	–	–	–	–	–	+
E. stellata	–	–	–	–	–	–	–	–	–	–	–	+
Bacillariophyceae												
Nitzschia palaea	+	+	+	–	–	–	–	–	–	–	+	+
Navicula gracilis	+	+	–	+	–	–	–	–	–	–	+	+
Fragilaria capucina	–	+	+	+	+	+	–	–	–	–	–	–
Gomphonema parvulum	–	+	+	+	–	–	–	–	–	–	–	–
Cymbella ventricosa	–	+	–	–	–	–	–	–	–	–	–	–
Achnanthes sp.	–	–	+	–	–	–	–	–	–	–	–	–
Synedra ulna	+	+	–	–	–	–	–	–	–	–	–	–
Xanthophyceae												
Tribonema bombycinum	–	–	–	–	+	–	–	–	–	–	–	–
Desmidaceae												
Cosmarium sp.	–	–	+	–	–	–	–	–	–	–	–	–

++: Bloom forming algae.

In May, 05 among cyanophyceae different species of *Oscillatoria*, a species of *Lyngbya* and *Stigeoclonium*, among chlorophyceae *Tetraedron muticum*, *Spirogyra fluviatilis* and one species of xanthophyceae, *i.e.*, *Tribonema bomycinum* were recorded.

After rains in August, 05 species of *Microcystis* (*M. aeruginosa* and *M. scripta*) developed which formed bloom in October, 05. According to Pathak (1979), free CO_2 will be limiting where there is a permanent bloom of BGA, especially *Microcystis* sp. Sabu and Azis (1995) emphasized on strong positive relationship between phosphorus and productivity. The foremost cause attributed to the triggering and spread of blooms is increasing pollution through anthropogenic activities (Bhat and

Matondkar, 2004). The abundance or *Microcystis* spp. might be responsible for the increased concentration of alkalinity and hardness in November–December, 05. Throughout the year, the alkalinity was found to be more than 100 ppm, which indicated highly productive water (Alikunhi, 1957). The blooms of *Microcystis* spp. are harmful as it manufactures microcystin, which yield 7 (or 14) amino acids on hydrolysis. It causes enlargement and congestion of the liver followed by haemorrhage and may also exhibit neurotoxic activity. In recent years, there has been considerable scientific attention towards harmful algal blooms (Blaxter and Southward, 1997; Chattopadhyay *et al.*, 2002b; Sarkar and Malchow, 2005). Harmful algal blooms refer to blooms, which cause discomforts or toxic effects on human health directly or indirectly. Algal blooms hurt the system in two ways. First, they cloud the water and block sunlight causing underwater grasses to die, which provide food and shelter for aquatic creatures. Second, when these algae die and decompose, oxygen is used up which is essential to most organisms living in the water.

Table 25.2: Selected Physico-chemical Characteristics of Lake Jaisamand in the Year, 2005

Sl.No.	*Parameters*	*Jan.*	*Feb.*	*Mar.*	*Apr.*	*May*	*Jun.*	*Jul..*	*Aug.*	*Sep.*	*Oct.*	*Nov.*	*Dec.*
1.	Air temperature	I	18.0	24.0	25.0	24.5	34.0	31.0	–	27.25	24.5	21.0	14.0
		II	19.0	24.0	25.0	24.0	36.0	31.0	–	27.5	27.5	21.0	15.0
2.	Water temperature	I	19.0	19.5	26.5	29.5	28.5	30.0	–	30.0	25.0	22.5	19.5
		II	20.0	19.0	26.5	29.0	29.5	30.0	–	29.59	26.5	23.0	19.5
3.	pH	I	8.39	8.69	8.71	8.64	8.22	8.59	–	9.15	8.38	8.04	8.06
		II	8.35	8.65	8.75	8.63	8.52	8.66	–	9.12	8.49	8.14	7.99
4.	D.O.	I	9.20	7.95	7.65	7.04	6.734	6.53	–	8.23	4.59	4.08	5.714
		II	8.90	8.05	7.96	7.35	6.53	6.99	–	8.299	5.71	4.49	5.714
5.	B.O.D.	I	1.56	1.43	5.002	2.55	4.49	6.326	–	4.558	1.531	1.041	1.63
		II	2.01	1.96	1.735	2.45	3.469	6.734	–	4.524	3.572	0.918	2.04
6.	Total alkalinity	I	148.0	166.0	168.0	210.0	208.0	188.0	–	131.0	138.0	160.0	163.0
		II	146.0	164.0	174.0	204.0	200.0	192.0	–	130.0	142.0	167.0	160.0
7.	Hardness	I	157.0	162.5	160.0	163.0	162.0	155.0	–	117.0	130.0	142.0	145.0
		II	155.0	202.0	164.0	162.0	162.0	155.0	–	115.0	132.0	141.0	141.0
8.	Sulphate	I	70.0	150.0	150.0	190.0	140.0	55.5	–	125.0	80.0	355.0	400.0
		II	90.0	101.0	101.0	220.0	115.0	115.0	–	230.0	195.0	350.0	340.0
9.	Nitrate	I	3.21	1.33	–	0.77	3.987	3.323	–	3.101	0.775	4.21	2.879
		II	4.02	6.31	–	2.75	3.101	6.756	–	7.42	0.332	12.105	2.215
10.	Phosphate	I	0.035	0.049	0.032	0.006	0.052	0.005	–	0.018	0.061	0.021	0.042
		II	0.038	0.043	0.028	0.008	0.049	0.011	–	0.024	0.098	0.021	0.042
11.	T.D.S.	I	320.0	517.0	749.0	317.0	287.0	313.0	–	2210	246.0	225.0	240.0
		II	326.0	527.0	766.0	320.0	289.0	311.0	–	223.0	248.0	2270	244.0
12.	Conductivity	I	0.640	1.03	1.49	0.633	0.579	0.627	–	0.443	0.493	0.450	0.992
		II	0.652	1.05	1.53	0.641	0.583	0.623	–	0.448	0.498	0.452	0.489

–: No sampling done, I and II are sampling stations.

All parameters are in ppm except for temperature (°C), pH and conductivity (mS/cm).

The highest number of species was recorded in December, 05 (Table 25.2). At the end it may be concluded that as the Lake Jaisamand is an important source of water supply therefore, care should be taken so that blooms of harmful algae may not develop. 4 per cent copper sulphate solution is useful to eradicate nuisance planktonic forms. Use of specific algicides will prove to be useful from the standpoint of fisheries management, as algae constitute the base of food pyramid. Thus, a monitoring of a water body from time to time is essential to maintain the water quality and its aesthetic value.

Acknowledgements

The author (P. Vikal) is grateful to UGC, New Delhi for giving research fellowship and to Prof. N.C. Aery, Head, Department of Botany, M.L.S. University, Udaipur for providing laboratory facilities. Author also acknowledges the encouragement of Prof. K. Kapoor. I am very much thankful to my father Shri P.D. Vikal and brothers Ghanshyam Singh and Rokesh Kumar Singh for their continuous help and support throughout.

References

Adesalu, T.A. and Nwankwo, D.I., 2005. Studies on the phytoplankton of Olero creek and parts of Benin river, Nigeria. *The Ekol.*, 3(2): 21–30.

Alikunhi, K.H., 1957. Fish culture in India. *Fram. Bull.*, 20, ICAR, New Delhi, pp. 144.

APHA, AWWA and WPCF, 1998. *Standard Methods for the Examination of Water and Wastewater*, 20th edn. American Public Health Association, New York.

Bhat, S.R. and Matondkar, S.G.P., 2004. Algal blooms in the seas around India: Networking for research and outreach. *Curr. Sci.*, 87(8): 1079–1083.

Blaxter, J. and Southward, A., 1997. *Advances in Marine Biology*. Academic Press, London.

Bold Harold, C. and Wynne Michael, J., 1978. *Introduction to Algae.* Prentice Hall of India Pvt. Ltd., New Delhi, pp. 706.

Chattopadhyay, J., Sarkar, R. and Mandal, S., 2002b. Toxin producing plankton may act as a biological control for planktonic blooms-field study and mathematical modeling. *J. Theor. Biol.*, 215: 333–344.

Das, A.K., 2002. Phytoplankton primary production in some selected reservoirs of A.P. *Geobios,* 29: 52–57.

Desikachary, T.V., 1959. *Cyanophyta.* Indian Council Agricultural Research, New Delhi, pp. 686.

Fasihuddin, S. and Puttaiah, E.T., 2004. Spatio-temporal trends in phytoplankton primary production in marikanave reservoir. *Indian J. Environ. and Ecoplan.*, 8(1): 179–184.

Kapoor, K. and Solanki, P.S., 1992. In water management problems related to water resources in Udaipur. *Nat. Sem. Strat. Env. Prot. Industrialization,* M.L.S. University, Udaipur, pp. 119–126.

Kodarkar, M.S., Muley, E.V. and Rao, V., 1991. Toxic algal blooms in the Lake Hussainsagar, Hyderabad. *J. Aquatic Biol.*, 6(1&2): 13–18.

Likens, G.E., 1975. Primary production of inland aquatic ecosystems. In: *The Primary Productivity of the Biospheres,* (Eds.) Lieth, H. and R.H. Whittaker. Springer-Verlag, New York.

Manivasakam, N., 1996. *Physico-chemical Examination of Water, Sewage and Industrial Effluents.* Pragati Prakashan, Meerut (India), pp. 257.

Palmer, C.M., 1980. *Algae and Water Pollution.* Castle House Publication Ltd., England, pp. 123.

Pathak, V., 1979. Evaluation of productivity in Nagarjunasagar reservoir (Andhra Pradesh) as a function of hydrological and limnological parameters. *J. Inland Fish. Soc.*, India, 11: 49–68.

Sabu, T. and Abdul Azis, P.K., 1995. A study on the primary production in Peppara reservoir. In: *The Proc. of Seventh Kerela Science Congress*, Palakkad, pp. 76–77.

Sarkar, R.R. and Malchow, H., 2005. Nutrients and toxin producing plankton control algal blooms: A spatio-temporal study in noisy environment. *J. Biosci.*, 30(5): 101–113.

Trivedi, R.K. and Goel, P.K., 1986. *Chemical and Biological Methods for Water Pollution Studies.* Environmental Publications, Karad, pp. 248.

Whitton, B.A., Rott, E. and Friedrick, G., 1991. *Use of Algae for Monitoring Rivers.* Institute of Botanik, Universitat Innsbruck, Germany.

Wong, S.L., 1989. Algal assays to interpret toxicity guidelines for natural waters. *J. Env. Sci. Health*, 24: 1001–1010.

Chapter 26

Fertigation Management in Cauliflower (*Brassica oleracea*)

U.S. Kadam[1], M.R. Deshmukh[2], S.A. Kadam[3] and H.M. Patil[3]
[1]Principal Investigator and Associate Professor of Agricultural Engineering;
[2]Assistant Professor of Horticulture;
[3]Junior Research Assistant (Agricultural Engineering)
Mahatma Phule Krishi Vidyapeeth, Rahuri – 413 722, Dist: Ahmednagar (M.S)

ABSTRACT

A field experiment was undertaken with a specific objective to find out the effect of different water-soluble fertilizer levels on the yield of cauliflower, FUE and WUE along with the effect of fertigation on field performance of drip irrigation system. The different levels of recommended dose of N,P, K nutrients for cauliflower were applied in the form of commercially available water soluble fertilizer (WSF) along with irrigation water through drip irrigation system as well as straight fertilizers by band placement. The study included 6 treatments and 3 replications. It was observed that 80 per cent of recommended dose (T_2) through WSF gave better results over the other treatments under study. The overall finding of the study indicated that when the quality of irrigation water is of C3S1 class and water soluble fertilizer which are acidic in nature are used then the system performance in terms of per cent reduction in initial discharge and the field emission uniformity was found influenced due to water quality and not due to fertigation. Further higher yields (554 q/ha) were recorded due to fertigation along with saving of 20 per cent quantity of fertilizers in T_2 (80 per cent of RD) when compared to conventional method of fertilizer application T_6 Control (N application through drip and P and K by band placement according to recommended dose of 150 : 75 : 75 kg/ ha) which gave a yield of 390 q/ha.

***Keywords**: Cauliflower, Fertigation.*

Introduction

Cauliflower is one of the important vegetable cash crops grown in Maharashtra. The area under cauliflower in Maharashtra state is 3500 ha with average yield and total production of 20 tones/ hactare and 71 tonnes/hactare respectively.

As the productivity of irrigated agriculture is more than twice the rainfed agriculture, it is very essential to bring the maximum area under irrigation. Therefore it is the need of time to use this input properly and judiciously to increase the productivity of land. Among sophisticated drip irrigation has prove its superiority due to direct application of water and fertilizer into the vicinity of root zone. By this system, the water is absorbed by the roots with high frequency than the conventional irrigation system, which results in better quality and quantity of the produce. It saves water to the extent of 60 per cent in different crops (Magar *et al.*, 1984). The research work on drip irrigation in the state is being carried out with the major aspect of water saving. Increased yields and efficient fertilizer application for crops such as sugarcane (Magar *et al.*, 1987), bhendi, bitter gourd and ridge gourd (Mane *et al.*, 1987).

Fertigation, which combines irrigation water with fertilizer, is well recognized as the most effective and convenient means of maintaining optimal fertility levels and water supply according to specific needs of each crop and type of soil. The technique applies both water and fertilizer at a low rate to the vicinity of plant root zone, resulting in higher yields and better quality of crops. There has been a continuous substantial growth in the use of fertilizers in liquid form in the developing countries.

With the introduction of water-soluble fertilizers, the farmers are fertigating the crops in majority cases for fertilizer economy. However it has been reported in research review committee report (Anonymous 1998) of Department of irrigation and drainage engineering MPKV, Rahuri that emitter-clogging problem was found aggravated. Therefore the present investigation on "Fertigation Management of Cauliflower" was undertaken.

Materials and Methods

The present investigation was carried out at the instructional farm of Department of irrigation and drainage engineering, Dr. A.S. College of Agricultural Engineering, M.P.K.V. Rahuri. The experiment was carried out in RBD with 6 treatments and three replications. The treatment details are given below.

T_1: Fertigation @ 60 per cent of RD
T_2: Fertigation @ 80 per cent of RD
T_3: Fertigation @ 100 per cent of RD
T_4: Fertigation @ 120 per cent of RD
T_5: Fertigation @ 140 per cent of RD
T_6: Control (N application through drip and P and K by band placement according to RD)
RD: Recommended dose (150 : 75 :75) kg/ha

Thirty days old healthy seedlings of variety golden –80 were transplanted on 15th November 2003 at a spacing of 0.6 × 0.4 m. Water-soluble fertilizers (Haisoul) of grade (19 : 19 : 19) was used for treatments T_1 to T_5. Urea (46.6 per cent), single super phosphate (16 per cent) P_2O_5), Murate of potash (60 per cent K_2O) were used for treatments T_6.

The average discharge and emission uniformity was computed by the formula suggested by Karmelli and Keller (1974). The recommended dose of fertilizer was given in 30 splits, the irrigations were scheduled on every alternate day. The depth of water to be applied per plant was calculated by using the equation:

$$D = PE \times Kc \times Kp$$

where,

D: Depth of water to be applied (mm)

Kc: Crop co-efficient (0.8, 1.0 and 1.2 for 3 months respectively)

Kp: Pan factor (0.7)

PE: Sum of pan evaporation for two days (mm)

The volume of water to be applied for treatment was computed by the equation:

$$V = D \times A \times N$$

where,

V: Volume of water (l)

D: Depth of water (mm)

A: Area of one plot

N: number of plots or replications per treatments.

The time of operation (hr) of drip irrigation system for each treatment was calculated by using equation:

$$To = V/q.\ Eu.n.\ N$$

where,

To: Time of application for the operation of drip irrigation unit for respective treatment (hr)

V: Volume of water to be applied for each application for all three replications of treatment (l)

q: Average discharge of emitters in respective treatments (lph)

Eu: Emission uniformity of drip irrigation unit n = No of emitters per treatment

N: No of plots per treatment.

Observations such as height of plant, no of leaves, weight of leaves, weight of curd, diameter of curd, days required for curd initiation, days required for maturation, curd to leaf ratio and specific gravity of curd were recorded. Fertilizer use efficiency was determined by the equation

$$FUE = \text{yield (kg/ha)} / \text{quantity of fertilizer applied (kg/ha)}$$

Results and Discussion

The results tabulated in Table 26.1 revealed that the average discharge in all the treatments was found to be 3.795 and there was very little variation within the treatments and replications. The treatments do not vary significantly from each other. It was observed that the average EU was closer to 94 per cent. The desired EU is expected to be around 90 per cent (Keller and Karmelli, 1974). Ahluwalia *et al.* (1987) studied the response of cauliflower to drip irrigation (3 irrigation levels) and conventional irrigation method. A uniformity co-efficient of 94 per cent could be achieved in drip irrigation. In the present studies the values are more than the desired. The treatments do not differ significantly from each other.

Table 26.1: Average Discharge and Emission Uniformity and Reduction Observed at 30, 60, and 90 Days Respectively

Treatment	Average Discharge (lph)	Average Eu (%)	% Reduction in Average Discharge After			% Reduction in Emission Uniformity After		
			30 Days	60 Days	90 Days	30 Days	60 Days	90 Days
T_1	3.788	94.31	1.71	3.270	4.883	0.882	1.351	2.285
T_2	3.826	94.08	0.914	2.090	3.450	0.588	1.116	1.615
T_3	3.840	94.40	0.520	1.380	2.421	0.463	0.784	1.255
T_4	3.791	94.54	0.466	1.112	1.905	0.356	0.516	0.871
T_5	3.749	94.58	0.328	0.693	1.226	0.134	0.404	0.571
T_6	3.780	93.84	2.469	4.285	6.375	1.137	1.526	2.385
SE±	0.021	0.474	0.185	0.182	0.438	0.083	0.058	0.082
CD at 5%	NS	NS	0.582	0.574	1.379	0.262	0.181	0.259

Observations on average emitter discharge and emission uniformity from all the three replicated plots before fertigation (initial) and from planting to the last fertigation at an interval of 30, 60 and 90 days are also presented in Table 26.1. The maximum per cent reduction (6.37 per cent) in emitter discharge was observed in treatment T_6 (Control) followed by treatment T_1 (4.88 per cent). It may be due to alkaline nature of water. The maximum per cent reduction (2.38 per cent) in Emission uniformity was observed in treatment T_6 (Control) followed by treatment T_1 (2.30 per cent). It is revealed that the reduction in EU was nearly same in control and fertigation treatment. Similar findings were reported by Anonymous (1998). They reported that the discharge was found reduced to nearly 72 per cent and 60 per cent of the initial average discharge and EU was reduced to 10 per cent from the initial value when commercial water soluble fertilizers and straight fertilizers (Urea) respectively were applied through drip.

The data tabulated in Table 26.2 indicated that the per cent plant survival was above 97 per cent in all the treatments, which was satisfactory. The number of days for curd initation and harvest was found to be 67 to 70 days and 13 to 15 days in all the treatments respectively.

Table 26.2: Effect of Various Treatments on Growth of Cauliflower

Treatments	Av. Plant Survival (%)	Days Required for Curd Initiation	Days to Harvest the Curd after Initiation	Average Weight of Leaves/ Plant (g)	Average Plant Height (cm)	No. of Leaves
T_1	97.54	70	14	1144	87.4	26
T_2	98.03	68	13	1543	95.9	28
T_3	97.54	69	15	1462	95.8	28
T_4	96.07	67	14	1361	94.1	27
T_5	97.05	69	15	1279	92.4	26
T_6	97.05	70	13	1216	89.3	26
SE±				105.14	1.357	0.803
CD at 5%				NS	4.213	NS

Maximum weight of leaves was observed in treatment T_2 where as it was minimum in treatment T_6. Observations in respective of plant height revealed that plant height was maximum in treatment T_2 whereas it was minimum in T_1 (60 per cent RD). Treatment T_1 and T_6 were at par with each other. Treatment T_2, T_3, T_4 and T_5 were at par with each other. But treatment T_2 was significantly superior over Treatment T_1 and T_6. The number of leaves per plant was more or less similar in all treatments and therefore the treatments did not differ significantly from each other.

The maximum yield of 554.15 q/ha was obtained in treatment T_2 followed by T_3 (55.1.65 q/ha), T_4 (506.51 q/ha), T_5 (480.65 q/ha), T_6 (389.57 q/ha) and T_1 (377.76 q/ha). From the statistical analysis it is revealed that treatment T_2 and T_3 were at par with each other. Treatment T_2 was significantly superior over the treatments T_1, T_5, T_5 and T_6. The increase in yield in treatment T_2 was 42.24 per cent over T_6 (Control). Padmakumari and Shivanappan (1978) reported that brinjal with drip irrigation required only 24 cm of water against 69 cm in conventional method and the yield obtained was 18.75 tonnes/ha which was above normal.

Water use efficiency observed in different treatments and given in Table 26.3 indicated that the maximum WUE (18.95 q/ha-cm) was observed in T_2 followed by T_3 (18.88 q/ha-cm), T_4 (17.34 q/ha-cm), T_5 (16.45 q/ha-cm), T_6 (13.34 q/ha-cm) and T_1 (12.93 q/ha-cm) whereas the FUE was found to be maximum in T_2 (230.92) followed by T_1 (209.86), T_3 (183.88), T_4 (140.69), T_6 (129.85) and T_5 (114.43). Gorantiwar *et al.* (1987) compared conventional furrow and improved irrigation method to cabbage in relation to yield, WUE and water saving. The yield in drip irrigation was higher (16.9 tonnes/ha) than yield in furrow irrigation method (14.67 tonnes /ha). Similarly the water saving to the extent of 46 per cent was observed in drip over furrow. WUE 6.35 q/ha-cm was obtained for drip irrigation method whereas 2.96 q/ha-cm for furrow irrigation.

Table 26.3: Water Use Efficiency Observed in Different Treatments (q/ha-cm)

Treatments	*Yield (q/ha)*	*Depth of Water Applied (cm)*	*Water Use Efficiency (q/ha-cm)*
T_1	377.76	29.23	12.93
T_2	554.15	29.23	18.95
T_3	551.65	29.23	18.88
T_4	506.91	29.23	17.34
T_5	480.65	29.23	16.45
T_6	389.57	29.23	13.34
SE±	8.838		0.242
CD at 5%	27.837		0.761

The data related with the cost economics of fertigation for different treatments indicated that Benefit: Cost ratio (2.15) was found maximum in treatment in T_2 followed by T_1 (2.00), T_6 (1.90), T_4 (1.72), T_1 (1.57) and T_5 (1.54). The net income was maximum from treatment T_2 (59,290 Rs/ha) closely followed by treatment T_1 (55,315 Rs/ha).

The overall findings of the present study indicated that when the quality of irrigation water was C3S1 and water soluble fertilizer that are acidic in nature the system performance in terms of per cent reduction in initial discharge and field emission uniformity was found influenced due to water quality and no due to fertigation. Further 42.24 per cent higher yields were recorded due to fertigation along with saving of 20 per cent in fertilizer. The maximum net returns were obtained when 80 per cent of recommended dose of water soluble fertilizers was applied through drip.

Table 26.4: Fertilizer Use Efficiency Observed in Different Treatments

Treatments	*Yield (q/ha)*	*Quantity of N, P, K Applied (kg/ha)*	*Fertilizer Use Efficiency*
T_1	377.76	180	209.86
T_2	554.15	240	230.92
T_3	551.65	300	183.88
T_4	506.91	360	140.69
T_5	480.65	420	114.43
T_6	389.57	300	129.85
SE±	8.838		2.505
CD at 5%	27.837		7.891

Table 26.5: Cost Economics of Fertigation for Different Treatments

Items	T_1	T_2	T_3	T_4	T_5	T_6
Cost of production (Rs/ha)	47,936	51,510	55,085	58,660	62,234	40,894
Gross monetary returns (Rs/ha)	75,600	110,800	1,10,400	1,01,400	96,200	78,000
Net Income (Rs/ha)	27,666	59,290	55,315	42,740	33,966	37,106
B : C ratio	1.57	2.15	2.00	1.72	1.54	1.90

References

Ahluwalia, M.S., Baldev Singh and Gill, B.S., 1993. Drip irrigation system. Its hydraulic performance and influence on cauliflower crop. *J. Water Management*, 1(1): 6–9

Anonymous, 1998. Research review committee report. Deptt. of irrigation and water management, M.P.K.V. Rahuri.

Gorantiwar, S.D., Pampattiwar, P.S., Suryawanshi, S.N. and Pingale, L.V., 1987. Comparative study of various irrigation methods for cabbage. *National Seminar*, 10–11: 142–149.

Magar, S.S. and Sonawane, B.V., 1987. Moisture distribution pattern and nitrogen saving in drip irrigation for cotton in vertisol. In: *Abstract of National Seminar on Drip and Sprinkler Irrigation Methods.* Adoption pp. 241–256.

Mane, T.A., Pampattiwar, P.S., Patil, B.R. and Khade, K.K., 1987. Comparative studies on drip and furrow methods of irrigation for bhendi crop. In: *Abstract of National Seminar on Drip and Sprinkler Irrigation Methods.* Adoption pp. 8.

Padmakumari and Shivanappan, R.K., 1978. Drip irrigation in brinjal, bhendi, tomato, water scarcity area. *Madras Agric. J.*, 65(9): 608–609.

Chapter 27

Uniformity in Micro-sprinkler Irrigation System

S.A. Kadam[1], U.S. Kadam[2], S.D. Gorantiwar[3] and H.M. Patil[1]
[1]Junior Research Assistant; [2]Principal Investigator; [3]Associate Professor
Department of Irrigation and Drainage Engineering,
Mahatma Phule Krishi Vidyapeeth, Rahuri – 413 722, Dist: Ahmednagar (M.S)

ABSTRACT

Field performance of micro-sprinkler in terms of uniformity as influenced by the operating pressure and micro-sprinkler spacing was evaluated at the Instructional Farm, Department of Irrigation and Drainage Engineering, MPKV, Rahuri (India). The results of the study indicated the uniformity decreased with the increase in micro-sprinkler spacing from 1.5 × 1.5 m to 3.75 × 3.75 m. It is also seen that as operating head increases from 0.75 to 2.0 kg/cm² uniformity increases. In general it was found that for the rotating head micro-sprinkler under consideration, the spacing of 1.5 × 1.5 m should be used for the operating head of 1.0 kg/cm² and 2.25 × 2.25 should be used for the operating head of 2.0 kg/cm².

Keywords: Irrigation, Micro-sprinkler, Uniformity.

Introduction

The total area under irrigation in Maharashtra is only 16.41 per cent of cultivable command area (CCA) and it is estimated that after utilisation of all the harnessable water resources, the irrigated area in the state may not exceed 34 per cent of cultivable command area (CCA) {Second Irrigation Commission Report, 1999). However this estimate is based on consideration that surface irrigation methods are adopted for irrigation. As the production of irrigated agriculture is more than 2–3 times the production of rainfed, agriculture, it is necessary to further increase the area under irrigation to produce the food forever increasing population. If the micro-irrigation methods such as drip and micro-sprinkler, which save water to the extent of 30 to 50 per cent over surface irrigation methods and increases yield upto 30 per cent (Pampattiwar *et al.*, 1991 and Magar and Firke, 1993), it is possible to bring more area under irrigation without going for creating new water resources which now need huge expenditure. Amongst the micro-irrigation methods, micro-sprinkler irrigation method is becoming popular as it overcomes the greatest demerit of clogging in drip irrigation. The use of micro-sprinkler irrigation for many crops has indicated an increase in crop yield and water saving (Singh and Singh, 1990, and Shinde and Firake, 1998). However due to more initial investment it is necessary to design and adopt micro-sprinkler irrigation method properly.

The important parameter in the design of micro-sprinkler irrigation is to optimize the spacing between the micro-sprinklers as spacing influences the discharge, uniformity and cost. Crop yield depends largely on the discharge and uniformity of water application. The study was therefore undertaken to decide the spacing between micro-sprinkles based on the uniformity of micro-sprinkler irrigation system. The information on uniformity of micro-sprinkler for their different spacing would also aid for irrigation scheduling.

Material and Methods

The experiment was conducted on the Instructional Farm of Mahatma Phule Krishi Vidyapeeth, Rahuri (India). The uniformity was evaluated for following four different micro-sprinkler spacing:

1. 1.5 × 1.5 m
2. 2.25 × 2.25 m
3. 3.0 × 3.0 m
4. 3.75 × 3.75 m

The rotating head micro-sprinklers of nozzle size 1.3 mm at stake height of 30 cm were used. The observations for uniformity were recorded for different operating heads (0.75, 1.00, 1.25, 1.50, 1.75, 2.00 kg/cm^2) for each of the micro-sprinkler spacing. The water in catch can was collected for 30 min and repeated for three times for each combination of spacing and operating head. The uniformity was estimated for the average depth of water collected in catch can by Christiansen's formula (Michael, 1978) which is given below:

$$UC = 100\,(I - \Sigma I \times I/m \times n)$$

where,

UC: Uniformity coefficient, per cent

IΣI: Sum of the absolute deviation of individual observation from the average (depth), mm

m: Average depth of all observations, mm

n: Number of observations

Results and Discussion

The uniformity coefficient values estimated for different combinations of spacing and operating heads are presented in Table 27.1. Table 27.1 indicates that as spacing increases, uniformity decreases and as operating head increases uniformity increases. These results are in close agreement with those obtained for conventional overhead sprinklers (Pampattiwar *et al.*, 1991). The minimum desirable uniformity in case of sprinkler irrigation should be 85 per cent (Michael, 1978). It is observed from the table that when the operating head is 0.75 kg/cm^2, the uniformity values are always below minimum desirable limit of 85 per cent. For the operating head of 1.00, 1.25, and 1.75 kg/cm^2 the uniformity values more than minimum desirable limit are obtained only for the spacing of 1.5 × 1.5 m which is the close spacing of all those tested. However, when operating head is increased to 2.0 kg/cm^2, the desirable uniformity is obtained for the spacing of 1.5 × 1.5 m and 2.25 × 2.25 m. Thus this study indicates that the micro-sprinkler under study be operated at 1.0 kg/cm^2 for the spacing of 1.5 × 1.5 m and at 2.0 kg/cm^2 for the spacing of 2.25 × 2.25 m. More micro-sprinkler spacing is desirable to reduce the cost of the system. However, if spacing of the micro-sprinkler is increased, they are to be operated at higher operating heads, needing more expenditure on energy cost. If the spacing are further increased and operating heads are further decreased, the uniformity is found to be reduced drastically. This may reduce eventually the crop yield.

Table 27.1: The Uniformity Coefficient Observed in Different Micro-sprinkler Spacing and Operating Heads

Pressure (kg/cm^2)	*Micro-sprinkler Spacing (m)*	*Uniformity Coefficient (%)*
0.75	1.5 × 1.5	81.66
	2.25 × 2.25	71.68
	3.0 × 3.0	65.59
	3.75 × 3.75	53.48
1.0	1.5 × 1.5	85.60
	2.25 × 2.25	75.47
	3.0 × 3.0	70.32
	3.75 × 3.75	56.77
1.25	1.5 × 1.5	86.93
	2.25 × 2.25	77.51
	3.0 × 3.0	75.91
	3.75 × 3.75	60.68
1.5	1.5 × 1.5	88.13
	2.25 × 2.25	80.95
	3.0 × 3.0	78.16
	3.75 × 3.75	64.01
1.75	1.5 × 1.5	89.88
	2.25 × 2.25	83.55
	3.0 × 3.0	81.63
	3.75 × 3.75	68.60
2.0	1.5 × 1.5	91.99
	2.25 × 2.25	86.20
	3.0 × 3.0	83.88
	3.75 × 3.75	73.12

References

Chitale Irrigation Commission Report, 1999.

Michael, A.M., 1978. *Irrigation: Theory and Practice.* Vikas Publishing House Pvt. Ltd., New Delhi, India, pp. 1–801.

Magar, S.S., Firake, N.N. and D.S. Salunkhe, 1993. Effect of uniformity coefficient of water application on spacing between micro-sprinklers in vegetables. *Maharashtra J. Hort.*, 7(1).

Pampattiwar, P.S., Firake, N.N. and D.S. Salunkhe, 1991. Evaluation of hydraulic performance of micro-sprinkler irrigation system. *Indian Journal of Agricultural Engineering (ICAR)*, 1(2): 141–144.

Singh, H. and B. Singh, 1990. Micro-sprinkler performance evaluation and constraints for its adoption. In: *Proc. XIth International Congress on the Use of Plastics in Agriculture*, held in New Delhi, B: 79–91.

Chapter 28

Micro-sprinkler Irrigation System for Onion

U.S. Kadam[1], S.A. Kadam[2], S.D. Gorantiwar[3] and H.M. Patil[3]

[1]Professor, [2]Junior Research Assistant, [3]Associate Professor and [4]Junior Research Assistant, Department of Irrigation and Drainage Engineering, CAE, Mahatma Phule Krishi Vidyapeeth, Rahuri – 413 722

ABSTRACT

The experiment entitled "Effect of Irrigation Scheduling on Yield and Water Use Efficiency of Onion Under Micro-sprinkler Irrigation System" was conducted at Instructional Farm, Department of Irrigation and Drainage Engineering, Dr. A.S. College of Agricultural Engineering, Rahuri. The results of the investigation revealed that maximum onion yield was recorded when the irrigations were scheduled at 20 per cent of available moisture depletion level which was found superior and statistically significant.

The water requirement and water use efficiency under T_1 treatment were as 514.40 mm and 961.82 kg/ha-cm for 20 per cent moisture depletion level. The water requirement in the control treatment (T_5) was found maximum (600 mm) resulting in lower WUE (416.15 kg/ha-cm) than any other micro-sprinkler irrigated treatments. However, it was observed that there was overall water saving to the extent of 19.56 per cent and simultaneously increase in yield to the extent of 37.86 per cent.

Micro-sprinkler method was found satisfactory for high density and low height crop like onion. It was possible because of optimum soil moisture aeration status enhancing the nutrient availability and soil microbial activities.

Keywords: Irrigation, Consumptive use, Water use efficiency, Micro-sprinklers.

Introduction

Onion (*Allium cepa* L.) is an important spice crop which is categorize under high renumarative cash crop in India. India has the second largest area under onion in the world. Agriculture Research Data Book, 2003 showed that India had 4.47 lakh ha area under onion and production was 47.21 lakh tones of bulbs (Anonymous, 2002). In Maharashtra area under onion is 1.18 1akh ha and production was 16.87 lakh tones. Apart from other input limitations soil moisture was one of the most important limitations. Hence micro-sprinkler method is considered the most appropriate method of irrigation for the crop like onion which is shallow rooted, high density and dwarf crop. The crop also needs frequent application of irrigation water based on its depletion. Soil water depletion is one of the criterions for scheduling of irrigation and maintenance of higher 75–80 per cent available soil moisture in effective root zone was possible due to micro-sprinkler irrigation system. Study was therefore aimed to locate the optimum point within the available range of soil moisture, when irrigation must be scheduled for maximum yield of onion bulbs and water use efficiency.

Materials and Methods

The experiment was conducted at the Instructional Farm, Department of Irrigation and Drainage Engineering, Dr. A.S. College of Agricultural Engineering, Rahuri. The soil of the experimental field was clayey type. The field capacity, permanent wilting point, bulk density and infiltration rate of the soil were 37.50 per cent, 19.13 per cent, 1.26 g/cm^3 and 0.42 cm/hr, respectively.

The experiment was laid out statistically with randomized block design (RBD) with five treatments based on per cent moisture depletion levels replicated for four times. The treatments T_1, T_2, T_3 and T_4 were based on different irrigation frequencies calculated on the basis of per cent moisture depletion levels for 20, 30, 40 and 50 per cent respectively over available soil moisture under micro-sprinkler irrigation system. However the control treatment T_5 irrigation water was applied as per recommendation of university considered as control treatment. The recommended cultivation practices were adopted for growing Phule Suverna variety of onion. Onion seedlings were transplanted in the month of December. In the control (check basin) treatment, water was applied by conventional method with the depth of 60 mm water applied at an interval of 10 days.

Total available soil moisture (ASM) was soil moisture constants and bulk density of soil profile indicating the effective root zone (Black *et al.*, 1957). It was estimated to the extent of 120 mm in the effective root zone of 60 cm.

However, the effective root zone depth of onion was considered 200 mm for initial stage upto 20 days and 450 mm for further crop growth period. Accordingly net irrigation depth (Di) was worked out as per depletion level (Dp) as indicated below:

$$Di = (TAW \times Dp)/100$$

For the present study the depletion levels (Dp) were taken as 20, 30, 40 and 50 per cent for the treatment T_1, T_2, T_3 and T_4 respectively,

$$Dg = Di/UC$$

where,

Dg: Gross depth of irrigation, mm

UC: Uniformity coefficient, per cent

$$To = Dg \times 60 / Pi$$

where,

To: Time of operation, min

Pi: Precipitation rate, mm/hr

The value of reference evapotranspiration is calculated by using equation:

$$Eto = PE \times F$$

where,

PE: Average pan evaporation over growth stage period, mm/day

F: Pan factor, taken as 0.80

The irrigation frequency is determined by considering net irrigation depth, potential evapotranspiration (ETo) for two days immediately after irrigation and crop evaporatranspiration (ETc), mm/day by following equations,

$$Dir = Di - (ETo \times 2)$$

where,

Dir: Net depth remaining after two days after irrigation, mm

Di: Net depth of irrigation to be applied, mm

ETo: Reference evapotranspiration, mm/day

The crop evapotranspiration (ETc) is calculated by using crop coefficient (Kc) as:

$$ETc = ETo \times Kc$$

The values of crop coefficients are as per the growth stages. They are taken as per FAO bulletin 1980 Vol. 33. (Doorenbose and Pruilt, 1975).

Hence, irrigation interval (IR) in days is calculated as:

$$IR = (Dir/ETc) + 2$$

These empirical equations and constants such as PF and Kc were studied and, tested for different crops under SAT climatic conditions at Rahuri (Magar *et al.*, 1993).

Results and Discussion

The average discharge of micro-sprinkler and average wetted diameter were obtained as 57.50 lph and 5.20 m, respectively. Similarly the average rate of precipitation and uniformity coefficient were obtained as 24.35 mm/hr and 92 per cent, respectively. The effective root zone depth at initial period upto 20 days and subsequently at full growth period from 21 days to maturity were 200 and 450 mm respectively. Moreover total available soil moisture was 46.4 and 104.50 mm respectively for the said period.

Irrigation depth data is the function of time pan factor and crop factor. The maximum gross depth (60 mm) per irrigation for initial period of 20 days after sowing was observed in control treatment whereas it was found minimum when irrigations were scheduled at 20 per cent depletion level.

The values of number of irrigations and total depth of irrigation in each treatment were reported in Table 28.1. The total depth of water given in each treatment was nearly same and the little variation in total depth of irrigation in treatment T_1 to T_4 which was due to different Kc factors, irrigation intervals and irrigation cut off dates. It is interesting to note that water requirement under micro-sprinkler irrigation system and number of irrigations has inverse relationship even though there was no much variation of water quantity. This has helped in maintaining excellent soil aeration and available soil moisture status. The effective root zone was under the dominance of field capacity during entire growth period.

Table 28.1: Number of Irrigations and Total Depth of Irrigation Applied in Different Treatments

Treatments	*Number of Irrigations*	*Total Depth of Irrigation (mm)*
T_1	26	514.00
T_2	17	502.40
T_3	12	467.70
T_4	9	446.40
T_5 (Control)	10	600.00

The average percentage soil moisture depletion over available soil moisture were 20.078, 29.972, 39.992, 50.003 and 46.565 per cent in treatments T_1, T_2, T_3, T_4 and T_5 respectively. Small variation was observed in each moisture depletion treatment which were scheduled at 20, 30, 40 and 50 per cent.

Table 28.2: Mean Yields of Onion Bulbs (q/ha), Yield of Stalks (q/ha) and Bulb to Stalk Ratio as Influenced by Different Treatments After Drying

Treatment	*Yield of Onion Bulbs (q/ha)*	*Yield of Stalks (q/ha)*	*Bulb to Stalk Ratio*	*Per cent Increase in Yield Over Control*	
				Yield of Bulbs	*Yield of Stalks*
T_1	494.375	60.625	8.154	97.99	49.23
T_2	363.125	49.375	7.356	45.43	21.54
T_3	298.125	47.500	6.281	19.40	16.92
T_4	220.938	37.813	5.884	– 11.51	– 6.92
T_5 (Control)	249.688	40.625	6.180	–	–
F test	Sig.	Sig.	Sig.	–	–
SE±	6.755	1.046	0.064	–	–
C.D. at 5%	20.822	3.223	0.198	–	–
General mean	325.250	47.188	6.771	–	–

However it is clear that fluctuations of available soil moisture content from depletion level of 50 per cent to saturation had cyclonic effect. The crop was under moisture stress condition before irrigation and suffered set backs for aeration status for a period of achieving field capacity after irrigation application. This has showed the dominance of micro-sprinkler system over traditional surface irrigation. The yield of onion bulbs obtained in each treatment was reported in Table 28.2. The yield of onion bulbs under treatment T_1 (494.375 q/ha) was significantly superior over all other treatments. However the treatment T_4 recorded the lowest yield of bulbs (220.938 q/ha) was due to inadequate water application and moisture stress. The per cent increase in the yield of onion bulbs over control

(T_5) was observed as 97.99 per cent, 45.43 per cent and 19.40 per cent in treatments T_1, T_2 and T_3 respectively. However, the yield of onion bulbs in treatment T_4 observed lower than that of control. This is because of the high available soil moisture depletion level (50.003 per cent) as compared to control (46.565 per cent).

The values water use efficiency in each treatment were reported in Table 28.3. Water use efficiency was maximum (961.82 kg/ha-cm) in treatment T_1 and decreased with increase in per cent moisture depletion levels. The WUE was minimum (416.15 kg/ha-cm) in control treatment T_5 than any other treatments. The per cent saving in water was observed to be 14.33, 16.26, 22.05 and 25.60 in treatment T_1, T_2, T_3 and T_4 respectively over control treatment.

Table 28.3: The WUE as Influenced by Different Irrigation Frequencies

Treatment	*Yield (q/ha)*	*Water Requirement (mm)*	*WUE kg/ha-cm*
T_1	494.375	514.40	961.82
T_2	363.125	502.40	722.78
T_3	298.125	467.70	637.43
T_4	220.938	446.40	494.93
T_5 (Control)	249.688	600.00	416.15

Conclusions

In brief it summarized that micro-sprinkler irrigation system should be scheduled below 30 per cent available soil moisture depletion level. The frequency and depth of irrigation water could be estimated from pan evaporation and crop coefficient data. There is a overall saving of significant water quantity to the extent of 19.56 per cent in addition to increase in the yield of garlic to the extent of 37.83 per cent. This was possible because of excellent soil moisture plant atmosphere relationship.

References

Anonymous, 2002. *The Hindu Survey of Indian Agriculture*, pp. 89.

Black, C.A. 1957. *Soil-Plant Relationship*. John Wiley and Sons, New York, 332 p.

Doorenbos, D.W. and W.C. Pruilt, 1975. *Guidelines for Predicting Crop Water Requirements*. Irrigation and Drainage Paper 24, Food and Agriculture Organisation of United Nations, Rome, Italy.

FAO, 1980. *The FAO Production Year Book*, 33: 109–111.

Magar, S.S., N.N. Firake and D.S. Salunkhe, 1993. Effect of uniformity coefficient of water application on spacing between micro-sprinklers in vegetables. *Maharashtra Journal of Hort.*, 7(1).

Chapter 29

Micro-sprinkler Irrigation System for Garlic

S.A. Kadam[1], U.S. Kadam[2] and H.M. Patil[3]
[1]Junior Research Assistant, [2]Principal Investigator and [3]Associate Professor, Department of Irrigation and Drainage Engineering, CAE, Mahatma Phule Krishi Vidyapeeth, Rahuri – 413 722

ABSTRACT

The experiment entitled "Effect of Irrigation Scheduling on Yield and Water Use Efficiency of Garlic Under Micro-sprinkler Irrigation System" was conducted at Instructional Farm, Department of Irrigation and Drainage Engineering, Dr. A.S. College of Agricultural Engineering, Rahuri. The results of the investigation revealed that maximum garlic yield was recorded when the irrigations were scheduled at 20 per cent of available moisture depletion level which was found superior and statically significant.

The water requirement and water use efficiency under T_1 treatment were as 802.40 mm and 84.20 kg/ha-cm for 20 per cent moisture depletion level. The water requirement in the control treatment (T_5) was found maximum (940 mm) resulting in lower water use efficiency (59.32 kg/ha-cm) than the other micro-sprinkler irrigated treatments. However, it was observed that there was water saving to the extent of 17.21 per cent and simultaneously increase in yield to the extent of 21 per cent.

Micro-sprinkler method was found satisfactory for high density and low height crop like garlic. It was possible because of optimum soil moisture aeration status enhancing the nutrient availability and soil microbial activities.

***Keywords**: Irrigation Scheduling, Micro-sprinkler, Water use efficiency, Yield, Water requirement.*

Introduction

Garlic (*Allium sativam* L.) is an important spice crops which is categorized under high renumarative cash crop in India. India has the second largest area under garlic in the world. Agriculture Research Data Book, 2003 showed that India had 1.44 lakh ha area under garlic and production was 4.953 lakh million tones (MT) of bulbs (Anonymous, 2002). In Maharashtra area under garlic is 6500 ha and production was 43600 MT. Apart from other input, limitations soil moisture is one of the most important limitation. Hence micro-sprinkler method is considered to be the most appropriate method of irrigation for the crop like garlic which is shallow rooted, high density and dwarf crop. The crop also needs frequent application of irrigation water based on its depletion. Soil water depletion is one of the criterion for scheduling of irrigation and maintenance of higher 75–80 per cent available soil moisture in effective root zone, which is possible due to micro-sprinkler irrigation system. Study was therefore aimed to locate the optimum point within the available range of soil moisture, when irrigation must be scheduled for maximum yield of garlic bulbs and water use efficiency.

Material and Methods

The experiment was conducted at the Instructional Farm, Department of Irrigation and Drainage Engineering, Dr. A.S. College of Agricultural Engineering, Rahuri. The soil of the experimental field was clayey type. The field capacity, permanent wilting point, bulk density and infiltration rate of the soil were 38.23 per cent, 19.80 per cent, 1.26 g/cm^3 and 0.42 cm/hr, respectively.

The experiment was laid out statistically with randomized block design (RED) with five treatments replicated for four times. The treatments T_1, T_2, T_3 and T_4 were based on different irrigation frequencies calculated on the basis of per cent moisture depletion levels for 20, 30, 40 and 50 per cent respectively under micro-sprinkler irrigation system. However for the control treatment T_5, irrigation water was applied as per recommendation of university. The recommended cultivation practices were adopted for growing G-41 variety of garlic. Garlic cloves were sown in the month of December. In the control (check basin) treatment, water was applied by conventional method with the depth of 70 mm water applied at an interval of 10 days.

Total available soil moisture (ASM) was soil moisture constants and bulk density of soil profile indicating the effective root zone. (Black et al, 1956). It was estimated to the extent of 140 mm in the effective root zone of 60 cm.

However, the effective root zone depth of garlic was considered 200 mm for initial stage upto 20 days and 450 mm for further crop growth period. Accordingly net irrigation depth (Di) was worked out as per depletion level (Dp) as indicated below:

$$Di = (TAW \times Dp)/100$$

For the present study the depletion levels (Dp) were taken as 20, 30, 40 and 50 per cent for the treatment T_1, T_2, T_3 and T_4 respectively.

$$Dg = Di/UC$$

where,

Dg: Gross depth of irrigation, mm

UC: Uniformity coefficient, per cent

$$To = Dg \times 60/Pi$$

where,

To: Time of operation, min

Pi: Precipitation rate, mm/hr

The value of reference evapotranspiration is calculated by using equation:

$$Eto = PE \times F$$

where,

PE: Average pan evaporation over growth stage period, mm/day

F: Pan factor, taken as 0.70

The irrigation frequency is determined by considering net irrigation depth, potential evapotranspiration (ETo) for two days immediately after irrigation and crop evapotranspiration (ETc), mm/day by following equations:

$$Dir = Di - (ETo \times 2)$$

where,

Dir: Net depth remaining after two days after irrigation, mm

Di: Net depth of irrigation to be applied, mm

ETo: Reference evapotranspiration, mm/day

The crop evapotranspiration (ETc) is calculated by using crop coefficient (Kc) as:

$$ETc = ETo \times Kc$$

The values of crop coefficients are as per the growth stages. They are taken as per FAO bulletin 1980 vol. 33. (Doorenbose and Pruilt, 1975).

Hence, irrigation frequency (IF) in days is calculated as:

$$IF = (Dir/ETc) + 2$$

These empirical equations and constants such as PF and Kc were studied and tested for different crops under SAT climatic conditions at Rahuri (Magar *et al.*, 1993).

Results and Discussion

The average discharge of micro-sprinkler and average wetted diameter were obtained as 37.83 lph and 4.64 m, respectively. Similarly the average rate of precipitation and uniformity coefficient were obtained as 12.34 mm/ha and 85 per cent, respectively. The effective root zone depth at initial period upto 20 days and subsequently at full growth period from 21 days to maturity were 200 and 450 mm respectively.

Irrigation depth data is the function of time pan factor and crop factor were depicted in Table 29.1. The maximum gross depth (70 mm) per irrigation for initial period of 20 days after sowing was observed in control treatment whereas it was found minimum when irrigations were scheduled at 20 per cent depletion level.

Table 29.1: Number of Irrigations and Total Depth of Irrigation Applied in Different Treatments

Treatments	*Number of Irrigations*	*Total Depth of Irrigation (mm)*
T_1	32	802.40
T_2	20	776.07
T_3	15	754.60
T_4	11	707.00
T_5 (Control)	12	940.00

The total depth of irrigation in each treatment were reported in Table 29.1. The total depth of water given in each treatment was nearly same and the little variation in total depth of irrigation in treatment T_1 to T_4 which was due to different Kc factors, irrigation frequencies and irrigation cut off dates. It is interesting to note that water requirement under micro-sprinkler irrigation system and number of irrigation frequency has inverse relationship even though there was no much variation of water quantity. This has helped in maintaining excellent soil aeration and available soil moisture status. The effective root zone was under the dominance of field capacity during entire growth period.

The average percentage soil moisture depletion over available soil moisture were 20.60, 30.76, 41.71, 50.95 and 45.62 per cent in treatments T_1, T_2, T_3, T_4 and T_5, respectively. Small variation was observed in each moisture depletion treatment which were scheduled at 20, 30, 40 and 50 per cent.

However it is clear that fluctuations of available soil moisture content from depletion level of 60 per cent to saturation had cyclonic effect. The crop was under moisture stress condition before irrigation and suffered set backs for aeration status for a period of achieving field capacity after irrigation application. This has showed the dominance of micro-sprinkler system over traditional surface irrigation.

The yield of bulbs in each treatment were reported in Table 29.2. The yield of bulbs under treatment T_1 (67.568 q/ha) was significantly superior over all other treatments except treatment T_2. However the treatment T_4 recorded the lowest yield of bulbs (55.767 q/ha) was due to inadequate water application and moisture stress. The per cent increase in the yield of garlic over control (T_5) was observed as 21.16 per cent, 14.24 per cent and 4.58 per cent in treatments T_1, T_2 and T_3 respectively. However, the yield of garlic in treatment T_4 observed lower than that of control. This is because of the high available soil moisture depletion level (50.95 per cent) as compared to control (45.62 per cent).

The water use efficiency in each treatment were reported in Table 29.3. Water use efficiency was maximum (84.20 kg/ha-cm) in treatment T_1 and decreased with increase in per cent moisture depletion levels. The water use efficiency was minimum (59.32 kg/ha-cm) in control treatment T_5 than any other treatments. The percentage increase in water use efficiency over control was observed to be 41.92, 38.37, 30.28 and 2.06 in treatment T_1, T_2, T_3 and T_4 respectively.

In brief it summarized that micro-sprinkler irrigation system should be scheduled below 30 per cent available soil moisture depletion level. The frequency and depth of irrigation, water could be estimated from pan evaporation and crop coefficient data. There is a saving of significant water quantity to the extent of 17.15 per cent in addition to increase in the yield of garlic to the extent of 21 per cent. This was possible because of excellent soil moisture plant atmosphere relationship.

Table 29.2: Mean Yields of Garlic Bulbs (q/ha), Yield of Stalks (q/ha) and Bulb to Stalk Ratio as Influenced by Different Treatments After Drying

Treatment	*Yield of Onion Bulbs (q/ha)*	*Yield of Stalks (q/ha)*	*Bulb to Stalk Ratio*	*Per cent Increase in Yield Over Control*	
				Yield of Bulbs	*Yield of Stalks*
T_1	67.568	14.96	4.51	21.16	15.69
T_2	63.710	13.28	4.79	14.24	2.70
T_3	58.325	12.13	4.80	4.58	11.37
T_4	42.810	10.33	4.14	– 23.23	– 3.94
T_5 (Control)	55.767	12.93	4.31	–	–
F test	Sig.	Sig.	Sig.	–	–
SE±	1.473	0.338	0.198	–	–
C.D. at 5%	4.539	1.043	0.611	–	–
General mean	57.636	12.726	4.51	–	–

Table 29.3: The WUE as Influenced by Different Irrigation Frequencies

Treatment	*Yield (q/ha)*	*WUE kg/ha-cm*
T_1	67.568	84.207
T_2	63.710	82.093
T_3	58.325	77.292
T_4	42.810	60.551
T_5 (Control)	55.767	59.326

Conclusions

Micro-sprinkler irrigation system should be scheduled below 30 per cent available soil moisture depletion level. The frequency and depth of irrigation water could be estimated from pan evaporation and crop coefficient data. There is a saving of significant water quantity to the extent of 17.15 per cent in addition to increase in the yield of garlic to the extent of 21 per cent. This was possible because of excellent soil moisture plant atmosphere relationship.

References

Anonymous, 2002. *The Hindu Survey of Indian Agriculture*, pp. 89

Black, C.A., 1957. *Soil-plant Relationship*. John Wiley and Sons, New York, 332 p.

Doorenbos, D.W. and W.C. Pruilt, 1975. *Guidelines for Predicting Crop Water Requirements*. Irrigation and Drainage Paper 24, Food and Agriculture Organisation of United Nations, Rome, Italy.

Magar, S.S., N.N. Firake and D.S. Salunkhe, 1993. Effect of uniformity coefficient of water application on spacing between micro-sprinklers in vegetables. *Maharashtra Journal of Hort.*, 7(1).

Chapter 30

Studies on the Physico-chemical Characteristics of Lake Pichhola

Premlata Vikal* and Sandhya Tyagi
Department of Botany, Mohanlal Sukhadia University, Udaipur – 313 002, Rajasthan, India

ABSTRACT

The article deals with the study of physico-chemical characteristics of Lake Pichhola from May, 2004 to April, 2005. The water was analyzed for various physico-chemical parameters, *e.g.*, water temperature, pH, dissolved oxygen (DO), bio-chemical oxygen demand (BOD), etc. During the entire study period the values of chloride, nitrate, phosphate, hardness and BOD were found quite high thus indicating eutrophic condition of the lake.

Keywords: *Lake Pichhola, Physico-chemical characteristics, Sewage, Eutrophication, Water quality.*

Introduction

Water being a good solvent, contains dissolved substances in varying concentrations. Water is the most essential commodity for the living beings; therefore, satisfactory maintenance of the quality of water is of prime concern. Physical and chemical nature of water affects the distribution of aquatic organisms too. With the increasing urbanization and industrialization, the quality of water is degrading day by day.

* E-mail: vikal16@yahoo.co.in

Udaipur City is known for a network of lake system. Among all the lakes, Lake Pichhola is an important one and its water is supplied to the inhabitants of Udaipur to fulfill their basic needs. It is situated between the parallels 24°33′55″ and 24°34′59″ north latitudes and 73°39′32″ and 73°41′08″ east longitudes (Mohan, 1990). The lake is about 7 kilometers long and 2 kilometers broad. River Sisarama chiefly supplies water to this lake.

Due to paucity of rains for the last seven to eight years, the water level of lake has been decreased a lot. Its main source of pollution is disposal of untreated sewage and other organic wastes. There are various ghats where people used to take bath and do washings. After rains, River Sisarama also adds pollutants by carrying agricultural run-offs and other wastes thrown by nearby villagers along with it. Since time immemorial, disposal of sewage and other organic refuse have been a problem and now it has become more complicated with increasing population and industrialization. This has resulted in increasing demand of water leading to its depletion in quantity and deterioration in quality. The polluted water has created many problems of taste, odour and hygiene. Therefore, the water should be of good quality along with its aesthetic value in the scenic environment.

To know the water quality of any water body, studies on its physico-chemical characteristics are essential. The aim of the present study was to assess the water quality of Lake Pichhola, which would help in providing us early signals of environmental damage and potential hazards to human health.

Materials and Methods

The physico-chemical characteristics of Lake Pichhola were studied from May, 2004 to April, 2005. For study two sampling stations, *i.e.*, Bansi Ghat and Maanji ka Mandir were selected. From each station about 1.5 litres of water was collected in polythene bottles and carried to laboratory for analysis as per standard methods (Manivasakam, 1996; Trivedi and Goel, 1986 and APHA, 1998). For the estimation of dissolved oxygen (DO) water was collected in BOD bottles of 300 ml capacity without bubbling and DO was fixed by adding 2–2 ml of Winkler A and B at the site itself. Water temperature was measured on the spot.

Results and Discussion

The solubility of minerals and other materials depends on water temperature to a great extent. It also effects the distribution of aquatic flora and fauna. The minimum and maximum temperature were 15°C (February, 05) and 28°C (June, 04), at station I and 17°C (January, 05) and 29°C (September, 04) at station II, respectively. Variations in pH of water are controlled by water chemistry and photosynthesis (Hutchinson, 1957). Since, H^+ is the most powerful ion among all the nutrient cations, its concentration has a direct role in releasing the absorbed nutrient cations especially, calcium, magnesium and potassium. Welch (1952) has shown the importance of pH on aquatic life. pH of drinking water should be between 6.5 to 8.5 and it was recorded to be within the limit (Table 30.1). pH showed a high correlation coefficient with water temperature at both the stations, *i.e.*, 0.83 and 0.77, respectively (Tables 30.2 and 30.3).

Carbon dioxide dissolves in water more readily than oxygen and its dissolution depends on temperature, pressure and mineral content of the water. Here free CO_2 displayed positive correlation with water temperature, phosphate and total alkalinity and negative correlation with other parameters. In fresh waters, chlorides are present as salts of sodium, calcium and magnesium (Ganapati, 1956). It imparts salty taste to water (Sohani *et al.*, 2001). The concentration of chloride in the lake ranged from 46.98 mg/l to 152.95 mg/l at station I and from 49.99 mg/l to 162.95 mg/l at station II. Hence, its high concentration indicated sewage pollution in the lake.

Table 30.1: Physico-chemical Characteristics of Lake Pichhola During May, 2004 to April, 2005

Sl.No.	*Parameters*	*Station*	*May' 04*	*June' 04*	*July' 04*	*Aug' 04*	*Sep' 04*	*Oct' 04*	*Nov' 04*	*Dec' 04*	*Jan' 05*	*Feb' 05*	*Mar' 05*	*Apr' 05*
1.	Water	I	27.0	27.5	28.0	26.5	26.0	26.0	–	16.5	15.5	15.0	18.5	23.5
	Temperature (°C)	II	27.0	27.5	28.0	240	29.0	25.0	–	18.5	17.0	17.5	21.5	25.0
2.	pH	I	8.15	8.32	8.27	8.15	7.95	7.68	–	7.69	7.70	7.52	7.73	7.93
		II	8.37	8.18	8.10	8.0	7.99	–	7.69	7.63	7.67	7.33	7.53	7.7
3.	Free CO_2	I	63.0	37.0	67.6	94.0	120.0	62.0	–	30.0	58.0	70.0	46.0	36.0
		II	38.0	40.0	66.6	106.0	116.0	106.0	–	26.0	48.0	60.0	61.0	28.0
4.	Chloride	I	76.98	81.97	88.97	64.98	70.98	52.98	–	46.98	86.97	100.9	110.96	152.95
		II	64.98	82.97	92.97	66.98	66.98	58.98	–	49.99	83.97	96.97	114.96	162.95
5.	Nitrate	I	2.55	7.31	5.09	2.44	3.69	9.52	–	9.86	8.86	11.08	8.19	6.98
		II	0.89	3.32	2.88	2.22	3.39	4.74	–	5.98	6.42	11.07	8.66	6.53
6.	Phosphate	I	0.14	0.073	0.171	0.116	0.346	0.309	–	0.263	0.092	0.098	0.116	0.229
		II	0.138	0.086	0.202	0.266	0.383	0.242	–	0.272	0.116	0.122	0.11	0.245
7.	Total	I	240.0	209.0	270.0	204.0	170.0	146.0	–	165.0	164.0	158.0	183.0	248.0
	alkalinity	II	200.0	217.0	250.0	206.0	182.0	154.0	–	167.0	168.0	184.0	197.0	242.0
8	Hardness	I	148.0	196.0	2260	206.0	166.0	167.0	–	266.0	280.0	350.0	305.0	313.0
		II	138.0	209.0	220.0	224.0	164.0	180.0	–	270.0	324.0	388.0	310.0	314.0
9.	DO	I	7.55	571	6.44	4.08	2.04	5.20	–	11.84	10.42	7.75	2.45	1.02
		II	10.2	6.53	4.89	5.71	3.26	6.40	–	11.43	9.58	4.48	133	0.82
10.	BOD	I	–	–	–	32.61	59.18	52.0	–	40.0	61.22	60.0	6.12	28.58
		II	–	–	15.5	28.57	61.22	34.0	–	32.0	81.63	42.0	12.24	30.61

Station I: Bansi Ghat, Station II: Maanji ka Mandir.

–: Not taken.

All the parameters are in mg/l except for water temperature and pH.

Nitrates and phosphates are the main nutrients for the growth of blue-green algae and are responsible for eutrophication of the lakes. The concentration of nitrate above 40 mg/l in water can cause methamoglobinaemia in infants. Phosphate is positively correlated with water temperature and free CO_2 to a considerable extent. Total alkalinity depends on pH, carbon dioxide and chloride. Phenolphthalein alkalinity is inversely proportional to free CO_2, *i.e.*, free CO_2 would be absent in its presence. The highly productive waters generally have total alkalinity above 100 mg/l (Alikunhi, 1957). In our study, it varied between 146–270 mg/l and 154–250 mg/l at station I and II, respectively. According to Kannan (1991), the water is classified as very hard if the values exceed 180 mg/l, therefore, water of the lake can be considered very hard as *per se.* Hardness showed significant positive correlation coefficient with chloride (0.68 and 0.51) and nitrate (0.43 and 0.89) at both the stations of the lake.

Table 30.2: Correlation Coefficient Among Various Parameters at Station I of Lake Pichhola (May, 04–April, 05)

Parameters	*Water Temperature*	*pH*	*Free CO_2*	*Chloride*	*Nitrate*	*Phosphate*	*Total Alkalinity*	*Hardness*	*DO*	*BOD*
Water temperature	+1.00									
pH	+0.831	+1.00								
Free CO_2	+0.305	+0.119	+1.00							
Chloride	–0.137	–0.016	–0.277	+1.00						
Nitrate	–0.615	–0.641	–0.455	–0.094	+100					
Phosphate	+0.212	–0.191	+0.272	–0.271	–0.187	+1.00				
Total alkalinity	+0.569	+0.769	–0.134	+0.442	–0.496	–0.213	+1.00			
Hardness	–0.785	–0.633	–0.407	+0.679	+0.426	–0.324	–0.129	+1.00		
DO	–0.480	–0.255	–0.308	–0.502	+0.191	–0.364	–0.266	+0.088	+1.00	
BOD	–0.086	–0.239	+0.412	–0.380	+0.132	+0.014	–0.482	–0.240	+0.442	+1.00

Table 30.2: Correlation Coefficient Among Various Parameters at Station II of Lake Pichhola (May, 04–April, 05)

Parameters	*Water Temperature*	*pH*	*Free CO_2*	*Chloride*	*Nitrate*	*Phosphate*	*Total Alkalinity*	*Hardness*	*DO*	*BOD*
Water temperature	+1.00									
pH	+0.770	+1.00								
Free CO_2	+0.328	–0.019	+1.00							
Chloride	–0.015	+0.050	–0.367	+1.00						
Nitrate	–0.005	–0.837	–0.225	+0.439	+1.00					
Phosphate	+0.341	+0.01	+0.550	+0.008	–0.307	+1.00				
Total Alkalinity	+0.506	+0.377	–0.262	+0.259	–0.2466	–0.113	+1.00			
Hardness	–0.796	–0.792	–0.383	+0.510	+0.899	–0.386	–0.011	+1.00		
DO	–0.279	+0.352	–0.285	–0.735	–0.331	–0.137	–0.468	–0.256	+1.00	
BOD	–0.286	–0.0006	+0.077	–0.237	+0.030	+0.004	–0.560	+0.046	+0.379	+1.00

DO support the aquatic life. Water saturated with oxygen gives pleasant taste to water. Hutchinson (1957) said "a skilful limnologist can probably learn more about the nature of a lake from a series of oxygen determinations than from any other kind of data". DO is inversely proportional to temperature. DO showed positive correlation with pH at station II and negative at station I. BOD helps in analyzing organically polluted water. A water body with BODs of 8 mg/l is considered to be moderately polluted. The values of BOD determined were quite high, indicating highly organically polluted water, which has resulted from the disposal of domestic wastes and sewages into the lake.

At the end it can be concluded that the water of Lake Pichhola is not suitable for drinking and other domestic purposes. Restricting strictly the disposal of sewage and other organic wastes in the lake can prevent the organic pollution. Bathing and washings should be banned. The lime soda treatment and ion exchanger is effective for hardness removal (Nawlakhe *et al.*, 1994).

Acknowledgement

The author is highly thankful to UGC, New Delhi for providing research fellowship. The author is also grateful to Prof. N.C. Aery, Head, Department of Botany, M.L.S. University, Udaipur for providing laboratory facilities and to Prof. K. Kapoor for his valuable guidance and encouragement. Encouragement and help from my father Shri P.D. Vikal and brothers Ghanshyam Singh and Rokesh Kumar Singh are also thankfully acknowledged.

References

Alikunhi, K.H., 1957. Fish culture in India. *Fram Bull.* 20, ICAR, New Delhi, pp. 144.

Ganapati, S.V., 1956. The limnology of two minor irrigation reservoirs near Madras I: The Errakupam reservoir. *Hydrobiol.*, 8: 365–438.

APHA, AWWA and WPCF, 1998. *Standard Methods for the Examination of Water and Wastewater*, 20th edition. American Public Health Association, New York.

Hutchinson, G.E., 1957. *A Treatise on Limnology I: Geography, Physics and Chemistry*. John Wiley and Sons. Inc. New York, London, pp. 1015.

Kannan, K., 1991. *Fundamentals of Environmental Pollution*. S. Chand and Company Ltd., New Delhi, pp. 189.

Manivasakam, N., 1996. *Physico-chemical Examination of Water, Sewage and Industrial Effluents*. Pragati Prakashan, Meerut, pp. 256.

Mohan, R., 1990. Studies on the eutrophication problem of Lake Pichhola. *Ph.D. Thesis*, Department of Botany, M.L.S. University, Udaipur, Rajasthan, India.

Nawlakhe, W.G., Govendwar, S.L., Ratnaparkhi, D.Y., Mehta, C. and Gokhe, B.H., 1994. Water quality of Palamu district, Bihar. *Indian Water Works Association*, 26: 9–12.

Sohani, D., Pande, S. and Srivastava, V.S., 2001. Groundwater quality at tribal town: Nandurbar (Maharashtra). *Indian J. Environ. and Ecoplan.*, 5(2): 475–479.

Trivedi, R.K. and Goel, P.K., 1986. *Chemical and Biological Methods for Water Pollution*. Enviro Media, Karad, pp. 248.

Welch, P.S., 1952. *Limnology*. Mc-Graw Hill Book Co., New York, pp. 538.

Chapter 31

Efficiency of Lactic Acid Bacteria Present in Locally Available Fermented Milks as Probiotic Cultures

R. Mallikarjuna Reddy
Associate Professor and Head, Department of Dairy Science, Jawahar Bharathi Degree and P.G. College, Kavali –524 201, Andhra Pradesh, India

ABSTRACT

Fermented milk samples, dahi and buttermilk, collected from the local house-holds and hotels are screened for the presence of lactic acid bacteria. Both the lactic *streptococci* and *lactobacilli* are identified in the samples. Both the lactic *streptococci* and *lactobacilli* exhibited equally good survival rate against bile salt concentration (0.1 per cent). *Lactobacilli* expressed greater cell surface hydrophobicity and greater antagonistic activity against pathogenic organisms compared to *streptococci*. The study indicates that the locally prepared fermented milks liave large number of probiotic *streptococci* and *lactobacilli* which improves the health of the consumers.

Keywords: Probiotics, Fermented milks, Lactic acid bacteria, Bile tolerance, Antagonistic activity.

Introduction

Probiotics are the microorganisms which upon ingestion in certain numbers exert health benefits beyond inherent basic nutrition. Some bacteria present in certain foods alter the balance of human intestinal flora so that the putrefactive types of fermentation occurring in intestines are suppressed. The members of *Lactobacilli, Streptococcus* and *Bifidobacterium* are the predominant probiotic organisms consumed by human beings commonly in India through fermented milks such as dahi, cultured butter milk etc. Further, these *Streptococci* and *Lactobacilli* constitute a significant proportion of human and animal gastrointestinal indigenous microflora (Johansson *et al.*, 1993). These organisms are known to help in maintaining health of the host by controlling proliferation of pathogenic microbes (Collins and Aramaki, 1980; Fernandes *et al.*, 1987). Further, these are responsible for the inhibition of carcinogenesis, improvement of lactose utilization, stimulation of immune system and synthesis of vitamins (Goldin and Gorbach, 1984). Besides this, probiotic bacteria act on the cholesterol in the intestines which ultimately lead to its reduction in absorption in the intestines. These characters lead to the use of these bacteria as probiotics for maintaining health as well as for normalizing gastrointestinal microflora in diseased and stressed condition.

An efficient probiotic culture, whether it is *Lactobacilli* or *Streptococci*, should survive well in normal gastrointestinal pH and in the presence of bile and phenolic compounds so that they can establish themselves in the intestines. At the same time they should have good adhesive ability to human intestinal mucosa for competitive exclusion of pathogens (Aftabuddin and Sinha, 2003). Colonization of these organisms in the human GIT to multiply and elaborate antimicrobial substances is necessary. Therefore, the population of probiotic microorganisms should be minimum of 10^6 to 10^7 cfu/g of the food product to compete with the gut flora.

The ability of the bacterial cells to adhere to intestinal epithelial cells as well as to hydrocarbon was significantly correlated (Rosenberg *et al.*, 1981; Prakash *et al.*, 1997). Various researchers reported different rate of survival to bile concentration (Overdahl and Zottala, 1991; Hoier, 1992; Kurmann, 1988). Thus, probiotic cultures have to be selected on the basis of bile tolerance and adhesion to intestinal wall. Therefore, an attempt is made to study the survival rate of *Lactobacilli* and *streptococci*, isolated from the local fermented milks, in the presence of bile and their adhesive capacity to intestinal wall for their potential application as probiotics. Though there are many studies on the efficacy of probiotic organisms, the studies regarding the presence and type of probiotic organisms in commonly used fermented milks like dahi, lassi etc are scanty. Therefore, an attempt is made to analyze the locally available fermented milks for the presence of probiotic organisms and their suitability as probiotic foods.

Materials and Methods

Materials

The lactic acid bacteria isolated from dahi samples (20 samples), collected from the selected households and local hotels are used in this study. Based on the microscopic examination and fermentation tests the *Streptococci*, *Lactobacilli* and other organisms present in the samples are identified. These bacteria are maintained in MRS broth (De Man *et al.*, 1960). Analytical grade chemicals such as ox bile (Hi Media), hexadecane, urea etc., are used in the study. To estimate the antibacterial activity of lactic acid bacteria against pathogenic organisms, pathogenic bacteria belonging to *Listeria monocytogenes* (ATCC 19111) *Salmonella typhimurium* (A TCC 14028) are used. These are obtained from the American Type Culture Collection, Maryland, USA.

Bile Tolerance

Survival of *Streptococcus* and *Lactobacillus* isolates in the presence of bile is determined according to the method suggested by Walker and Gilliland (1993). All the isolates are grown in MRS broth containing 0.1 to 0.5 per cent of bile salt (ox bile). The MRS broth without bile salt is used as control.

The absorbance is measured at 600 nm after 48 hrs of incubation at 37°C. To measure the survival rate of *Streptococcus* and *Lactobacilli* in the presence of bile salts, the following formula is used.

$$\text{Survival rate (\%)} = \frac{\text{(Absorbance of isolates in MRS broth containing bile salt)}}{\text{(Absorbance of isolates in MRS broth without bile salt)}} \times 100$$

Adhesive Capacity of Lactic Acid Bacteria to Intestinal Wall

The ability of lactic acid bacteria to adhere to the human intestinal wall is measured through their cell surface hydrophobicity. The cell surface hydrophobicity of lactic acid bacterial isolates is measured in 18 hours old cultures as per the method described by Prakash *et al.* (1997). The cells are harvested by centrifugation @ 12,000 rpm for 10 minutes and washed twice and suspended in Phosphate Urea Magnesium Sulphate (PUM) buffer ($K_2HPO_4.2H_2O$ = 22.2 gm; KH_2PO_4 = 7.76 gm; Urea = 1.8 gm and $MgSO_4.7H_2O$ = 0.2 gm in one liter of distilled water; pH should be 71). Exactly 1.2 ml of bacterial suspension having absorbance of 1.4 at 650 nm (A) is taken in acid washed test tubes and to which 0.2 ml of hexadecane is added. These tubes are incubated at 30°C for 10 minutes and the two contents (phases) are mixed by vortexing for 2 minutes. The content is allowed to separate for 15 minutes at room temperature. The aqueous phase is carefully pipetted out and its absorbance (B) is measured by spectrophotometer. Per cent cell surface hydrophobicity is calculated according to the following fomlula.

$$\text{Cell Surface Hydrophobicity (\%)} = \frac{A - B}{A} \times 100$$

Antibacterial Activity

Antibacterial activity is determined by agar well diffusion method as per Toba *et al.* (1991). The lactic acid bacterial cultures are grown in 200 ml of MRS broth in a conical flask. The cultures are centrifuged @ 10,000 rpm in a high speed centrifuge (REMI). The culture supernatants are neutralized to pH 6.5 with the addition of sterile 1.0 N NaOH so as to neutralize the effect of organic acids produced during the growth of lactic acid bacteria (LAB) in MRS broth. The neutralized as well as un-neutralized culture supernatants are tested for antibacterial activity. The culture supernatants are treated with catalase (5 mg/ml culture) and incubated at 37°C for 3–4 hours in order to decompose the hydrogen peroxide present if any. Controls are run without the supernatant. The inhibition zone are measured in the agar plates. The difference in the inhibition zone diameter is a measure of the effect due to organic acid or hydrogen peroxide or both.

Results and Discussion

Bacterial Profile of Fermented Milks

The predominant fermented milk products available in the local market are dahi (curd) and butter milk. These fermented milks are commonly used by the people of this area as part of their diet. Therefore, about 20 samples of dahi and butter milk are collected from the market (hotels, dairies and private vendors) and from selected house holds. The samples are tested for titratable acidity, lactic acid bacterial counts and total plate counts. All the samples could be divided into three groups based on their acidity. The samples having acidity up to 0.70 per cent lactic acid have shown 4.23×10^5 lactic acid bacteria out of total bacterial count of 4.42×10^6 (Table 31.1). *Lactic streptococci* appears to be the predominant lactic acid bacteria in these samples having low acidity. The samples having acidity of 1.0 per cent to 1.4 per cent lactic acid have the lactic acid bacterial population of 3.1×10^6 to 4.4×10^7

per gram indicating that these samples are better probiotic foods because they have sufficient bacterial population required to suppress the growth of pathogens in the human intestines (Fernandes *et al.*, 1987). The fermented milk samples having lactic acid up to 1.0 per cent showed the mixture of *Streptococci* and *Lactobacilli* as their predominant organisms. The samples having acidity of 1.4 per cent or more contains large proportion of *Lactobacilli.* This variation could be attributed to the tolerance level of these organisms to acidity of the product. *Lactic streptococci* can tolerate acidity up to 1.0–1.2 per cent, whereas *Lactobacilli* can grow in the medium having acidity up to 2–2.4 per cent lactic acid. Therefore, *Lactobacilli* could be more potential probiotic organisms compared to *Lactic streptococci.* However, all the fermented milks tested showed higher proportion of lactic acid bacteria among total bacterial count. Pathogenic bacteria like *E. coli* is observed in the samples having lower level of acidity. The samples having low level of acidity belong to hotels and restaurants where as the samples having higher acidity belong to individual households. Thus, it could be inferred that the dahi or butter milk consumed by households are better probiotic foods compared to that of hotels and restaurants.

Table 31.1: Bacterial Species Identified in the Locally Available Fermented Milks

Sl.No.	Average Acidity of the Sample	Average TPC/g	Average LAB Count/g	Cultures Identified
1.	Up to 0.70 per cent lactic acid	4.42×10^6	4.23×10^5	*S. lactis*
				S. cremoris
				L. plantarum
				L. acidophilus
				E. coli
				S. faecalis
2.	0.94 per cent lactic acid	7.25×10^7	4.4×10^7	*S. lactis*
				S. cremoris
				L. brevis
				L. bulgaricus
				L. acidophilus
3.	1.4 per cent lactic acid and more	3.65×10^6	3.11×10^6	*L. bulgaricus*
				L. acidophilus
				L. casei
				S. thermophilus
				L. plantarum

TPC: Total Plate Count; LAB: Lactic Acid Bacteria

Tolerance to Bile Salts

A total of 9 lactic acid bacterial isolates are tested against their bile tolerance. All the. isolates used showed decreased survival rate with increase in the bile salt concentration (Table 31.2). Thus, the results indicate that the survival rate of LAB declines with the increase in bile salt concentration. However, *S. lactis, L. plantarum, L. acidophilus* and *L. bulgaricus* showed higher survival rate ranging from 85–98.2 per cent in the presence of 0.1 per cent bile salts. The other strains exhibited moderate survival even in lower concentration of bile salts. *S faecalis* showed very poor survival rate at 0.1 per cent bile concentration indicating that it is unfit to be a probiotic organism. Further, very few isolates have shown moderate survival rate of 20 to 33 per cent in the presence of 0.2 per cent bile concentration. Further increase in bile salt concentration still reduced the survival rate of all the lactic acid bacteria. Bile salt concentration in the intestines of the humans is 2 per cent at a maximum level, which is equivalent to 0.22 per cent bile salt (ox bile) used in this study (Davenport, 1977). Taking this intestinal concentration of bile salts into consideration, it could be noted that *S. cremoris, S. lactis, L. bulgrnicus, L. acidophilus, L. plantarum* and *L. brevis* showed better survival rate and thus these could be considered as useful probiotic organisms and the remaining are not potential probiotics.

Table 31.2: Survival Rate of Lactic Acid Bacterial Isolates in Different Bile Salt Concentrations

Sl.No.	*Isolate*	*Bile Salt Concentration (per cent)*				
		0.1	*0.2*	*0.3*	*0.4*	*0.5*
1.	*S. latis*	92.4	33.2	16.4	5.2	4.0
2.	*S. cremoris*	69.2	20.3	10.4	3.5	1.9
3.	*S. faecalis*	19.4	6.4	5.7	1.6	0.8
4.	*S. thermophilus*	36.4	15.9	5.2	1.4	0.7
5.	*L. casie*	23.0	5.6	4.6	1.4	0.7
6.	*L. brevis*	75.4	23.0	10.5	2.1	1.6
7.	*L. plantarum*	85.6	24.3	9.4	4.3	2.2
8.	*L. acidophilus*	98.2	22.9	12.2	2.4	2.1
9.	*L. bulgaricus*	90.4	20.5	8.1	2.1	1.2

Cell Surface Hydrophobicity

The ability of lactic acid bacteria to adhere to the intestinal wall varies from 10.8 to 94.1 per cent (Table 31.3). Among the nine isolates examined *L. acidophilus, L. bulgaricus* and *L. plantarum* have shown greater adherence to intestinal wall while *S. lactis, S. thermephilus* and *L. casei* exhibited moderate (42–51 per cent) adherence capacity. Aftabuddin and Sinha (2003) reported still higher values of cell surface hydrophobicity, while Bhowmick *et al.* (1985) reported 50 per cent and Prakash *et al.* (1997) reported 40 per cent cell surface hydrophobicity in the case of *L. acidophilus.* These variations in cell surface hydrophobicity are thought to be due to bacterial cell surface components, their type and amount of lipoteichoic acid-protein-complex (Christensen *et al.*, 1985; Sherman and Savage, 1986). The variation could also be due to the presence of cell surface hydrophobic protein moiety (Henriksson *et al.*, 1991). Greater surface hydrophobicity of cells results in greater attractive force and thereby higher level of adhesion to epithelial cells (Rijnaarts *et al.*, 1993; Prakash *et al.*, 1997).

Antagonistic Activity of Lactic Acid Bacteria Against Pathogens

The antagonistic activities of the lactic acid cultures isolated from the locally available curd samples are tested against two pathogenic bacteria namely, *Listeria monocytogens* and *Salmonella typhimurium,* The antibacterial activity of lactic acid cultures could be mainly due to the production of organic acids, hydrogen peroxide and bacteriocins produced by the cultures. Lactic acid is the predominant organic acid followed by acetic acid produced by these lactic acid cultures present in fermented milk products.

Table 31.3: Cell Surface Hydrophobicity of Lactic Acid Bacterial Isolates

Sl.No.	Isolate	Cell Surface Hydrophobicity (%)
1.	*S. latis*	51.5
2.	*S. cremoris*	35.4
3.	*S. faecalis*	12.4
4.	*S. thermophilus*	42.3
5.	*L. casie*	53.2
6.	*L. brevis*	34.1
7.	*L. plantarum*	84.2
8.	*L. acidophilus*	72.8
9.	*L. bulgaricus*	74.4

The culture supernatants of the isolates are tested after treating with alkali and catalase. Neutralization of supernatants with alkali decreased the antibacterial activity to some extent, which shows that the antagonistic activity is partially due to lactic acid production (Nair and Surendran, 2004). The zone of inhibition caused by the neutralized samples is less than 50 per cent of that caused by the original culture supernatants of the isolates indicates that about 50 per cent of the antagonistic activity caused by the organism is due to the production of organic acids (Table 31.4). The inhibitory action was not affected by catalase treatment indicating that it is not caused by the action of hydrogen peroxide.

Table 31.4: Inhibition of Pathogenic Bacterial Growth by Probiotic Cultures (zone in mm)

Sl.No.	Isolates	*Listeria monocytogenes*		*Salmonella typhimurium*	
		Un-neutralized	Neutralized	Un-neutralized	Neutralized
1.	*Streptoccus lactis*	14	12	8	6
2.	*Streptococcus cremoris*	8	10	5	4
3.	*Streptococcus faecalis*	2	NZD	2	NZD
4.	*Streptococcus thermephilus*	12	13	8	7
5.	*Lactobacillus bulgaricus*	18	14	10	8
6.	*Lactobacillus casei*	12	12	6	4
7.	*Lactobacillus brevis*	16	14	7	8
8.	*Lactobacillus plantarum*	24	20	14	12
9.	*Lactobacillus acidophilus*	18	14	10	8

NZD: No zone detected.

The antibacterial activity of the isolates against selected pathogens indicates all the isolates are able to inhibit the growth of *L. monocytogenes* to varying degrees (Table 31.4). The highest inhibitory action is observed in the case of *L. plantarum* followed by *L. bulgariccus* and *L. acidophilus.* While *Streptococcus faecalis* exhibited very low antibacterial activity, other *Streptococci* showed moderate antagonistic activity against *L. monocytogenes.* In the case of *S. typhimurium* also similar antagonistic activities are observed. The antibacterial activity exhibited by these lactic acid bacteria is probably due to a combination of factors such as lactic acid and bacteriocin production by the cultures (Gillikend and Speck, 1977). Bacteriocins or antibiotics such as nicin, plantaricin, bulgaricin, acidophillin etc are produced by *S. lactis, L. plantarum, L. bulgaricus* and *L. acidophilus* respectively. These antibiotics suppress the growth of pathogens in the human intestines.

The foregoing analysis indicates that the locally available fermented milk products have probiotic organisms which promotes the health of the consumers. Among the probiotic cultures identified *Lactobacilli* exhibited higher bile salt tolerance, greater adherence to intestinal wall and antagonistic effect against pathogenic organisms compared to *Lactic streptococci*. However, the results indicate that *Lactic streptococci* could also be used as potential probiotic cultures along with *Lactobacillus* cultures. Further, the study indicates that taking fermented milks such as dahi (curd), butter milk etc., would improve the health of the consumers.

Acknowledgements

The author is thankful to the authorities of Department of Biotechnology, Jawahar Bharati Degree and P.G College, Kavali for the opportunity given to utilize the laboratory for this study.

References

Aftabuddin and Sinha, 2003. Survival of *Lactobacilli* in the presence of bile salts and their cell surface hydrophobicity for use as probiotics. *Indian J. Dairy Sci.*, 56(6): 363–366.

Bhowmick, T., Johnson, M.C. and Ray, B., 1985. Isolation and partial characterization of surface proteins of *Lactobacillus acidophilus* strains. *Int. J. Food Microbiol.*, 12: 311–321.

Christensen, G.D., Simpson, W.A. and Beachey, E.H., 1985. Adhesion of bacteria to animal tissues: Complex mechanism. In: *Bacterial Adhesion: Mechanism and Physiological Significance*, (Eds.) Savage, D.C. and Fletcher, M. Plenum Press, New York, pp. 279–305.

Collins, E.B. and Aramaki, K., 1980. Production of hydrogen peroxide by *Lactobacillus acidophilus. J. Dairy Sci.*, 63: 353–357.

Davenport, H.W., 1977. *Physiology of Digestive Tract*, 4th edn. Year Book. Medical Publisher Inc., Chicago, IL, pp. 232–234.

De man, J.C., Rogosa, M. and Sharpe, M.E., 1960. A medium for cultivation of *Lactobacilli. J. Appl, Bacteriol.*, 23: 130– 135.

Fernandes, C.F., Sahani, K.M. and Amer, M.A. 1987. Therapeutic role of dietary *Lactobacilli* and *Lactobacilli* fermented dairy products. *FEMS Microbiol. Rev.*, 46: 343–356.

Gilliland, S.E. and Speck, M.L., 1977. Antagonistic action of *Lactobacillus acidophilus* toward intestinal and food borne pathogens in associative cultures. *J. Food. Prot.*, 40: 820–823.

Goldin, B.R. and Gorbach, S.L., 1984. Importance of bile tolerance of *Lactobacillus acidophilus* used as dietary adjunct. *J. Dairy Sci.*, 67: 3045–3051.

Henriksson, A., Szewzyk, R. and Conway, P.L 1991. Characteristics of the adhesive determinants of *Lactobacillus fermentum* 104. *Appl. Environ. Microbial* 57(2): 499.

Hoier, E 1992. Acid and bile tolerance of *Lactobacillus acidophilus* and *Bifldobacteria. Dairy Sci. Abstr* 54(12): 1036.

Johansson, M.L., Molin, G., Jeppsson, B., Nobaek, S., Ahrne, S. and Bengmark, S 1993. Administration of different *Lactobacillus* strains in fermented oatmeal soup: *In vivo* colonization of human intestinal mucosa and effect on the indigenous flora. *Appl. Environ. Microbiol* 57(1): 15–20.

Kumann, J.A 1988. In: *Starters for Fermented Milks, Section 5: Starters with Selected Intestinal Bacteria.* IDF Bulletin Fil., 227: 41–55.

Nair, Seema P. and P.K. Surendran, 2004. Antibacterial activity of lactic acid bacteria from frozen fish against *Listeria monocytogenes* and *Salmonella typhimurium. Asian J. of Microbiol. Biotech. Env. Sci.*, 6(3): 437–442.

Overdahal, B.J. and Zottola, E.A., 1991. Relationship between bile tolerance and the presence of ruthenium red staining layer on strains of *Lactobacillus acidophilus*. *J. Dairy Sci.*, 74(d): 1196–1200.

Prakash, R., Sinha, P.R., Sinha, R.N. and Singh, B., 1997. Adherence of *Lactobacilli* to epithelial cells and hexadecane for use as probiotics. *Indian J. Dairy Sci.*, 50(1): 60–65.

Rijnaarts, H.M.M., Norde, W., Bouwer, E.J., Lyklema, J. and Sehnder, A.J.B., 1993. Bacterial adhesion under static and dynamic conditions. *Appl. Environ. Microbiol.*, 59(10): 3255–3265.

Rossenberg, M., Perry, A., Bayer, E.A., Gutnick, D.L., Rosenberg, E. and Ofek, I., 1981. Adherence of *Acinetobacter calacoaceticus* RAG–1 to human epithelial cells and to lexadecane. *Infect. Immun.*, 33: 29–33.

Sherman, L.A. and Savage, D.C., 1986. Lipoteichoic acids in *Lactobacillus* strains that colonize the mouse gastric epithelium. *Appl. Environ. Microbiol.*, 52: 302–304.

Toba, T., Samant S.K. and Itoh, T., 1991. Assay system for detecting bacteriocin in microdilution. *Lett. Appl. Microbiology*, 70: 473–478.

Walker, D.K. and Gilliland, S.E., 1993. Relationships among bile tolerance, bile salt deconjugation and assimilation of cholesterol by *Lactobacillus acidophilus*. *J. Dairy Sci.*, 76(4): 950–961.

Chapter 32

Water Scarcity Zones of Jamshedpur, Jharkhand

Gouri Suresh[1], Neena Sharma[2], G. Rajalakshmi[2] and Rajni Kumari[2]
[1]Reader in Botany, Jamshedpur Women's College, Jamshedpur
[2]Guest Faculty, Environment and Water Management, Jamshedpur Women's College, Jamshedpur

ABSTRACT

The present article is based on a survey conducted by Jamshedpur Women's College Eco-Club, of which the author is the moderator. The city of Jamshedpur was divided into 9 segments and questionnaires were distributed to households in a random fashion in all the segments. Based on the information collected, a water scarcity map of Jamshedpur was prepared; this map will definitely be of use to the city planners. It was also found that awareness activities are essential in the city.

Keywords: Water source, Water quality, Awareness, Water scarcity map, Water conservation.

Introduction

Jamshedpur is a picturesque city situated in the state of Jharkhand, at the confluence of two rivers, Subarnarekha and Kharkhai, and surrounded by the rugged hills of the Chotanagpur plateau. 64 sq.kms in area, the city is located at 532 metres above sea level. The Mapping of Water Scarcity Zones of Jamshedpur was a project, sponsored by the Ministry of Environment and Forests, Government of India, undertaken by the Eco-Club of Jamshedpur Women's College. Eco-Clubs of different schools of Jamshedpur were also involved in this massive venture.

Materials and Methods

Jamshedpur was divided into 9 zones of approximately equal areas based on the main roads separating these zones. Copies of a questionnaire prepared were distributed among the members of the Eco-Clubs of our college and also the participating schools. The members were divided into 9 groups based on the localities they lived in and a leader was selected for each group. Thereafter, the groups fanned out into the 9 zones and collected answers for the questions from the local people. A very large quantity of data was obtained by this method. The questionnaire:

1. Name of the zone:
2. Information collected from:
3. Main source of water–Supply Water/Surface Water/Ground Water

   Any specification:
4. If ground water source water is available, at what depth:

   Maximum depth Minimum depth
5. Quality of water:
6. Is there any scarcity of water in your area? Yes/No

   If yes, throughout the year/for a few months only

   Specifications, if any:
7. Cleaning procedure practised:

   Self/Community/Company or Municipality Details:
8. Awareness regarding water problems among the local people:
9. Is there any water body in your zone? Lakes/Ponds/River/Any other
10. Any other water related problem faced by the local people of the zone:
11. Suggestions to improve the prevailing condition:

    (*a*) At individual level

    (*b*) At community level

    (*c*) At Government level
12. Suggestions by the Eco-Club members:

Results and Discussion

Based on the collected data, a detailed map of Jamshedpur with regard to its Water scarcity Zones was prepared. This map is sure to help the authorities in chalking out solutions to the problem of water scarcity in parts of Jamshedpur.

From the map of the city, it is clear that Tata Steel Works is situated almost in the centre of the city, towards the southern part and it occupies a considerable area. This area was excluded for the purpose of the study. The 9 zones studied are all outside the company area.

Tata, the biggest company of Jamshedpur, does maintenance of most parts of the city through JUSCO (Jamshedpur Utility Services Company). In these area, there is ample water supply. However, some other areas (about 20 per cent of the total) face water shortage in summer months from April to July. Yet other such areas (about 10 per cent of the total) face the problem throughout the year.

Some parts of the city where there is acute water shortage all the year round are located away from water bodies; however, quite contrary to expectations, some such areas are very near to either River Subarnarekha or River Kharkhai. To know the exact reason for this, hydro-geological studies will have to be conducted.

The survey has also thrown up some facts regarding the approach of the public in dealing with water supply and crisis. A majority of the people feel that supplying water is purely the responsibility of the government. It is true that water, like land and forests, is a state subject and it is the duty of the state governments to enforce legislations for surface and subsurface water. Yet people's participation is a must for the redressing of any local scarcity problem. The present survey showed that local people in the affected areas are either non-committal or ignorant about water conservation methods. From this, it is obvious that there is an urgent need of educating the people and creating awareness in them regarding different methods of water conservation.

In many suburban areas of Jamshedpur, where acute water scarcity was seen clusters of residential complexes have been built after draining wetlands. Inland wetlands help to replenish groundwater sources (Miller, Jr., 1996). Therefore draining them is suicidal to water conservation efforts and not many people know this.

It has also been observed that the public is using some water bodies as dumping grounds. These can be cleaned and maintained by the same public–what is needed is motivation. Success stories like that of how such a dump was transformed back to the original 9,000 square meter *jheel* of clear water in Kolkata (Dasgupta, 2003) can be an inspiration.

In areas outside the command of JUSCO, people use mainly hand pumps for all their water needs. But bore wells would yield water only if they are recharged over along period of time (Kumar, 2001). Therefore people must realise that as many types of water harvesting techniques as possible must be implemented. The 3 main methods are roof harvesting, run off harvesting and floodwater harvesting (Suresh, 2004).

Of the several avenues to raising water productivity, the key is pricing water at its market value (Brown, 2001). However, there is a contrary view that privatisation of water resources implies that water is an economic commodity and scarcity is a business opportunity (Jayararnan, 4003). Thus, making the local populace realise the essentiality of practising water conservation measures is the need of the hour. Motivating them to do so is the duty of all those who are aware of the grimness of the situation.

Acknowledgement

The authors are thankful to the Ministry of Environment and Forests, Government of India for sponsoring the project and to Gram Vikas Kendra, Jamshedpur Regional Resource Agency for forwarding the same to the ministry. Thanks are due to all the Eco-Club members who participated in the project. The authors are also grateful to the Principal, Jamshedpur Women's College for providing the facilities needed.

References

Brown, Lester R., 2001. *State of the World*. W.W. Norton and Company, New York, London, pp. 54–55.

Dasgupta, Rajashri, 2003. Kolkata: Clean waters again. *Survey of the Environment*, (Ed.) N. Ravi. The Hindu, pp. 165–170.

Jayaraman, Nityanand, 2003. Privatisation of water. *Survey of the Environment*, (Ed.) N. Ravi. The Hindu, pp. 83–86.

Kumar, Dinesh M., 2001. Drinking water: A scarce resource. In: *Survey of the Environment*, (Ed.) N. Ravi. The Hindu, pp. 93–101.

Miller, G. Tyler, Jr., 1996. *Living in the Environment: Principles, Connections and Solutions*. Wadsworth Publishing Company, New York, pp. 160–162.

Suresh, R., 2004. *Soil and Water Conservation Engineering*. Standard Publishers Distributors, Delhi, pp. 523–527.

Chapter 33

Bhitarkanika Mangrove Forest: A Sensitive Fragile Ecosystem

Lakshman Nayak and Durga Prasad Behera
P.G. Department of Marine Sciences, Berhampur University,
Bhanja Bihar – 760 007, Orissa

ABSTRACT

Out of 110 species of mangrove identified in the globe, 63 species are found in Bhitarkanika. The mangrove supports the marine life of 60 genera and 64 species of microfauna, 34 genera and 55 species of meiofauna, 19 species of amphibians, 42 species of reptiles including the rare and endangered sea turtle, more than 200 species of birds, 60 species of mammals and 6 species of marine mammals. 300 species of fishes and 229 species of crustacea are also inhabiting this sanctuary. The wild animals such as leopard, cat, fishing cat, jungle cat, hyena, wild boar, spotted dear, sambar, porcupine, pythons, dolphins, king cobra, saltwater crocodiles, water monitor lizard, terrapins, common mangoose, Indian field mouse, Indian mole rat are living in this sanctuary. The water temperature of the sanctuary varied from 25°C to 32°C. The air temperature varied from 28°C to 36°C. The pH value varied from 6.98 to 8.7. Salinity fluctuated from 9 ppm to 33.8 ppm. The dissolved oxygen varied from 2.325 ml/l to 7.250 ml/l. The carbonate alkalinity value varied from 25 ppm to 60 ppm and bicarbonate alkalinity varied from 65 ppm to 120 ppm.

Keywords: Bhitarkanika, Sensitive, Fragile, Ecosystem.

Introduction

Mangroves are the tidal forests of coastal wetlands, existing in the intertidal zones of sheltered shores, estuaries, tidal creeks, backwaters, lagoons, marshes and mud-flats of the tropical and subtropical regions of the world. The Bhitarkanika mangrove forest is the second largest mangrove in India. Mangroves of Bramhani and Baitarani delta of Kendrapara district have been declared as Bhitarkanika Wildlife Sanctuary in April 1975 covering an area of 672 sq.km. The core area of the sanctuary has been declared as Bhitarkanika National Park in September 1998 covering an area of 145 sq.km. Bordering the Bhitarkanika Sanctuary/National Park, an area of 1435 sq.km. has been declared as Gahirmatha Marine Wildlife Sanctuary in September 1997. Bhitarkanika is endowed with very complex sub-ecosystem of freshwater, marine and terrestrial, which are intricately mixed with each other. The maintenance of such an ecosystem depends on regular influence of freshwater from the adjoining land and tidal inflow from the sea. The horizontal and vertical zonation of the mangrove plant communities influenced by freshwater influx, tidal inflow, degree of inundation, seasonal rainfall and salinity gradient etc. have its influence on the wildlife and their occurrence. Mangrove swamps are one of the richest and most productive area and form the base of the food chain in the sea and coastal waters. The mangroves, wetlands and coral reefs, which are fragile ecosystems face threat due to various anthropogenic activities. Its over exploitation will certainly hamper the existence of many communities of plants and animals in this fragile ecosystem. The mangrove ecosystem play a very vital role in contributing to the food web in general and detritus food chain in particular supporting rich estuarine and adjacent marine fisheries (Chadha and Kar, 1999). The litter fall of mangrove trees amount to a large quantity of nutrients which are circulated both within the mangrove forest and in the surrounding marine ecosystem. The leaf material is utilised by bacteria and fungi and forms the base of an aquatic food web (Odum and Heald, 1975; Snedker, 1978). The mangroves play a very significant role in protecting the hinterland against cyclones and the ingress of seawater during tidal surge. Mangroves stabilise coastal landmass against sea erosion. They are repositories of immense biological diversity and are also the nursery and breeding ground of several marine life forms like species of shrimps, crabs, fishes and molluscs. Mangroves sustain the ecological security of the coastal zones. They also provide livelihood to thousands of fishermen and other villagers who resides in and around these areas. Mangrove forest acts as a buffer against any natural disasters. Depending on the recommendations of National Committees on Wetlands, Mangroves and Coral Reefs, 24 wetlands, 33 mangroves and four coral reefs in the country have been identified for conservation and management by the Ministry for Conservation and Management (Mandal and Rastogi, 2004).

Several works have been done on the distribution of mangroves in India (Sidhu, 1963; Nayak, 1993 and Pande, 2002). Some works have been done on the floral characteristics of Bhitarkanika by Banerjee and Das (1972), Mishra and Panigrahy (1987), Pattnaik and Choudhury (1989) and Banerjee and Rao (1990). Little work has been carried out on the hydrographical features of Mahanadi estuary by Upadhyay (1988) and Raju *et at.* (2000). No work has been done on the hydrological characteristics of Bhitarkanika mangrove forest. Therefore, the present work is an attempt to study the hydrological parameters of Bhitarkanika.

Materials and Methods

Bhitarkanika Sanctuary is one of the spectacular sanctuaries lies between 20°4′ and 20°8′ N latitude and 86°45′ and 87°5′E longitude covering an area of about 650 sq km of which about 150 sq km are under mangrove forest (Figure 52.1). The study was carried out during January 2004 to December 2004. The monthly water samples were collected from three different stations. The samples on hydrological parameters such as temperature, pH, salinity, dissolved oxygen and alkalinity were collected regularly. The air and water temperature were recorded using a thermometer (±0.1°C). The water samples were brought to the laboratory and analysed for pH, salinity, dissolved oxygen and alkalinity following the standard method (APHA, 1995). During the water sampling the flowering and fruit bearing periods of different species of mangrove trees were observed and recorded. Station 1 is chosen at Nalitapatia, which is located near to Dhamara river. Station 2 is Dangmal, located in between Nalitapatia and Bhitarkanika estuary. Station 3 is chosen at Bhitarkanika estuary, which is towards the sea, and the hydrological parameters are highly fluctuative.

Results and Discussion

The hydrological data collected from three stations of Bhitarkanika were represented in Tables 33.1 to 33.3. The temperature was highest being 36°C during May and lowest temperature was being 28.10°C during January at Station-1 *i.e.*, Nalitapatia. The surface water temperature was highest being 32°C in the month of May and the lowest temperature was observed to be 25.3°C in the month of December. In Station-2 (Dangamal), the highest water temperature was observed to be 32°C during April and the lowest water temperature was observed during the month of December being 25.10°C. The highest atmospheric temperature was observed being 36°C during April and lowest atmospheric temperature was observed to be 28.5°C during January. In Station-3 (Bhitarkanika estuary), the highest atmospheric temperature was observed during the month of May being 35°C and the lowest atmospheric temperature was observed during the month of January being 27°C. The highest and lowest water temperature in this station was observed being 32°C and 25°C in the month of May and December respectively. There has been gradual decrease in the temperature from Station-l to Station-3 towards the down stream along the river system. The finding are in general agreement with that of Upadhyay (1988) and Raju *et al.* (2000).

Table 33.1: Monthly Average Value of Hydrological Parameters of Bhitarkanika During January 2004 to December 2004 at Station-1

Month	*Temperature (°C)*		*pH*	*Salinity (ppm)*	*Dissolved Oxygen (ml/l)*	*Alkalinity in ppm*	
	Air	*Water*				*CO_3*	*HCO_3*
January	28.10	26.0	8.20	31.40	6.355	40	100
February	30.50	27.0	7.92	29.30	6.532	25	090
March	32.80	28.0	8.50	30.80	6.500	30	085
April	35.72	29.2	7.05	23.00	7.012	45	090
May	36.00	32.0	7.02	33.80	5.268	35	075
June	34.80	29.7	7.00	15.00	5.360	45	105
July	33.70	29.2	8.00	10.00	6.330	40	110
August	32.50	29.0	7.60	19.00	2.580	25	075
September	32.10	28.7	7.20	17.00	6.130	35	080
October	31.80	28.2	7.05	28.00	7.160	40	100
November	31.00	27.0	7.80	26.00	7.210	30	090
December	30.50	25.3	8.70	30.00	7.250	40	085

The highest pH value was observed in the month of December being 8.70 and lowest pH value was observed during the month of June being 7.00 at Station-1 (Nalitapatia). In Station-2 (Dangmal), the highest pH was observed during the month of November being 8.4 and the lowest pH value was observed during the month of July being 8.00. Similarly at Station-3 (Bhitarkanika estuary) the highest pH was observed during the month of December being 8.5 and the lowest pH value was recorded during the month of June being 6.98. The higher value of pH may be due to influence of seawater penetration. The lower value of pH was quite low during monsoon period, which may be due to the influence of fresh water influx and decomposition of organic matter carried out by the floodwater into the riverine system (Raju *et al.*, 2000 and Shriadah, 2000).

Table 33.2: Monthly Average Value of Hydrological Parameters of Bhitarkanika During January 2004 to December 2004 at Station-2

Month	*Temperature (°C)*		*pH*	*Salinity (ppm)*	*Dissolved Oxygen (ml/l)*	*Alkalinity in ppm*	
	Air	*Water*				CO_3	HCO_3
January	28.50	26.00	8.30	31.00	5.525	35	105
February	29.70	28.00	8.10	28.00	5.120	30	080
March	32.00	29.00	8.20	31.00	4.140	25	070
April	36.00	32.00	8.17	32.00	5.585	35	085
May	35.80	31.00	8.20	18.00	5.023	30	080
June	34.20	30.00	8.12	16.00	3.792	30	070
July	33.00	29.25	8.00	09.00	3.315	45	090
August	32.50	27.00	8.20	23.00	2.325	25	070
September	32.20	29.00	8.30	27.00	2.790	30	085
October	31.80	28.50	8.31	25.00	4.682	40	060
November	31.25	27.00	8.40	28.00	5.104	30	090
December	30.00	25.10	8.30	29.00	6.788	35	110

Table 33.3: Monthly Average Value of Hydrological Parameters of Bhitarkanika During January 2004 to December 2004 at Station-3

Month	*Temperature (°C)*		*pH*	*Salinity (ppm)*	*Dissolved Oxygen (ml/l)*	*Alkalinity in ppm*	
	Air	*Water*				CO_3	HCO_3
January	27.00	25.25	8.20	28.00	6.296	45	110
February	29.50	27.00	8.30	29.00	6.564	35	105
March	31.80	29.00	8.02	30.00	5.255	25	080
April	34.50	31.00	8.13	32.0	5.012	35	075
May	35.00	32.00	8.15	28.00	5.181	38	085
June	33.00	29.00	6.98	28.00	5.190	40	090
July	32.50	29.50	8.00	10.00	5.182	35	075
August	32.20	29.60	7.16	12.00	5.162	30	085
September	32.00	28.07	7.26	16.00	5.590	25	075
October	31.50	28.00	7.80	20.00	6.280	45	085
November	31.00	27.00	8.40	22.00	7.091	50	110
December	30.00	25.00	8.50	25.00	7.210	60	120

Seasonal and regional fluctuation in salinity was observed at all the Stations (Tables 33.3 to 33.3). The salinity values increased both in post-monsoon and pre-monsoon months. The highest value of salinity was observed in the month of May being 33.8 ppm and lowest in the month of July being 10 ppm at Station-1. In Station-2 the lowest salinity value was observed during the month of July being 9 ppm and highest in the month of April being 32 ppm. Similarly the higher and lower values of salinity were observed in the month of April and July respectively being 32 ppm and 10 ppm at Station-3. The salinity of the estuary did not go usually above 33.8 ppm or below 9 ppm although during the monsoon season there is freshwater influx into the estuary. The high salinity values may be due to high evaporation and low intensity of river run off into the estuarine system. In the present study the annual variation of salinity was 9 ppm to 33.8 ppm. Salinity values were maximum during peak summer month at all the stations. However, drastic reduction of salinity was noticed during the monsoon period, which is in conformity with the observation of Achuthankutty (1987) from the estuary of Goa.

The dissolved oxygen concentration showed a wide range of variation throughout the year (Tables 52.2 to 52.4). In the month of December the dissolved oxygen concentration was very high being 7.25 ml/lit and minimum being 2.58 ml/lit in the month of August at Station-1. The highest and lowest value was observed during December and August being 6.788 ml/lit and 2.325 ml/lit at Station-2. Likewise the maximum and minimum values were observed during December and August respectively at Station-3 being 7.21 ml/lit and 5.162 ml/lit. Comparatively high dissolved oxygen values were associated with high pH values along the estuarine system during post-monsoon period, which suggest the abundant growth of phytoplankton and related zooplankton. Low values of dissolved oxygen concentration correspond to monsoon period, which may be due to bio-degradation of organic matter in the water column (Raju *et al.*, 2000).

Alkalinity generally determines the magnitude of dial fluctuation of pH of water. Water with low alkalinity (less than 15 mg/l) results in wide fluctuation in pH value from 7.5 at dawn to 10 or even highest in the afternoon. Alkalinity value was generally higher at all the stations. The highest value was observed in the month of December and lowest value was observed in the month of April. The bicarbonate alkalinity varied from station to station. Bicarbonate alkalinity was more in the month of June and July, which may be due to high influx of fresh water and high pH in estuarine system of Bhitarkanika. This observation is in conformity with the observation of Nayak and Behera (2004) from Chilika lagoon.

References

APHA, 1995. *Standard Methods for the Examination of Water and Wastewater*. American Public Health Association. AWWA and WPCF, Washington D.C.

Achutankutty, C.T., 1987. Commercially important penaeid shrimp larvae in the estuaries of Goa. *Mahasagar Bull. of Nat. Inst. Oceanog.*, 20(4): 217–224.

Banerjee, L.K. and Das, G.C., 1972. New distributional records from Orissa coast. *Bulletin of Botanical Survey of India*, 14(1–4): 184–186.

Banerjee, L.K. and Rao, T.A., 1990. *Mangroves of Orissa Coast and their Ecology*. Bishen Singh Mahendra Pal Singh, Dehera Dun, p. 1–114.

Chadha, S. and Kar, C.S. 1999. *Bhitarkanika Myth and Reality*. Natraj Publisher, Dehra Dun, p. 55–170.

Misra, S.C. and Panigrahi, G., 1987. Studies on the mangrove flora of Orissa with particular reference to the Rhizophoraceae. R.Br. *Journal of Economic and Taxonomic Botany*, 11(1): 121–132.

Mandal, N. and Rastogi, M., 2004. Living wealth: Conservation is the key–the preservation of biodiversity. *Science Reporter*, June: 24–27.

Nayak, S.T., 1993. Remote sensing application in the management of wetland ecosystem with special emphasis on management of mangrove ecosystem and coastal protection, Visakhapatnam. India, pp. 21.

Nayak, L. and Behera, D.P., 2004. Seasonal variation of some physio-chemical parameters of Chilika lagoon (east coast of India) after opening of new mouth, near Sipakuda. *Indian Journal of Marine Sciences*, 33(2): 206–208.

Patnaik, S.N. and Choudhury, B.P., 1989. Present status and future development of mangrove vegetation in Bhitarkanika. In *INDO-US Workshop on Wetlands, Mangroves and Biosphere Reserve*, pp. 89–95. Government of India, Ministry of Environment and Forests, New Delhi.

Raju, V.A., Nayak, L. and Das, R.C., 2000. Seasonal abundance of post larvae of penaeid prawns in the Mahanadi estuary, Orissa. *Asian J. Zool. Science*, 9: 15–23.

Shriadah, A.M., 2000. Chemistry of the mangrove waters and sediments along the Arabian Gulf Shoreline of the United Arab Emirates. *Indian Journal of Marine Sciences*, 29: 224–229.

Sidhu, S.S., 1963. Studies on the mangroves of India. In: *Proceedings of Indian Academy of Sciences*, 33(8): 129–136.

Upadhyay, S., 1988. Physico-chemical characteristics of the Mahanadi estuarine ecosystem, East coast of India. *Indian Journal of Marine Sciences*, 17: 19–23.

Chapter 34

Characterization of Conducting Polymers by MALDI-TOF MS Method

R. Palanivel, K. Balamurugan and M.E. Raja saravanan
Annamalai University, Annamalainagar – 608 002, Tamil Nadu, India

ABSTRACT

Polymers, the synthetic materials have truly unique properties. The polymers are a very special kind of compounds, which do not behave like small molecules do, but the properties of polymers depend on molecular weight and molecular weight distribution (MMD). The molecular characterization of PANi base synthesized chemically by condensation method is attempted using MALDI-TOF MS technique. Further the study is extended on doped PANi with two different strengths (0.01 N, 1 N) of dopant $HClO_4$. The detailed descriptions of the method of TOF (Time of Flight) and the molecular weight determination of the PANi base and its doped form as conducting polymers have been studied quantitatively and strength of the dopant on the molecular characterization is also identified.

Keywords: Polyaniline, Doped PANi, MALDI-TOF MS, Molecular weight.

Introduction

In recent times the narrow perspective about the carbon based polymers as insulators have been changed to as a new class of polymer called intrinsically conductive polymers or electro active polymers. The potential uses of conductive polymers are quite significant today. Their characterization study in determining the molecular weight is vital part of the present study.

Materials and Methods

Aniline is polymerized to polyaniline (PANi) by chemical method with electrical conducting properties. Their conductivity was also enhanced upon doping with an oxidizing or reducing agent (Skotheim, 1986). The molecular weights of the new class of synthetic metals are in the range between thousand to hundreds of thousand with their structural units of monomer recurring structurally (Trivedi, 1994). Dopant induces bond alterations resulting in increased electrical conductivity (Mool *et al.*, 1994) between 10 to 100 S/cm.

The MALDI-TOF MS instrument from Broker Daltonics (Germany) used the ultraviolet laser (Nitrogen) at the wavelength of 337 nm. The Operating pulse rate is 50 Hz and ions accelerated at 20 kv. The samples were analyzed by operating in linear mode. Then ions were detected using micro channel plate (MCP). Dihydrobenzoic (DHB) acid was used as matrix and the samples were dissolved in Ethanol.

The molecular weight and molar-mass distribution of a polymer sample is a critical determinant of its material characteristics (Vreeland *et al.*, 2001). The determination of the molecular weight (MW) and molecular weight distribution (MWD) (Mool *et al.*, 1994) using MALDI-TOF-MS requires only pico moles of samples, comparing with Gel Permeation technique (GPC). Also a time of 10 minute is sufficient to sensitively give the detailed information of the sample. The single monomer resolution of the samples up to ~20kDa in molar mass is also possible in MALDI-TOF-MS (Itoda, 2003).

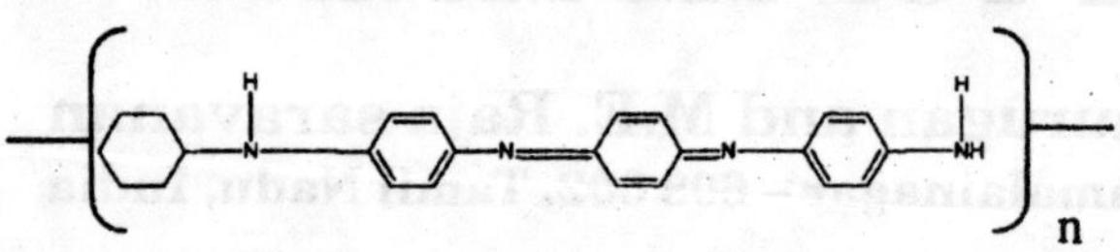

Figure 34.1: Molecular Structure of Polyaniline (Base)

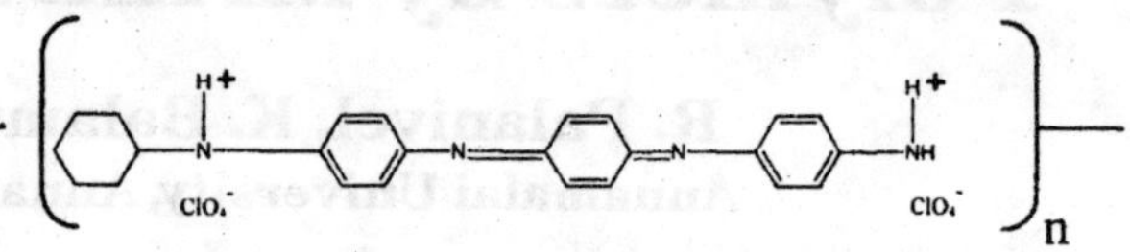

Figure 34.2: Molecular structure of Polyaniline⁺ $HclO_4$-(1N and 0.01N)

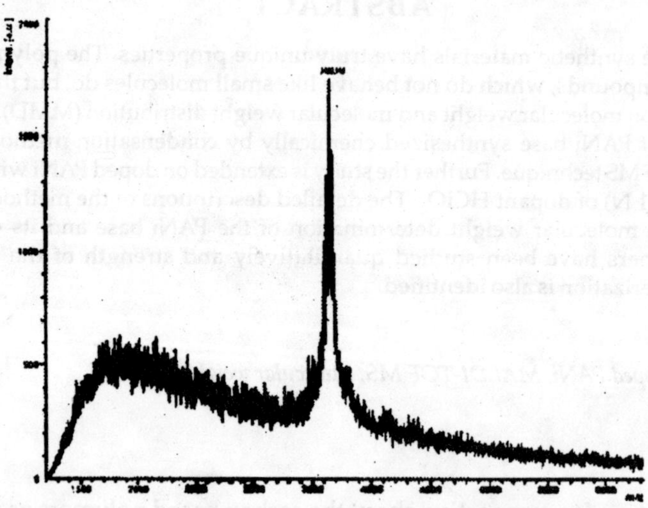

Figure 34.3: MALDI-TOF MS PANI BASE

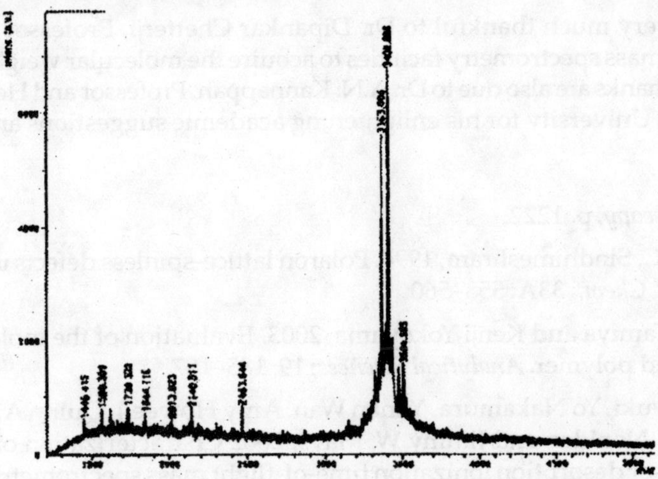

Figure 34.4: MALDI-TOF MS PANI⁺ $HClO_4$(1N)

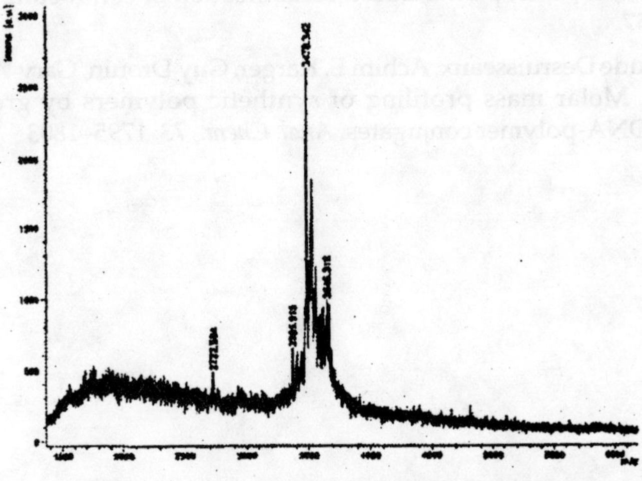

Figure 34.5: MALDI-TOF MS PANI⁺ $HClO_4$(0.01N)

Results and Discussion

Time of flight mass Spectrometry (TOF-MS), a separation-in-time technique, is one in which ions are spatially separated. The time needed by the ions pockets to travel the length of the drift region (the time of flight) can be related to the mass of the ions. The molecules of 1,000,000 amu and greater have been analyzed on this technique (Bush, 1997; Haining Ji *et al.*, 2002). It is found that the molecular weight of the base and its doped polymer as 3645.116 da 3406.886 da and 3479.342 da respectively. The decrease in molecular weight on doping may be attributed to the link breakages or the solubility of monomer in the acid doping.

Acknowledgements

The authors are very much thankful to Dr. Dipankar Chetterji, Professor of Bio Physics, IISC, Bangalore for offering mass spectrometry facilities to acquire the molecular weight data of the samples studied in this article. Thanks are also due to Dr. A.N. Kannappan, Professor and Head of the Department of Physics, Annamalai University for his enlightening academic suggestions and advices.

References

Bush, K., 1997. *Spectroscopy*, p. 1222.

Gupta, Mool C. and D.C. Sindhimeshram, 1994. Polaron lattice-spinless defects in conducting organic polymers. *Indian J. Chem.*, 33A: 558–560.

Itoda, Kazuya, Eiichi Tamiya and Kenji Yokoyama, 2003. Evaluation of the molecular recognition of peptide-conjugated polymer. *Analytical Sciences*, 19: 185–187.

Sato, Haining Ji Nobuyuki, Yo Nakamura, Yunan Wan, Amy Howeell, Quinn A. Thomas, Robson, F., Storey, William K. Nonidez and Jimmy W. Mays, 2002 Characterization of polyisobutytene by matrix-assisted laser desorption ionization time-of-flight mass spectrometry. *Macromolecules*, 35: 1196–1199.

Skotheim, Terje A., 1986. *Hand Book of Conducting Polymers*, Vol. 1, p. 362.

Trivedi, Dinesh Chandra, 1994. Dopant induced solubilization of conducting polyaniline. *Indian J. Chem.*, 33A: 552–557.

Vreeland, Wyatt N., Claude Desruisseaux, Achim E. Karger, Guy Drouin, Gary W. Slater, and Annelise E. Barronm, 2001. Molar mass profiling of synthetic polymers by gree-solution capillary electrophoresis of DNA-polymer conjugates. *Anal. Chem.*, 73: 1795–1803.

Chapter 35

Application of Clustering Algorithms in Robotic Manipulators

A.K. Dash
Assistant Professor, School of Mechanical Engineering, SASTRA (Deemed University), Thanjavur – 613 402, Tamil Nadu, India

ABSTRACT

Clustering analysis is a primary method for database mining. It is used as a separate tool to get insight into the distribution of a data set, mainly used for feature recognition, classification, and pattern identification in various fields. In this article, a novel application of this technique is discussed in the filed of robotic manipulators. Robotic manipulators encounter singular points (where their behaviour become uncertain) at many places within its workspace. Clustering algorithm is used, as shown in this article, in grouping these singular points into several clusters. Subsequently, clusters are modeled as obstacles and a path is determined avoiding these obstacles. Thus, the robotic manipulator traverses from home position to goal position avoiding these singular points.

Keywords: *Clustering algorithm, Robotic manipulators, Singularity avoidance.*

Introduction

Clustering algorithms are attractive for tasks of class identification. There are two main types of clustering algorithms: partitioning and hierarchical algorithms. The partitioning algorithms construct a partition of a database into a set of clusters and they use an input parameter for these algorithms are based on some domain knowledge that is unfortunately not available for many applications. On the other hand, hierarchical algorithms create a hierarchical decomposition of the database. The hierarchical decomposition is represented by a *dendrogram,* a tree that splits the database into smaller subsets until each subset consists of only one object.

Density Based Notion of Clusters

Looking at a sample set of points in Figure 35.1, clusters of points and noisy points not belonging to any of the clusters are easily detected. The main reason of recognition of these clusters is that within each cluster there is atypical density of points which is considerably higher than outside of the cluster. The density within the areas of noise is lower than the density in any of the clusters. Based on this notion of density, many density based clustering algorithms are developed. DBSCCAN (Density Based Spatial Clustering of Applications with Noise) is one such algorithm.

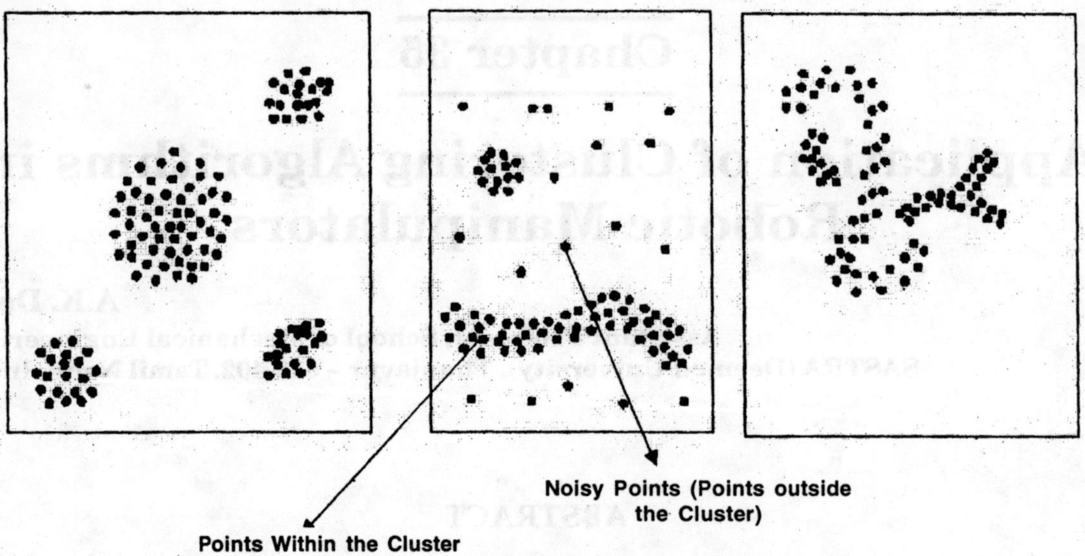

Figure 35.1: Density Based Clustering
(Points Within the Cluster are Closely Spaced than the Noisy Points)

Many definitions are proposed before DBSCAN is introduced (Sander *et al.*, 1998) in the scientific community. DBSCAN is designed to discover the clusters and the noise in a spatial database. Ideally, the parameters *Eps* and *MinPts* of each cluster and at least one point from each cluster are required. Then points those are density-reachable from the given point can be retrieved using the correct parameters. There is no easy way to get this information in advance for all clusters of the database. However, there is a simple yet effective heuristic to determine the parameters *Eps* and *MinPts* of the thinnest, *i.e.*, the least dense, cluster in the database. Therefore, DBSCAN uses global values for *Eps* and *MinPts, i.e.*, the same values for all clusters.

Robotic Manipulators

Robotic manipulators are broadly classified as (*i*) serial manipulators and (*ii*) parallel manipulators. Figure 35.2(*a*) is a 7-degrees of freedom serial manipulator and that in Figure 35.2(*b*) is 6 degrees of freedom parallel manipulator. These manipulators are suitably applied in different fields

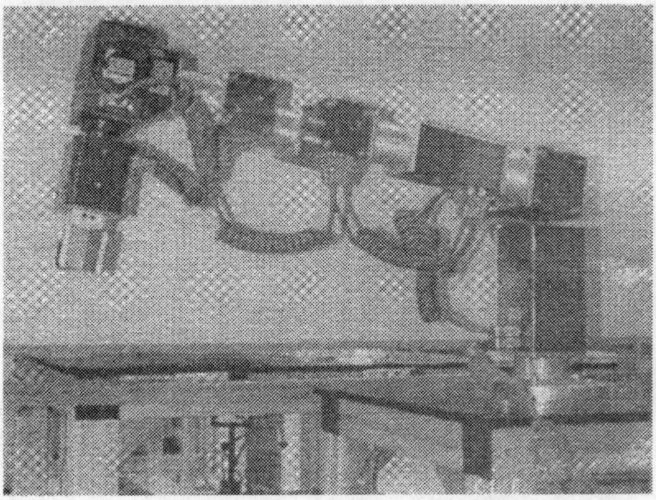

(*a*) (*b*)

Figure 35.2: (*a*) 7-DOF Serial Manipulator (NTU, Singapore); (*b*) 6-DOF Parallel Manipulator (Hexa robot, Braunschweig, Germany)

of applications. However, during an application, when the manipulator moves from one point to another following a particular trajectory, sometimes it encounters points at which its mobility ceases or its mobility increases (becomes uncontrollable). In robotic notation, the ceasing of mobility is known as reduction in degrees of freedom and increase in mobility is known as addition of degrees of freedom. These points are mathematically called as singular points because at these points, the mathematics of the robots becomes underfined.

In case of serial manipulators, these singular points are less and occur at the boundary of the workspace of the manipulator; however in case of parallel manipulators, these singular points are more and they occur within the workspace of the manipulator. Thus, one of the most widely studied areas in the research of parallel manipulators is the avoidance of singular points in the workspace of parallel manipulators. In this article, it is shown how clustering algorithms can be used in avoidance of singularities from the workspace.

Materials and Methodology

In this article, a three legged, six degrees of freedom, parallel manipulator (Yang *et al.* 2001, A. K. Dash *et al.*, 2004) is studied. The prototype and line diagram of this manipulator is shown in Figure 35.3.

In this manipulator, as shown in Figure 35.3(*a*), there are two platforms-a mobile platform being supported by three legs upon base platform. Each leg has four joints (shown in Figure 35.3(*b*) for leg-1), out of which two are 1-DOF actuators, one is 1-DOF passive rotary joint and the last is a 3-DOF spherical joint.

The methodology adopted to determine a path avoiding the singularity points is as below:

Step 1: Determination of singularity points by determining the Jacobian matrix of this manipulator (Angeles, 1997. and Tsai Lung Wen, 1999).

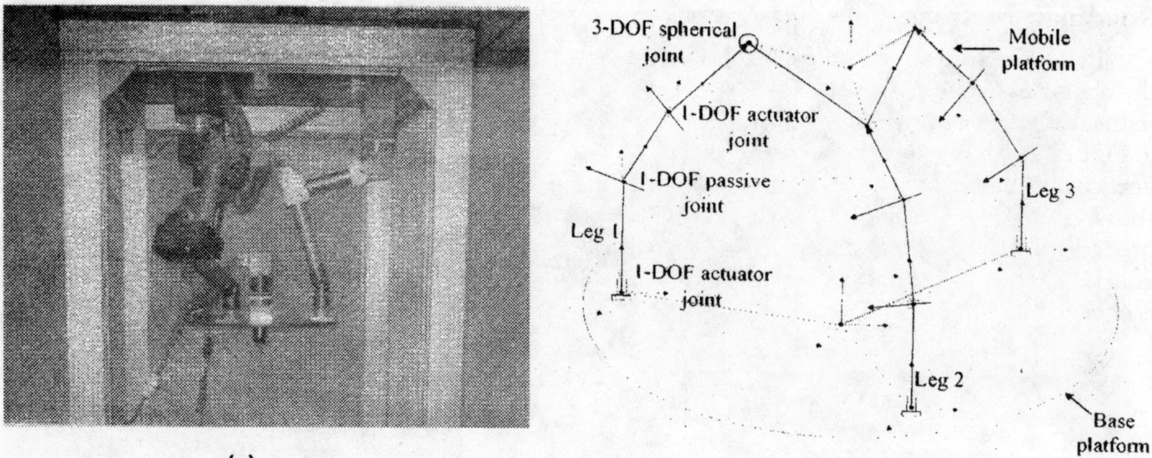

(*a*) (*b*)

Figure 35.3: (*a*) Prototype of Three-legged, 6-DOF Parallel Manipulator; (*b*) Line Diagram of the Same (upside down)

Step 2: Application of DBSCAN clustering algorithm to the above determined data set of singularity points to classify them into clusters and noisy points.

Step 3: Circumscribing each of these clusters by a polygon.

Step 4: Modeling of each of the clusters thus obtained into obstacles.

Step 5: Application of a path planning algorithm to move the manipulator from initial pose to final pose avoiding these obstacles.

Step 6: If the path thus determined avoids all the obstacles and the noisy points, then the algorithm terminates; in case it encounters any noisy point on the path, then a local method of avoiding the noisy points based on Grassmann line geometry is to be implemented.

Step 2 gives a number of clusters of singular points determined in step 1. In step 3, each of the dataset of clusters is sorted by 2D sorting method (in case singularity points are determined in a 2-D space). This sorting gives the points lying on the vertices of the cluster under analysis. Joining these vertices gives a polygon (or polyhedron in case of 3D space) circumscribing the cluster. Thus as many clusters are determined, that many number of polygons are determined. In step 4, each of these polygons is modeled as an obstacle. This method is based on determining the geometrical center of a polygon, connecting the vertices of the polygon with the center, and then increasing the radial distance of the vertices. Joining the new vertices produces an enlarged polygon which acts as an obstacle to be avoided. The idea is that in the next step an obstacle avoidance algorithm can be used to determine a path free from the singularity points within these polygons. Thus all the polygons are modeled as obstacles. Next, in step 5, a Road map method (Latombe, 1991) of obsfacle avoidance algorithm is implemented to determine a path that avoids all these obstacles and, thus, the clusters of singular points.

Roadmap Method

In Roadmap method, a term *free space* is defined as the space in the workspace of a robot excluding the obstacles. In this free space, a connectivity diagram is determined which is the set of all possible paths from the beginning point of a robot (shown as q_{init} in Figure 35.4) to the goal point (shown as q_{goal} in Figure 35.4). These paths are determined by connecting the beginning and goal points and all the vertices of the obstacles with other such points which it can see from its position. For example, as shown in Figure 35.4, q_{init} can only be connected to points 1, 2 and 5; not to any other points. This procedure is done for all other points. From this, a set of possible paths joining the beginning and the goal points are determined (as can be observed in Figure 35.4). Out of all these paths, the optimal path is chosen as the desired path (in Figure 35.4, it is found to be q_{init}-2-9-q_{goal}).

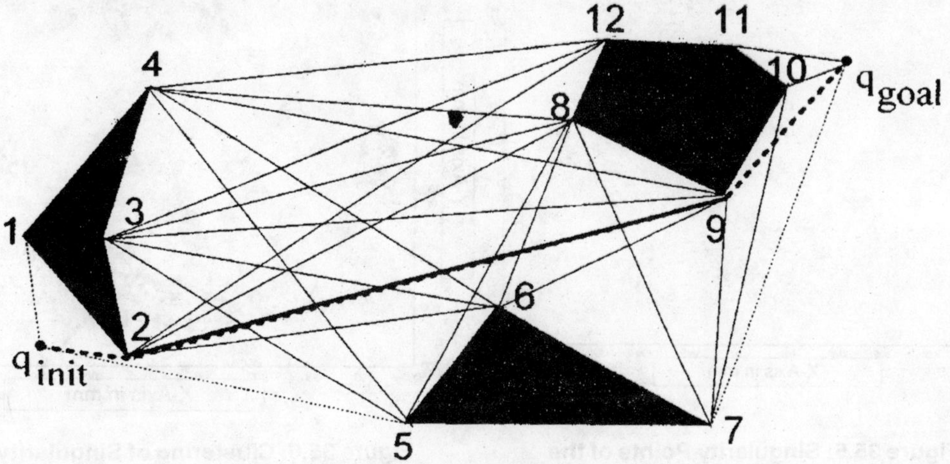

Figure 35.4: Road Map Method of Finding Optimal Path

The path determined in Step 5 avoids all the obstacles and, in step 7, it is checked if it also avoids the noisy points in the workspace. If it avoids, then the algorithm terminates successfully; otherwise a *local method* based on Grassmann line geometry is used.

Local Method

As discussed in the introduction, at a singular point within the workspace of a parallel manipulator, the mobile platform gains a degree-of-freedom and the required joint torque becomes, mathematically, infinite at that point. Grassmann line geometry (Merlet, 1989) can determine the direction of the degree-of-freedom gained and the directions of the other degrees-of-freedom, along which the parallel manipulator can have stable motion. In this local method, as the mobile platform approaches the noisy point (which is a singular point), a point is determined at a *safe distance* (where the joint torque is well within the limits, Gosselin and Angeles, 1991) away from the singular point along a line the direction of which is the constrained degree-of-freedom. In stead of going through the singular point, the platform traces this new point which is not a singular point. Again it joins back the optimal path after coming a safe distance away from the singular point. Thus the manipulator avoids all the singularities within the workspace.

Results and Discussion

As outlined above, the proposed methodology can be applied in the 30 space of a parallel manipulator. For simplicity purpose, the algorithm is tested for singularity avoidance in one plane of the manipulator. When the mobile platform is at given height from the base, the singularity points are determined by using Jacobian matrix and they are as shown in Figure 35.5. DBSCAN algorithm is applied on this dataset. The results are shown in Figure 35.6. The figure shows a total of 4 clusters and some noisy points. A 2-D sorting algorithm is applied to each cluster of this dataset of singularity points, by which all the vertices of a cluster are determined. Joining the vertices produces a polygon; for four clusters of singular points, four polygons are determined as shown in Figure 35.7. Each of these polygons is modeled as an obstacle. Thus in Figure 35.8, there are 4 obstacles and some noisy points which do not belong to any cluster. Now, the objective is to determine an optimal path avoiding

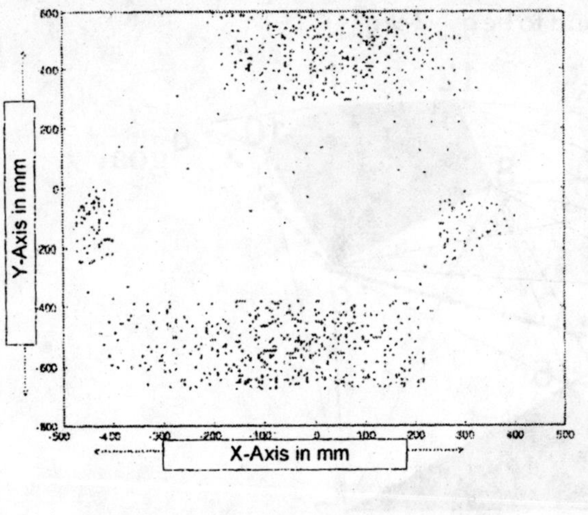

Figure 35.5: Singularity Points of the Manipulator Shown in Figure 35.3

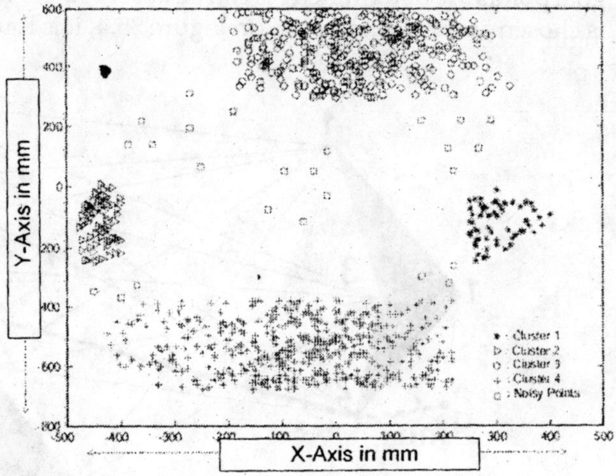

Figure 35.6: Clustering of Singularity Points

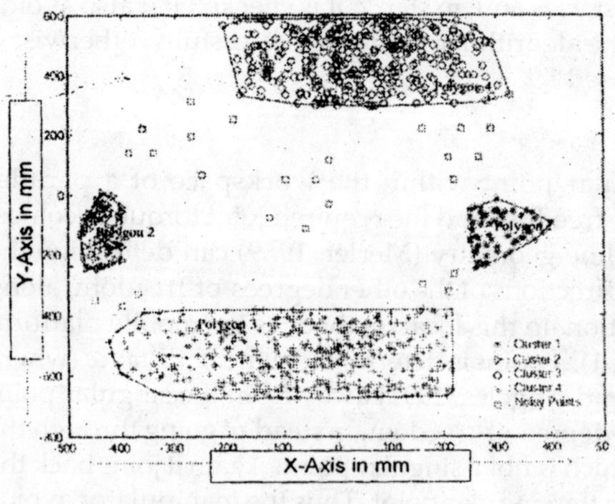

Figure 35.7: Obstacles Made Around the Clusters of Singular Points

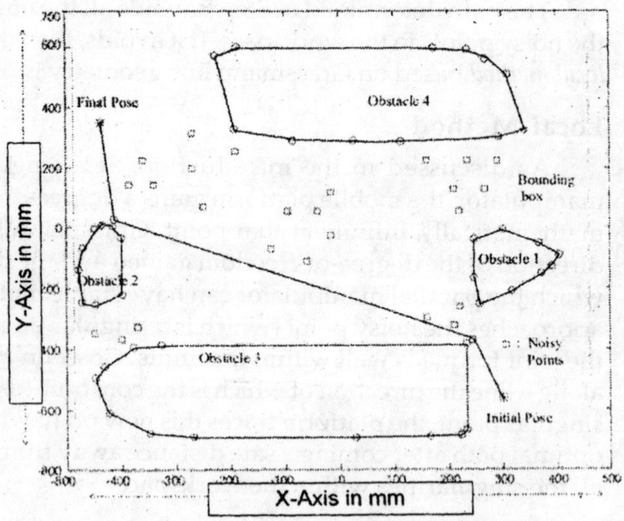

Figure 35.8: Optimal Path Avoiding the Singular Points

these obstacles and the noisy points. To avoid the obstacles, the Roadmap method of path finding is adopted. Adopting this method, it is shown in Figure 35.7 that the optimal path is free from any singularity point of the workspace.

This article presents a novel application of clustering techniques (DBSCAN in this case) in determining a singularity-avoidance path inside the workspace of parallel robotic manipulators. The algorithm first determines a path connecting the beginning and end poses. This path avoids the bulk of the singularity points where the bulk of the singularity points are modeled as polyhedron obstacles. To determine this path, a numerical program using the Jacobian matrices is used to determine the singularity points inside the workspace. These singularity points are grouped into several clusters of singularity points and some isolated singularity points. Then clusters of singularity points are modeled as obstacles. Subsequently, a path-planning algorithm is used to determine an optimal path avoiding these obstacles. If any singularity point lies on or near the path, this path is restructured by a local method. In this local method, Grassmann's line geometry is used to restructure the path until it rejoins the first path. In a case study of a parallel manipulator, this algorithm is verified.

Acknowledgement

The author is thankful to Nanyang Technological University, Singapore and in particular to Prof. (Dr.) I-Ming Chen of School of mechanical and aerospace engineering of the same university to support this research work carried during his PhD study during 1999–2002.

References

Angeles, J., 1997. *Fundamental of Robotic Material Systems: Theory, Methods and Algorithms.* Springer Verlag, USA.

Das, Anjan Kumar, I. Ming Chen, Song Huat Yeo, and Guilin Yang, 2004. Instantaneous kinematics and singularly analysis of three legged parallel manipulators. *Robotica,* 22: 189–203.

Gosselin, C. & J. Angeles, 1991. Singularity analysis of closed-loop kinematic chains. *IEEE Transactions on Robotics and Automation,* 6(3): 281–291.

Latombe, Jean-Claude, 1991. *Robot Motion Planning.* Kluwer Academic Publisher, USA.

Merlet, J.P., 1989. Singular configurations of parallel manipulators and grassmann geometry. *International J. Robotics and Res.,* 8(5): 45–56.

Sander, J., Ester, M., Kriegel, H. and Xu, X., 1998. Density-based clustering in spatial databases: the algorithm (GDBSCAN) and its applications. *Data Mining and Knowledge Discovery,* 2(2): 169–194.

Wen, Tsai Lung, 1999. *Robot Analysis: The Mechanics of Serial and Parallel Manipulators.* John Wiley and Sons, NY.

Yang, G., I.M. Chen, W. Lin and J. Angeles, 2001. Singularity analysis of three-legged parallel robots based on passive-joint velocities, In: *Proceedings of IEEE Conference on Robotics and Automation, S. Korea,* p. 2407–2412.

Chapter 36

Effects of Dietary Supplementation of *Spirulina maxima* on the Feeding Energetics, Growth and Carcass Composition of Common Carp *Cyprinus carpio* (*Linnaeus*) frys.

T.A. Sethuramalingam, K. Alangara Ahila and J. Mary Benny
Centre for Aquafeed and Nutrition (CAFeN),
Department of Advanced Zoology and Biotechnology,
St. Xavier's College (Autonomous), Palayamkottai – 627 002

ABSTRACT

Three types of feeding experiments were conducted to assess the feeding energetics and growth performance of *Cyprinus carpio* frys (420 ± 0.67 mg) fed with three types of formulated diets *i.e.*, carbohydrate rich (type–I) protein rich (type–II) and fat rich (type–III) diets. All the three types of feed were incorporated separately with unicellular algae (*Spirulina maxima*) 1, 25, 50, and 100 per cent replacing ground nut oil cake / soybean meal of the control diet and fed to fishes for 45 days. A separate control diet was maintained for each type of feed without probiont incorporation. There were significant differences observed in feeding rate (FR), absorption rate (AR), and growth rate (SGR) and food conversion ratio. Fish fed with 1 and 25 per cent of algae in protein and fat rich diets showed low food conversion ratio (FCR) as compared to control diet. The carbohydrate rich diet fed fish showed higher FR, absorption efficiency (AE) and SGR with low FCR at 50 per cent *Spirulina* incorporated diets. Protein intake (PI), protein efficiency ratio (PER) and protein utilization (ANPU) of the fish were greatly influenced by 1 and 25 per cent of *Spirulina* contained protein and fat rich diets. The bioavailability of nutrients from the feed as well as digestive capability of the fish were found higher in that concentrations of algal feed. In the present study it was evident that *S. maxima* incorporated in specific concentrations were found more effective to specific nutrients present in diets. So addition of *S. maxima* at specific concentrations in to the food paved way to produce cost effective diets for carps.

Keywords: Spirulina supplementation, Carbohydrate, Protein and Fat rich diets.

Introduction

In the context of increasing food production on a sustainable basis, freshwater aquaculture has come out to be a major option not only for producing quality protein, but also to enhance higher fish production. Aquaculture is a feed based industry with over 60 per cent of operational cost coming only from feed source. In intensive farming or in polyculture system use of nutritionally balanced feed is an obvious approach to achieve optimum growth and production of fishes (Mukhopadhyay, 1998). An understanding of the relationship between the rates of feeding and conversion of the food is of utmost importance in the system as it helps to avoid wastage of a costly inputs of feed, both in the form of unutilized feed and non assimilated fraction of the ingested diet.

Carbohydrate is an inexpensive and immediate source of energy in fish diets and serves as a precursor for the various intermediary metabolic functions. Though it improves the pellet quality of feed, carbohydrate utilization with additive supplementation and bioavailability of nutrients from feed remains unclear in many fish species (NRC, 1993; Sethuramalingam and Haniffa, 2002). On the other hand the cost of feed is largely influenced by the level and source of protein (dietary fish meal) which is the most expensive component in fish diet. However, because of rapid expansion of aquaculture industry and increasing cost of good quality fish meal, it becomes inevitable to find alternative protein sources to reduce feed cost. Most of the informations available for protein utilization in fishes were limited only to the nutrients as a whole (Mohanty *et al.*, 1990; Jena *et al.*, 1998). But specific nutrient bioavailability to fish by addition of additives in the form of probiotics or other micronutrients is fully needed. In recent times feed stuffs that supply lipids primarily receive less attention than ones that supply protein because the latter are more expensive. Yet lipids provide essential fatty acids, phospholipids and fat soluble vitamins, as well as concentrated energy which can spare protein for growth (NRC, 1993). In addition, lipids are often supplemented in commercial fish feeds to increase their buoyancy, prevent nutrient leaching and increase palatability (Lovelll, 1989). Though more reports are available on nutrient basis, works on the lipid sources with probiotic influence are still to be studied.

The use of probiotics such as unicellular algae, beneficial bacteria or yeast in fish feed is now in the early stage of development. Research on the evaluation of potential strains of probiotic organisms for fish species is very scanty and improvement on this process is urgently needed.

On this view, the present study was focused (*a*) to assess the feeding energetics and growth performance of *Cyprinus carpio* frys fed with carbohydrate rich, protein rich and fat rich diets containing different concentrations of enriched *Spirulina maxima*, and (*b*) to find out the changes in the tissue chemistry of fish by analyzing the carcass composition.

Materials and Methods

Frys of *Cyprinus Carpio* were bought from Fish Farmer's Development Agency, (FFDA), Manimuthar and transported to the Center for Aquafeed and Nutrition (CAFeN) laboratory St. Xavier's College,

Palayamkottai and acclimatized for 10 days in rectangular cement tanks (400 l Capacity). During this period fish were fed with groundnut oilcake and rice bran (1 : 1) Sixty fish (420 + 0.67 mg) were recruited from the stock and divided into 3 groups and each group in turn divided into 4 sets, each set having 5 fish and were introduced into rectangular plastic troughs (20 l). All the troughs were interconnected with PVC pipes and a water flow-through system was established using aerated well water with a flow rate of 4 l/hr. Three types of test diets namely carbohydrate rich (Type–I) protein rich (Type–II) and fat rich (Type–III) were prepared using ingredients as given in the Table 36.1 and to each test diet unicellular algae *Spirulina maxima* was added at 1, 25, 50 and 100 per cent replacing GOC/SBM of control diet. For each type of test diets a control feed was prepared and fed to fish for 41 days. Throughout the experimental period fish were fed with 5 per cent of initial body weight of feed twice a day at 9 AM and 4 PM. The unfed and faeces were removed from each setup daily by manual siphoning, dried and stored. After the experimental period fish from each treatment were removed, weighed, dried and stored for biochemical analyses (AOAC, 1990). The feeding energetics was calculated by using IBP formulae (Petrusewicz and Macfadyen, 1970) and growth performance was assessed by following the method of Jauncey (1982b). The data were subjected to one way ANOVA (Zar, 1984) and the difference in the means from each treatment were analyzed by Duncan's multiple range fests (Snedacor and Cochran, 1967).

Results and Discussion

Among the three types of feeds tested for *C. carpio* frys the feeding rate (FR) and absorption efficiency (AE) were higher in fat rich diet (36.89 mg/g fish/day) and (71.56 per cent) followed by protein rich diet 17.25mg/g fish/day and 84.97 per cent at 1 per cent *spirulina* incorporation levels. At the same concentration the carbohydrate rich diet fed fish showed a lower feeding rate and absorption efficiency (14.12 mg/g fish/day and 70.98 per cent). The feeding rate of fish fed carbohydrate rich diet increased from 14.12 to 16.64 (mg/g fish/day) but the protein and fat rich diet fed fish showed a decreased feeding rate from 17.25 to 10.08 (mg/g fish/day) and 36.89 to 14.91 (mg/g fish/day) respectively. The AE also declined with the increase *spirulina* concentration in the three types of diets. The gross and net conversion efficiencies of *C. carpio* frys fed with carbohydrate rich diets were higher at 1 and 25 per cent *Spirulina* concentration (14.19, 15.69 per cent and 16.87 and 17.49 per cent) and decreased thereafter. The same trend was seen in protein (14.92 to 8.21 and 16.27 to 9.67 per cent) and fat rich diet fed fishes (9.49 to 6.69 per cent and 10.36 to 7.38 per cent) respectively. Of the three types of diets fed to *C. carpio* fingerlings, higher FR and AE were observed at low concentration (1 and 25 per cent) of algae incorporated protein rich feed, followed by fat rich diets. The GCE and NCE were more or less same in carbohydrate and protein rich diets than fat rich diet fed fishes (Table 36.2).

The growth of *C. carpio* frys was highly influenced by fat (1 per cent) and carbohydrate (50 per cent) rich diets followed by protein rich diet (1 per cent). This was evidenced by higher specific growth rate of fish (1.55, 1.38 and 1.25) respectively. The lowest (best) food conversion ratio was also attained by the fish (1.15, 1.09 and 1.24) in the respective diets. This result indicated that 1 per cent of *Spirulina* incorporated fat rich diet influenced the fish to convert the food to flesh more efficiently than higher incorporation levels. The same with 1 per cent as well as 50 per cent spirulina incorporated protein and carbohydrate rich diets respectively.

It has been widely accepted that omnivores and herbivores could utilize higher dietary levels of carbohydrate, than carnivorous fishes, since carnivores appears to utilize carbohydrate poorly at both digestive and metabolic levels (Shimeno *et al.*, 1981; Furuichi and Yone, 1982) (Lovell, 1989). As *C. carpio* is an omnivore, the carbohydrate rich diets at 75 per cent, spirulina incorporation level produced

Table 36.1: Percentage and Proximate Composition of Formulated Diets Incorporated with *Spirulina maxima*

Ingredients	*Type–I*					*Type–II*					*Type–III*				
	C_1	*1*	*2*	*3*	*4*	C_2	*1*	*2*	*3*	*4*	C_3	*1*	*2*	*3*	*4*
Fish Meal (FM)	10	–	–	–	–	22.7	–	–	–	–	24	30	24	21	16
Prawnhead Waste (PW)	–	–	–	–	–	11.9	35	31	28	23	–	12	14	16	18
Groundnut Oil Cake (GOC)	12	7.25	6	4	–	17	21	29.	31	39	20	13	15	15	15
Rice Bran (RB)	16	23	22	20	18	–	18	16.3	19	14	–	–	–	–	–
Broken Rice (BR)	12	19	15	15	14	–	–	–	–	–	15	12	14	15	18
Wheat Meal (WM)	11	11	13	14	15	10	10	7.7	6	8	–	–	–	–	–
Corn Flour (CF)	16	24	25	27	26	14.73	–	–	–	–	–	–	–	–	–
Soy Bean Meal (SBM)	16	10	12	11	14	13.67	7.25	6.0	4	–	14	7.25	6	4	–
Cod Liver Oil (CLO)	–	–	–	–	–	3	3	3	3	3	12	12	12	12	12
Tapioca Flour (TF)	5	5	5	5	5	5	5	5	5	5	9	9	9	9	9
Vit. And Min. Mix*	2	–	–	–	–	2	–	–	–	–	2	–	–	–	–
Egg yolk	–	–	–	–	–	–	–	–	–	–	4	4	4	4	4
Spirulina maxima	–	0.75	2	4	8	–	0.75	2	4	8	–	0.75	2	4	8
Proximate composition															
Crude protein (%)	32.11	29.34	27.49	24.12	21.32	44.62	45.08	45.19	44.83	44.92	42.30	41.80	41.54	41.34	41.07
Crude lipid (%)	4.73	4.82	4.76	4.66	4.43	7.79	7.53	7.68	7.91	7.96	8.16	11.32	12.89	13.68	16.28
Nitrogen Free Extract (NFE)	32.29	34.27	39.29	42.36	44.28	24.67	22.71	22.96	23.09	23.46	19.42	17.23	16.64	14.28	11.82
Crude Fibre (%)	6.71	6.70	6.54	6.39	6.18	9.79	11.31	12.49	13.74	14.21	8.38	9.51	9.24	8.91	8.64
Ash (%)	7.82	6.61	6.48	5.84	5.66	7.82	9.23	9.12	8.84	8.13	6.21	7.16	7.22	7.49	7.27
Caloric Content (K cal./100 g)	398.6	394.21	386.41	390.68	393.43	412.59	410.84	409.32	408.83	406.89	416.32	412.73	414.83	410.39	416.83

* Supradiin tablets–Nicholas Pharma, Pithampur, M.P. India.

better growth. The effect of *Spirulina* on the feeding efficiency and growth was evident from this study, Even at 1 and 25 per cent incorporation level the protein and fat rich diet fed *C. carpio* fingerlings produced a reasonable growth. The results of Jena *et al.* (1998) in *C. carpio* frys revealed a marked difference in the growth, SGR and FCR between control and experimental groups showing the essential requirement of animal protein in the diet to supplement with essential amino acids and fatty acids. In the present study the protein and fat rich diet supplemented with 1 and 25 per cent spirulina influenced the fish to get essential amino acids and fatty acids from the feed which was well evidenced by higher SGR and lower FCR.

Table 36.2: Overall Energy Budget of *Cyprinus carpio* Fingerlings Fed with Formulated Diets

Feed Types Feed No.	*Food Consumed (mg)*	*Food Absorbed (mg)*	*Feeding Rate (FR) (mg/g fish/day)*	*(AE) (%)*	*Gross Conversion Efficiency (GCE) (%)*	*Net Conversion Efficiency (NCE) (%)*
Carbohydrate Rich Diets (Type–I)						
Control–1	1950	1436	15.76[a]	63.64[a]	13.16[bc]	14.30[bc]
1–1%	1761	1074	14.12[a]	70.98[ab]	14.19[bc]	16.87[c]
2–25%	1849	1109	14.72[a]	63.98[a]	15.69[c]	17.49[c]
3–50%	1945	1205	15.95[a]	61.96[a]	11.06[b]	13.86[b]
4–100%	2084	1289	16.64[ab]	61.85[a]	10.39[ab]	12.33[b]
Protein Rich Diets (Type–II)						
Control–2	2304.89	1798.63	14.39[a]	78.04[c]	12.55[b]	14.44[bc]
5–1%	2307.81	2214.06	17.25[ab]	84.97[c]	14.92[c]	16.27[c]
6–25%	1596.88	1421.88	12.07[a]	77.04[b]	12.24[b]	13.17[b]
7–50%	1636.25	1483.13	12.07[a]	70.65[b]	8.53[a]	11.84[ab]
8–100%	2259.68	2150.32	10.08[a]	65.22[a]	8.21[a]	9.67[a]
Fat Rich Diets (Type–III)						
Control–3	3896.64	3742.88	28.89[c]	66.05[a]	8.09[a]	9.59[a]
9–1%	2752.31	2518.99	36.89[c]	71.56[ab]	9.49[ab]	10.36[ab]
10–25%	1930.116	1697.47	19.27[b]	67.9[ab]	7.33[a]	8.27[a]
11–50%	1569.72	1371.32	15.03[a]	62.36[a]	8.31[a]	9.53[a]
12–100%	1600.25	1313.6	14.91[a]	57.63[a]	6.69[a]	7.38[a]

The mean values in the same column having different superscripts differ significantly ($p < 0.05$).

Corollary results are also obtained in *Rhabdosargus sarba* fed with 15 per cent *Spirulina* meal produced higher feeding efficiency, protein utilization (El-sayed, 1994). Out of 5 protein rich ingredients used for South African abalon; *Spirulina* based diet produced higher growth than other algal diets (Britz, 1996). Raw *Spirulina* at lower concentration appears to be more effective for larval tilapia because higher concentration hinders the process of digestion because the plant cell wall may act as a barrier to the digestive enzymes (Lu *et al.*, 2002). Supplementation of micro algae was also found to improve the growth and survival of larval turbot (Reitan *et al.*, 1997). In Japanese flunder (*Paralichthys Olivaceus*) fed with *Chlorella* enhanced growth and protein utilization (Thomas and Sung, 2001) (Koo *et al.*, 2001). In the present study *Spirulina* at lower concentrations improved the intestinal flora of the fish by breaking down the food components there by extracting more nutrient from the food. *Spirulina* has less intestinal compaction but improves the growth very well, stimulates the enzyme production that transport fat within the body. This may be one of the reasons for higher growth of C.*carpio* fed with 25 per cent spirulina incorporated fat rich diets. This fish has utilized the fat to improve growth instead of descipating it Cannon, 2002).

Table 36.3: Overall Growth Performance and Protein Budget of *C. carpio* Fingerlings Fed with Formulated Diets

Control	*Initial Weight (IW) (mg)*	*Final Weight (FW) (mg)*	*Growth Rate (GR)*	*Specific Growth Rate (SGR)*	*Food Conversion Ratio (FCR)*	*Protein Efficiency Ratio (PER)*	*Apparent Net Protein Utilization (ANPU) (%)*
Carbohydrate rich diets (Type–I)							
Control	353.53±2.56	415.19±2.72	6.11±0.29	0.46[a]	2.94[a]	0.87[a]	19.6[a]
1–1%	356.25±1.96	43..23±2.71	7.25±0.01	0.54[a]	2.22[b]	1.01[a]	22.47[a]
2–25%	358.75±1.29	463.75±2.66	10.26±0.12	0.73[b]	1.63[c]	1.39[b]	28.17[ab]
3–50%	348.33±2.08	563.5±2.77	21.62±0.24	1.38[c]	1.09[d]	1.79[c]	36.44[b]
4–100%	337.52±2.16	470.13±1.96	13.81±0.13	0.95[b]	1.45[c]	1.35[b]	26.41[ab]
Protein rich diets (Type–II)							
Control–2	457.3±2.61	797.63±3.84	21.24±20.62	1.09[a]	1.08[a]	1.74[b]	36.01[a]
5–1%	461.74±3.16	701.62±3.13	18.37±0.66	1.25[b]	1.24[ab]	1.58[c]	29.96[ab]
6–25%	452.35±2.87	647.64±4.62	15.11±0.51	1.03[a]	1.59[b]	1.46[a]	26.67[b]
7–50%	474.49±3.08	665.06±3.19	14.25±0.83	0.95[a]	1.79[c]	1.38[a]	24.48[b]
8–100%	462.81±2.88	648.41±3.55	14.01±0.65	0.94[a]	1.83[c]	1.35[a]	22.83[b]
Fat rich diets (Type–III)							
Control–3	350.66±3.85	491.28±2.86	14.26±1.26	0.96[a]	1.81[b]	1.36[b]	33.42[a]
9–1%	361.72±2.16	622.93±2.09	21.64±1.92	1.55[b]	1.15[a]	1.69[c]	35.69[a]
10–25%	351.14±1.83	491.43±3.44	13.98±1.28	0.96[a]	1.82[b]	1.35[b]	34.09[a]
11–50%	365.79±2.81	482.15±2.48	12.47±0.88	0.89[a]	2.19[c]	0.86[a]	24.84[b]
12–100%	354.16±1.87	474.47±2.71	11.89±1.01	0.86[a]	2.67[d]	0.77[a]	21.62[b]

The mean values in the same column having different superscripts differ significantly ($p < 0.05$).

Among the three types of feed tested, the PER was higher in protein rich diets than carbohydrate or fat rich diets, whereas the ANPU was high in 1 per cent and 25 per cent fat rich diets followed by 50 per cent carbohydrate rich diet. Though the PER was higher in protein rich diet, the result of ANPU indicated that higher flesh production was resulted by 1 per cent and 25 per cent *Spirulina* incorporated fat rich diet, the reason being the fish could mobilize the fat more effectively than other diets. The PER and ANPU of fish fed with carbohydrate rich diets increased linearly upto 75 per cent incorporation level and decreased there after but, the protein and fat rich diet fed fish showed slight decrease as the incorporation level or *Spirulina* increased in the after 25 per cent respective diets.

Table 36.4: Carcass Composition of *Cyprinus carpio* Fingerlings Fed with Formulated Diets

Feed Types Feed No.	*Carbohydrate (%)*	*Protein (%)*	*Lipid (%)*
Carbohydrate rich diets (Type–I)			
Initial Fish	15.43	47.37	14.16
Control–1	16.71	55.51	15.46
1–1%	17.67	54.71	15.08
2–25%	17.73	54.28	15.23
3–50%	17.84	54.21	15.29
4–100%	17.96	54.18	15.36
Protein rich diets (Type–II)			
Control–2	16.84	55.76	16.74
1–1%	16.89	56.84	16.53
2–25%	16.92	56.37	16.61
3–50%	17.06	56.22	16.74
4–100%	17.11	56.13	16.79
Fat rich diets (Type–III)			
Control–3	15.26	55.56	16.98
1–1%	15.52	55.71	17.67
2–25%	15.67	55.36	17.84
3–50%	15.73	54.82	17.91
4–100%	15.78	54.13	17.97

Higher body carbohydrate and lower fat was found in carbohydrate rich diet fed fishes. Higher protein and moderate body fat was evidenced in protein rich feed fed fishes. A higher body fat and moderate body protein were quantified in fat rich diet fed fishes. Lower body lipid and higher body protein might be due to the effect of *Spirulina* concentration on the respective diets, which mobilized the fat to become protenacious than fatty fish. Supplementation of spirulina mobilized the fat and led to positive effects of liver function. Higher carcass of fat in fat diet fed fishes might be due to inability of either spirulina (or) other dietary components to regulate the liver function thereby reduce (or) degrade the muscles (Nakagawa *et al.*, 1984; Nematipour *et al.*, 1990; Xu *et al.*, 1993; and Koo *et al.*, 2001). The increase in the visceral fat content indicated the preferential utilization of lipid in the diet, thereby the change in the tissue chemistry explicitly revealed the incorporation of spirulina influenced through diet composition (Helland and Helland, 1998).

From this study it was evident that the individual effect of nutrients in the diet was enhanced by supplementation of *Spirulina* in specific concentrations. It could be presumed that not only the specific nutrient but also specific concentrations of probiont like *Spirulina* could be used to formulate cost effective diet for carps.

References

Annon., 2002. 'Spirulina' super food for ornamental fish. Technical information shrimp larval and enrichment feed. *Artemia International*, LL C: 1–4.

AOAC, 1990. *Official Methods of Analysis Association of Official Analytical Chemists*, Washington D.C., pp. 69–90.

Britz, J., 1996. The suitability of selected protein sources for inclusion in formulated diets for the South African abalone, *Haliotis midae. Aquaculture*, 140: 63–73.

El Sayed, A.M., 1994. Evaluation of soybean meal, *Spirulina* meal and chicken offal meal as protein sources for silver sea bream (*Rhabdosargus sarba*) fingerlings. *Aquaculture*, 127: 169–176.

Furuichi, M. and Yone, Y., 1982. Availability of carbohydrate in nutrition of carp and red sea bream. *Bull. Jpn. Soc. Sci. Fish*, 48: 945–948.

Helland, S.J. and Grisdale Helland, B., 1998. The influence of replacing fish meal diet with fish oil on growth, feed utilization and body composition of Atlantic salmon (*Salmon salar*) during the smoltification period. *Aquaculture*, 162: 1–10.

Jauncey, K., 1982b. Carp (*Cyprinus carpio*) Nutrition: A review. In: *Recent Advance in Aquaculture*, (Eds.) Ceds. J.F. Muir and R.S. Roberts. Croom Helm Ltd., London, pp. 215–263.

Jena, J.K., Mukhopadhyay, P.K. and Aravindakshan, P.K., 1998. Dietary incorporation of meat meal as a substitute for fish meal in carp fry rearing. *Indian J Fish*, 45(1): 43–49.

Koo, J., Sungchul Bai, C., Kangwoong Kim and Sekwon Kim, 2001. Optimum dietary level of *Chlorella* powder as a feed additive for growth performance of juvenile olive flounder *Paralichthys olivaceus* pukyong. National University 599–1, Daeyeon Dong, Nam, Korea, pp. 608–737.

Lovell, R.J., 1989. *Nutrition and Feeding of Fish.* AVI Van Nostrand Rein Hold Co., New York.

Lu, J., Yoshizaki, G., Sakai, K. and Takeuchi, J., 2002. Acceptability of raw *Spirulina* to larval tilapia the *Oreochromis niloticus. Fish. Sci.*, 68: 51–58.

Mohanty, S.N., Swamy, D.N. and Tripathi, S.D., 1990. Growth, nutritional indices and carcass composition of the Indian carp fry *Catla catla, Labeo rohita* and *Cirrhinus mrigala* fed four dietary protein levels. *Aquaculture Hung*, 6: 217–221.

Mukhopadhyay, P.K., 1998. Recent advances in carp nutrition and diet development studies in India: An overview. In: *Current and Emerging Trends in Aquaculture*, (Ed.) P.C. Thomas. Daya Publishing House, New Delhi, India, pp. 272–277.

Nakagawa, H., Kasahara, S., Tsujimura, A. and Akira, K., 1984. Changes of body composition during starvation in chlorella extract fed agu. *Bulletin of Japanese Society of Scientific Fisheries*, 50L: 665–671.

NAS-NRC, 1993. *Nutrient Requirements of Fish.* National Academy of Sciences, Washington, D.C., p. 128.

Nematipour, G.R., Nakagawa, H. and Ohya, S., 1990. Effect of *chlorella* extract supplement to diet on *in vitro* lipolysis in agu. *Nipon Suisan Gakkaishi*, 56: 777–782.

Petrusewicz, K. and Macfadyen, A., 1970. *Productivity of Terrestrial Animals.* IBP Handbook No. 13. Blackwell Scientific Publications, Oxford.

Reitan, Rainuzzo, Gunvoroie, and Yngvarolsen, 1997. A review of the nutritional effects of algae in marine fish larvae. *Aquaculture*, 155: 207–221.

Sethuramalingam, T.A. and Haniffa, M.A., 2002. Optimum carbohydrate to lipid ratio in diets of *Labeo rohita* fry. *Environment and Ecology*, 19(1): 152–157.

Shimeno, S., Takeda, M., Takayama, S., Fukut, A., Saski, H. and Kajiyama, M., 1981. Adaptation of hepatopancreatic enzymes to dietary carbohydrate in carp. *Bull. Jpn. Soc. Sci. Fish.*, 47: 71–77.

Snedacor, G.W. and Cochran, W.G., 1967. *Statistical Methods*. Oxford and IBH publishing Company, New Delhi, pp. 135.

Tomas Cebrera and Sung Bum Hur, 2001. The nutritional value of live foods on the larval growth and survival of Japanese flounder, *Paralichthys olivaceus*. *Journal of Applied Aquaculture*, 11.

Xu, B.T., Yamasaki, S. and Hirata, H., 1993. Supplementary *Ulva* sp. Var. meal level in diet of Japanese flounder *Paralichthys olivaceus*. *Suisan Zoshoku*., 41: 461–468.

Zar, J.H., 1984. *Biostatistical Analysis*, 2nd Edition. Prentice Hall Eaglewood Cliffs, New Delhi, Jersey, pp. 718.

Chapter 37

Soils of Mandhala Watershed in Shiwalik Foot Hills of Himachal Pradesh–I: Their Characterisation, Classification and Mapping

J.C. Sharma and Sanjeev K. Chaudhary
Department of Soil Science and Water Management,
Dr. Y.S. Parmar University of Horticulture and Forestry,
Nauni, Solan – 173 230, Himachal Pradesh

ABSTRACT

A detailed soil survey was conducted to study the soils of Mandhala watershed situated in Shiwalik foot hills of Himachal Pradesh. These soils, developed on sandstone and alluvial/colluvial deposits are well drained, extremely shallow to deep, slightly acidic to neutral in reaction, low to very high in organic carbon and have low to moderate CEC. Sand is the dominant mechanical fraction. On the basis of morphological and physico-chemical properties eight tentative soil series representing six sub groups *i.e.* Typic Hapludalfs, Dystric Eutrudepts, Lithic Udorthents, Typic Udorthents, Typic Udifluvents and Lithic Udipsamments under three orders *viz.* Entisol, Inceptisol and Alfisol were identified and mapped into twenty eight phases.

***Keywords**: Soil classification, Mapping, Shiwalik foot hills, Himachal Pradesh.*

Introduction

An understanding of the status of natural resources is crucial for their efficient management. Soil, a basic natural resource is a vital component of life support system. Documentation of soil properties in systematic form is a pre-requisite in formulating effective land use planning programme. Soil survey provides the means to define the nature and properties of soils in well defined taxonomic system, map their location and extent and provide basic soil and site characteristics for multipurpose interpretations (Bali, 1985). Such soil database is of limited availability in hilly states like Himachal Pradesh due to climatic, topographical and financial constraints. Although the soils of Himachal Pradesh have been mapped at 1 : 0.5 m scale (NBSS and LUP, 1997) and major sub orders identified are Orthents, Ochrepts, Fluvents, Psamments, Udolls and Udalfs. However, since most of the land in Himachal Pradesh is non arable (an association of rock outcrops and Orthents alone represents about half of the geographical area of the state) with scattered arable pockets, soil mapping at a larger scale is needed to harness the full potential of limited soil resource. Besides, most of the earlier work has been done on individual profile basis without mapping the soils (Singh *et al.*, 1991). A necessity is therefore, always felt for more soil database on watershed basis for further interpretation. Keeping this in view, a study was undertaken to characterise, classify and map the soils of Mandhala watershed in Shiwalik foot hills of Himachal Pradesh.

Materials and Methods

The study area (Mandhala watershed) is located between 76°50′ to 76°54′ E longitude and 30°53′45″ to 30°56′15″ N latitudes at an elevation ranging from 330 m to 900 m above msl in Kasauli development block of Solan district in Himachal Pradesh. The watershed comprises a total geographical area of 1453.53 ha. Physiographically, the study area represents Shiwalik foot hills of Himalayan region.

Geological structure of the region comprises of Shiwalik formations. The rock consists of massive sandstone bands with alternating clay layers with NW–SE trend lines and dipping moderately in the NE direction. Downstream area is mostly plain whereas upstream area in general, is highly rugged, dotted by frequent elevations and depressions which are characterized by moderately steep to very-very steep slopes. The area is well drained with sub-humid sub-tropical climate characterized by extreme summers and severe winters. Precipitation is received mainly in the form of summer rainfall, most of which is concentrated during monsoon season (mid June to mid September). Average annual rainfall is about 1000 mm. The area has udic moisture and hyperthermic temperature regimes. Natural vegetation mainly consists of mixed deciduous forests of khair (*Acacia catechu*) with its various associates. While grasses like *Heteropogon contortus, Eulaliopsis binata, Chrysopogon fulvus etc.* occupy waste / scrub lands, scattered fruit trees like mango (*Mangifera indica*), papaya (*Carica papaya*), lemon (*Citrus* spp.) etc. alongwith fodder trees occur on field bunds.

A detailed soil survey of the watershed was conducted using SOI toposheet as base map. Soil studies were based on pedons, auger-bore and mini-pit observations and spatially delineating the soil boundaries with phases of soil series as map units by traversing and following standard procedures (IARI, 1971; Sehgal *et al.*, 1987). Horizon-wise soil samples from representative soil profiles of the series were air dried in shade and ground to pass through 2 mm sieve. The processed samples were analysed in the laboratory for their physical and chemical properties following standard procedures (Jackson, 1973; Sarma *et al.*, 1987). The soils were classified as per the Keys to Soil Taxonomy (Soil Survey Staff, 1998). The data generated (land and soil) both spatial and attribute, were processed using a Geographical Information System (GIS) for the generation of thematic maps at 1 : 12,500 scale.

Results and Discussion

Based on detailed soil survey, eight tentative soil series *viz.*, Johranpur, Mandhala I, Mandhala II, Mandhala III, Mandhala IV, Majri, Dhaular and Haripur were identified and spatially delineated

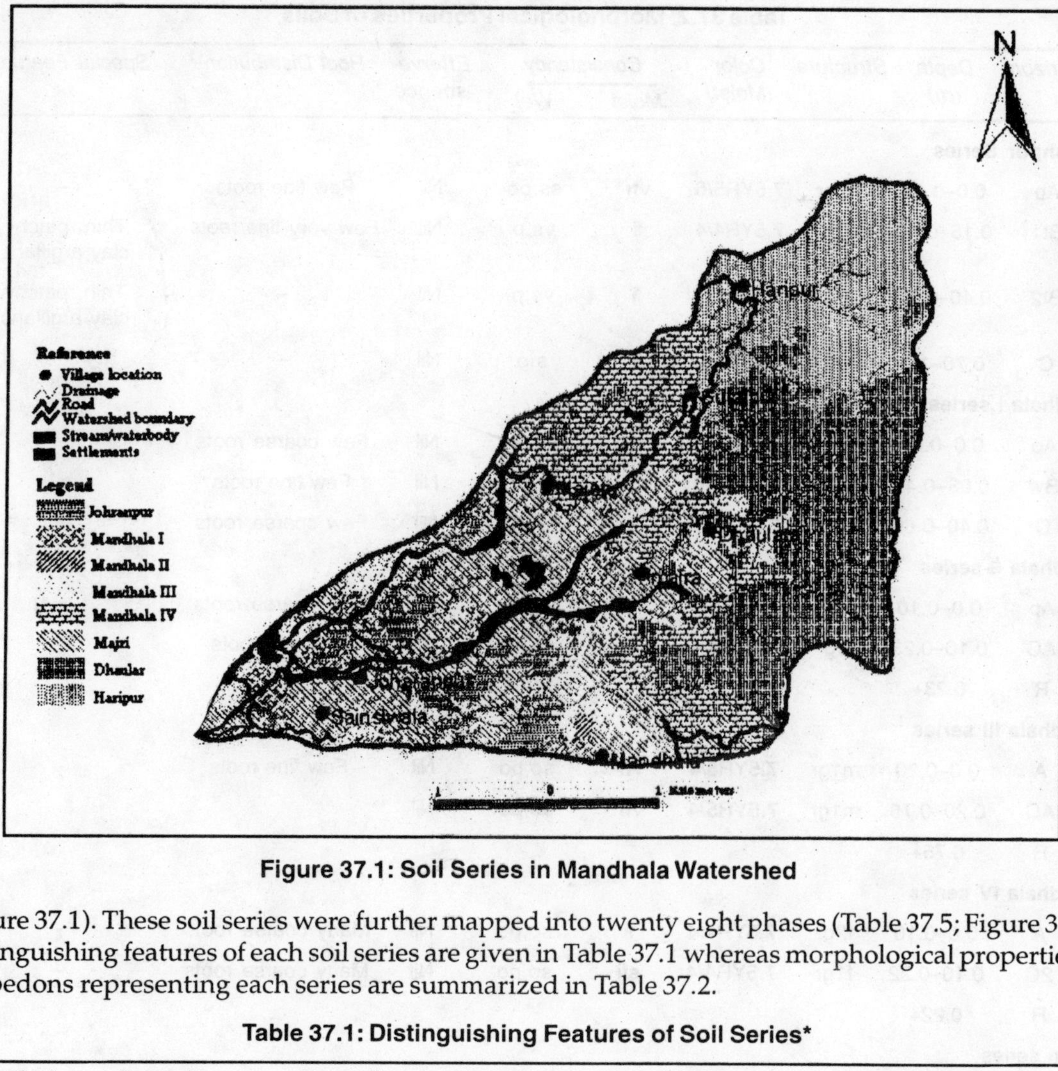

Figure 37.1: Soil Series in Mandhala Watershed

(Figure 37.1). These soil series were further mapped into twenty eight phases (Table 37.5; Figure 37.2). Distinguishing features of each soil series are given in Table 37.1 whereas morphological properties of the pedons representing each series are summarized in Table 37.2.

Table 37.1: Distinguishing Features of Soil Series*

Soil Series	*Depth Range (m)*	*General Slope (%)*	*Coarse Fragments (%)*	*Parent Material*	*Major Land Use*
Johranpur	0.75–1.20	0–10	1–13	Old alluvium	Agriculture
Mandhala I	0.52–0.60	0–10	7–28	Old alluvium	Agriculture, scrub, grassland
Mandhala II	0.20–0.23	0–5	7–27	Shallow alluvium	Agriculture
Mandhala III	0.55–0.75	0–10	29–43	Shallow alluvium	Grassland, scrub, agriculture
Mandhala IV	0.22–0.40	15–33	40–65	Sandstone	Forest, scrub
Majri	0.07–0.25	0–10	21–38	Shallow colluvium	Scrub, grassland, agriculture, plantations
Dhaular	0.15–0.40	mostly > 25	50–60	Sandstone	Forest
Haripur	0.21–0.24	mostly > 25	48–55	Sandstone	Forest

*All soils are well drained.

Table 37.2: Morphological Properties of Soils

Horizon	*Depth (m)*	*Structure*	*Color (Moist)*	*Consistency*		*Efferve-scence*	*Root Distribution*	*Special Features*
				Moist	*Wet*			
Johranpur Series								
Ap	0.0–0.15	m2gr	7.5YR5/6	vfr	ss,po	Nil	Few fine roots	–
Bt1	0.15–0.40	vf3sbk	7.5YR4/4	fi	vs,p	Nil	Few very fine roots	Thin, patchy clay argillans
Bt2	0.40–0.70	vf3sbk	7.5YR4/4	fi	vs,p	Nil	–	Thin, patchy clay argillans
C	0.70–1.0	m2sbk	7.5YR4/4	fi	s,p	Nil	–	
Mandhala I series								
Ap	0.0–0.08	sg	7.5YR5/4	l	so,po	Nil	Few coarse roots	–
Bw	0.08–0.40	f1sbk	7.5YR5/4	fr	so,po	Nil	Few fine roots	
C	0.40–0.60	m	7.5YR5/4	fi	so,po	Nil	Few coarse roots	–
Mandhala II series								
Ap	0.0–0.10	sg	7.5YR4/4	l	so,po	Nil	Few coarse roots	–
AC	0.10–0.23	sg	7.5YR4/4	l	so,po	Nil	Few fine roots	–
R	0.23+							
Mandhala III series								
A	0.0–0.20	m1gr	7.5YR5/4	vfr	so,po	Nil	Few fine roots	–
AC	0.20–0.75	m1gr	7.5YR5/4	vfr	so,po	Nil		
R	0.75+							
Mandhala IV series								
A	0.0–0.10	f2gr	7.5YR4/4	fi	so,po	Nil	Many coarse roots	–
2C	0.10–0.22	f1gr	7.5YR4/4	sfi	so,po	Nil	Many coarse roots	–
R	0.22+							
Majri series								
A	0.0–0.15	m2gr	7.5YR4/4	fr	so,po	Nil	Few coarse roots	–
R	0.15+							
Dhaular series								
A	0.0–0.17	m1gr	10YR5/3	fr	ss,ps	e	Many fine/coarse roots	–
Cr	0.17–0.40	Semi weathered sandstone						
Haripur series								
A	0.0–0.10	m2gr	10YR7/4	fr	so,po	Nil	Many coarse roots	–
AC	0.10–0.21	m2gr	10YR7/4	fr	ss,ps	Nil	Many coarse roots	–
R	0.21+							

Table 37.3: Physico-chemical Characteristics of Soils

Horizon	*Depth (m)*	*Gravel*	*Sand (%)*	*Silt (%)*	*Clay*	*Texture*	*BD (mg m⁻³)*	*pH*	*EC (dSm⁻¹)*	*OC (g kg⁻¹)*	*CaCO₃ (%)*	*CEC [cmol (p⁺) kg⁻¹]*	*BSP*
Johranpur series													
Ap	0.0–0.15	10	66	26	8	ls	1.14	6.95	0.37	6.6	0.0	7.8	82.86
Bt1	0.15–0.40	11	52	26	22	sil	1.20	7.29	0.25	5.1	0.0	12.1	84.53
Bt2	0.40–0.70	11	48	32	20	sil	1.25	7.20	0.29	3.9	0.0	13.9	80.86
C	0.70–1.0	4	69	13	18	l	1.32	7.10	0.34	3.9	0.0	13.5	78.43
Mandhala I series													
Ap	0.0–0.08	20	90	2	8	gs	1.43	6.25	0.29	2.7	0.0	7.5	61.73
Bw	0.08–0.40	14	80	10	10	ls	1.44	6.82	0.25	1.5	0.0	10.2	66.56
C	0.40–0.60	17	78	11	11	ls	1.39	6.64	0.21	1.8	0.0	11.2	66.78
Mandhala II series													
Ap	0.0–0.10	7	84	2	14	si	1.37	6.80	0.32	3.9	0.0	5.2	73.69
AC	0.10–0.23	8	82	4	14	si	1.34	6.90	0.27	2.1	0.0	4.7	82.04
R	0.23+												
Mandhala III series													
A	0.0–0.20	29	82	12	6	gls	1.49	6.38	0.17	15.9	0.0	5.8	71.63
AC	0.20–0.75	37	74	18	8	gls	1.39	6.59	0.11	13.9	0.0	7.8	77.70
R	0.75+												
Mandhala IV													
A	0.0–0.10	40	58	38	4	gsil	1.22	6.80	0.36	10.2	0.0	15.4	68.22
2C	0.10–0.22	41	85	32	2	gsil	1.28	6.90	0.34	4.2	0.0	13.5	75.49
R	0.22+												
Majri series													
A	0.0–0.15	21	64	26	10	gls	1.26	6.76	0.4	4.5	0.0	11.47	71.57
R	0.15+												
Dhaular series													
A	0.0–0.17	50	72	20	8	gls	1.24	6.55	0.09	12.0	0.1	6.6	71.57
Cr	0.17–0.40	Semi weathered sandstone											
Haripur series													
A	0.0–0.10	38	78	16	6	gls	1.32	6.74	0.24	10.5	0.0	12.5	80.10
AC	0.10–0.21	40	80	8	12	gls	1.35	6.98	0.21	7.5	0.0	14.0	81.20
R	0.21+												

Table 37.4: Classification of Studied Soils

Series	*Order*	*Sub-order*	*Great Group*	*Sub Group*	*Family*
Johranpur	Alfisol	Udalf	Hapludalfs	Typic Hapludalf	Fine loamy, mix, hyperthermic, Typic Hapludalfs
Mandhala I	Inceptisol	Udept	Eutrudepts	Dystric Eutrudepts	Coarse loamy, mixed, hyperthermic, Dystric Eutrudepts
Mandhala II	Entisol	Orthent	Udorthents	Lithic Udorthents	Coarse loamy, mixed, hyperthermic, Lithic Udorthents
Mandhala III	Entisol	Orthent	Udorthents	Typic Udorthents	Sandy, mixed, hyperthermic, Typic Udorthents
Mandhala IV	Entisol	Fluvent	Udifluvent	Typic Udifluvents	Coarse loamy skeletal, mixed, hyperthermic, Typic Udifluvents
Majri	Entisol	Psamment	Udipsamments	Lithic Udipsamments	Sandy, mixed, hyperthermic, Lithic Udipsamments
Dhaular	Entisol	Psamment	Udipsamments	Lithic Udipsamments	Sandy skeletal, mixed, hyperthermic, Lithic Udipsamments
Haripur	Entisol	Orthent	Udorthents	Lithic Udorthents	Sandy skeletal, mixed, hyperthermic, Lithic Udorthents

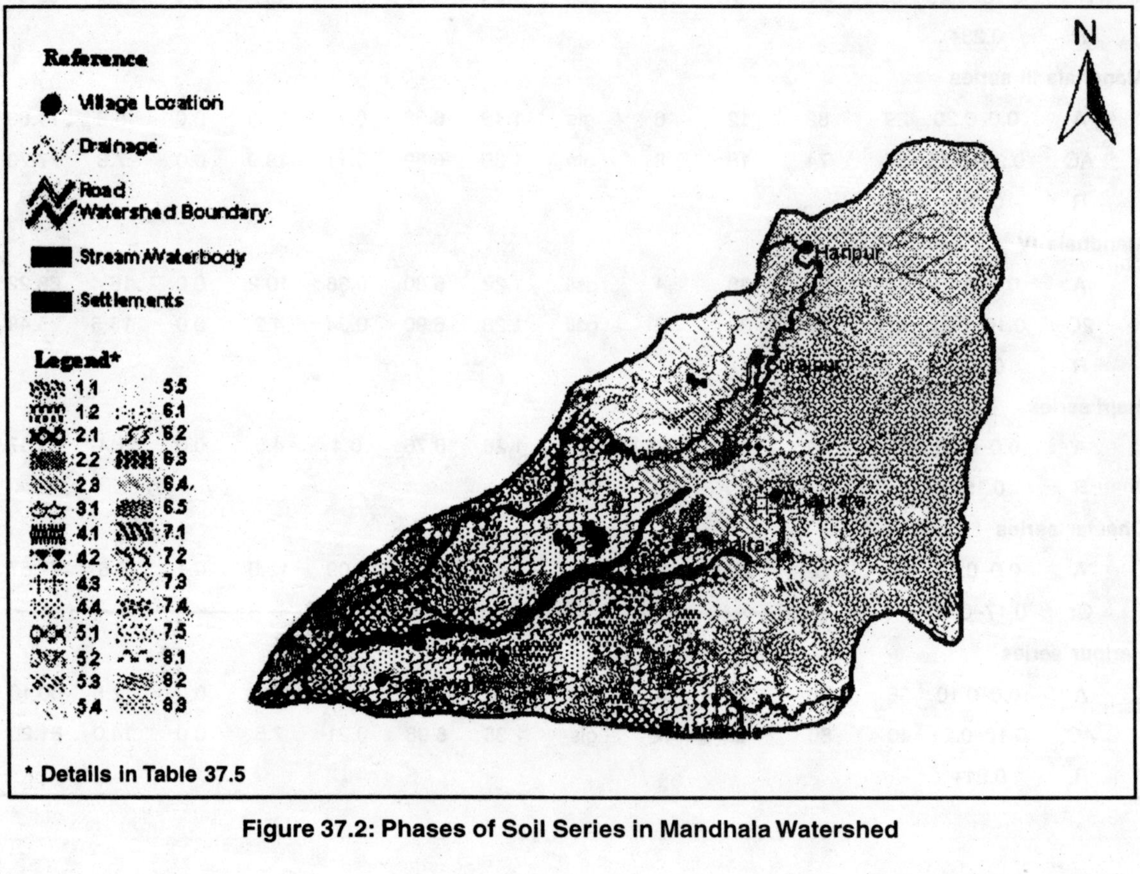

Figure 37.2: Phases of Soil Series in Mandhala Watershed

Table 37.5: Phases of Soil Series in Mandhala Watershed

Legend No.*	*Phases of Soil Series*	*Area (ha)*	*% TGA*
Johranpur Series: Fine loamy, mixed, hyperthermic, Typic Hapludalfs		**83.24**	**5.73**
1.1	Nearly level to v. gently sloping, deep to mod. deep, slightly eroded	46.41	3.19
1.2	Gently to mod. sloping, deep to mod. deep, slightly eroded	36.83	2.54
Mandhala I Series: Coarse loamy, mixed, hyperthermic, Dystric Eutrudepts		**109.44**	**7.53**
2.1	Nearly level to gently sloping, shallow to slightly deep, gravelly, sev. to mod. eroded	69.00	4.75
2.2	Mod. to strongly sloping, slightly deep, gravelly, sev. to mod. eroded	27.22	1.87
2.3	Mod. steep to steep, slightly deep, gravelly, mod. eroded	13.22	0.91
Mandhala II Series: Coarse loamy, mixed, hyperthermic, Lithic Udorthents		**3.46**	**0.24**
3.1	Nearly level to gently sloping, v. shallow to shallow, sandy, slightly eroded	3.46	0.24
Mandhala III Series: Sandy, mixed, hyperthermic, Typic Udorthents		**119.21**	**8.20**
4.1	Nearly level to gently sloping, shallow to slightly deep, gravelly, slightly to sev. eroded	63.63	4.37
4.2	Mod. to strongly sloping, shallow to slightly deep, gravelly, slightly to sev. eroded	35.42	2.44
4.3	Mod. steep to steep, shallow to slightly deep, gravelly, slightly eroded	9.16	0.63
4.4	Steep to v. steep, shallow, gravelly, slightly eroded	11.00	0.76
Mandhala IV Series: Coarse loamy skeletal, mixed, hyperthermic, Typic Udifluvents		**187.40**	**12.89**
5.1	Nearly level to gently sloping, v. shallow to shallow, gravelly, slightly eroded	6.08	0.42
5.2	Mod. sloping, v. shallow to shallow, gravelly, slightly eroded	18.67	1.28
5.3	Strongly sloping, v. shallow to shallow, gravelly, slightly eroded	18.00	1.24
5.4	Mod. steep to steep, v. shallow to shallow, gravelly, mod. eroded	95.24	6.55
5.5	Steep to v. v. steep, v. shallow to shallow, gravelly, mod. eroded	49.41	3.40
Majri Series: Sandy, mixed, hyperthermic, Lithic Udipsamments		**293.15**	**20.17**
6.1	Nearly level to v. gently sloping, v. shallow to mod. deep, gravelly, slightly eroded	65.78	4.51
6.2	Gently to mod. sloping, v. shallow to shallow, gravelly, slightly to sev. eroded	167.78	11.52
6.3	Strongly sloping, extr. shallow to shallow, gravelly, mod. to sev. eroded	18.89	1.29
6.4	Mod. steep to steep, extr. shallow to shallow, gravelly, mod. to sev. eroded	39.96	2.74
6.5	V. steep to v. v. steep, v. shallow, gravelly, mod. eroded	1.74	0.11
Dhaular Series: Sandy skeletal, mixed, hyperthermic, Lithic Udipsamments		**361.11**	**24.84**
7.1	Gently to mod. sloping, v. shallow to shallow, stony, mod. eroded	4.64	0.33
7.2	Strongly sloping, v. shallow to shallow, stony, sev. eroded	7.95	0.55
7.3	Mod. steep to steep, v. shallow to shallow, stony, sev. eroded	61.85	4.24
7.4	Steep to v. steep, v. shallow to shallow, stony, sev. eroded	260.23	17.90
7.5	v.v. steep, v. shallow, stony, sev. eroded	26.44	1.82
Haripur Series: Sandy skeletal, mixed, hyperthermic, Lithic Udorthents		**241.64**	**16.62**
8.1	Mod. sloping, v. shallow, stony, mod. eroded	9.62	0.66
8.2	Mod. steep to steep, v. shallow, stony, mod. eroded	33.12	2.28
8.3	Steep to v.v. steep, v. shallow, stony, mod. eroded	198.90	13.68

v.: Very; mod.: Moderately; sev.: Severely; extr.: Extremely; TGA: Total geographical area; *Legend No. in Figure 37.2.

Note: A 3.78 per cent of TGA (54.88 ha) is under streams/water bodies/settlements/roads etc.

Majority of the soils had 7.5 YR as the common hue of the soil matrix except in Dhaular and Haripur series where it was 10 YR. High colour values ranging from 4 to 7 denote well drained nature of these soils. Surface horizons had predominately granular structure except in Mandhala I and Mandhala II series where it was single grained. Sub-surface horizons varied in soil structure from sub angular blocky to massive. Consistency of moist soils of surface horizons was very friable/friable except in case of Mandhala I, Mandhala II and Mandhala IV where it was observed to be loose, loose and firm respectively. However in sub-surface horizons it varied from loose to firm. The consistency of wet soils varied from non-sticky and non-plastic to very sticky and plastic. The soils are non calcareous in nature as indicated by the absence of any effervescence upon dilute HCl treatment except in Dhaular soils. Few to many roots occurring in different horizons varied in size from very fine to coarse. Thin, patchy clay argillans were observed in subjacent horizons of Johranpur soils only.

Physico-chemical Characteristics

The data on physical properties pertaining to representative profiles of each series is presented in Table 37.3. Sand constituted the major soil separate both in surface and sub-surface horizons (48–90 per cent) while silt content varied from 2 to 38 per cent. The coarse texture of these soils may be ascribed to their origin from sandstone/ alluvial parent material. The clay content was low (2–22 per cent). Generally, sub-surface horizons exhibited higher clay content as compared to surface ones. This may be due to the illuviation process occurring during soil development. It was, however sufficient enough to give rise to argillic horizons only in Johranpur soils. Bulk density in surface and sub-surface horizons varied from 1.14 to 1.49 mg m^{-3} and 1.20 to 1.44 mg m^{-3} respectively. Bulk it density was related directly to sand fraction ($r = 0.65^{**}$) and inversely to silt ($r = -0.50^{*}$). Its negative relationship with clay ($r = -0.24$) however, failed to attain the level of significance.

The soils under study were found to be slightly acidic to neutral in reaction (pH range 6.25–7.29). The surface horizons had lower pH as compared to sub-surface soils. However, organic carbon content (1.5–15.9 g kg^{-1}) was recorded to be higher in surface horizons as compared to sub-surface ones. This may be attributed to leaf fall/decay and manurial applications. This might also explain the lower pH values observed in surface than in sub-surface horizons. The results are further substantiated by negative and significant correlation ($r = -0.48^{*}$) observed between organic carbon and pH. Similar results were reported by Sharma *et al.* (2004) while studying North-West Himalayan soils. CEC values ranged from 4.7 to 15.4 cmol (p^+) kg^{-1}. Free $CaCO_3$ was absent in all except Dhaular soils where it was found to be as low as 0.1 per cent. BSP varied from 61.73 to 84.53 per cent. The variations in CEC and BSP may be ascribed to the variations in pH values, type and/or content of soil colloids. Significant positive correlations of pH with CEC ($r = 0.52$) and BSP ($r = 0.69^{**}$) were observed in these studies. Biswas and Mukherjee (1995) also explained the effects of soil pH and soil colloids on these characteristics.

Soil Classification

All the soils were characterized by ochric surface diagnostic horizon (epipedon). Johranpur soils with *Bt* as argillic sub-surface diagnostic horizon (endopedon) and a base saturation more than 35 per cent were placed in order Alfisol. Mandhala I soils being in the stage of inception showed very little profile development with only cambic B as sub-surface diagnostic horizon due to insufficient illuviation. As such, these were placed in order Inceptisol. Rest of the soils (Mandhala II, Mandhala III, Mandhala IV, Majri, Dhaular and Haripur) being young, stratified without any sub-surface diagnostic horizon were placed under order Entisol. Since area qualifies for udic moisture regime, Johranpur soils were placed under great group Hapludalf. As these represented the central concept of great group, these were further classified as Typic Hapludalfs at sub group level. A base saturation of more than 60 per cent in sub-surface horizons within 75 cm of the surface coupled with the presence of diagnostic cambic B horizon enabled Mandhala I soils to be put under great group Eutrudepts. Further, because of absence of carbonates within one meter of surface, these were placed under sub group Dystric Eutrudepts. Mandhala II, Mandhala III and Haripur soils representing sub order Orthent were placed under Udorthents great group due to udic moisture regime. Since Mandhala III represented the central concept of great group, it was keyed out as Typic Udorthents at sub group level. However, in Mandhala II and Haripur soils depth is only 23 cm and 21 cm respectively and there exists lithic contact at a depth less than 25 cm which enabled these to be placed under sub group Lithic Udorthents. Absence of sub-surface diagnostic horizon coupled with udic moisture regime and irregular decrease in OC content with depth as evidenced by lithological discontinuity at 10 cm depth enabled Mandhala IV soils to be classified as Typic Udifluvents at sub group level. The control section of Majri and Dhaular soils is characterized by sandy or even coarser texture (upto 60 per cent coarse fragments). Accordingly, these qualified for Psamments sub order and Udipsamments great group. Further, a lithic/paralithic contact at a depth less than 25 cm enabled these soils to be classified as Lithic Udipsamments at sub group level. The mineralogy of the soils was considered to be mixed for classifying them at family level. In light of the area and extent given in Table 37.5 the studied series may be arranged in the order: Dhaular (361.11 ha) > Majri (293.15 ha) > Haripur (241.64 ha) > Mandhala IV (187.40 ha) > Mandhala III (119.21 ha) > Mandhala I (109.44 ha) > Johranpur (83.24 ha) > Mandhala II (3.46 ha) covering an area of 24.84, 20.17, 16.62, 12.89, 8.20, 7.53, 5.73 and 0.24 per cent of the total geographical area of the watershed, respectively.

A great variety of soils were found to occur during the systematic survey of the watershed. The studies revealed that these soils are in initial stage of development with O, mostly A-C profile sequence. The dominance of sand in soil separates, high bulk density values and poor structural status suggested greater susceptibility of these soils to erosion. This calls for provision of elaborate soil and water conservation measures to conserve the fragile ecosystem and arrest any further degradation of these lands. Fewer amounts of clay and low CEC values are indicative of low retentivity of native/added nutrient cations in these soils. The data reported here may serve as the basis/benchmark for further interpretation for land use planning at watershed level.

Acknowledgements

Authors are grateful to Dr. Karan Singh, Professor, Soils (Retired) for technical guidance.

References

Bali, Y.P., 1985. Soil Surveys and soil conservation for agricultural development. *J. Indian Soc. Soil Sci.*, 33: 740.

Biswas, T.D. and Mukherjee, S.K., 1995. *Textbook of Soil Science*. Tata McGraw-Hill Publishing Company Ltd., New Delhi, pp. 153–221.

I.A.R.I. 1971. *Soil Survey Manual*, AIS&LUS Organization, IARI, New Delhi, pp. 121.

Jackson, M.L., 1973. *Soil Chemical Analysis*. Prentice Hall of India Pvt. Ltd., New Delhi.

NBSS and LUP, 1997. *Soils of Himachal Pradesh for Optimizing Land Use*. Publ. 57, NBSS and LUP, Nagpur, India.

Sehgal, J.L., Sexena, R.K. and Vadivelu, S., 1987. *Field Manual*, Publ. 13, NBSS and LUP, Nagpur, pp. 73.

Sarma, V.A.K., Krishnan, P. and Budihal, S.L., 1987. *Laboratory Methods*. Publ. 14, NBSS&LUP, Nagpur, pp. 49.

Sharma, V.K., Sharma, P.D., Sharma, S.P., Acharya, C.L. and Sood, R.K., 2004. Characterization of cultivated soils of Neogal watershed in North-West Himalayas and their suitability for major crops. *J. Indian Soc. Soil Sci.* 52: 63–68.

Singh, K., Bhandari, A.R. and Tomar, K.P., 1991. Morphology, genesis and classification of some soils of North-West Himalayas. *J. Indian Soc. Soil Sci.*, 39: 139–146.

Soil Survey Staff, 1998. *Keys to Soil Taxonomy*, 8th Edn. USDA, Washington DC, USA, pp. 326.

Chapter 38

Soils of Mandhala Watershed in Shiwalik Foot Hills of Himachal Pradesh–II: Available S and Suitability for Mustard

J.C. Sharma and Sanjeev K. Chaudhary
Department of Soil Science and Water Management,
Dr. Y.S. Parmar University of Horticulture and Forestry,
Nauni, Solan – 173 230, Himachal Pradesh

ABSTRACT

Soils of a watershed occurring under Siwalik foot hill conditions of Himachal Pradesh were analysed for available S content and evaluated for suitability of growing mustard crop. Available S was positively and significantly correlated with silt, clay and CEC and negatively with soil pH. S deficiency was observed to be pravelent in these soils with major part of the watershed amounting to about 79 per cent of total area being permanently not suitable for mustard growing. Other categories identified *viz.*, moderately suitable, marginally suitable and temporarily not suitable occupied relatively smaller proportion of the watershed area.

Keywords: *S content, Soil-site suitability, Mustard, Himachal Pradesh.*

Introduction

Indian mustard (*Brassica juncea* L.) is an important oil seed crop grown in Shiwalik foot hills of Himachal Pradesh. However, its productivity is quite low. Major factors contributing to low oilseed production identified are its cultivation under marginal lands in the absence of any crop specific soil-site suitability recommendation and low rates of fertilizer application. Whatever little fertilizers being used by the farmers mainly consist of urea, DAP and MOP which are sulphur (S) free. Problem is further exaggerated as the soils in Shiwalik foot hills have traditionally been low in fertility due to their coarse texture and high rates of leaching and erosion. S is often required by plants in quantities equal to P. In fact, S is now being recognised as fourth major essential plant nutrient element after N, P and K (Sarkar *et al.*, 2000). Special role played by S in formation and development of protein rich oil storage organs in crucifers is well documented (Aulakh and Pasricha, 1988). The fact that oil seed crops have a high S requirement is adequately reflected by the remarkable response of oilseed crops to S applications frequently obtained in various studies. It has been reported that 100 kg of mustard seeds remove on an average 6.8, 6.2 and 2.5 kg of S, N and P_2O_5. Among its various functions S is an important element for increasing oil content, proteins and essential amino acids (Aulakh *et al.*, 1980).

Plant roots absorb S in the form of SO_4^{2-} ions from the soil solution. Therefore sulphate-S is considered to be the most important form from the point of view of plant nutrition and may prove to be a suitable index in evaluating the S availability to plants. The magnitude of S adsorption in soil depends (apart from surface area, its configuration, intensity of electrical charges on the adsorbent, anionic composition of soil solution), upon pH, organic matter *etc.* An attempt was made to study available S and delineate sites suitable for the cultivation of Indian mustard under Shiwalik foot hill conditions of Himachal Pradesh.

Materials and Methods

The study area mapped (as detailed in another communication) was further generalized into eighteen soils cape (physiography cum soil) mapping units for sake of simplicity by overlaying visually interpreted IRS imagery (November 10, 2001) supported by ground truth verification. Physico-chemical characteristics of surface (0–15 cm) soils falling in each mapping unit were determined by following standard analytical procedures. Available S in soils was extracted with 0.15 per cent $CaCl_2$ solution (Williams and Steinbergs, 1959). Sulphur in the soil extract was determined terbidimetrically by Chesnin and Yein's method (1951). In order to find out the influence of various soil properties on available S simple correlations as well as multiple linear regression equations were computed between available S and soil properties. Land was given a suitability rating in accordance with the framework of land evaluation (FAO, 1976; 1993; Sys *et al.*, 1993) depending upon how well its properties meet the requirements of the crop.

Results and Discussion

Salient characteristics of the studied soilscape units have been presented in Table 38.1.

Available S and Relationship with Soil Properties

Available S content varied from 11.2–38.2 kg ha^{-1} with a mean value of 22.7 kg ha^{-1} (Table 38.1). Considering 10 mg kg^{-1} (22.4 kg ha^{-1}) as the threshold value for S deficiency (Mehta *et al.*, 1988), H11, H23, H24, P11, P12, P23, FP2 and FP3 mapping units can be rated to be deficient in available S suggesting a strong need for S application in these soils. Leaching might have caused S deficiency in these units as sulphur is a mobile nutrient element in the soil system which moves down to the lower soil layers with frequent irrigations and rains (Tripathi and Hazra, 2000).

Table 38.1: Salient Characteristics and Available S (kg ha⁻¹) of Soil Scape Units

Soil Scape Unit	Soil Association	Slope Class	Soil Erosion	Soil Depth (cm)	Soil Texture	Surface Stoniness (%)	NPK Rating	Available S	
								Range	Mean
H11 (Shiwalik foot hills-upper)	Lithic Udorthents	D–I	e2–e3	15–33	gsl–gls	30–45	MHL	19.1–22.9	20.3
	Lithic Udipsamments								
H12 (Shiwalik foot hills-upper)	Lithic Udipsamments	C–I	e2–e3	18–35	gsl–gls	35–50	MHL	18.8–37.0	25.7
	Lithic Udorthents								
H13 (Shiwalik foot hills-upper)	Lithic Udipsamments	F–I	e2–e3	21–55	gsl–gls	30–50	LHL	18.2–38.2	26.2
	Lithic Udorthents								
	Typic Udorthents								
H21 (Shiwalik foot hills-lower)	Typic Udifluvents	A–I	e1–e3	22–40	gsil–ls	> 40	MHL	23.8–27.4	24.1
H22 (Shiwalik foot hills-lower)	Lithic Udipsamments	F–H	e3	20–40	gls	> 40	LHL	19.5–35.2	27.2
H23 (Shiwalik foot hills-lower)	Dystric Eutrudepts	A–G	e1–e3	9–105	l–gs	2–20	LHL	11.2–31.9	15.4
	Typic Hapludalfs								
	Lithic Udipsamments								
H24 (Shiwalik foot hills-lower)	Lithic Udipsamments	A–G	e1–e3	7.35	gsil–gls	> 35	LHL	12.6–35.8	18.7
	Typic Udifluvents								
H25 (Shiwalik foot hills-lower)	Typic Udorthents	D–H	e1–e3	10–75	ls–gls	> 40	LMM	12.8–30.4	23.9
	Lithic Udipsamments								
	Dystric Eutrudepts								
P11 (Piedmonts-upper)	Lithic Udipsamments	A–D	e1–e3	15–110	l–gis	2–15	LHL	12.5–31.1	21.7
	Typic Hapludalfs								
	Lithic Udorthents								
P12 (Piedmonts-upper)	Lithic Udipsamments	A–I	e1–e3	18–55	ls–gs	25	LHL	11.8–33.2	22.0
	Dystric Eutrudepts								

Contd...

Table 38.1–Contd...

Soil Scape Unit	Soil Association	Slope Class	Soil Erosion	Soil Depth (cm)	Soil Texture	Surface Stoniness (%)	NPK Rating	Available S	
								Range	Mean
P13 (Piedmonts-upper)	Lithic Udipsamments	A–G	e1–e3	21–52	ls–gs	40	LMM	12.6–30.3	23.3
	Typic Udorthents								
	Dystric Eutrudepts								
P21 (Piedmonts-lower)	Typic Hapludalfs	A–D	e1–e3	25–120	l–gis	2–13	LHL	11.5–35.0	27.1
	Lithic Udipsamments								
P22 (Piedmonts-lower)	Lithic Udipsamments	E	e2–e3	8–26	gls	25	LHL	12.7–32.8	23.4
P23 (Piedmonts-lower)	Lithic Udipsamments	A–G	e1–e3	15–76	ls–gs	30	LHL	12.9–30.8	18.6
	Typic Udorthents								
	Dystric Eutrudepts								
P24 (Piedmonts-lower)	Typic Udorthents	A–G	e1–e3	9–74	ls–gs	36	LMM	11.9–31.9	23.9
	Dystric Eutrudepts								
	Lithic Udipsamments								
FP1 (Flood plains)	Typic Udorthents	B–D	e1–e3	19–56	gls–gs	32	LHL	12.6–30.1	22.5
	Lithic Udipsamments								
	Dystric Eutrudepts								
FP2 (Flood plains)	Dystric Eutrudepts	A–D	e1–e3	23–60	ls–gs	38	LMM	12.3–31.7	21.8
	Lithic Udipsamments								
FP3 (Flood plains)	Lithic Udipsamments	A–C	e1–e3	24–100	l–gls	2–10	LHL	11.2–32.5	22.2
	Typic Hapludalfs								
	Typic Udorthents								

A: 0–1 per cent; B: 1–3 per cent; C: 3–5 per cent; D: 5–10 per cent; E: 10–15 per cent; F: 15–25 per cent; G: 25–33 per cent; HL 33–50 per cent; I: > 50 percent; e1: Slight; e2: Moderate; e3: Severe; g: Gravelly; ls: Loamy sand; sil: Silt loam; sl: Sandy loam, s: Sand; L: Low; M: Medium, H: High.

Available S was positively and significantly correlated with silt (r = 0.24**), clay (r = 0.42**) and CEC (r = 0.40**). Sharma and Jaggi (2001) observed significant positive correlation of available S with silt + clay and CEC in acid Alfisols of Himachal Pradesh. These relationships are obvious as texture (clay and silt) is also one of the major factors regulating the S status of a soil as the amount of clay and silt determines the number of edge adsorption sites for SO_4^{2-}. CEC, an important soil characteristic in soil fertility and plant nutrition, is the manifestation of OC and clay content as these provide surface area for the exchange of cations. The significant negative correlation (r = –0.26**) between soil pH and available S observed in the present investigation is in line with the results of Aggarwal and Nayyar (1998). This relationship is attributed to the decrease in adsorption of SO_4^{-2} in soils and its loss through leaching. Clay was found to be the strongest determinant of available S as is evident from the highest r value. The predictability of available S can be described by following multiple regression equation:

$$\text{Available S} = 29.364 + 0.888\,(\text{clay})^* - 2.007\,(\text{pH}) + 5.103\,(\text{OC}) + 0.788\,(\text{CEC}) - 0.54\,(\text{BSP}),\ R^2 = 46.31$$

Inclusion of EC improved the predictability (R^2 to nearly 49 per cent as per equation:

$$\text{Available S} = 8.258 + 0.728\,(\text{clay})^* - 7.299\,(\text{pH}) + 3.513\,(\text{OC}) + 0.749\,(\text{CEC}) - 0.628\,(\text{BSP}) - 19.454\,(\text{EC}),\ R^2 = 49.31$$

Inclusion of silt further appreciated R^2 value to nearly 55 per cent according to the equation:

$$\text{Available S} = 59.183 + 0.245\,(\text{silt}) + 0.903\,(\text{clay})^* - 1.835\,(\text{pH}) + 0.098\,(\text{OC}) + 0.612\,(\text{CEC}) - 0.487\,(\text{BSP}) - 26.116\,(\text{EC}),\ R^2 = 55.02$$

Soil Site Suitability for Mustard

These soils have been evaluated to asses their suitability for mustard growing according to criteria given in Table 38.2. The ratings have been presented in. The soils are grouped into five suitability categories (Table 38.3) *i.e.* highly suitable (S^1) *i.e.* lands having no significant limitations or slight limitations (P11, P21 and FP3 units), moderately suitable (S2) *i.e.* lands having moderate limitations or aggregation of slight limitations (P21 unit), marginally suitable (S3) *ie.* lands having aggregation of severe limitations (H23 unit), currently not suitable (N1) *i.e.* lands having limitations which cannot be corrected by accepted methods (P24 and FP2 units) and lands permanently not suitable (N2) *viz.* H11, H12, H13, H21, H22, H24, H25, P12, P13, P22 and FP1 units). The factors limiting the use of these soils have also been indicated in suitability sub-classes. A close scrutiny of Figure 38.1 shows that major part of the watershed amounting to 74.94 per cent (1089.36 ha) was found not suitable for mustard crop while only 8.56 per cent (124.38 ha) was found highly suitable. Other categories included moderately suitable 4.94 per cent (71.80 ha), marginally suitable 3.08 (44.76 ha) and temporarily not suitable 4.70 per cent (68.35 ha).

It may be concluded that available S content in these coarse textured soils is low with fairly wide spread S deficiencies. Hence there is a need for adequate S fertilization on deficient soils for attaining higher productivity levels of oilseeds and cereals grown in the area. Soil properties (especially clay, silt, CEC and pH) are the major determinants of available S in these soils. Majority of the soils are not suitable for mustard growing owing to limitations of slope, soil depth, texture, erosion and fertility.

Table 38.2: Criteria for Rating Soil Site Suitability for Mustard*

Soil Site Characteristics	*Degree of Limitation*				
	0 (None)	*1 (Slight)*	*2 (Moderate)*	*3 (Severe)*	*4 (Very Severe)*
Rain fall (mm)	> 400	300–400	200–300	100–200	< 100
MAT (°C)	15–18	18–22	> 22	–	–
Slope (per cent)	0–1	1–3	3–8	> 8	–
Drainage	Well	Moderately well	Imperfect	Poor	Very poor
Flood hazards	Nil	Slight	Slight	Moderate	Severe
Erosion	None	Slight	Moderate	High	Very high
Soil depth (cm)	> 80	50–80	20–50	< 20	–
Soil texture	sil, sicl, cl	sc, scl, l, sl	sic, c, ls	fs, s	–
Surface-stoniness (%)	< 1	1–3	3–10	> 10	–
NPK Rating	HHH	MMM	MLL	LLL	–
Base saturation (%)	> 80	50–80	35–50	< 35	–

*Modified and developed from Sys *et al.* (1993).

Table 38.3: Soil-site Suitability Ratings for Mustard

Soilscape Unit	*Rain-fall*	*Temperature*	*Slope*	*Flood Hazard*	*Texture*	*Depth*	*Erosion*	*Stoni-ness*	*NPK*	*BSP*	*Suitability Sub-class*
H11	1	1	4	0	2	3	2	2	1	1	N_2stde
H12	1	1	4	0	2	3	2	3	1	1	N_2stde
H13	1	1	4	0	2	2	3	3	2	1	N_2stedf
H21	1	1	4	0	2	2	2	3	1	1	N_2sted
H22	1	1	4	0	2	2	3	3	2	1	N_2stedf
H23	1	1	2	0	1	0	1	2	2	1	S_3stef
H24	1	1	2	0	2	3	2	2	2	1	N_2sdtef
H25	1	1	2	0	2	1	2	2	2	1	N_2stef
P11	1	1	1	0	1	1	1	1	2	1	S_1stef
P12	1	1	2	0	2	2	2	2	2	1	N_2stdef
P13	1	1	2	0	2	2	2	2	1–2	1	N_2stdef
P21	0	1	1	0	1	1	1	1	2	1	S_1tf
P22	0	1	1	0	2	3	2	2	2	1	N_2dtf
P23	1	1	1	0	2	1	1	2	1	1	S_2
P24	1	1	1	0	3	1	1	2	1–2	1	N_1tf
FP1	1	1	1	1	2	2	2	2	2	1	N_2dtef
FP2	1	1	1	1	2	2	2	2	1–2	1	N_1tef
FP3	0	1	1	1	1	1	1	1	2	1	S_1tf

s: Slope; t: Texture; d: Soil depth; e: Erosion; f: Fertility.

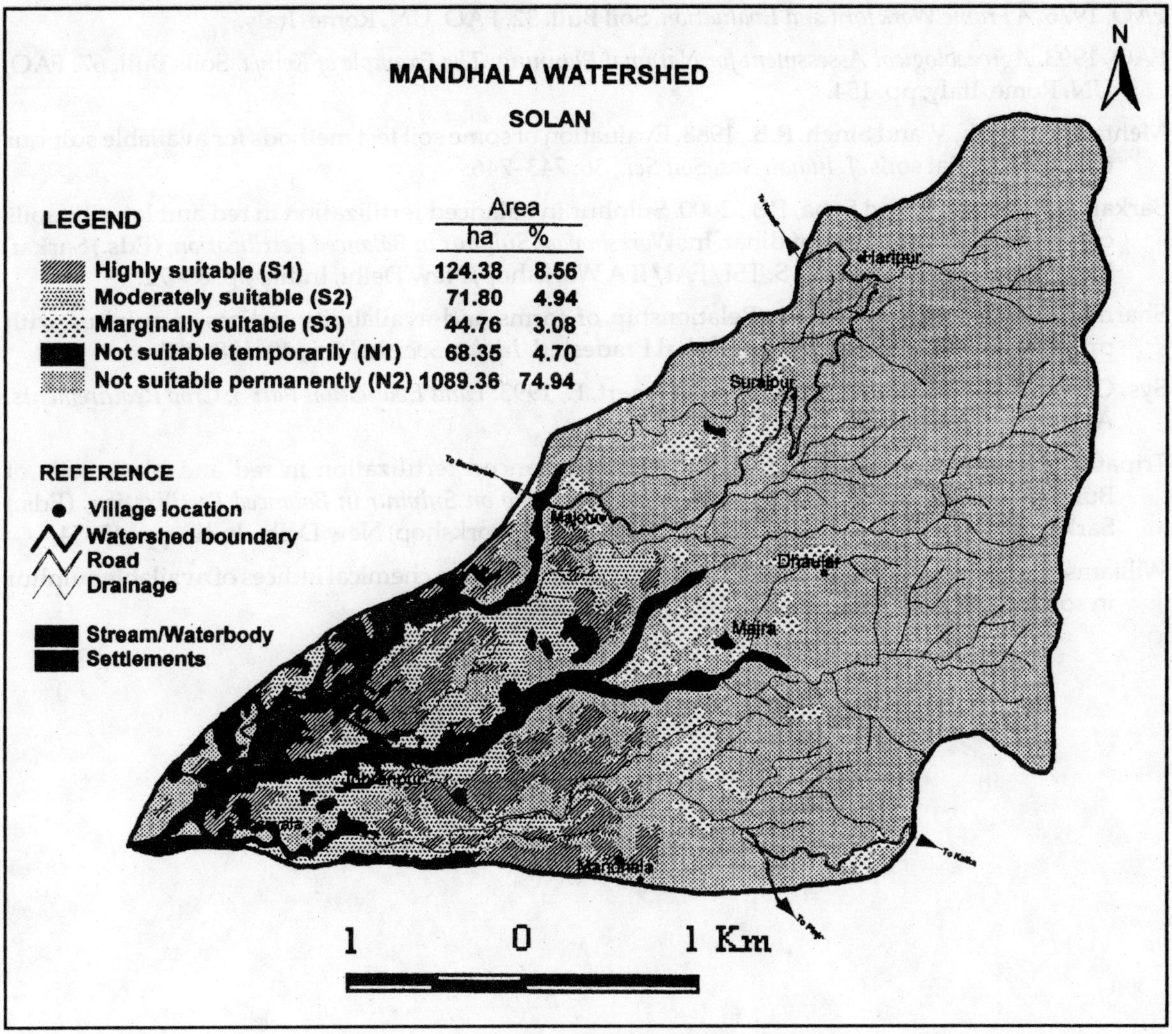

Figure 38.1: Soil-site Suitability Classes for Mustard

References

Aggarwal, V. and Nayyar, V.K., 1998. Available soil sulphur status and sulphur nutrition of wheat crop. *J. Indian Soc. Soil Sci.*, 46: 71–75.

Aulakh, M.S. and Pasricha, N.S., 1988. *Sulphur in Indian Agriculture, Section II*, 3. The Sulphur Institute, Washington DC and FAI, New Delhi, pp. 1–14.

Aulakh, M.S., Pasricha, N.S. and Sahota, N.S., 1980. Yield, nutrient concentration and quality of mustard crop as influenced by nitrogen and sulphur fertilizers. *J. Agric. Sci.*, 94: 545–549.

Chesnin, L. and Yien, C.H., 1951. Turbidimetric determination of available sulphur. *Soil Sci. Soc. Amer. Proc.*, 15: 149–151.

FAO, 1976. *A Frame Work for Land Evaluation*. Soil Bull. 32. FAO, UN, Rome, Italy.

FAO, 1993. *Agroecological Assessment for National Planning: The Example of Kenya*. Soils Bull. 67. FAO, UN, Rome, Italy, pp. 154.

Mehta, V.S., Singh, V. and Singh, R.B., 1988. Evaluation of some soil test methods for available sulphur in some alluvial soils. *J. Indian Soc. Soil Sci.*, 36: 743–746.

Sarkar, A.K., Singh, S. and Saha, P.B., 2000. Sulphur in balanced fertilization in red and lateritic soils of Chhotanagpur plateau of Bihar. In: *Workshop on Sulphur in Balanced Fertilization,* (Eds.) Sarkar, M.C., Biswas, B.C. and Das, S. TSI/FAI/IFA Workshop, New Delhi, India, pp. 65–72.

Sharma, R. and Jaggi, R.C., 2001. Relationship of forms and availability indices of sulphur with properties of soils of Kangra, Himachal Pradesh. *J. Indian Soc. Soil Sci.*, 49: 698–702.

Sys, C., Van Ranst, E., Debaveye, J. and Beemaert, F., 1993. *Land Evaluation Part 3. Crop Requirements*. Agric. Publ. 7, Belgium.

Tripathi, S.B. and Hazra, C.R., 2000. Sulphur in balanced fertilization in red and black soils of Bundelkhand region of Utter Pradesh. In: *Workshop on Sulphur in Balanced Fertilization,* (Eds.) Sarkar, M.C., Biswas, B.C. and Das, S. TSI/FAI/IFA workshop, New Delhi, India, pp. 43–54.

Williams, C.H. and Steinbergs, A., 1959. Soil sulphur fractions as chemical indices of available sulphur in some Australian soils. *Aust. J. Agric. Res.*, 10: 340–352.

Chapter 39

Effect of Different Planting Methods and Irrigation Schedules on Water Use and Water Use Efficiency of Soybean (*Glycine max* L.)

C.S. Maan, S.S. Mahal, R.K. Mahey and K.K. Vashist
Department of Agronomy, Agrometeorology and Forestry, Punjab Agricultural University, Ludhiana – 141 004

ABSTRACT

A field experiment was conducted during the *kharif* season of 2005 on loamy sand soil, to assess the effect of different planting methods *i.e.* bed planted with roller with mulch (M_1), bed planted with roller without mulch (M_2), bed planted without roller with mulch (M_3), bed planted without roller without mulch (M_4), conventional (flat) with mulch (M_5 and conventional (flat) without mulch (M_6) and three irrigation schedules viz. irrigation applied at 50 mm CPE (I_1), 65 mm CPE (I_2) and 80 mm CPE (I_3). The results revealed that maximum water use efficiency (WUE) was obtained with M_1 planting method (45.63 kg ha^{-1} cm^{-1}) and was 5.24, 5.96, 7.01, 8.14 and 10.18 kg ha^{-1} cm^{-1} higher over M_2, M_3, M_4, M_5, and M_6 methods of planting, respectively. The highest WUE (41.92 kg ha^{-1} cm^{-1}) was observed in I_3 irrigation schedule followed by 38.74 and 38.18 kg ha^{-1} cm^{-1} in I_2 and I_1 irrigation schedules, respectively.

Keywords: *Soybean, Planting method, Irrigation schedule, Water use efficiency.*

Introduction

In Punjab the prevailing rice-wheat system resulted in yield plateauing and also an adverse affect on natural ground water resources which is declining at the rate of 54 cm per year (Anonymous, 2006). Soybean is considered as potential alternative to rice due to its wider adaptability under varied soil and climatic conditions and multiple industrial uses. Its cultivation would replace rice to diversify rice-wheat rotation and also help to bridge the widening gap between demand and supply of edible oil and to overcome protein related malnutrition to great extent. So there may be some technological gaps in its management, which need to be identified and corrected. Soybean cultivation is gaining importance and in view of this proper agronomic practices are to be worked out for successful cultivation. One among them on which not much has been reported is the proper method of planting, scheduling of irrigation, facilitating optimum crop stand, which is pre-requisite for high yields. In view of this, the present experiments were planned to study the effect of land management system and scheduling of irrigation on water use, water use efficiency of soybean.

Materials and Methods

The present investigations were carried out during *Kharif* season of 2005 at student's farm, Department of Agronomy, Agrometeorology and Forestry, Punjab Agricultural University, Ludhiana on loamy sand soil, to study the effect of planting methods and irrigation schedules on water use and water use efficiency of soybean. The available nutrient status of the soil was low in organic carbon and available nitrogen, medium in available phosphorus and potassium. The field capacity and wilting point the soil were 27.40 and 12.01 per cent respectively. The bulk density was 1.56 g cc^{-1}; the soil pH and EC were 7.9 and 0.23 dSm^{-1}, respectively. Soybean (SL525) with duration of 144 days was used, as the test crop. Irrigation scheduling was based on the cumulative pan evaporation (CPE) recorded for USWB open pan evaporimeter. The total rainfall received during the crop season was 589.9 mm. The experiment was laid in split plot design with four replications comprising six planting methods *i.e.* bed planted with roller with mulch (M_1), bed planted with roller without mulch (M_2), bed planted without roller with mulch (M_3), bed planted without roller without mulch (M_4), conventional (flat) with mulch (M_5), and conventional (flat) without mulch (M_6) in main plots while three irrigation schedules *viz.*, irrigation applied at 50 mm CPE (I_1), 65 mm CPE (I_2) and 80 mm CPE (I_3) were kept in sub plots. Recommended package of practices was adopted. The consumptive use of water under different treatments was computed as per the procedure of Dastane (1967). The water use efficiency in kg ha^{-1} cm^{-1} for a given treatment was calculated dividing the seed yield by respective total consumptive use for the crop period.

Results and Discussion

Seed Yield

The treatment M_1 being (Table 39.1) at par with M_2 both treatments produced significantly higher seed yield of soybean as compare to all other planting treatments. The treatment M_2 differed non-significantly with other planting treatments except M_6, which produced lowest seed yield. Soybean sown with method of planting M_1 produced 10.8, 14.1, 14.9, 20.7 and 26.7 per cent higher seed yield as compare to M_2, M_3, M_4, M_5, and M_6 methods of planting, respectively. These results are contradictory to the findings of Singh *et al.* (1999) and Raut *et al.* (2000) who reported that bed planting gave better yield than flat sowing due to better aeration and more transfer of assimilates from source to sink which ultimately contributed towards higher seed yield in bed sown crop. Different scheduling of irrigation at I_1, I_2 and I_3 resulted non-significant differences as all the growth and yield contributing characters varied markedly. The numerical increase in seed yield under I_1 irrigation schedule was 3.3 and 4.5 per cent over I_2 and I_3 irrigation schedules, respectively. Jagathambae and Karivaratharaju (1998) also observed similar results with respect to different irrigation schedules. The interactional effects of different planting methods and irrigation schedules were found to be non-significant.

Table 39.1: Water Use and Water Use Efficiency of Soybean as Influenced by Different Planting Methods and Irrigation Schedules

Treatments	*No. of Irrigations*	*Irrigation Depth (cm)*	*Effective Rainfall (cm)*	*Water Use (cm)*	*Seed Yield (q ha^{-1})*	*WUE (kg ha^{-1} cm^{-1})*
Planting Methods						
M_1	–	5.0	33.93	55.40	25.28	45.63
M_2	–	5.0	33.93	56.48	22.81	40.39
M_3	–	5.0	33.93	55.83	22.15	39.67
M_4	–	5.0	33.93	56.96	22.00	38.62
M_5	–	7.5	33.93	55.85	20.94	37.49
M_6	–	7.5	33.93	56.31	19.96	35.45
CD (P = 0.05)	–	–	–	–	2.55	–
Irrigation Schedules						
I_1	6	–	31.7	59.61	22.76	38.18
I_2	5	–	32.1	56.86	22.03	38.74
I_3	3	–	37.8	51.95	21.78	41.92
CD (P = 0.05)	–	–	–	–	NS	–
Interaction	–	–	–	–	NS	–

Water Use

The highest seasonal (Table 39.1) water use (56.96 cm) was obtained with M_4 method of planting, which was 0.48, 0.65, 1.11, 1.13 and 1.56 cm more than that of M_2, M_6, M_5, M_3 and M_1 methods of planting, respectively. The water use of different planting methods and irrigation levels showed that the mulched treatments had less water use as compare to the non-mulched plots. The frequent showers during the growing season might have vitiated the soil-moisture status and thus caused marginal differences in the soil moisture. Different irrigation schedules also influenced the water use of soybean. Maximum water use (59.61 cm) was recorded with I_1 irrigation schedule, which is followed by I_2 (56.86 cm) and 13 (51.95 cm), respectively. Frequent irrigation application might be the cause of more water use in I_1 (six post sowing irrigation) than I_2 (five post sowing irrigation) and I_3 (three post sowing irrigation). Dubey *et al.* (1995) reported that maximum water use was obtained when irrigation was applied at 1.00 IW : CPE ratio compared to 0.50 and 0.75 ratios of IW and CPE.

Water Use Efficiency (WUE)

The maximum WUE was obtained (Table 39.1) with M_1 planting method (45.63 kg ha^{-1} cm^{-1}) which was 5.24, 5.96, 7.01, 8.14 and 10.18 kg ha^{-1} cm^{-1} more over M_2, M_3, M_4, M_5 and M_6 methods of planting, respectively. Bharambe *et al.* (1999) observed that the water use efficiency was more in a broad bed and furrow layout compared to flat layout. Among the irrigation schedules maximum WUE (41.92 kg ha^{-1} cm^{-1}) was observed in I_3 followed by 38.74 and 38.18 kg ha^{-1} cm^{-1} in hand I_1 irrigation schedules, respectively which was at the tune of 3.18 and 3.74 kg ha^{-1} cm^{-1} more than I_2 and I_1 irrigation schedules. Dubey *et al.* (1995) reported that maximum water use was obtained when irrigation was applied at 0.75 IW : CPE ratio compared to 0.50 and 1.00 ratios of IW and CPE.

References

Anonymous, 2006. Punjab facing groundwater crisis. *The Hindu*, Mar 04, 2006.

Bharambe, P.R., Oza, S.R., Shelke, D.K. and Jadhav, G.S., 1999. Effect of irrigation on crop productivity and soil-plant water relation in soybean-rabi sorghum cropping system. *J. Indian Soc. Soil Sci.*, 47: 689–694.

Dastane, N.G., 1967. *A Practical Manual for Water Use: Research in Agriculture*, 2nd edn. Navbharat Prakashan, Pune.

Dubey, O.P., Dixit, J.P., Tiwari, K.P. and Khan, R.A., 1995. Influence of irrigation on growth, yield and water-use efficiency of spring soybean (*Glycine max*) genotypes. *Indian J. Agron.*, 40: 616–619.

Jegathambal, R. and Karivaratharaju, T.V., 1998. Influence of irrigation and phosphorus levels on seed quality characteristics in soybean. *Madras Agric. J.*, 85: 177–178.

Raut, V.M., Taware, S.P., Halvankar, G.B. and Varghere, P., 2000. Comparison of different sowing methods in soybean. *J. Maharashtra Agric. Univ.*, 25: 218–219.

Singh, P., Alagarswamy, G., Hoogenboom, G., Pathak, P., Wani, S.P. and Virmani, S.M., 1999. Soybean-chickpea rotation on vertic inceptisols II. Long-term simulation of water balance and crop yields. *Fld Crops Res.*, 63: 225–236.

Chapter 40

A Study on the Selection of Habitat by the Post Larval and Juvenile Banana Shrimp, *Penaeus merguiensis* in and Around Kali Estuary, Karwar

S. Sampath Kumar and Kusumia Neelakantan
School of Ocean Sciences, Kodibag, Karwar – 581 303

ABSTRACT

The distribution of post larval and juvenile stages of *Penaeus merguiensis* in four different habitats adjoining the Kali estuary having different abiotic characteristics were examined for a period of one lunar month during the pre-monsoon season, when maximum availability of this species occurred. The results indicate that the variation in the density was due to the difference in substrate features such as pH, Organic content, sediment fraction type and vegetation cover. The study shows that slightly low pH might make the immigrating post larval forms to move out of a particular habitat. The present investigation provides evidence to suggest that though muddy substratum is the most preferred habitat, it can tolerate other substrate types, if the habitat provides sufficient amount of food and protection. The occurrence and density of post larvae and juvenile of *P. merguiensis* in a specific habitat may be dependent on the particular environmental factor like muddy substratum, optimal pH of water and sediments, water temperature, oxygen content, vegetation covers etc.

Keywords: Habitat selection, Banana shrimp, Penaeus merguiensis.

Introduction

The influence of environmental factors on the process of habitat selection by penaeids in general and *Penaeus merguiensis* in particular is not well understood. Most species utilize habitats, which encompass ecologically hospitable abiotic and biotic parameters, guided by their physiological limits of tolerance. The habitat selection is essentially the relationship between behavior and environment which determines the local distribution (Meadows and Campbell, 1972). *P. merguiensis* like most penaeids make use of mangrove lined estuary as the nursery feeding ground (Gunter, 1967; Staples and Vance, 1979; Staples *et al.*, 1985; Dall *et al.*, 1990; Vance *et al.*, 1990). Studies have been carried out in the Kali estuary Karwar to assess the seasonality in the distribution and abundance of penaeid prawns in general (Gunaga *et al.*, 1989) and *P. merguiensis* in particular (Sampath Kumar, 1996). Some information is available on the habitat requirement of *P. merguiensis* from Australian waters (Staples *et al.*, 1985; Vance *et al.*, 1990; Robertson, 1988) and also from the waters of Goa (Achuthankutty, 1988).

The present study was an extension of a work on the seed abundance of *P. merguiensis* in relation to some abiotic and biotic factors in Kali estuary for a period of two years (Sampath Kumar, 1996). During the study it was noted that even though immigration was occurring into all the creeks during the flood tides, the density of post larvae and juveniles in different creeks adjoining the estuary during low tide were different. As a result of this observation, this particular study was planned and carried out, during the peak immigration period, to know the environmental conditions prevailing in the adjoining locations and the factors limiting the survival in these areas, which in turn might help in elucidating the habitat preference of *P. merguiensis* and to identify the natural causes of variation in density of post larval and juveniles of *P. merguiensis* in the Kali estuary.

Materials and Methods

The habitat preference and selection were calculated on the basis of the number of individuals obtained per unit area and is expressed as no/m^2.

The study was carried out during the pre-monsoon season, when the immigration was frequent and the abundance of the post larvae and juvenile stages of *P. merguiensis* in the estuary was at its peak (Sampath Kumar, 1996) at four study stations, referred to as Habitat 1, Habitat 2, Habitat 3 and Habitat 4.

Weekly sampling was carried out at all the four habitats, during low tide for a period of one lunar month. The sampling was done using a standard velon screen hand held drag net, having an area of 3 × 1 m and 2 mm mesh size. The sampling was done at each sampling area, by dragging the net over a distance of 10 mts, near the edge of the water, by holding the two ends of the net by two people and counting the number of post larvae and juveniles obtained per drag. The net was dragged 3 times and the average number was calculated. Concomitantly, surface water and sediment (core) samples were collected for analysis.

Temperature was measured using a centigrade thermometer. Salinity was determined by Mohrknudson's method (Strickland and Parson, 1956), dissolved oxygen was estimated by Winkler's method (Strickland and Parson, 1956), and nutrients like Nitrate, Nitrite, Phosphate and silicate were determined by following the methods of Strickland and Parson (1956). Organic matter in sediments was estimated by the method of El Wakeel and Riley (1956) and pH was measured using pH meter (Systronic make, model 7024). Size fraction was determined by sieve analysis and pipette analysis (Holme and Intyre, 1984). Suspended load was determined by filtering 1 liter of water sample collected from the sampling site, through preheated weighed G.f/C filter paper and weighing it after drying till constant weight was attained and is expressed as mg/l.

Study Stations

Habitat 1: Kodibag Creek

This creek joins the Kali estuary along the southern bank at a distance of 0.3 km from the mouth. Even though only about 1 km long, the link with the estuary is lost during low tide as major portion of the creek bottom lies exposed during low tide, except the far end where the water depth is around 1.5 m. It appears like a pond, when the connection with the estuary is lost, during low tide. The area is bordered by a few mangrove shrubs. The initial survey conducted revealed that immigration of post larvae of *P. merguiensis* does occur during the flood tide, but sometimes no shrimps were caught during the ebb tide.

Habitat 2: Mavinahole Creek

This creek joins the estuary along the northern side, nearly opposite to Kodibag creek and is approximately 6 kms long. The creek near the study area is bordered by fringing mangroves. The creek bottom is covered by sea grass and water is shallow and sediment is sandy.

Habitat 3: Sandy Backwater Area

This is located on the southern side of the Kharland bund near the mangrove free area of the Kadwad backwaters of the Kali estuary. It is approximately 6 kms from the mouth of the estuary and has a sandy bottom with no vegetation cover, which gets exposed during the low tide.

Habitat 4: Muddy Mangrove Area

This is located in the backwater area of Kadwad, near to Habitat 3 on the northern side of the Kharland bund. This area includes a steep intertidal bank adjoining the fringing mangrove forest.

Statistical Analysis

The data acquired from the investigation was subjected to correlation analysis (Snedecor and Cochran, 1967) to test the significance of the relationship.

Results and Discussion

From Table 40.1, it can be seen that only post larvae and early stages of juveniles were seen at Habitat 1 and Table 40.2 shows that during 2nd and 4th collection they were totally absent at Habitat 1 thus suggesting that even though post larvae entered the creek, they did not remain or survive there as some of them were seen to move out during the following ebb tide along with the water and were collected in the drag net tide against the receding waters. This may be attributed to the slightly low pH of the water.

Table 40.1: Percentage Availability of Different Size Groups Collected from Each Habitat

	Size Group	Habitat 1	Habitat 2	Habitat 3	Habitat 4
Percentage Availability of Different Size Groups Collected from Each Habitat	60	0	0	0	21
	50 mm	0	4	0	42
	40 mm	0	33	0	92
	30 mm	0	21	0	105
	20 mm	18	40	0	147
	10 mm	7	6	0	12

Table 40.2: Variation in the Density of Post Larval and Juveniles of *P. merquiensis* at Different Habitats

	Collections	Habitat 1	Habitat 2	Habitat 3	Habitat 4
Post larval and	1	19	83	0	234
Juvenile Density	2	0	121	0	583
(no/m²)	3	84	144	0	302
	4	0	117	0	148
	5	21	56	0	827

William's had indicated that hydrogen ion concentration of the substrate may influence attraction to a substrate, but the present observation suggests that slightly low pH can induce repulsion from a particular substrate type, when it is not tolerable. Pedini (1981) has shown that acidic sediments can decrease growth rates and increase mortality in aquaculture ponds.

The results of hydrological analysis shows that, the pH of water. pH of the sediment and dissolved oxygen was less in comparison to all other habitats (Table 40.3). Even though there was no significant relationship existing between seed abundance and any particular environmental factor statistically, at Habitat 1 there existed significant relationship between water temperature and sediment temperature (at $P < 0.05$), water temperature and dissolved oxygen (at $P < 0.01$), which might suggest that as the soil is exposed to the sun during day time at low tide it gets heated up and in turn increasing the temperature of the inflowing flood tide waters, which might result in decreasing the oxygen content of water.

Table 40.3: Variation in the Abiotic Factors and Seed Abundance at Different Habitats During the Study Period

Parameters	Habitat 1	Habitat 2	Habitat 3	Habitat 4
Seed Abundance (no/m²)	124	521	0	2094
Water Temperature (°C)	30.1	31.3	30.4	30.4
Salinity (per cent)	27.5	28.6	19.6	20.2
pH	6.4	7.3	7.8	7.8
Suspended Load (mg/g)	0.009	0.009	0.007	0.007
Dissolved Oxygen (ml/l)	2.9	3.9	4.6	5.2
Sediment Temperature (°C)	29.7	31.2	31.1	30
pH	6.6	7.6	7.7	7.7
Organic Carbon (mg/g)	4.9	5.4	0.4	6.3
Sand (per cent)	99.5	86.3	99.5	11.8
Silt (per cent)	0.5	10.5	0.5	33.7
Clay (per cent)	0	3.2	0	54.5

At Habitat 2 the density was (56–144 no/m²) greater than Habitat 1 (0–84 no/m²). This habitat recorded almost all the size groups except the larger size group (Table 40.1), even though the density was low in comparison to Habitat 4 (148–827 no/m²). The continued presence of early stages suggests regular immigration during flood tides (Staples and Vance, 1979). The amount of organic content of the sediment implies richness in terms of food availability for the shrimps. The post larvae appear to survive even though the substrate was predominated by sand (Table 40.3). The lack of muddy substrate, which is its preferred habitat (Robertson and Duke, 1987) may be outweighed by the protection rendered by vegetation cover and the organic richness as a food source of food.

Habitat 3 was sandy and without any vegetation cover and no post larvae were collected. This might be attributed to the lack of food (absence of organic matter) and lack of protection by way of the absence of vegetation cover and no soft muddy substratum. The presence of coarse sand makes it difficult to bury quickly, unlike muddy substrate. Somers *et al.* (1987) inferred from their study that low organic carbon becomes a limiting factor in the distribution of the commercial species, during their estuarine phase. Variation in shrimp abundance has been linked to the organic carbon content of the sediment both in Gulf of Mexico (Gardy, 1971) and in the Sudanese Red Sea (Branford, 1981).

Habitat 4 recorded very high densities (148–827 no/m^2) and all the size groups were represented. This may be due to factors like, increased oxygen content, high organic detritus (Table 40.3), floral and faunal abundance in the water and sediments (Sampath Kumar, 1996) and protection offered by the soft muddy substratum.

Correlation analysis did not reveal any significant relationship between seed abundance and environmental factors at any Habitats and hence the data is not presented.

Sick *et al.* (1972) have inferred that a suitable substratum is not only important for prawn species to maintain their normal burrowing habits but also in promoting high survival and growth rates in culture systems. The present result apparently agrees with that of Young and Carpenter (1977) who determined that the final selection of a nursery ground probably depends on habitat features such as depth, substratum and types of plant or other cover for early juveniles of various species.

The greater densities in muddy mangrove area are mainly due to the aggregation of immigrating population (Sampath Kumar, 1996). Vance *et al.* (1996) observed during their study that large catch of *P. merguiensis* post larvae were only recorded from mangrove lined banks of the estuary, although small numbers of post larvae settled in all parts of the estuary.

The present observation at Habitat 4 provides further evidence for the hypothesis suggested by Robertson and Duke (1987) that mangrove habitat is more productive and has greater food resource for consumers. However they further added that the structural heterogeneity provided by mangrove tree, roots and mangrove snags in creeks afford greater protection from predators. In the present study it was observed that during the low tide the water recedes to such an extent that it reaches below the level of any growing mangrove trees, and has no prop roots or stilt roots to provide protection to the inhabiting organism. The only way of self defense in this habitat of high density seems to be by burying themselves in the muddy substratum and disappearing from its predators. Hence it shows that the degree of protection that the shrimps gains from the mangrove roots is less significant in comparison to the muddy substratum.

The observation recorded at Habitat 1 and Habitat 2 is similar to the influence drawn by Vance *et al.*, 1990. From their study in the Embley estuary they concluded that, if the habitat in which the post larvae and small juveniles settles, bears less than optimal environmental conditions, then either they move back into the main estuary to be carried by subsequent flood tides (Habitat 1) or remain where they first settle and be subjected to increased predation and decreased growth (Habitat 2). The difference in the densities of post larvae and juveniles of *P. merguiensis* recorded at Habitat 2 and Habitat 4 might help us in suggesting that, Habitat 2 does not bear very optimal environmental condition for the species, but remains there as it gets food and protection. In addition the temperature recorded is relatively higher than all other habitats.

As the environmental condition of the estuary during the study in general was appropriate for immigration and proper growth (Gunaga *et al.*, 1989, Sampath Kumar, 1996), the differences in densities at each habitat was due to the differences in the prevailing environmental conditions (Table 40.4). Thus it is possible to conclude that, muddy areas provide the most protection to these prawns and it may be their most preferred habitat and if they land up in a sandy habitat, they might remain there if the area provides protection by means of vegetation cover. If the inundating waters take the post larvae to an area with unfavorable pH, then they do not remain there and move out along with the receding waters.

Table 40.4: Probable Abiotic Factors that might be Having Some Influence on Seed Abundance, at Different Study Sites

Factors	*Habitat 1*	*Habitat 2*	*Habitat 3*	*Habitat 4*
Seed Abundance (no/m²)	124	521	0	2094
pH of Water	6.4	7.3	7.8	7.8
Dissolved Oxygen (ml/l)	2.9	3.9	4.6	5.2
Sediment pH	6.6	7.6	7.7	7.7
Organic Carbon (mg/g)	4.9	5.4	0.4	6.3
Sand (per cent)	99.5	86.3	99.5	11.8
Clay (per cent)	0	3.2	0	54.5

References

Achuthankutty, C.T., 1988. Biology of the banana prawn, *Penaeus merguiensis* de Man from Goa waters. *Journal of Aquatic Organisms*, 21: 1–10.

Branford, J.R., 1981. Sediment and distribution of penaeid shrimp in Sudanese Red Sea. *Estuar. Coast. Shelf Sci.*, 13: 349–354.

Dall, W., Hill, B.J., Rothlishberg, P.C. and Staples, D.J., 1990. The biology of the penaeidae. *Adv. Mar. Biol.*, 27: 1–489.

El-Wakeel, S.K. and Riley, J.P., 1956. Determination of organic carbon in marine muds. *J. Cons. Perm. Int. Explor. Mer.*, 22: 180–183.

Gardy, J.R., 1971. The distribution of sediment properties and shrimp catch on two shrimp shrimping grounds on the continental shelf of Gulf of Mexico. *Proc. Gulf and Caribbean Fish. Inst.*, 23 (Ann. Sess.): 139–148.

Gunaga, V.R., Neelakantan, K. and Neelakantan, B., 1989. Seasonal abundance of penaeid post larvae in the Kali estuary, Karwar. *Mahasagar*, 22(4): 193–197.

Gunter, G., 1967. Some relationships of estuaries to the fisheries of the Gulf of Mexico. In: *Estuaries*, (Ed.) G.H. Lauff. American Association for the Advancement of Science, Washington D.C, 621.

Holme, N.A. and McIntyre, A.D., 1984. *Methods for the Study of Marine Benthos*, Blackwell Scientific Publication, Oxford, 172 pp.

Meadows, P.S. and Campbell, J.I., 1981. Habitat selection by aquatic invertebrates. In: *Adv. Mar. Biol.*, (Eds.) Russel, F.S., Young, M. Academic Press, London, 10: 271.

Pedni, M., 1981. Penaeid shrimp culture in tropical developing countries. *FAD. Fish Cil.*, 732 pp.

Robertson, A.I., 1988. Abundance, diet and predators of juvenile banana prawn, *Penaeus merguiensis* in a tropical mangrove estuary. *Aust. J. Mar. Freswat. Res.*, 39: 467–478.

Robertson, A.I. and Duke, N.C., 1987. Mangroves are nursery sites: Comparison of the abundance and species composition of fish and crustaceans in mangroves and other near shore habitats in tropical Australia. *Mar. Biol. (Berl.)*, 96: 193–206.

Sampath Kumar, S., 1996. Some eco-physiological aspects of *Penaeus merguiensis* from Karwar waters. *Ph.D. Thesis*, Karnatak University, 265 pp.

Sick, L.V., Andrews, J.W. and White, D.B., 1972. Preliminary studies of selected environmental and nutritional requirements for the culture of penaeid shrimps. *Fish. Bull. US Fish. Wildlife Ser.*, 70: 101.

Sivabalan, A. 1993. Studies on the ecology of the mangrove of Kali estuary, Karwar. *Ph.D. thesis*, Karnatak University. 390 pp.

Snedecor, G.W. and Cochran, W.C., 1967. *Statistical Methods*, 6th edition. Oxford and IBH Publishing Co., New Delhi, 172 pp.

Somers, I.F., Croaces, P.J. and Hill, B.J., 1987. Distribution and abundance of tiger prawns *Penaeus esculentus* and *Penaeus semisulcatus*. *Aust. J. Freshwat. Res.*, 38: 63.

Staples, D.J. and Vance, D.J., 1979. Effects of changes in catchability on sampling of juvenile and adolescent banana prawns *Penaeus merguiensis* de Man. *Aust. J. Mar. Freshwat. Res.*, 30: 511–519.

Staples, D.J., Vance, D.J. and Heals, D.S., 1985. Habitat requirement of juvenile penaeid prawn and their relationship to offshore fisheries, In: *Sec. Aust. Prawn Sem.*, (Eds.) Rothlisberg, P.C., Hill, B.J. and, Staples, D.J. NPS_2 Clevland, Australia, 47.

Strickland, J.D.H. and Parson, T.R., 1956. A practical handbook of seawater analysis. *Bull. Fish. Res. Bd. Canada*, 167: 310.

Vance, D.J., Haywood, M.D.E. and Staples, D.J., 1990. Use of a mangrove estuary as a nursery area by post larval prawns, *Peneus merguiensis* de Man, in northern Australia. *Estuar. Coastal Shelf Sci.*, 31: 689–701.

Williams, A.B., 1958. Substrate as a factor in shrimp distribution. *Limnol Oceanogr.*, 3: 283–291.

Young, P.C. and Carpenter, S.M., 1977. Recruitment of post larval prawns to nursery areas in Moreton Bay, Qld. Australia. *Aust. J. Mar. Freshwat. Res.*, 28: 745–773.

Chapter 41

Impact of Ethanol and *Mucuna pruriens* Seed on Rat Testicular Phosphomonoesterases

G. Krishnamoorthy, A. Sivamady and A. Antoniraj
Department of Zoology, K.M. Centre for P.G. Studies, Pondicherry – 605 008, India

ABSTRACT

Oral administration of ethanol induce structural damage to the seminiferous tubules and reduce the population of spermatozoa in testis. Changes in the activities of acid phosphatase and alkaline phosphatase in response to ethanol, ethanol + *Mucuna pruriens* seed treatment indicates the functional and structural status of testis. The importance of *Mucuna pruriens* seed on seminiferous tubular regeneration and fertility improvement in alcohol fed rat has been discussed.

Keywords: *Ethanol, Mucuna pruriens, Acid phosphatase, Alkaline phosphatase.*

Introduction

Alcoholics often have fertility disturbances like impotence, loss of libido and sterility (Boyden and pampenter, 1983). Semen samples of alcoholics shows decreased sperm count, impaired motility and morphologically abnormal spermatozoa (Gomathi *et al.*, 1993; Villata *et al.*, 1997). Ethanol administration not only disturbs the testicular biochemical and metabolic pathways and also damage its histoarchitecture (Reddy *et al.*, 1997; Oh *et al.*, 2002). Further intake of ethanol inhibits the secretions of epididymis and retard the sperm maturation (Willis *et al.*, 1983; Haider *et al.*, 1985). Ethanol act directly at different level and causes many hormonal changes which induce gonad functional impairment and fertility problems (Calleja *et al.*, 1997). Chronic alcohol consumption exerts an inhibitory effect on gonodotrophs in the pituitary (Somer *et al.*, 1988; Ciero *et al.*, 1997) and decrease the gonodotropin release (Rettori and Mc Cann, 1997).

Interest on herbal medicine research has been increasing recently. Review of literature also enlights the value of medicinal plants on the fertility improvement with less side effects. *Mucuna pruriens* recognized as an aphrodiasic in ayurveda, is known to enhance the functional activities of gonad and sex accessory organs (Saxena and Dixit, 1987). Further, the *Mucuna pruriens* seed extract treatment bring about a noteworthy increase of sperm population in experimental rats (Amin, 1996). Therefore, the present study was aimed to elucidate the augumentary role of *Mucuna pruriens* seed on testes in ethanol treated rats.

Materials and Methods

Fifteen adult male albino rats (*Rattus norvegicus*) were selected and randomly divided into three groups of five each. They were given balanced diet and water *ad libitum* and maintained in temperature controlled room. Group I includes control rats (Received an isocaloric quantity of sucrose) and Group II includes Ethanol treated (25 per cent of ethanol (1 ml), twice daily at regular intervals for 30 days). Group III includes Ethanol (1 ml of 25 per cent) + *Mucuna pruriens* seed extract (15 mg per kilogram body weight for 30 days, twice daily at regular intervals). At the end of experimental period (30 days) the animals were sacrificed. The testes were removed immediately and processed for further analysis. Acid phosphatase was estimated following the method of Tenniswood *et al.* (1976) and the Alkaline phosphatase was assayed by the method of Bessey *et al.* (1946) using reagent kits (Span diagnostics).

Results and Discussion

In the present study ethanol treatment severely damaged the histoarchitecture of the testes in rats. Shrunken seminiferous tubules with narrowed lumen and vacoulated space in the interstitial region was observed in ethanol treated rats. Earlier studies also reported the testicular damage following alcohol treatment (Haider *et al.*, 1985; Zhu *et al.*, 1997). The structural integrity of seminiferous tubule was drastically disturbed by alcohol intake. The testicular level of lipid peroxide was increased by alcohol, which may be involved in part in testicular dysfunction and damage (EI-Sokkazy *et al.*, 1999). In addition the rupture and atrophy of seminiferous tubules in ethanol treated rats may be due to microtubular disruption of the differentiating germ and sertoli cells (Nakai and Hess, 1997). Further, the increased activities of Acid phosphatase and Alkaline phosphatase in ethanol treated rats can be attributed to destruction of cell membrane and testes damage (Saxena and Sarin, 1980). These degenerative changes in testes due to raised phosphomonoesterases action are very well correlated to the histological findings of the present study.

Table 41.1: Effect of Ethanol and *Mucuna pruriens* Seed on Phosphomonoesterases in Rat Testis

Group	*Alkaline Phosphatase*			*Acid Phosphatase*		
Control	0.35		0.02	0.33		0.02
Ethanol treated (E)	0.77	±	0.03***	0.70	±	0.03***
Ethanol + Mucuna pruriens seed treated (EM)	0.38	±	C 0.03	0.39	±	C 0.02

Each value is Mean ± SEM of five animals.

Enzyme activities are expressed as μ moles of p-nitrophenol formed/h/μg Protein.

***: $P < 0.001$ = Control Vs Ethanol treated (E); C: $P < 0.001$ = Ethanol treated + *Mucuna pruriens* seed treated (EM).

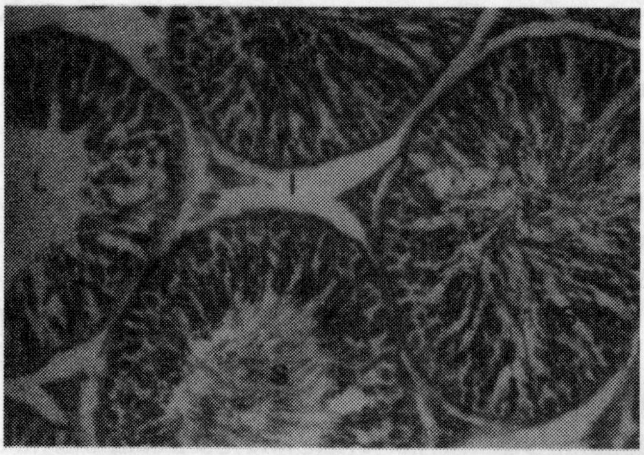

(A) Intact seminiferous tubules (ST) with wide lumen (L), spermatozoa (S) and interstitial cells (I) of control rat

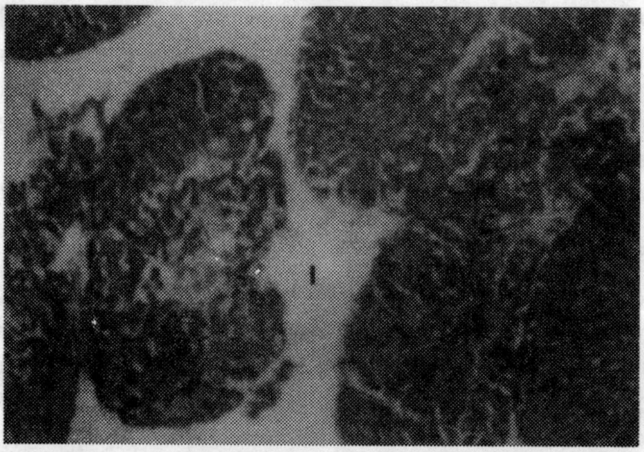

(B) Shrunken seminiferous tubules (ST) and reduced in interstitial cells (I) of ethanol treated rat

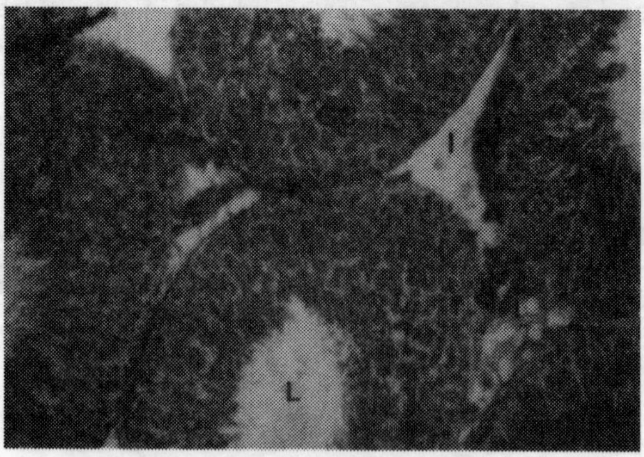

(C) Undamaged seminiferous tubules (ST) with more germ cells (GC) and spermatozoa (S) of ethanol + Mucuna treated rat

Figure 41.1: Histology of Rat Testis

In the combined treatment the Mucuna seed extract not only prevents the ethanol, induced structural damage and also act as a promotor for proliferation of sertoli, germ and leydig cells. The enrichment of cells within the seminiferous tubules reveals the augumentary role of Mucuna seed on spermatogenesis. The *Mucuna pruriens* seed extract itself exhibits androgen activity (Saxena and Dixit, 1987) and also induce the synthesis of testicular androgen (Amin, 1996). The activities of Acid phosphatase and Alkaline phosphatase were brought to normal in ethanol + Mucuna treatment. This restoration may be due to the stabilization of testicular lysosome. The bioactive principles like steroid from Mucuna along with testicular androgen would have improved the stability of lysosome. So, the *Mucuna pruriens* seed extract may be beneficial to improve the testicular function in alcoholics.

References

Amin, K.M.Y., 1996. Ayurvedic properties of Mucuna. *Fitoterapia,* 67: 53–58.

Bessey, C.A., Lowry, C.H. and Brock, M.S., 1946. A method for the rapid determination of alkaline phosphatase with five cubic millimeters of serum. *J. Biol. Chem.*, 164: 321–330.

Boyden, T.W. and Pampernter, R.W., 1983. Effects of ethanol on the male hypothalamic pituitary gonadal axis. *Endocr. Rev.*, 4: 384–395.

Calleja, E.J., Rodriguez, T.L.A., Fernandez, H.B., and Uaquerom, P.C., 1997. Testicular changes produced by alcohol. *Urologica Espanolas.* 21: 337–342.

Cicero, T.J., Greenwald, J., Nock, H.C. and Connor, L., 1990. Castration–induced changes in the response of the hypothalamic pituitary axis to alcohol in the male rat. *J. Pharmacol. Exp. Ther.*, 252: 456–461.

El-Sokkary, G.R., Reiter, R.J., Tan, D.X., Kim, S.J., and Cabrera, J., 1999. Inhibitory effect of melatonin on products of lipid peroxidation resulting from chronic ethanol administration. *Alcohol Alcoholism,* 34: 842–850.

Gomathi, C., Balasubramanian, K., Vijayabanu, N., Srikanth, V. and Govindarajulu, P., 1993. Effects of chronic alcoholism on semen studies on lipid profiles. *Int. J. Androl.*, 16: 175–181.

Haider, R., Grignon, G., and Guedenet, J.C., 1985. Development of the seminal vesicles. *Prog. Reprod. Bio. Med.*, 12: 4–7.

Nakai, M. and Ress, R.A., 1997. Effects of carbendazim on meiotic spermatocytes and subsequent spermiogenesis in the rat testis. *Anat. Rec.*, 247: 379.

Oh, S.L., Lee, M.S., Kim, C.L., Sang, K.Y. and Park, S.C., 2002. Asparate modulaterthe ethanol-induced oxidative stress and glutathione utilizing enzymes in rat testes. *Exp. Mol. Med.*, 34: 47–52.

Rettori, V. and Mc Cann, S.M., 1997. The mechanism of alcohol to suppress gonadotropin secretion. *Mol. Psychiatry,* 2: 350–354.

Saxena, A.K. and Sarin, K., 1980. Pathological and biochemical changes in the liver and testis of the desert gerbil. Effect of a single intra peritoneal injection of phorate. *Ind. J. Exp. Biol.*, 18: 1001–1004.

Saxena, S. and Dixit, V.K., 1987. Androgenic properties of Mucuna pruriens. *Ind. Nat. Prod.*, 3: 3–7.

Somer, L., Wrobel, K.R. and Schimnael, M., 1988. Long-term alcohol consumption. *Acta Anat.*, 131: 41–46.

Tenniswood, M., Biri, C.F. and Clark, A.F., 1976. Acid phosphatase androgen dependent markers of rat prostate. *Can. J. Biochem.*, 54: 350–355.

Villata, J., Ballesca, J.L., Cicolas, J.M., Martinez-de-Osaba, M.J., Antumez, E. and Pimental, C., 1997. Testicular function in asymptomatic chronic alcoholics: Relation to ethanol intake. *Alcohol Clin. Exp. Res.*, 21: 128–133.

Willis, B.R., Anderson, R.A., Jr., Oswald, C. and Zaneveld, L.J.D., 1983. Ethanol, induced male reproductive tract pathology as a function of ethanol dose and duration of exposure. *J. Pharmacol. Exp. Ther.*, 225: 470–478.

Zhu, Q., Van Thiel, D.H. and Cravaler, J.S., 1997. Effects of ethanol on rat Sertoli cell function. *Studies in vitro*, 21: 1409.

Chapter 42

Evaluation of Basmati Rice Based Crop Sequences for Productivity, Economic Returns and Soil Health

M.S. Sidhu* and R. Sikka**

***Senior Agronomist and **Soil Chemist**

Department of Agronomy and Agrometerology, Punjab Agricultural University, Ludhiana – 141 004, Punjab

ABSTRACT

A field experiment was conducted consecutively for three years at research farm of Punjab Agricultural University, Ludhiana, to evaluate the productivity of different basmati rice based crop sequences and their effect on soil health. Maximum mean grain yield of basmati rice was recorded in fallow-basmati rice sequence (22.9 q/ha) and minimum grain yield was recorded in green manure-basmati rice sequence (14.3 q/ha). However, the yield of subsequent crops, wheat, celery, field peas and sunflower increased substantially where green manure was practiced. Maximum basmati rice equivalent yield was recorded in *bajra* (f)-basmati rice-sunflower crop sequence (53.8 q/ha) and minimum was obtained in green manure-basmati rice-celery sequence (25.7 q/ha). The gross returns also followed the same trend. The mean net returns were, however, maximum in *bajra*-basmati rice crop sequence (Rs. 19900 /ha) and minimum in green manure-basmati rice sequence (Rs. 8800 /ha) but there was not much variation in the fallow-basmati and *bajra* (*f*)-basmati rice crop sequence. Incorporation of *Sesbania aculeata* increased the organic carbon from 0.33 per cent as its initial value to 0.49 per cent and available N from 110 kg/ha to 127 kg/ha after three crop cycles. The content of available P increased than its initial status under all the treatments. However, a significant decrease in available K content of the soil was observed after three crop cycles under all treatments.

Keywords: *Diversification, Green manure, Basmati rice, Productivity, economic returns, soil fertility*

Introduction

Growing demand for food in the country due to burgeoning human population on one hand and progressive shrinking of agricultural land available per capita on the other hand led to intensification of time and space dimensions in Indian agriculture. This has been made possible through adoption of multiple cropping where more than two crops are grown one after another during a year on the same piece of land.

Rice-wheat crop sequence is the backbone of India's food security. This system occupies an area of 10.5 m ha in the Indo-Gangetic plains of India. There are reports of grain yield stagnation and decline in soil productivity (Yadav *et al.*, 1998), due to continuous adoption of this system. Rapid spread of rice-wheat system in the post-green revolution era in this region caused an eclipse on sustainability of soil productivity (Rattan and Singh, 1997) and low economic returns.

The decline in crop yields and economic returns could be minimized by introduction of alternate crops and crop sequences and inclusion of green manures (Chauhan, 2000). Basmati rice is one such alternative known for its aroma and cooking quality and has great export potential. Generally wheat is grown in sequence with basmati rice but wheat gives poor production due to its delayed sowing. Moreover, both the crops are cereals, which create problem of mining of nutrients from soil. Sunflower another such crop has also been found remunerative than late sown wheat

Therefore, a need is felt to diversify the rice-wheat system by introducing celery, field peas sunflower and *bajra* in rotation with basmati rice and including organic source of fertilizer like green manure crop May and June before planting basmati rice. Accordingly, different basmati rice based crop sequences were tried to evaluate their productivity, economic returns and their effect on soil health.

Materials and Methods

A field experiment was conducted at the research farm of Punjab Agricultural University, Ludhiana for 3 consecutive years. The experiment was conducted on a sandy loam soil, low in organic carbon (0.33 per cent) and available N (110.2 kg/ha), high in available P (42.8 kg/ha) and low in available K (132.6 kg/ha) content. The pH and EC of the soil (1 : 2) were 7.62 and 0.23 dS/m, respectively. The experiment was laid out in split plot design. There were three main treatments *viz.*, fallow (where no crop was grown before basmati), *bajra* (a fodder grown before basmati) and green manure (*Sesbania aculeata* grown before basmati) which were kept as main plots and four crop sequences followed thereafter *viz.*, basmati rice (*Oryza sativa*)-wheat (*Triticum aestivum*), basmati rice-field peas (*Pisum sativum*) basmati rice-celery (*Apium graveolens*) and basmati-sunflower (*Helianthus annuus*) which were kept as sub plots. *Bajra* was grown as fodder and harvested after 8 to 7 weeks and green manure dhaincha (*Sesbania aculeata*) was grown for about 60 days and incorporated in soil a day before transplanting basmati rice in the respective treatments.

In all there were twelve treatments, which were replicated thrice. All recommended package of practices were followed to grow different crops. The varieties and fertilizer applied to different crops are given in Table 42.1. The crops were harvested at proper maturity and yield data were recorded and the yield of basmati rice, wheat, celery, field peas and sunflower were adjusted at 18, 10, 10, 10 and 14 per cent moisture respectively.

The productivity of different crop sequences were compared by converting the yield of all the component crops of a crop sequence into basmati equivalent yield on the basis of minimum support price and gross and net returns were also worked out. After the completion of three cycles of crop rotations surface soil samples (0-15 cm) were taken and analyzed for pH (1 : 2), EC (1 : 2), OC (Walkley and Black 1934), available N (Subbiah and Asija 1956) available P (Olsen *et al.*, 1956) and available K (Meernin and Peach 1950) to determine the changes in soil fertility.

Table 42.1: Varieties of Crops and Nutrients Applied to Different Crops

Crop	*Variety*	*Nutrient Applied (kg/ha)*		
		N	P_2O_5	K_2O
Sesbania aculeate (GM)	Local	0	0	0
Bajra	Local	50	0	0
Basmati rice	Basmati–370/386	60	30	0
Field peas	Field peas-48	30	40	0
Wheat	PBW–373	120	60	30
Celery	Local	100	40	0
Sunflower	MSFH–8	60	30	30

Results and Discussion

Crop Productivity

The data on yield of component crops of different crop sequences are given in Table 42.2. The grain yield of basmati rice ranged from 22.0 to 24.1 q/ha. in fallow-basmati rice sequence and from 18.2 to 19.4 q/ha in different *bajra* (*f*)-basmati rice sequence and the corresponding values ranged from 14.2 to 14.4 q/ha in green manure-basmati rice crop sequences. When the mean values were compared it was found that maximum mean grain yield of basmati was recorded in fallow-basmati rice sequence (22.9 q/ha) and minimum mean grain yield of basmati was recorded in green manure-basmati rice sequence (14.3 q/ha).

The reduction in grain yield in green manure-basmati rice sequence may be due to severe lodging which occurred in two out of the three years of experimentation. As basmati rice grown after green manured crop also received recommended N which might have made the crop more succulent and, therefore, resulted in lodging. In the *bajra* (*f*)-basmati rice crop sequences the mining of nutrients by *bajra* (fodder) might have resulted in the low yield of following basmati rice crop as is also evident from nutrient status of soils particularly N and K after harvest of the crop (Table 42.4).

Earlier also it has been reported by Sidhu *et al.* (2004) that where green manuring was practiced along with application of recommended level of N the basmati crop suffered from severe lodging which resulted in low grain yield. On the other hand the yield of wheat, celery, field peas and sunflower succeeding basmati rice increased substantially where green manure was practiced. The mean grain yield of wheat was 44.9, 43.9 and 47.1 q/ha, respectively, in fallow-basmati rice-wheat, *bajra* (*f*)-basmati rice-wheat and green manure basmati rice-wheat crop sequence. The seed yield of celery, field peas and sunflower were also higher in crop sequences with green manuring, indicating there by that there was considerable residual effect of green manure in the form of increasing the available nutrients in the soil (Table 42.4) and the other associated beneficial effects.

There was not much variation in the seed yield of celery and sunflower in fallow-basmati rice and *bajra* (fodder)-basmati rice crop sequences. However, the seed yield of field peas was high in fallow-basmati rice sequence as compared to *bajra* (*f*)-basmati rice-field pea sequence. Green manure of *dhaincha* not only benefit the crop before which incorporated but also leave sufficient residual effect to benefit the succeeding crop.

Table 42.2: Grain Yield of Component Crops and BREY of Basmati Rice Based Crop Sequences (3 year mean)

Crop Sequence	*Grain Yield (q/ha)*		*BREY (q/ha)*
	I	*II*	
Fallow			
Basmati rice-wheat	22.0	44.9	41.4
Basmati rice-celery	24.1	5.6	34.2
Basmati rice-field peas	22.2	14.6	35.5
Basmati rice-sunflower	23.4	27.0	46.7
Mean	22.9		39.5
Basmati rice-wheat	(472) 18.5	43.9	50.3
Basmati rice-celery	(455) 19.4	5.7	42.1
Basmati rice-field peas	(455) 18.7	11.7	41.7
Basmati rice-sunflower	(443) 18.2	27.2	53.8
Mean	18.7		47.0
Green manure			
Basmati rice-wheat	14.2	47.1	34.6
Basmati rice-celery	14.2	6.3	25.7
Basmati rice-field peas	14.3	16.2	29.1
Basmati rice-sunflower	14.4	28.2	38.8
Mean	14.3		32.1

Figures in parentheses are fresh biomass of *bajra* (fodder) q/ha.

Basmati Rice Equivalent Yield (BREY) and Economic Analysis

The grain yield of component crops of different crop sequences was converted into basmati rice equivalent yield (BREY) and the data are presented in Table 42.2. The maximum BREY was obtained in *bajra (f)*-basmati rice-sunflower crop sequence (53.8 q/ha) and minimum was obtained in green manure-basmati rice-celery system (25.7 q/ha) which could be due to inclusion of *bajra* crop in the former and in the later crop sequence it could be because of low grain yield of basmati rice. Maximum mean BREY was obtained in *bajra (f)* based crop sequence (47.0 q/ha) followed by fallow (39.0) and minimum was obtained in green manure based crop sequence (32.1).

As expected gross returns (Table 42.3) also followed the same trend and the mean gross returns were maximum (Rs. 51,700 /ha) in *bajra*-basmati rice sequence and minimum in green manure-basmati rice sequence, which could also be due to low yields of component crops in these sequence (Table 42.2).

On the other hand when net returns (Table 42.4) were compared it was found that the mean net returns obtained from green manure-basmati rice were significantly below as compared to fallow-basmati rice and *bajra (f)*-basmati rice which may be due to poor yield of basmati rice obtained in this crop sequence. However there was not much variation in the mean net returns obtained from fallow-basmati rice and *bajra (f)*-basmati rice (Rs. 19200 and 19900/ha). But when sub crop sequences were compared fallow-basmati rice-sunflower and *bajra* (*f*)-basmati rice-sunflower gave the highest net returns, which were significantly better than fallow-basmati rice-wheat and *bajra* (*f*)-basmati rice-wheat sequence.

Table 42.3: Gross Returns from Basmati Rice Based Crop Sequences (mean of 3 years)

Crop Sequence	*Fallow*	*Bajra (fodder)*	*Green Manure*	*Mean*
Basmati rice-wheat	45.6	55.3	38.0	46.6
Basmati rice-celery	37.6	46.3	28.3	37.4
Basmati rice-field peas	39.0	45.9	31.9	38.9
Basmati rice-sunflower	51.5	59.2	42.7	51.1
Mean	43.4	51.7	35.2	

CD at 5% Main plot: NS, Sub plot: 5.7, Interaction: NS.

Table 42.4: Net Returns from Basmati Rice Based Crop Sequences (mean of 3 years)

Crop Sequence	*Fallow*	*Bajra (fodder)*	*Green Manure*	*Mean*
Basmati rice-wheat	20.8	23.0	11.0	18.3
Basmati rice-celery	15.2	16.4	3.7	11.8
Basmati rice-field peas	14.4	13.8	5.1	11.1
Basmati rice-sunflower	26.3	26.5	15.3	22.7
Mean	19.2	19.9	8.8	

CD at 5% Main plot: 4.7, Sub plot: 2.8, Interaction: NS.

Nitrogen Addition by Incorporating Green Manure

About 60 days old dhaincha was incorporated for green manure during all the three years of cropping which produced 51.6, 54.5 and 62.0 q/ha of dry weight during 1996, 1997 and 1998 respectively. The N content of green manure during the three years was 2.2, 1.97 and 1,87 per cent, respectively and that added 113.5,107.4 and 115.9 kg N/ha during different years. Earlier, Meelu *et al.* (1994) had also reported that incorporation of green manure resulted in addition of 75–100 kg N/ha. Therefore, the improvement in soil fertility and BREY was observed in green manure compared to other treatments due to addition of biomass, nitrogen and other associated beneficial effects.

Soil Health

After three cycles of cropping there was a reduction in the soil pH than the initial value (Table 42.5). The mean soil pH decreased from 7.62 to 7.45, 7.48 and 7.46 in fallow-basmati rice, *bajra* (*f*)-basmati rice and green manure-basmati rice crop sequence. Although within the mean values there were marginal changes but when compared with initial pH (7.62) there was considerable decline in soil pH. This could be due to addition of urea which is known to be acid forming fertilizer and release of some organic acids produced during the decomposition of roots of different crops remaining in the soil.

Table 42.5: Influence of Basmati Rice Based Crop Sequences on Soil Health

Crop Sequence		*pH (1 : 2)*	*OC (%)*	*Available N (kg/ha)*	*Available P (kg/ha)*	*Available K (kg/ha)*
Fallow						
Basmati rice-wheat		7.52	0.34	114.4	47.6	119.0
Basmati rice-celery		7.56	0.43	125.4	55.0	108.9
Basmati rice-field peas		7.36	0.36	122.3	55.0	119.7
Basmati rice-sunflower		7.36	0.35	101.8	52.0	88.6
Mean		7.45	0.37	115.9	52.4	109.0
Bajra (fodder)						
Basmati rice-wheat		7.50	0.36	112.9	47.6	89.9
Basmati rice-celery		7.46	0.38	109.7	51.7	99.7
Basmati rice-field peas		7.49	0.39	106.6	55.0	59.3
Basmati rice-sunflower		7.49	0.41	103.4	54.6	106.1
Mean		7.48	0.38	108.1	52.0	95.0
Green manure						
Basmati rice-wheat		7.54	0.51	130.1	64.4	119.7
Basmati rice-celery		7.53	0.52	126.9	57.2	115.2
Basmati rice-field peas		7.39	0.49	128.5	59.5	123.2
Basmati rice-sunflower		7.38	0.45	122.2	58.2	116.6
Mean		7.46	0.49	127.0	59.2	118.7
CD at 5%	Main plot	0.05	0.06	NS	5.6	7.9
	Sub plot	NS	NS	NS	NS	NS
	Interaction	NS	NS	NS	NS	NS
Initial value		7.62	0.33	110.2	42.8	132.6

The mean organic carbon content of the soil increased over its initial content but the extent of increase was higher where green manure was incorporated and it increased from its initial status of 0.33 per cent to 0.49. Mahapatra *et al.* (2002) reported beneficial effects of *Sesbania aculeate* green manuring in rice-wheat sequence on soil organic carbon and available N. There was not much variation in the available N content of the soil than its initial content except in green manuring treatment where it increased to 127 kg/ha over its initial content of 110 kg/ha.

On the other hand mean available P of the soil increased than its initial status under all the treatments but again the extent of increase was maximum under green manure incorporation. This may be due to addition of P fertilizers to crops and because of lesser P uptake by the crops than the amounts applied to different crops. Earlier Sikka and Sidhu (1997) had reported that lesser P uptake in spite of much higher P application to different crop sequences resulted in build up of available P in soils.

However, the available K content of the soil decreased significantly than its initial status in all the crop sequences but the magnitude of reduction was little less in green manure incorporation. The decrease in available K status may be due to exceedingly higher K uptake by the crops and no addition of K fertilizer to the crops as is evident from the data in Table 42.1. The mean available K content was minimum under *bajra*-basmati rice sequence, which may be due to greater removal by *bajra* crop, which received no K application.

It may be concluded that fallow basmati rice-sunflower sequence was more viable both economically and biologically as it produced substantially higher rice equivalent yield and higher returns over variable cost than the prevalent basmati rice-wheat sequence and also sustained the nutrient availability in the soil except K. It may also be mentioned that green manuring did not prove beneficial for basmati rice as it resulted in severe lodging of basmati causing a reduction in its grain yield.

References

Chauhan, D.S., Sharma, R.K. and Kharub, A.S. 2000. Diversification of rice-wheat system for sustainability and profitability. In: *Extended Summary, International Conference on Managing Natural Resources,* held during 14–19 Feburary at IARI, New Delhi.

Mahapatra, B.S., Kumar, Ajay, Shukla, D.K. and Shukla, R K. 2002. Summer legumes in relation to productivity and fertility in rice-wheat cropping system. In: *Paper Presented at the International Conference on Managing Natural Resources,* held during 14–19 Feburary at IARI, New Delhi, 1: 155–156.

Meelu, O.P., Singh, B. and Singh, Y. 1994. Effect of green manure and crop residue cycling on N economy, organic matter and physical properties of soil in rice-wheat cropping system. In: *Proceeding of Symposium on Sustainability of Rice-wheat System in India,* held during 7–8 May at Regional Station, Karnal, p. 15–24.

Merwin, H.D. and Peech, M. 1950. Exchangeability of soil potassium in the sand, silt and clay fractions as influenced by the nature of the complementary exchangeable cations. *Soil Sci. Soc. Am. Proc.,* 15: 125–128.

Olsen, S.R., Cole, C.Y., Watanabe, F.S. and Dean, L.A. 1954. Estimation of available phosphorus by extraction with sodium bicarbonate. *USDA Circular,* 939: 19.

Rattan, R.K. and Singh, A.K. 1997. Role of balanced fertilization in rice-wheat cropping system. *Fertilizer News,* 42(4): 592–600.

Sidhu, M.S., Thakar Singh and R.K. Sikka. 2002. Response of basmati rice to nitrogen in different cropping systems. In: *Extended Summary of 2nd International Agronomy Congress,* New Delhi, p. 402–403.

Sidhu, M.S., R. Sikka and Thakar Singh. 2004. Performance of transplanted Basmati rice in different cropping systems as affected by nitrogen application. *IRRI Notes,* p. 62–65.

Sikka, R. and M.S. Sidhu. 1997. Evaluation of summer groundnut based cropping systems for nutrient uptake and changes in soil fertility. In: *Proceedings of the 62nd Annual Convention of the Indian Society of Soil Science* held from 18–21 October at Kolkata, pp. 209.

Subbiah, B.V. and Asija, G.L. 1956. A rapid procedure for estimation of available nitrogen in soils. *Curr Sci.,* 25: 259–260.

Walkley, A. and Black, I.A. 1934. An estimation of degtareff method for determining soil organic matter and a proposed modification of the chromic acid titration method. *Soil Sci.,* 4: 51–56.

Yadav, R.L., Yadav, D.S., Singh, R.M. and Kumar, A. 1998. Long term effects of inorganic-inputs and crop productivity in rice-wheat cropping system. *Nutrient Cycling in Agroecosystems,* 51: 193–200.

Chapter 43

Puberty Acceleration Effects of Male Urine in Female Lesser Bandicoot Rat *Bandicota bengalensis* (Gray and Hardwicke)

Gurpreet Kaur* and V.R. Parshad
Department of Zoology and Fisheries,
Punjab Agricultural University, Ludhiana – 141 004, India

ABSTRACT

Juvenile females (30 days old) of the lesser bandicoot rat, *Bandicota bengalensis,* exposed individually daily to adult male urine soaked in filter paper pieces for nasal contact during bedding, attained the onset of puberty 21 days earlier than female juveniles exposed to water soaked filter paper pieces. Females treated with male urine showed vaginal opening (VO) at age of 141.0±30.0 days and first oestrous (FO) at 152.0±28.0 days with corresponding body weights of 155.0±6.6 g and 168.0±9.0 g compared to VO at 162.0±29.3 days and FO at 171.0±29.1 days of age with corresponding body weights of 160.0±2.0 g and 165.0±6.8 g in untreated females. Individual variations in these parameters occurred irrespective of the origin of females from the colonies and uniformity in initial body weights and experimental conditions of treated and untreated females.

Keywords*: Puberty acceleration, Urinary pheromones, Bandicota bengalensis.*

---

* Corresponding Author.

Introduction

In mammals, the timing of puberty is a finely tuned event dependent on genetic heritage, social and environmental factors (Vandenbergh, 1967). Pheromonal acceleration of puberty in the laboratory mouse was first described by Vandenbergh (1967) and has been termed as the Vandenbergh effect (Vandenbergh, 1999). Male stimulus accelerates puberty in females, either by the presence of adult male as in *Rattus norvegicus* (Vandenbergh, 1969) hopping mouse *Notomys alexis* (Breed, 1976) or by its urinary pheromones as in house mouse *Mus musculus* (Vandenbergh, 1969) meadow vole *Microtus pennsylvanicus* (Baddaloo and Chulow, 1981) and prairie deer mouse *Peromyscus maniculatus bairdii* (Teague and Bradley, 1978). The lesser bandicoot rat, *Bandicota bengalensis,* which is the most predominant agricultural pest in South Asia (Parshad, 1999) breeds throughout the year with large litters and maintains its large populations (Parshad *et al.*, 1989). However, little is known about the factors regulating its puberty, reproduction and population dynamics. This article describes the puberty acceleration effects of urine of adult males in females of *B. bengalensis.*

Materials and Methods

Adult *B. bengalensis* caught from the fields were paired in laboratory breeding cages (each of 76 × 38 × 40 cm size) with food and water *ad libitum.* Food consisted of a mixture of cracked wheat, skimmed milk powder, powdered sugar and groundnut oil (85 : 5 : 5 : 5). Green leafy vegetables and vitamin B complex syrup (Glaxo India Limited) in water were also provided 2 to 3 times in a week. Breeding cages for the experiment were kept in separate rooms so as to avoid the effect of odours of strange males or females. The males were separated from females at about 12 days of pregnancy. Young ones from three colonies were used for the experiment (Table 43.1).

Table 43.1: Composition of 3 Colonies and Number of Juvenile Females Used in Control and Treatment Groups

Colony Number	Litter Size	Number of Males	Number of Females	Female Used	
				Control	Treatment
C_1	10	6	4	2	2
C_2	4	2	2	1	1
C_3	9	5	4	2	2
Total		**13**	**10**	**5**	**5**

On day 30, the young-ones were weaned from their mothers. Female young ones were separated, weighed and divided into 2 groups of 5 each (one group was kept as control group and other as treatment group). Rats were transferred in individual cages. Before it, the cages were washed and deodorized with a detergent. Mean body weight of 30 day old rats in each group was almost similar 42.5 g in control and 43.0 g in treatment group. Cages of both groups were placed in separate rooms in which no other animal was housed so as to avoid the effect of strange odours on puberty.

Urine Donors

Laboratory born 6–12 months old male rats were used for collection of urine on filter paper pieces. The bottom of the cages of individually housed male rats were lined with filter paper which absorbed excreted urine. Parts of the filter paper which had soaked urine were cut into small pieces and placed inside the cage of each individual female rat of treated group (n = 5), while water soaked pieces of filter paper were placed in cages (n = 5) of control group.

Exposure of Young Females to Male Urine Odour

Urine and water soaked pieces of filter paper were presented daily to individually caged female rats from day 30 of age onwards until they attained puberty by the method as earlier adopted by Massey and Vandenbergh (1981). The recipient female rats torned the pieces of filter paper and used these as bedding material. The filter paper pieces were changed daily so that the rats of the treatment group had a fresh source of urinary odour. Five recipient rats received odour from filter paper pieces soaked in urine of 6 adult male rats.

Observations

In both control and treated groups, daily feed intake and body weights at weekly intervals were recorded. All females were examined daily for vaginal opening and vaginal smears were prepared from each female daily after introitus. Day of first oestrous was recorded from the presence of fully formed cornified cells in the vaginal smear of rats.

Statistical Analysis

The data of body weights, daily food consumption, day of vaginal opening and day of first oestrous of treated and control groups were compared by student's t-test (Gupta, 1991).

Results and Discussion

Exposure of female rats to the urine of adult male rats on the filter paper pieces, which the females used as bedding material, from day 30 of age continuously till the day of first oestrous affected the food consumption, growth of rats as well as their sexual maturity. Though 5 females in each group derived from three different colonies had the same initial body weight on day 30, but variations in different parameters related to puberty were evident during the present study (Tables 43.2 and 43.3).

Table 43.2: Effects of Male Urine on Puberty Acceleration of Female *B. bengalensis*

	Control (Water)	*Treated (Male Urine)*	*t-value*	*Level of Significance*
Body weight (g) of female rats (n = 4)				
Day 30	42.5±2.5	43.0±3.0	0.30^{NS}	
Day 60	83.5±10.6	93.0±7.05	1.20*	P < 0.40
Day 90	121.2±9.0	129.0±6.6	1.24*	P < 0.40
Body weight (g) on day of vaginal opening (n = 4)**	160.0±2.0	155.0±6.6	0.79*	P < 0.50
Body weight (g) on day of first oestrous (n = 4)**	165.0±6.8	168.0±9.0	0.63^{NS}	
Mean daily food consumption g/100g body weight*	8.91±0.13 (n = 716)**	8.03±0.13 (n = 716)**	2.38	P < 0.05
Day of vaginal opening (n = 4)**	162.0±29.3	141.0±30.0	3.11	P < 0.13
Day of first oestrous (n = 4)**	171.0±29.1	152.0±28.0	2.33	P < 0.20

*: Based on the actual number of rats till the day of first oestrous in rats from different colonies.

**: One female from the control group did not show vaginal opening till 309 days this was excluded from the data.

Table 43.3: Body Weight (g) Differences in Individual Rats of Different Colonies on the Day (Age) of Vaginal Opening and First Oestrous

			Colony C_1			Colony C_2			Colony C_3	
			Body Weight	Day		Body Weight	Day		Body Weight	Day
Control	Vaginal opening	UR_1	160	119	UR_3	155	231	UR_4	160	190
		UR_2	165	108						
	First oestrous	UR_1	160	124	UR_3	155	237	UR_4	160	203
		UR_2	185	120						
Treated	Vaginal opening	TR_1	153	108	TR_3	165	231	TR_4	137	117
		TR_2	165	108						
	First oestrous	TR_1	180	119	TR_3	165	236	TR_4	145	137
		TR_2	185	119						

R_1 to R_4 are the number assigned to individual rats and U represent untreated and T represented treated rats.

Food Consumption

Mean daily food consumption of female rats exposed to urine of adult males was 8.03±0.13g/100g body weight which was significantly less than 8.91±0.13 in the untreated rats as calculated on the basis of daily food consumption from day 30 to day of first oestrous which ranged from 108 to 237 days in different rats. However, the mean daily food consumption of urine treated rats was less upto day 192. But from day 193 the mean daily food consumption increases in treated rats as compared to female rats provided with water soaked filter paper.

Body Weight

The female rats in the control and treatment groups were derived in equal numbers from the same 3 colonies. Therefore, the composition of both the groups was genetically similar.

The mean body weight at vaginal opening of treated rats was 155.0±6.6g which was significantly less than the 160.0±2.0g in untreated rats (Table 43.2) and there were individual variations even they were derived from the same colony (Table 43.3). From day 30 of age, the increase in body weight was more in rats provided with urine soaked filter paper up to 24 weeks. But after 24 weeks the body weight of the treated female rats decreased upto 30 weeks. After 30 weeks the body weight of the treated female rats again increased and it remained as such upto 34 weeks or until first oestrous. The body weight at first oestrous was 168.0±9.0g in treated and 165.0±6.8g in untreated rats and this difference was statistically non-significant as also observed in *P. maniculatus* (Teague and Bradley, 1978). However, according to Colby and Vandenbergh (1974) mice exposed to male urine were 1.2g lighter in weight at first oestrous than mice not exposed to urine.

Vaginal Opening and First Oestrous

The mean age at vaginal opening of treated rats was 141.0±30.0 days which was significantly less than 162.0±29.3 days in untreated rats (Table 43.2). Treated female rats exhibited vaginal opening 21 days earlier than untreated female rats and individual variations were evident even within the juveniles of the same colony (Table 43.3). According to Colby and Vandenbergh (1974) them mice exposed to urine exhibited vaginal opening 1.4 days earlier than the mice exposed to tap water. The mean age at first oestrous of treated rats was 152.0±28.0 days which was significantly less than 171.0±29.1 days in untreated rats (Table 43.2). Treated female bandicoot rats attained first oestrous about 18 days earlier than the untreated females as compared to 10 to 20 days in mice (Vandenbergh *et al.*, 1975) 16 days in meadow voles (Baddaloo and Chulow, 1981) and 14 days in Prairie Vole (Carter *et al.*, 1980). In the present study one of the female provided with water soaked filter paper did not show vaginal opening and oestrous.

Previous studies showed that male stimulation mediated through urinary pheromones affects endocrinological events related to puberty. Male stimulation resulted in the elevation of luteinized hormone (LH) in the prepubertal female mice (Bronson and Desjardins, 1974). By the second day after exposure to male, FSH levels were somewhat lower and prolactin levels began to rise (Bronson and Maruniak, 1976). Studies on adult female Prairie voles, *Microtus ochrogaster* showed a 185 per cent rapid increase in luteinizing hormone releasing factor in the posterior portion of the olfactory bulb of the female with 60 minute exposure to a single drop of male urine (Dluzen *et al.*, 1981). Present studies showed that the differences in day of vaginal opening and day of first oestrous between littermates were less but urinary signals from the adult males accelerated puberty in them. It is concluded that heredity plays an important role in establishing the approximate time of puberty in *B. bengalensis*, but the social stimulus from male conspecifics also play role in scheduling the time of puberty within the range of ages limited by heredity.

Acknowledgements

Thanks are due to Professor Dr. J.G. Vandenbergh, Department of Zoology, N.C. State University, Raleigh, NC 27650, USA for providing literature and to Indian Council of Agricultural Research, New Delhi for providing support to centre of All India Network Project on rodent control under which this work was carried out.

References

Baddaloo, E.G.Y. and Chulow, F.V., 1981. Effects of the male on growth, sexual maturation and ovulation of young female meadow voles, *Microtus pennsylvanicus*. *Can. J. Zool.*, 59: 415–421.

Breed, W.G., 1976. Effect of environment on ovarian activity of wild hopping mice *Notomys alexis*. *J. Reprod. Fertil.*, 47: 395–397.

Bronson, F.H. and Desjardins, C., 1974. Circulating concentrations of FSH, LH, estradiol and progesterone associated with acute, male-induced puberty in female mice. *Endocrinology*, 94: 1658–1668.

Bronson, F.H. and Maruniak, J.A., 1976. Differential effects of male stimuli on follide-stimulating hormone, luteinizing hormone and prolactin secretion in prepubertal female mice. *Endocrinology*, 98: 1101–1108.

Carter, C.S., Getz, L.L., Garvish, L., McDermott, J.L. and Arnold, P., 1980. Male-related pheromones and the activation of female reproduction in the prairie vole, *Microtus ochrogaster*. *Biol. Reprod.*, 23: 1038–1045.

Colby, D.R. and Vandenbergh, J.G., 1974. Regulatory effects of urinary pheromones on puberty in the mouse. *Biol. Reprod.*, 11: 268–279.

Dluzen, D.E., Ramirez, V.D., Carter, C.S. and Getz, L.L., 1981. Male vole urine changes luteinizing hormone releasing hormone and norepinephrine in female olfactory bulb. *Science*, 212: 573–575.

Gupta, S.P., 1991. *Statistical Methods*. Sultan Chand and Sons, New Delhi.

Massey, A. and Vandenbergh, J.G., 1981. Puberty acceleration by a urinary cue from male mice in feral *populations. Biol. Reprod.*, 24: 523–527.

Parshad, V.R., 1999. Rodent Control in India. In: *Pest. Mgmt. Rev.*, 4: 97–126.

Parshad, V.R., Kaur, P. and Guraya, S.S., 1989. Reproductive cycles of mammals: Rodentia. In: *Reproductive Cycle of Indian Vegetables*, (Ed.) S.K. Saidapuri, pp. 347–408, Allied Publishers Limited.

Teague, L.G. and Bradley, E.L., 1978. The existence of a puberty accelerating pheromone in the urine of the male prairie deermouse *Peromyscus maniculatus bairdii. Biol. Reprod.*, 19: 314–317.

Vandenbergh, J.G., 1967. Effect of the presence of a male on the sexual maturation of female mice. *Endocrinology*, 8: 345–349.

Vandenbergh, J.G., 1969. Male odour accelerates female sexual maturation in mice. *Endocrinology*, 24: 658–660.

Vandenbergh, J.G., 1999. Puberty acceleration. In: *Encyclopedia of Reproduction*, p. 125–128.

Vandenbergh, J.G., Whitsett, J.M. and Lombardi, J.R., 1975. Partial isolation of a pheromone accelerating puberty in female mice. *J. Reprod. Fertil.*, 43: 515–523.

Chapter 44

Medicinal Wealth of District Pauri Garhwal

Upma Dobhal and Sneh Lata Bhandari
Department of Botany, H.N.B Garhwal University, Campus Pauri, Pauri (Garhwal), Uttaranchal – 246 001

ABSTRACT

Medicinal plants constitute nearly 25 per cent of the prescribed drugs Herbal drugs are being increasingly preferred in medicine science today. Pauri Garhwal a district of Uttaranchal has a remarkable wealth of medicinal herbs. The field study in the remote localities of different tehsils was carried out and information on the folk utilization of medicinal plants was gathered from inhabitants. Periodical visits were conducted to different villages located around the district. The vaidyas and local inhabitants were interviewed to obtain information on medicinal uses of different plants. The present study illustrates 15 plant species reported to be medicinally important, used for curing various ailments with their uses, action, preparation and dosages.

Keywords: *Aromatic, Herbs, Vaidyas, Ailments.*

Introduction

Plants have been used by man in various ways. The very existence of mankind depends upon plants. Besides many uses, plants are the important source of various drugs. Uttaranchal, has a great wealth of medicinal and commercial plants of economic importance.

Pauri Garhwal, a district of Uttaranchal encompasses an area of 5,440 sq km and is situated between 29°45′ to 30°15′ latitude and 78°24′ to 79°23′ longitude. The district has unique ecosystem that comprises forests, meadows, savannah grasslands, marshes and rivers as well, as wildlife geology and several other geographically distinctive peculiarities. Due to diverse geographical and climatic conditions, there is a remarkable biodiversity exhibited by flora and fauna of district Pauri Garhwal. Forests form the major part of district, in these forests is present the number of plant species which are rich source of medicines and drugs.

Material and Methods

The field study was made from August to December 2005 in various tehsils of district Pauri Garhwal, in order to find the plants of medicinal values for research work. The survey was conducted among local people, rural folks and vaidyas to know the local names and medicinal importance of various plants, of the various localities. The plants with medicinal values, as r.nown from local people and rural folks were collected and studies were made to know their medicinal uses. The notable contribution to medicinal plants of Pauri Garhwal have also been reported by some workers (Maheshwari *et al.*, 1984: Ansari *et al.*, 1985 and Dhasmana, 1987).

Results and Discussions

The data collected after field study show about 15 species of plants, which are of medicinal importance and are in use by local people and rural folks since ancient times, for traditional remedies of various diseases.

Owing to the rapid rate of destruction of forests precious herbal medicines are becoming rare day by day. Some of these medicinal plants are much threatened and others are vanishing rapidly. So, it is indispensable to take necessary steps to protect the plant from degeneration. These plants can be conserved for benefit of mankind. *In situ* conservation of these plants in their natural habitat can be made by involving local people of the district. These plants can also be conserved by *ex situ* means. Nurseries of these plants can be prepared and plants can be multiplied rapidly by tissue culture technique. Uttaranchal has been declared as a herbal state. Therefore the local inhabitants can get the financial assistance and grants for establishing such nurseries. Local people of this area should be made aware of their heritage of these important medicinal plants.

There is a need to establish a systematic and organized collection of valuable medicinal plants of the district. Cottage industries should be established for the preparation of crude medicines, powders, shampoos etc.

Research priority should be given to the survey of medicinal plants in their natural habitats. Many important plants of this district have not been screened so far for their medicinal importance. Therefore studies should be carried out for their screening.

It can be concluded from the present study that district Pauri hoods treasure of many medicinal plants. By correct identification of these plants, many diseases can be cured without side effects though research work is still needed for such plants. On account of ignorance and unemployment, the valuable herbs of this area are disappearing at an alarming rate. Uncontrolled forest fires during summer destroy natural regeneration of these herbs. Greedy herb contractors with the help of locals also play great role in the exploitation and depletion of this great wealth.

List of Medicinal Plants of District Pauri Garhwal

Achellia melliofolium (Family: Asteraceae)

Local name: Gandrain

Common name: Yarrow

Part used: Aerial part

Actions: Diaphoretic, Hyposensitive, Astringent, Anti-inflammatory, Piuretic, Anti-microbial, Bitter, Hepatic.

Preparation and Dosage: Infusion: Pour a cup of boiling water into 1–2 tablespoon of a dried herb and left for infusion for 10–15 min. This should be drunk hourly.

Argimonia eupatoria (Family: Rosaceae)

Local name: Lesukuria

Common name: Argimony

Part used: Aerial parts

Action: Astringent, Tonic, Bitter, Diuretic, Vulnerary, Anti-spasmodic, Carminative, Hepatic, Cholagogue.

Preparation and Dosage: Infusion: Pour a cup of boiling water into 1–2 teaspoonfuls of the dried herb and leave to infuse for 10–15 min. This should be drunk three times a day.

Artemisia vulgaris (Family: Asteraceae)

Local name: Kunjaa

Common name: Mugwort

Part used: Leaves and root

Action: Bitter tonic, Stimulant, Nervine tonic, Emmenagogue.

Preparation and Dosage: Infusion: Pour a cup of boiling water into 1–2 to teaspoonfuls of dried herb and leave to infuse for 10–15 min in covered container. This should be drunk three times a day. *Combination*: May be used widely where bitter action is needed.

Astragulus candolleans (Family: Fabaceae)

Local name: Rudrawati

Common name: Milk vetch

Action: Immunomodulator

Preparation and dosage: Decoction: Put one teaspoonful of the root into a cup of water, bring to boil and simmer for 10–15 min. This should be drunk three times a day.

Berberis aristata (Berberidaceae)

Local name: Kingoor

Common name: Barberry

Part used: Bark of root and stem

Action: Cholagogue, Hepatic, Anti-emetic, Bitter, Laxative.

Preparation and Dosage: Decoction: Put one teaspoonful of the bark into a cup of cold water and bring to boil. Leave for 10–15 min. This should be drunk three times a day.

Hypericum perforatum (Family: Hypericaceae)

Local name: Choli phulya

Common name: St. John's wort

Part used: Aerial parts

Action: Anti-inflammatory, Astringent, Vulnerary, Nervine, Anti-microbial.

Preparation and Dosage: Infusion: Pour a cup of boiling water into 1–2 teaspoonful of dried herbs and leave the infuse for 10–15 min. This should be drunk three times a day.

Nepeta ciliaris (Family: Lamiaceae)

Local name: Catmint

Common name: Nueet

Part used: Leaves and flowering tops.

Action: Carminative, Anti-spasmodic, Diaphoretic, Nervine, Astringent.

Preparation and Dosage: Infusion: Pour a cup of boiling water into two teaspoonful of dried herbs and leave to infuse for 10–15 min. This should be drunk three times a day.

Combination: It may be used with Yarrow (*Achellia melliofolium*) in cold.

Plantago erosa (Family: Oleaceae)

Local name: Isabgol

Common name: Great plaintain

Part used: Leaves and aerial parts.

Action: Vulnerary, Expectorant, Demulcent, Anti-inflammatory, Astringent, Diuretic, Anti-microbial.

Preparation and Dosage: Infusion: Pour a cup of boiling water into two teaspoonful of dried herbs and leave to infuse for 10 min. This should be drunk three times a day. An ointment can be prepared to aids cuts.

Polygala arvensis (Family: Polygalaceae)

Local name: Sanjeevani

Common name: Milkwort

Action: Expectorant, Diaphoretic, Sialogogue, Emetic.

Preparation and Dosage: Infusion: Pour a cup of boiling water into half teaspoonful of dried herbs and leave to infuse for 5–10 min. This should be drunk three times a day, 1–2 ml of tincture should be taken three times a day.

Smilax aspera (Family: Smilacaceae)

Local name: Kukardara.

Common name: China root.

Part used: Root and Rhizome.

Action: Alterative, Anti-rheumatic, Diuretic, Diaphoretic.

Preparation and Dosage: Decoction: Pour 1–2 teaspoonful of root in cup of water, bring to boil and simmer 10–15 min. This should be drunk three times a day.

Stellaria media (Family: Caryophyllaceae)

Local name: Badyalu

Common name: Chick weed

Action: Anti-rheumatic, Vulnerary, Emollient

Preparation and Dosage: Infusion: Pour a cup of boiling water into two teaspoonfuls of dried herbs and leave to infuse for 5 min. This should be drunk three times a day. Fresh chickwood may be made into green drink by placing handful of fresh plant into blender with pineapple juice, blend and strain. To ease itching, a strong infusion of a fresh plant makes a useful addition to a bath water.

Taraxucum officinale (Family: Asteraceae)

Local name: Kanphuliya

Common name: Dandelion

Part used: Root and leaf

Action: Diuretic, Hepatic, Cholagogue, Anti-rheumatic, Laxative, Bitter toxic.

Preparation and Dosage: Decoction: Put 2–3 teaspoonful of root into one cup of water, bring to boil and gently simmer for 10–15 min. this should be drunk three times a day. Its combination with barberry (*Berberis aristata*) may be used for liver and gall bladder problem. It may be used with yarrow (*Achellia melliofolium*) for water retention.

Trifolium Prantense (Family: Fabaceae)

Local name: Mithla

Common name: Red clover

Part used: Flower heads

Action: Alternative, Expectorant, Anti-Spasmodic.

Preparation and Dosage: Infusion: Pour a cup of boiling water into 1–2 teaspoonful of dried herbs and leave to infuse for 10–15 min. This should be drunk times a day. Its combination with nettle used for skin diseases.

Utrica dioca (Utricaceae)

Local name: Kandali

Common name: Stinging nettle

Part used: Aerial parts

Action: Diuretic, Astringent, Hypotensive, Diuretic.

Preparation and Dosage: Infusion: Pour a cup of boiling water into 1–2 teaspoonful of a dried herb and leave to infuse for 10–15 min. This should be drunk three times a day. 1–2 ml tincture can be taken three times a day.

***Viola odorcta* (Family: Violaceae)**

Local name: Banfsa

Common name: Sweet violet

Action: Expectorant, Alterative, Anti-Inflammatory, Diuretic.

Preparation and Dosage: Infusion: Pour a cup of boiling water into one teaspoonful of the herb and let infuse for 10–15 min. This should be drunk three times a day.

References

Ansari, A.A. and Ghana Nand., 1985. Some medicinal plants of Pauri Garhwal. *Himalayan Chem. Pharm. Bull.*, 2: 42–44.

Dhasmana, H., 1987. Medicinal plants of Pauri town and adjacent forest region (Pauri Garhwal). *J. Sci. Res. Pl. Med.*, 8: 1–8.

Maheshwari, J.K. and J.P. Singh, 1984. Contribution to the ethnobotany of Bhoxa tribe of Bijnor and Pauri Garhwal district, U.P. *J. Econ. Tax. Bot.*, 5(2): 251–259.

Gaur, R.D., 1999. *Flora of the District Garhwal North West Himalaya*. Transmedia, Srinagar (Garhwal), p. 50–744.

Chapter 45

Soil-Water-Plant Relations of Sunflower Under Various Tillage and Irrigation Regimes in Rice Fallow Alfisol of Semiarid Tropics

P. Gurumurthy[1]* and M. Singa Rao[2]

[1]Assistant Professor, Department of Soil Science and Agricultural Chemistry, Agricultural College, Naira, Srikakulam Dist., Andhra Pradesh

[2]Professor of Soil Physics, Department of Soil Science and Agricultural Chemistry, College of Agriculture, ANGR Agricultural University, Rajendranagar, Hyderabad – 500 030, India

ABSTRACT

Field experiments were conducted on sandy clay loam soil (Typic Haplustalt) to evaluate the influence of different tillage practices and irrigation regimes, their interaction on soil-plant-water relations of sunflower (*Helianthus annuus* L.) crop under rice fallow condition. Farmers' tillage practice of two passages of bullock drawn desi plough + two passages of three tyne bollock drawn cultivator was compared with four tillage practices *viz.*, one passage of tractor drawn disc plough + rotavator; one passage of tractor drawn MB plough + two passages of tractor drawn cultivator; two passages of bullock drawn desi plough + one passage of power tiller drawn rotavator, and zero tillage (seed dibbling). Four irrigation regimes consisting of 60 mm irrigation at IW/CPE = 0.6, 0.8, 1.0 and 1.2 were superimposed on tillage treatments. The soil physical changes were measured with respect of tillage depth, fineness of seedbed expressed as mean weight diameter (MWD) of dry clods, dry bulk density, penetration resistance and soil

* Corresponding Author: E-mail: murthypeddinti@rediffmail.com.

water retention and release were measured. Plant water stress indices (XWP and LDR) and crop growth parameters were also recorded. Primary tillage with tractor drawn M.B plough (T_4) or tractor drawn disc plough (T_5) resulted 26 cm deep tillage and significantly reduced soil dry bulk density, penetration resistance and increased water retention in upper 30 cm soil layer, but had no influence on these measurements beyond 30 cm depth. Tillage with bullock drawn desi plough was resulted to shallow tillage (11 cm) and influenced the above physical properties only in top 15 cm soil layer. Secondary tillage with rotavator, either power tiller drawn or tractor drawn made finer tilth measuring low MWD of dry clods and recorded significantly higher water retention compared to other tillage practices. Irrigation regimes did not show any significant effect on soil physical properties except soil penetration resistance, which is lower with intensity of irrigation. These favourable soil physical conditions induced in deep tillage caused higher XWP and lower LDR in the sunflower crop, thus significantly increased crop growth measuring higher LAI, head diameter land dry matter production.

Keywords: *Tillage-irrigation regimes, Soil-plant-water relations, Rice fallows-sunflower.*

Introduction

Plant water status controls the physiological processes and conditions, which determine the quantity and quality of its growth (Kramer, 1969). Morphological, physiological and biochemical modification caused by water stress in the plant may substantially reduce crop yield (Hsiao, 1973). Tillage modifies soil physical properties and influences water retention and release in soil. Rice fallows are characterized by massive soil structure which pose problems for the following crop through impervious for emerging seedlings, plant roots, air and water (Reddy and Reddy 1989 and Singa Rao *et al.*, 1995). The problem is ameliorated to a considerable extent by loosening of the soil achieved by tillage (Bennie and Botha, 1986). Different indices have been used to evaluate the plant water status. These indices are affected by various factors such as soil and water management practices, plant and atmosphere parameters. Certain stages of plant growth are comparatively more sensitive than others to moisture stress. Effect of tillage practices on corn (Reicosky *et al.*, 1976) and cotton (Young and Browning, 1977) and water management practices in sunflower (Singh *et al.*, 1995) has been reported on the effect of moisture stress. But the combined effect of tillage and moisture has not been reported hither to. The present study was therefore undertaken to study the individual and interaction effects of tillage and irrigation regimes on soil-water-plant relations of sunflower and its growth rice fallow Alfisols.

Materials and Methods

Field experiments were conducted on transplanted low land rice fallow at the University Experimental Farm, Acharya N.G. Ranga Agricultural University, Hyderabad (17°19′ North Latitude and 78° 28′ East longitude at an altitude of 535 m above MSL), during *rabi* for two consecutive years (1998 and 1999). The soil (Typic Haplustalf) was medium deep (50 cm deep), sandy clay loam in texture with 66.00, 12.33, and 21.67 per cent sand, silt and clay, respectively with an initial bulk density of 1.47 (0–15 cm), 1.55 (15–30 cm), 1.59 (30–45 cm) Mg m^{-3}. The moisture per cent at –0.03 MPa and –1.5 MPa pressure were 16.65 and 5.40 by weight, respectively. The soil was neutral (pH 7.54), non-saline (E.C. 0.174 d Sm^{-1}), medium in organic carbon (0.55 per cent), low in available nitrogen (223 kg/ha) and phosphorus (21.4 kg P_2O_5/ha) and medium in available potassium (287 kg K_2O/ha).

The experiment was laid in strip plot (6.6 m × 4.5 m) design with five tillage treatments *viz.*, T_1–zero tillage (dibbling of seeds), T_2–two passages each of bullock drawn desi plough + bullock drawn three type cultivator-farmers' practice, T_3–two passages of bullock drawn desi plough + one passage of power tiller drawn rotavator (Modified farmers' practice), T_4–one passage of tractor drawn MB plough + two passages of tractor drawn cultivator and T_5–one passage each of tractor drawn disc plough + rotavator and four levels of irrigation regimes of 60mm irrigation at IW/CPE ratio of 0.6 (I_1), 0.8 (I_2), 1.0 (I_3) and 1.2 (I_4) superimposed on tillage treatments. The treatments were replicated thrice. Tillage treatments were carried out at around 10 per cent soil moisture content thereby avoiding compaction during tillage operations. The crop (Morden cultivar) with plant to plant and row to row distance of 45 cm and 30 cm received 60 kg N (urea in two equal splits as basal dose and at buttoning stage), 60 kg P_2O_5 (SSP as basal) and 30 kg K_2O (MoP as basal).

Recording of soil physical properties were carried out at sowing, 10 DAS (emergence) 30 DAS (Vegetative stage) and 60 DAS (maturing stage) and at harvest. Depth of tillage was measured by gently pushing a scale in to seedbed until it hit a hard ground. After the tillage operation, the entire soil in pockets of 50 cm × 50 cm was removed to a depth of 15 cm from three randomly selected sites, dry sieved and the mean weight diameter (MWD) estimated by the procedure of Youker and Mc Guinness (1967) to quantify the fineness of seed bed. Total porosity was measured up to 45 cm depth at 15 cm interval by core sampler (Black, 1965). The soil penetration resistance was determined with cone penetrometer as described by Davidson (1965). Seedling emergence was determined by counting the number of seedlings in gross plot after their emergence.

The parameters of water stress *viz.*, leaf water potential (LWP) using Pressure bomb apparatus (Scolander *et al.*, 1965) and leaf diffusion resistance (LDR) using Steady State Porometer (Model, LT–1600) were measured before riving scheduled irrigation at regular intervals of 30, 45 and 60 DAS coinciding vegetative, flowering and filling stages, respectively during crop growth period for both the years of study. The above parameters were measured before irrigation schedule around stipulated time. Crop growth was recorded in the form of leaf area index (LAI), sunflower head diameter and total shoot dry matter produced at regular intervals.

Soil water retention characteristics were estimated by the procedure outlined by Jalota *et al.* (1998). Irrigation water applied was measured with V-notch weirs. Soil water availability at different soil depths was monitored by determining soil moisture content gravimetrically up to 45 cm soil layer in three depths using core sampler (Michael *et al.*, 1977) just before and 24 h after irrigation for each irrigation cycle. The crop ETa was as computed by the following relation

$$\text{Crop ETa} = \sum_{n=1}^{i=N} \frac{M_1 i - M_2 i}{10} \times \text{pbi} \times \text{Di}$$

where,

ETa: Crop evapotranspiration (mm) from the effective root zone depth within an irrigation cycle.

n: Number of soil layers sampled in the root zone depth.

M_1i: Mass water percentage 24 hr after n[th] irrigation in the ith layer, cm

M_2i: Mass water percentage just before n[th] irrigation in the ith layer

pbi: Bulk density of the soil in the ith layer, cm

Σ: Summation

i: Integer representing soil layer.

The experimental data were subjected to 'Analysis of Variance' (Panse arid Sukathme, 1965).

Results and Discussion

Depth of Tillage

Primary tillage with heavy draught machinery such as MB plough (T_4) and Disc plough (T_5) tilled the soil to deeper soil depth of 26 cm, while tillage with bullock drawn wooden desi plough tilled to 11 cm soil depth (Table 45.1), however, secondary tillage implements have no effect on depth of tillage. These results are in agreement with the results obtained by Rautaray (1983) and AICRP on SPCIP (1987).

Table 45.1: Effect of Tillage Practices and Irrigation Regimes on Plant stand, Leaf Area Index (LAI) and Head Diameter and of Sunflower in Rice Fallows

Treatment	*Depth of Tillage*		*Seedbed Tilth (MWD of Clods)*		*Initial Crop Stand (%)*		*Leaf Area Index (at Flowering)*		*Head Diameter (cm)*	
	1998	*1999*	*1998*	*1999*	*1998*	*1999*	*1998*	*1999*	*1998*	*1999*
Tillage practice (T)										
T_1	–	–	–	–	58.64	56.36	1.78	1.74	11.5	10.7
T_2	11.1	10.9	38.3	35.7	66.36	65.36	2.09	2.07	13.5	13.1
T_3	11.0	11.4	23.8	21.4	79.10	80.45	2.50	2.36	14.1	13.7
T_4	26.2	25.7	37.0	34.3	68.18	66.00	2.74	2.60	16.4	16.0
T_5	25.6	25.4	20.5	20.4	80.91	82.72	3.02	2.86	16.1	15.7
Irrigation regime (M)										
I_1	–	–	–	–	69.42	69.72	1.93	1.87	12.6	11.8
I_2	–	–	–	–	71.65	72.33	2.35	2.23	13.9	13.2
I_3	–	–	–	–	71.29	70.29	2.67	2.55	15.2	14.8
I_4	–	–	–	–	70.23	68.38	2.77	2.67	15.7	15.6
CD (P = 0.05)										
T	1.20	1.08	3.2	2.1	4.88	4.52	0.17	0.26	0.63	1.00
M	*NS*	*NS*	*NS*	*NS*	NS	NS	0.21	0.18	0.53	0.91
T × M	*NS*	*NS*	*NS*	*NS*	NS	NS	0.38	0.41	1.36	2.04

Note: Tillage depth and seedbed tilth were measured after imposition of tillage treatments, hence effect of irrigation regimes on these parameters were not presented.

T_1: Zero tillage (dibbling of seeds)

T_2: Two passages each of bullock drawn desi plough + bullock drawn three type cultivator-farmers' practice

T_3: Two passages of bullock drawn desi plough + one passage of power tiller drawn rotavator (modified farmers' practice)

T_4: One passage of tractor drawn MB plough + two passages of tractor drawn cultivator

T_5: One passage each of tractor drawn disc plough + rotavator.

Irrigation regimes of 60 mm irrigation at IW/CPE ratio of 0.6 (I_1), 0.8 (I_2), 1.0 (I_3) and 1.2 (I_4).

Clod Size Distribution and Mean Weight Diameter (MWD) of Dry Clods in the Seed Bed

The soil fraction under the tillage treatment of one passage of tractor drawn disc plough + rotavator (T_5) contained highest percentage of smaller clods and lower percentage of larger ones (Table 45.1). It also revealed that different tillage treatments especially secondary tillage implements could induce considerable differences in fineness of tilth in expressed as clod MWD (mean weight diameter). The MWD of clods was higher in case of secondary tillage with cultivator (either bullock drawn or tractor drawn) and lower under rotavator either power tiller drawn or tractor drawn irrespective of primary tillage. High-speed types in rotavator broke large size clods to small clods thus measuring low clod mean weight diameter finer seedbed tilth. Khan (1984) and Singa Rao *et al.* (1995) with rotary tillage reported similar observations. Bhushan and Ghildyal (1971) reported larger clods under mould board plough due to the increase in radius of curvature of the mould board plough. The soil was hard, compact and undisturbed in T_1 (zero tillage treatment).

Porespace

The variations in total porosity with the lapse of time during various sunflower growth stages are presented in Table 45.2. The higher values of porosity were observed in plots received tillage once with tractor drawn disc plough followed by twice with rotavator (T_5) and lowest in untilled plots (T_1). In general, the porosity values of rotavator treated plots were higher than cultivator treated plots for both the years. Since the porosity is directly related to the bulk density, any change in the later should produced corresponding variation in the former. After the first irrigation, there was a rapid decrease in porosity in all the tilled plots, but thereafter the decrease was gradual and reached a value close to that of untilled plots by the end of the experiment. Higher values of porosity obtained in tilled plots agree with the findings of Allamaras *et al.* (1977). Among the tillage treatments, higher porosity recorded under tillage once with disc plough + rotavator, both tractor drawn (T_5) and once with mould board plough + cultivator, both tractor drawn (T_4) maintained for longer period.

Table 45.2: Effect of Tillage Treatments on Total Pore Space of Soil at 0–15, 15–30 and 30–45 cm During Sunflower Crop Growth Periods

Treatment	*Pore Space (per cent)*							
	1998 (Crop Growth Period)				*1999 (Crop Growth Period)*			
	Sowing	*30 DAS*	*60 DAS*	*Harvest*	*Sowing*	*30 DAS*	*60 DAS*	*Harvest*
Tillage (T)	**0–15 cm**							
T_1	45.2	45.2	44.7	44.6	43.8	43.3	43.6	42.5
T_2	49.3	48.4	47.3	46.9	48.3	46.6	45.8	44.0
T_3	49.9	48.7	47.9	47.5	49.1	49.9	46.5	45.0
T_4	49.8	48.8	48.7	47.2	51.1	48.0	47.4	45.4
T_5	51.6	50.1	49.4	48.4	53.0	49.8	48.4	46.1
CD (P = 0.05)	0.25	1.05	0.57	0.37	0.80	0.76	0.55	0.30
	15–30 cm							
T_1	41.1	42.2	42.3	40.1	42.4	41.3	40.8	40.0
T_2	40.5	41.5	41.7	40.8	41.9	40.8	40.4	39.9
T_3	41.0	41.4	41.9	40.7	42.4	41.3	40.3	39.6
T_4	48.4	45.7	44.6	43.4	48.5	46.3	45.0	43.2
T_5	49.0	45.6	45.7	44.1	50.1	46.5	45.0	43.8
CD (P = 0.05)	0.20	1.71	1.11	1.30	1.47	0.64	1.54	1.08

Soil Penetration Resistance (SPR)

Soil penetration resistance of surface layer influenced by various tillage practices and irrigation regimes during emergence (10 DAS), flowering (50 DAS) and seed filling stages (75 DAS) of sunflower is shown in Figure 45.1. It is observed that SPR is significantly higher in untilled plots (T_1) than tilled plots. The resistance offered by soil tilled with tractor drawn Disc plough (T_5) and tractor drawn MB plough (T_4) is significantly lower compared to that with two passages of bullock drawn desi plough (T_2 and T_3). Mechanical impedance of soil is a function of both bulk density and soil moisture content. Higher penetration resistance in zero tillage plots could be attributed to their high dry bulk density of soil. Chaudhary *et al.* (1985) and Gajri *et al.* (1997) reported similar relation between bulk density and penetration resistance under different tillage practices. SPR at emergence (10 DAS) positively correlated with dry bulk density (r = 0.805**). Irrigation regimes also influenced the penetration resistance in the soil. At seed filling stage the penetration resistance was measured in all the plots just before giving irrigation, which coincided with cumulative pan evaporation at 300 mm in all the treatments. Lower soil penetration resistance values were recorded in irrigation regime IW/CPE = 1.2 and higher SPR in IW/CPE = 0.6 irrigation regime. Reduction in penetration resistance with increase in water application ($I_1 > I_2 > I_3 > I_4$) may be attributed to higher soil moisture content in IW/CPE = 1.2. Soil moisture content and SPR recorded significantly negative correlation (r = –0.558*).

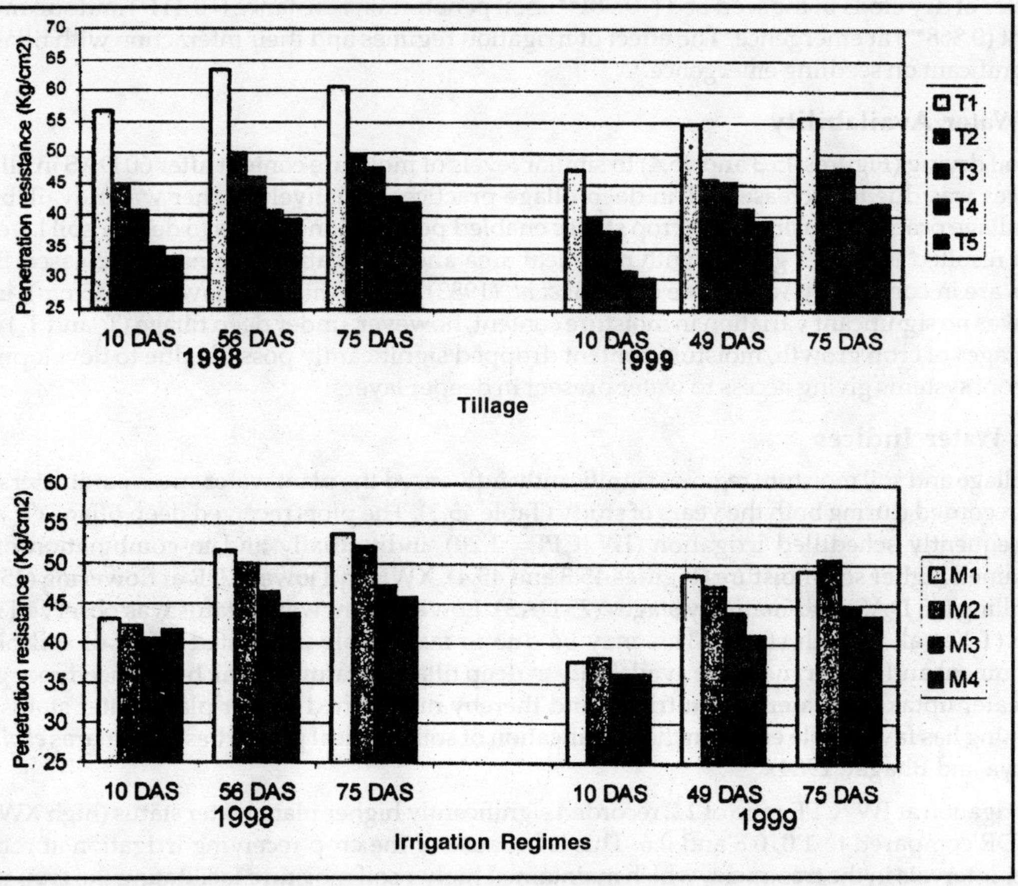

Figure 45.1: Effect of Tillage Practices and Irrigation Regimes on Soil Penetration Resistance

Soil Water Retention

Tillage practices increased soil water retention (Figure 45.2). Deep tillage (T_5 and T_4) caused increase in water retention up to 30 cm soil depth; however tillage with bullock drawn country plough influenced water retention in surface soil layer only. Secondary tillage with rotavator (T_3 and T_5) formed fine tilth with more number of capillary pores in surface soil layer is attributed for high water retention in such tillage practices. Similar results of higher soil moisture content with deep tillage under rainfed conditions owing to higher water storage capacity of soil profile were reported by Vittal *et al.* (1983). The soil under zero tillage was massive and compact with low pore space hence low water retention.

Initial Crop Stand

High initial crop with rotavator tillage (as secondary tillage) may be attributed to formation of fine tilth resulting in improved seed-soil-water contact and soil physical conditions that are prerequisites for germination and emergence (Table 45.1). Similar was the case with rotavator used as secondary tillage (81.81 per cent incase of T_5 and 79.73 incase of T_3) compared to cultivator (67.1 per cent incase of T_4 and 65.86 per cent in case of T_2) and zero tillage plots (57.5 per cent). Seedling emergence per cent, was significantly correlated with other soil physical properties *viz.*, mean mass diameter of dry clods in the seed bed (–0.961**), soil penetration resistance (–0.411*) and soil moisture content (0.868**) at emergence. The effect of irrigation regimes and their interaction with tillage was not significant on seedling emergence.

Soil Water Availability

Soil drying (Figures 45.3 and 45.4) to similar levels of moisture content after 60 DAS in all tillage practices was due to increased ET in deep tillage practices. Relatively higher water availability in deep tillage practices during early crop stages enabled penetration of roots to deeper soil layers that inturn resulted in higher growth with more leaf area and dry matter causing an increased ET. The results are in conformity with those of Vittal *et al.* (1983). In the third soil layer (30–45 cm), initially there was no significant variation in moisture content, however, under deep tillage (T_4 and T_5) during later stages of crop growth, moisture content dropped significantly possibly due to development of deep root systems giving access to water present in deeper layers.

Plant Water Indices

Tillage and soil moisture regimes significantly influenced the plant water stress parameters (XWP, LDR) recorded during both the years of study (Table 45.3). The plots received deep tillage (T_5 and T_4) and frequently scheduled irrigation (IW/CPE= 1.20), individually and in combination of both, maintained higher soil moisture (Figures 45.3 and 45.4), XWP and lower LDR at flowering (45 DAS), seed filling (65 DAS) and maturity stages (75 DAS). Lowest plant water status was observed in zero tillage (T_1) at all growth stages. This may be due to favourable integrated effect of soil physical environment and higher moisture availability in deep tillage treatment which enhanced root growth for greater uptake of water and nutrients and thereby maintained higher plant water status. Deep ploughing has favourable effect on the modification of soil physical properties (Allamaras *et al.*, 1977; Acharya and Bhagat, 1984).

Irrigation at IW/CPE ratio of 1.2 recorded significantly higher plant water status (high XWP) and low LDR compared to 1.0, 0.8 and 0.6. This is because of the crop receiving irrigation at relatively shorter intervals in the treatment, which maintained higher soil moisture facilitating the crop to draw required amount of moisture.

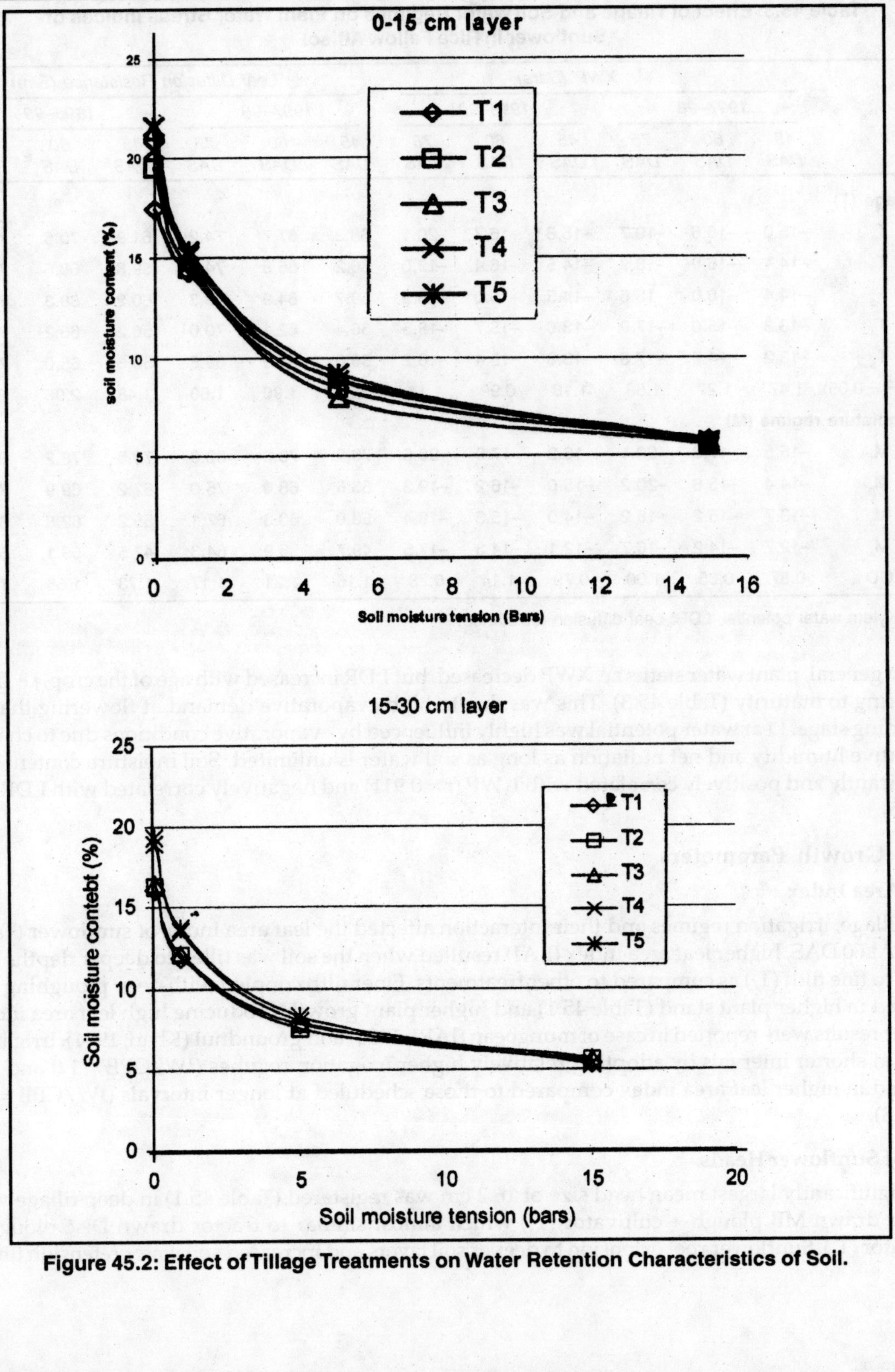

Figure 45.2: Effect of Tillage Treatments on Water Retention Characteristics of Soil.

Table 45.3: Effect of Tillage and Soil Water Regimes on Plant Water Stress Indices of Sunflower in Rice Fallow Alfisol

	XWP (Bars)						*Leaf Diffusion Resistance (S/m)*					
	1997–98			*1998–99*			*1997–98*			*1998–99*		
	45 DAS	*60 DAS*	*75 DAS*	*45 DAS*	*60 DAS*	*75 DAS*	*45 DAS*	*60 DAS*	*75 DAS*	*45 DAS*	*60 DAS*	*75 DAS*
Tillage (T)												
T_1	–15.0	–16.6	–19.7	–15.6	–16.7	–20.1	63.3	67.7	74.2	61.9	70.5	77.9
T_2	–14.4	–16.0	–18.9	–14.5	–16.4	–19.0	60.2	65.8	74.3	59,8	68.6	76.7
T_3	–14.4	–16.0	–18.6	–14.5	–16.0	–18.9	59.7	64.9	73.3	60.2	68.3	75.8
T_4	–13.3	–15.0	–17.9	–13.6	–15.7	–18.5	56.4	62.1	70.0	56.2	65.2	72.5
T_5	–13.2	–14.7	–17.8	–13.6	–15.4	–18.2	56.0	61.9	69.5	56.3	65.0	72.0
CD (P = 0.05)	1.47	1.27	1.58	1.18	0.99	1.14	2.25	1.90	1.60	1.48	2.96	2.53
Soil moisture regime (M)												
M_1	–15.5	–17.3	–20.1	–16.2	–17.9	–20.6	70.2	73.8	83.0	70.5	78.2	83.7
M_2	–14.4	–15.8	–20.2	–15.0	–16.2	–19.3	63.6	68.0	75.0	62.2	69.9	78.9
M_3	–13.7	–15.2	–18.2	–14.0	–15.3	–18.4	53.0	60.3	67.1	55.2	62.8	69.8
M_4	–12.7	–14.2	–16.7	–12.1	–14.3	–17.5	49.7	55.9	64.3	47.5	59.1	67.5
CD	0.87	0.85	1.00	0.79	1.14	0.78	1.16	1.21	1.17	1.73	1.55	1.68

XWP: Xylem water potential; LDR: Leaf diffusion resistance.

In general, plant water status *i.e.* XWP decreased, but LDR increased with age of the crop, *i.e.* from flowering to maturity (Table 45.3). This was due to high evaporative demand at flowering than at preceding stage. Leaf water potential was highly influenced by evaporative conditions due to change in relative humidity and net radiation as long as soil water is unlimited. Soil moisture content had significantly and positively correlated with L WP (r = 0.911) and negatively correlated with LDR (r = –0.885).

Crop Growth Parameters

Leaf Area Index

Tillage, irrigation regimes and their interaction affected the leaf area index of sunflower (Table 45.1). At 60 DAS, higher leaf area index (LAI) resulted when the soil was tilled to deeper depths and made to fine tilth (T_5) as compared to other treatments. Finer tilth coupled with deep ploughing (T_5) resulted in higher plant stand (Table 45.1) and higher plant growth producing high leaf area index. Similar results were reported in case of mungbean (IARI, 1984) and groundnut (Khan, 1984). Irrigation given at shorter intervals by adopting relatively higher irrigation regimes (IW/CPE = 1.0 and 1.2) resulted in higher leaf area index compared to those scheduled at longer intervals (IW/CPE = 0.6 and 0.8).

Size of Sunflower Heads

Significantly largest mean head size of 16.2 cm was registered (Table 45.1) in deep tillage with tractor drawn MB plough + cultivator (T_4) which almost similar to tractor drawn Disc plough + rotavator (T_5). Sunflower root extension to deeper soil layers and increased soil water retension under

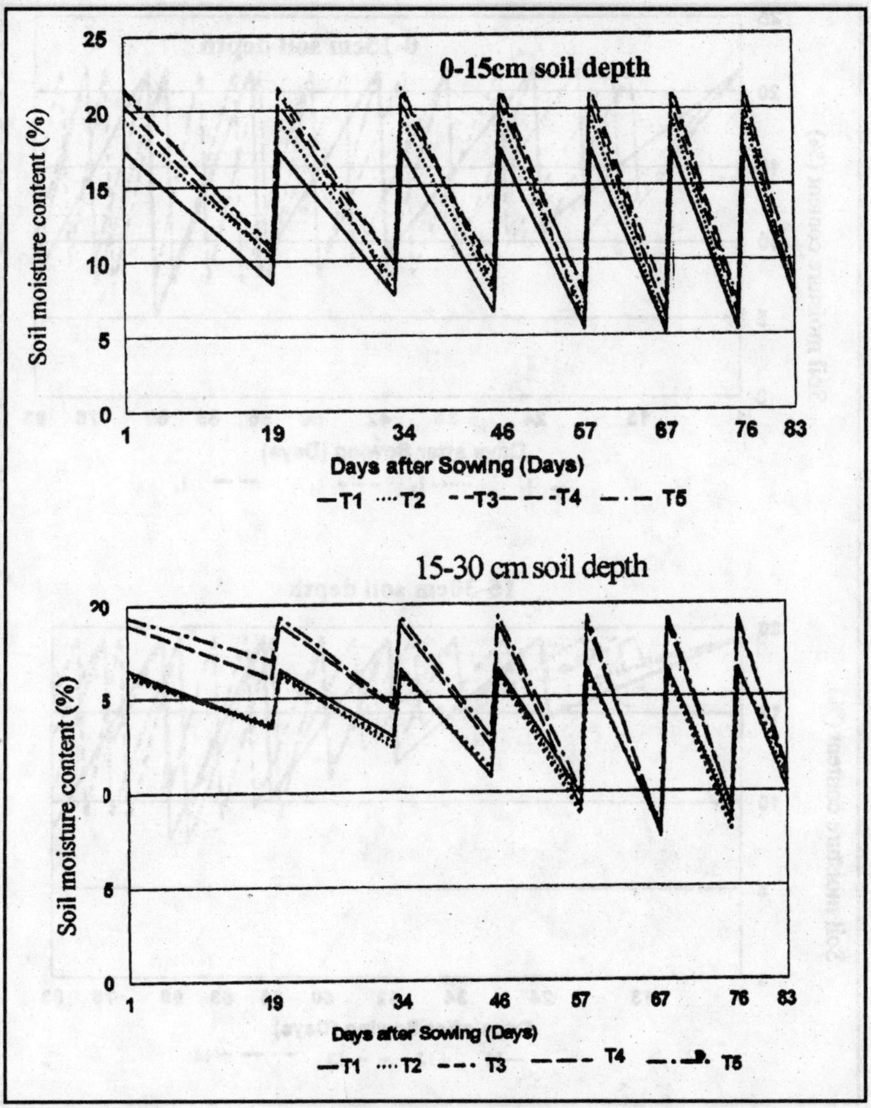

Figure 45.3: Effect of Tillage on Soil Moisture Availability Between Irrigation Cycles (Drying Cycles) of IW/CPE = 0.8 Regime During Crop Period

deep tillage practices, which inturn resulted in higher crop growth is ascribed for the production of larger heads. Irrigation scheduled at IW/CPE= 1.2 (I_4) registered significantly large heads is due mainly to optimum water supply in the particular irrigation schedule. Increase in sunflower head size with intensity of irrigation frequency was also earlier reported by Singh *et al.*, (1995).

Drymatter Produced

Favorable soil conditions created with tractor drawn disc plough once + rotovator twice (T_5) supported the crop with higher relative leaf water content and leaf water potential, thus resulting in

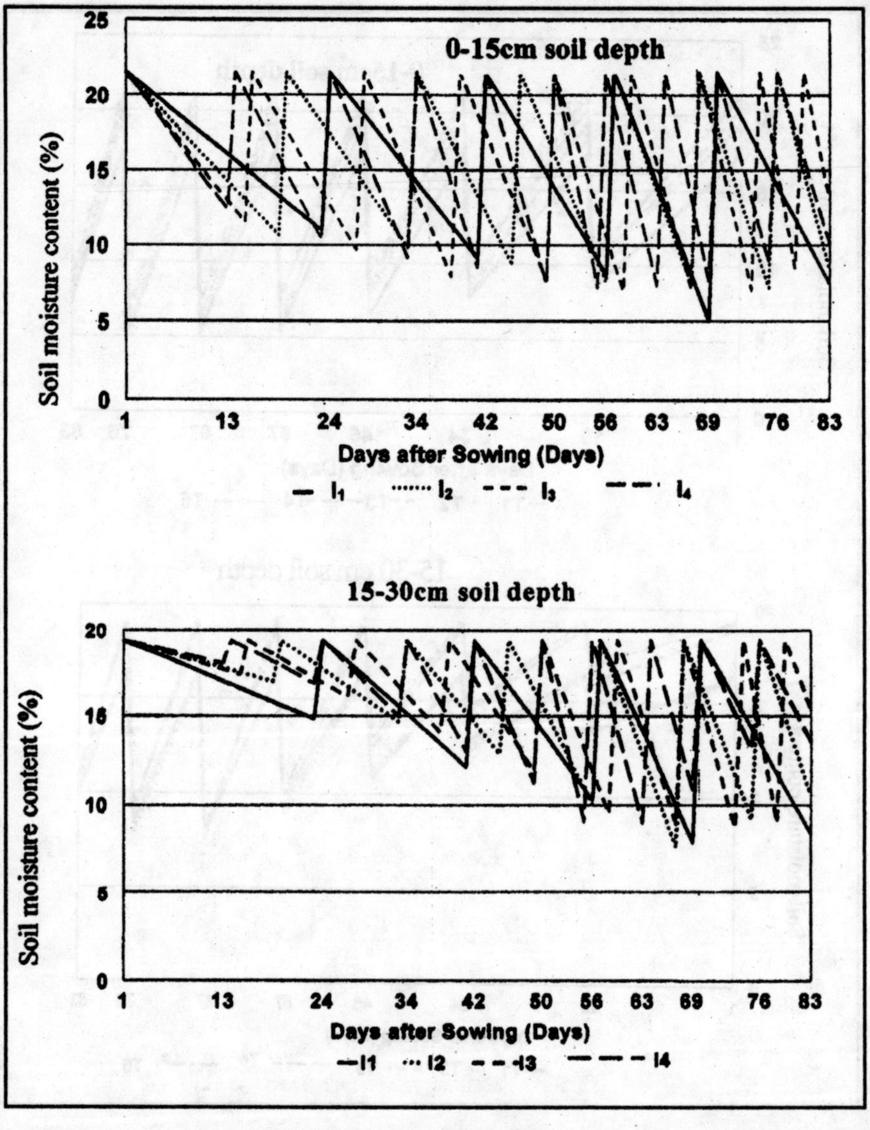

Figure 45.4: Effect of Irrigation Regimes on Soil Moisture Availability Between Irrigation Cycles (Drying Cycles) of Deep-fine Tillage Practice (T_5) during Crop Period

higher dry matter accumulation (Figure 45.5). Increase in soil moisture regime significantly increased the dry matter accumulation at harvest. This could be attributed to lower soil moisture stress prevailing before next irrigation in the plots irrigated with IW/CPE ratio of 1.0 and 1.2. On the other hand in IW/CPE = 0.8 and IW/CPE= 0.6 moisture regimes, plants experienced relatively higher soil water stress thereby lower plant growth hence low dry matter accumulation. The, interaction effects of tillage and soil moisture regimes significantly affected the dry matter production, which appeared due to high seedling emergence and relatively better growth in the initial stage of crop.

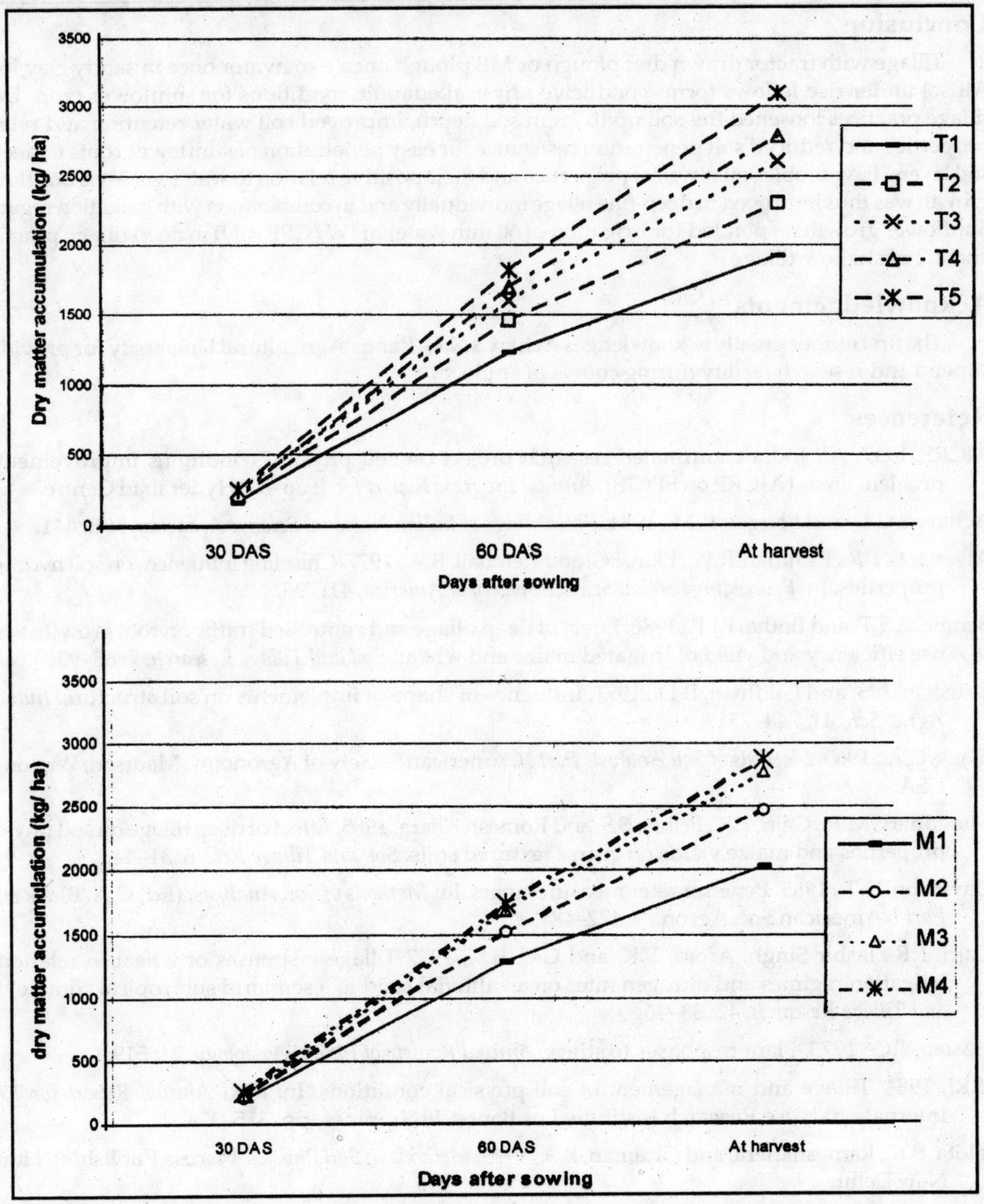

Figure 45.5: Effect of Tillage and Soil Water Regimes on Total Dry Matter Accumulation of Post Rainy Season Sunflower (Means 1998 and 1999)

Conclusion

Tillage with tractor drawn disc plough or MB plough once + rotavator once in sandy clay loam Alfisol under rice fallows forms conducive physical edaphic conditions for sunflower crop. Deep tillage practices loosened the soil up to 26cm soil depth, improved soil water retention and release properties and reduced soil penetration resistance for easy penetration of sunflower roots to deeper soil layers. Favourable soil physical properties and their positive relation to water vis-à-vis Sunflower growth was thus improved in deep-fine tillage individually and in combination with irrigation regimes. Sunflower growth responded for irrigation of 60 mm water at IW/CPE = 1.0 in deep tillage practices and 1.2 in shallow tillage.

Acknowledgements

The first author greatly acknowledges Acharya N.G. Ranga Agricultural University for providing stipend and research facility during course of study.

References

AICRP, 1987. All India coordinated research project on soil physical conditions improvement in problem areas (AICRP on SPCIP). *Annual Progress Report for 1986–87*, Hyderabad Centre.

Acharya, C.L. and Bhagat, R.M., 1984. *Proceedings of Indian National Science Academy*, 50B: 441.

Allamaras, P.R., Richman, R.W., Ekin, I.G. and Kemball, B.A., 1977. Chiseling influences on soil hydraulic properties. In: *Proceedings of Soil Science Society of America*, 41: 796.

Bennie, A.T.P. and Botha, F.J.P., 1986. Effect of deep tillage and controlled traffic on root growth, water use efficiency and yield of irrigated maize and wheat. *Soil and Tillage Research*, 7: 85–95.

Bhushan, L.S. and Gildhyal, B.D., 1971. Influence of shape of implements on soil structure. *Indian J. Agric. Sci.*, 41: 744–751.

Black, C.A., 1965. *Methods of Soil Analysis Part II*. American Society of Agronomy, Madison, Wisconsin, USA.

Chaudhary, M.R., Gajri, P.R., Prihar, S.S. and Romesh Khera, 1985. Effect of deep tillage on soil physical properties and maize yields on coarse textured soils. *Soil and Tillage Res.*, 6: 31–44.

Davidson, D.T., 1965. Penetrometer measurements. In: *Methods of Soil Analysis*, (Ed) C.A. Black *et al.*, *Part I*. American Soc. Agron., 9: 472–484.

Gajri, P.R., Jasbir Singh, Arora, V.K. and Gill, B.S., 1997 Tillage responses of wheat in relation to irrigation regimes and nitrogen rates on an alluvial sand in a semiarid subtropical climate. *Soil and Tillage Research*, 42: 33–46.

Hasiao, T.C., 1973. Plant responses to stress. *Annual Review of Plant Physiology*, 24: 519.

IRRI, 1984. Tillage and management of soil physical conditions. In: *IRRI Annual Report for 1983*, International Rice Research Institute, Los Banos, Philippines, pp. 315–324.

Jalota, S.K., Rameshkhera, and Ghuman, B.S., 1998. *Methods in Soil Physics*. Narosa Publishing House, New Delhi.

Khan, A.R., 1984. Studies on tillage induced physical edaphic properties in relation to peanut crop. *Soil and Tillage Research*, 4: 225–236.

Kramer, P.J., 1969 *Plant and Water Relationships: A Modern Synthesis*. McGraw Hill, New York, pp. 183.

Michael, A.M., Hukkeri, S.B. and Singh, N.P., 1977. *Estimating Water Requirement of Crops*. Indian Agricultural Research Institute, Monograph No. 4, New Delhi, pp. 91–162.

Panse, V.G. and Sukhatme, P.V., 1965. *Statistical Methods for Agricultural Workers*. Indian Council of Agricultural Research Press, New Delhi.

Rautaray, S.K., 1983. Tillage systems for production of sorghum in black soils. *J. Agric. Engg.*, 20: 27–38.

Reddy, G.H.S. and Reddy, A.A., 1989. Cause of low yields of groundnut in six districts of Andhra Pradesh. A Survey Report submitted to Andhra Pradesh Agricultural University, Hyderabad, India.

Reicosky, D.C., Campbell, R.B. and Doty, C.W., 1976. Corn plant water stress as influenced by Chiseling, irrigation and Water table depth. *Agronomy Journal*, 68: 499.

Scholandar, P.F., Hamnael, H.T., Badstreet and Hemmingsen, E.A., 1965. Sap pressure in vascular plants. *Science*, 148: 339–346.

Singa Rao, M., Prabhu Prasadini, R., Nagender Rao, C. and Jayasree, G., 1995. Soil physical constraints and their management for increasing crop production in Andhra Pradesh. *Highlights of Research*, 1967–1994.

Singh, V., Sharma, S.K., Verma, B.L. and Singh, B., 1995. Response of rainy season sunflower (*Helianthus annuus* L.) to irrigation and nitrogen under North Western Rajasthan. *Indian Journal of Agronomy*, 40(2): 239–242.

Vittal, K.P.R., Vijayalakshmi, K. and Rao, U.M.B., 1983. Effect of deep tillage on dryland crop production in red soils of India. *Soils and Tillage Res.*, 3(4): 377–384.

Youker, R.E. and McGunnis, K., 1967. A short method of obtaining mean weight diameter value of aggregate analysis of soils. Soil Sci., 83: 291–294.

Young, R.E. and Browning, D., 1977 Water stress influence on cotton. *Trans. American Society of Agricultural Engineering*, 20: 673.

Chapter 46

Correlation Coefficient (r) Value Between LCC Vs SPAD as Influenced by Hybrids and Nitrogen Management Practices

S. Ramesh, S. Ravi and B. Chandrasekaran
Department of Agronomy, Tamil Nadu Agricultural University, Coimbatore – 641 003, Tamil Nadu

ABSTRACT

Field experiments were conducted at wetland research farm of Tamil Nadu Agricultural University, Coimbatore during *Kharif* 2000 and 2001 to study the relationship between the LCC values and SPAD reading, and the LCC values Vs Grain yield as influenced by nitrogen management practices in rice hybrids. Treatments consisted of two hybrids in the main plot and seven levels of N management practices during *Kharif* 2000 and eight sub-plot treatments during *Kharif* 2001. Application of green manure @ 6.25 tha^{-1} combined with LCC cv5 based N application (N_4) registered, higher level of LCC and SPAD values than the other treatments. LCC Vs SPAD values had relatively stronger relationship in both the years at all the stages of crop growth and also confirmed the existence of strong positive correlation co-efficient between LCC Vs grain yield at all the stages of crop except at 14 days after transplanting.

Keywords: Hybrids, Nitrogen management, Leaf colour chart, SPAD.

Introduction

To sustain the self-sufficiency in the coming decades the rice production needs to be geared up every year by almost 2 million tones. This is a daunting task, in view of plateauing trend observed in yield potential of high yielding varieties, and decreasing and declining natural resource base. Among the many genetic approaches being explored to break the yield barrier in rice, hybrid rice technology appears to be the most feasible and readily adoptable one. Swaminathan (1998) emphasized precision farming to reduce the cost of production and improve the productivity on an ecologically sustainable basis. Matching the time course N demand of the rice crop with N supply leading to maximum utilization is the precision N management. Assessing the leaf N concentration is one of the techniques to maintain specific N concentration at different physiological stages by which the yield potential can be exploited. This is essentially because N concentration in leaves is highly correlated with concentration of the CO_2 fixing enzyme in photosynthesis (Greenwood *et al.*, 1991) and hence, the potential maximum rate of photosynthesis could be linearly related to the N concentration in the leaf (Black, 1993). One of the recently introduced N management approach for estimating the leaf N concentration was by the measurement of leaf greenness. Leaf green colour was calibrated very precisely to the leaf N content (Tanno, 1988). Among the different tools available to measure the leaf greenness, the non destructive measurement of leaf green colour intensity using leaf green colour charts (LCC) and the instrument namely, chlorophyll meter (SPAD-502) are gaining much importance. Hence in this context, it becomes imperative to conduct this experiment to study the relative accuracy of LCC to assess the leaf N status and it was compared and correlated with chlorophyll meter reading (IRRI, 1996) which is highly essential to determine the time of N topdressing for specific cultivar groups such as hybrids.

Materials and Methods

Field experiments to study the effect of green manuring and fertilizer N application through LCC grades on rice hybrids were conducted during *Kharif* 2000 and 2001 in wetland research farm of Tamil Nadu Agricultural University, Coimbatore. The soil of the experimental field was moderately drained, deep clay loam, classified taxonomically as typic haplustalf. The soils were low in available soil N, medium in available soil P and high in available soil K respectively. The experiment was laid out in split plot design with three replications. The main plot treatments consisted of two hybrids PA6201 and CoRH2 and the sub-plots with seven levels of N management practices during *Kharif* 2000 whereas during *Kharif* 2001 one more sub-plot treatment was included which is the existing recommended practice under Integrated nitrogen management practices (N_8). The rice hybrid PA6201 was released in 2000 by Central Release Committee (CVRC). It is a medium duration (125–130 days) semi dwarf, erect type having long panicle with long slender grain. The rice hybrid $CoRH_2$ released during 1998 by Paddy Breeding Station, Tamil Nadu Agricultural University, Coimbatore is also a medium duration semi-dwarf type having compact panicle with medium grain type.

The details of the treatments and notations used are as under:

A. Main plot: Hybrid rice

H_1: PA6201

H_2: CoRH2

B. Sub-Plot: Nitrogen management

N_1: Basal + 3 splits of recommended dose (200 kgha^{-1})

N_2: Green manure @6.25 tha^{-1} + N application as per LCC cv 3

N_3: Green manure 6.25 tha^{-1} + N application as per LCC cv 4

N_4: Green manure 6.25 tha^{-1} + N application as per LCC cv 5

N_5: 20 kg Nha^{-1} as basal + N as per LCC cv 3

N_6: 20 kg Nha^{-1} as basal + N as per LCC cv 4

N_7: 20 kg Nha^{-1} as basal + N as per LCC cv 5

N_8: Green manure @ 6.25 tha^{-1} + 3 splits of recommended dose (200 kgha^{-1})

An absolute control was maintained separately in the experimental field to find out the efficiency of soil applied nitrogen. The above ground biomass of *Sesbania aculeata* at the age of 40 days was harvested, weighed and applied to the respective plots as per the treatment schedule. The green manure was spread uniformly and incorporated one week before transplanting. The fertilizer nitrogen was applied to the rice crop as per the treatments. In LCC based N management treatments, the LCC values were recorded as per the standard procedure (IRRI, 1996) at weekly intervals starting from 14 DAT to flowering. Leaf colour chart consists of six strips from light yellowish green (No. 1) to dark green (No. 6). The rice leaf colour was compared with standard LCC under the same environmental condition. The colour of the single leaf was measured by holding the leaf, colour chart vertically and placing the middle part of leaf, 1 cm in front of colour strip for comparison (IRRI, 1996). Ten readings were taken for each plot and average was computed by rounding off to the nearest 0.5 to determine the need for N top dressing. Whenever the LCC values fall below the fixed critical level the amount of N to be applied at different growth stages was as per the schedule given below (IRRI, 1996).

Crop Growth Stages	*N to be Applied (kgha^{-1})*
Early growth stage (14–21 DAT)	20
Rapid growth stage (28–42 DAT)	30
Late growth stage (45-Flowering)	20

Table 46.1: Leaf Colour Chart Based N Application at Weekly Intervals-*Kharif* 2000

Time of Fertilizer Application (DAT)		H_1N_1	H_1N_2	H_1N_3	H_1N_4	H_1N_5	H_1N_6	H_1N_7	H_2N_1	H_2N_2	H_2N_3	H_2N_4	H_2N_5	H_2N_6	H_2N_7
Basal		50	–	–	–	20	20	20	50	–	–	–	20	20	20
Early growth stages (EGS)	14	–	–	20	20	–	20	20	–	–	20	20	–	20	20
	21	50	20	–	20	20	–	20	50	20	–	20	20	–	20
Rapid growth stages (RGS)	28	–	–	30	–	–	30	–	–	–	30	–	–	30	–
	35	–	30	–	30	30	–	30	–	30	–	30	30	–	30
	42	50	–	30	–	–	30	–	50	–	30	–	–	30	–
Late growth stages (LGS)	49	–	20	–	20	20	–	20	–	20	–	20	20	–	20
	56	–	–	20	20	–	20	20	–	–	20	20	–	20	20
	63	50	20	–	20	20	–	20	50	20	–	20	20	–	20
	70	–	–	20	20	–	20	20	–	–	20	20	–	20	20
Total kgha^{-1}		**200**	**90**	**120**	**150**	**110**	**140**	**170**	**200**	**90**	**120**	**150**	**110**	**140**	**170**

The total quantity of N applied as per the treatments are furnished in Tables 46.1 and 46.2. All the treatments including control received a uniform dose of 50 kg P_2O_5 and 50 kg K_2Oha^{-1}. The entire dose of P as Single Superphosphate (16 per cent P_2O_5) was applied as basal before transplanting. Potassium in the form of Muriate of Potash (60 per cent K_2O) was applied in three equal splits *viz.*, 50 per cent as basal and 25 per cent each at active tillering and panicle initiation stages.

Chlorophyll meter (SPAD 502) readings were recorded from 14 DAT in all the treatments at 7 days interval until flowering stage. The top most fully opened leaf was chosen for SPAD measurement and SPAD readings were taken on one side of the midrib of the leaf blade, mid way between leaf base and leaf tip. At early growth stages, measurements were taken near the tip of the leaf and even on the midrib, since the leaf blade was too narrow for SPAD measurements. Ten readings were taken at random for each plot and average was arrived. The correlation coefficient studies were made between LCC Vs SPAD values and between LCC V s grain yield.

Results and Discussion

Leaf Colour Chart (LCC) Values

Leaf colour chart (LCC) readings were recorded from 14 DAT to flowering stage in all the treatments at an interval of seven days (Tables 46.3 and 46.4). The two rice hybrids *viz.*, PA 6201 and CoRH2 had not exhibited any distinct colour difference among themselves at every point of LCC reading in both the seasons. This is evidenced by the observed LCC values for H_1 and H_2, their corresponding number of split doses and the total quantity of N applied to each treatment. However, variations in the LCC values were observed due to N management practices. The number of N splits across N_1 to N_7 was 4, 5, 7, 5, 6 and 7 and their corresponding quantity of N was 200, 90, 120, 150, 110, 140 and 170 kgha^{-1} for both PA6201 and CoRH2 hybrids in both the seasons. Considering the growth and yield performance of the hybrids, the N_4 treatment *i.e.*, green manure @ 6.25 tha^{-1} followed by inorganic N at LCC critical value 5 based N applications was significantly superior to all other treatments. Therefore, N_4 treatment is found to be the optimum N management practice which requires seven splits of top dressing, amounting to 150 kg Nha^{-1}. This treatment resulted in a saving of 50 kg Nha^{-1} as compared to the recommended dose of 200 kg Nha^{-1} alone (N_1) besides the yield advantage.

SPAD Values

The SPAD measurements were taken in the same leaves used for LCC measurements at the same time and intervals. The observations were made across the hybrids and N application practices. The recorded SPAD values were used for the correlation studies with LCC values (Tables 46.5 and 46.6).

Correlation Coefficient Between LCC Vs SPAD Values

The correlation coefficient studies were made between LCC Vs SPAD values and between LCC Vs grain yield. The relationship between LCC Vs SPAD values revealed the existence of a strong positive correlation in both the seasons at all the stages of crop growth (Table 46.7). The r values ranged from 0.53 and 0.59 at 14 DAT to 0.95 at 49 DAT respectively for *Kharif* 2000 and 2001. Also it indicated the closeness during 28, 35, 42 and 49 DAT in both the years. However, the relationship between LCC Vs SPAD values at 14 DAT was exhibited significantly. With regard to correlation coefficient between LCC Vs Grain yield, there was a significant positive correlation coefficient (0.37 to 0.82) at all the time of observations except at 14 DAT (Table 46.8). The relationship between LCC Vs Grain yield was almost similar to that of LCC Vs SPAD in both the seasons. In case of SPAD Vs Grain yield (0.86 to 0.96) higher positive correlation was registered in all the stages of observations. Johnkutty and Palaniappan (1996) reported that the higher positive correlation coefficient between SPAD Vs Grain yield (0.75 to 0.90) in lowland rice at Coimbatore, Tamil Nadu.

Table 46.2: Leaf Colour Chart Based N Application at Weekly Intervals-*Kharif* 2001

Time of Fertilizer Application (DAT)		H_1N_1	H_1N_2	H_1N_3	H_1N_4	H_1N_5	H_1N_6	H_1N_7	H_1N_8	H_2N_1	H_2N_2	H_2N_3	H_2N_4	H_2N_5	H_2N_6	H_2N_7	H_2N_8
Basal		50	–	–	–	20	20	20	–	50	–	–	–	20	20	20	–
Early growth	14	–	–	20	20	–	20	20	–	–	–	20	20	–	20	20	–
stages (EGS)	21	50	20	–	20	20	–	20	50	50	20	–	20	20	–	20	50
Rapid growth	28	–	–	30	–	–	30	–	–	–	–	30	–	–	30	–	–
stages (RGS)	35	–	30	–	30	30	–	30	–	–	30	–	30	30	–	30	–
	42	50	–	30	–	–	30	–	50	50	–	30	–	–	30	–	50
Late growth	49	–	20	–	20	20	–	20	–	–	20	–	20	20	–	20	–
stages (LGS)	56	–	–	20	20	–	20	20	–	–	–	20	20	–	20	20	–
	63	50	20	–	20	20	–	20	50	50	20	–	20	–	–	20	50
	70	–	–	20	20	–	20	20	–	–	–	20	20	20	20	20	–
Total $kgha^{-1}$		**200**	**90**	**120**	**150**	**110**	**140**	**170**	**150**	**200**	**90**	**120**	**150**	**110**	**140**	**170**	**15**

Table 46.3: LCC Values Recorded at Different Days After Transplanting in the Two Hybrids as Influenced by Different N Regimes. Mean of 10 observations in *Kharif* 2000

Hybrid/ N-Regime	*Leaf Colour Chart (LCC) Values at Different Days After Transplanting*								
	14	*21*	*28*	*35*	*42*	*49*	*56*	*63*	*70*
H_1N_1	3.5	4.0	5.0	4.5	4.0	4.5	4.5	4.5	4.5
H_1N_2	3.5	3.0	4.0	3.0	3.0	4.0	4.0	3.0	3.5
H_1N_3	3.5	4.5	3.5	4.5	4.0	4.5	4.0	4.5	4.0
H_1N_4	3.5	4.5	5.0	4.5	5.0	4.5	4.5	4.5	4.5
H_1N_5	3.0	3.0	3.5	3.0	4.0	3.0	3.5	3.0	3.5
H_1N_6	3.0	3.5	3.5	4.0	3.5	4.0	3.5	4.0	3.5
H_1N_7	3.0	3.5	5.0	4.5	5.0	4.5	4.0	4.5	4.5
H_2N_1	3.5	3.5	5.0	4.5	4.0	4.5	4.5	4.0	4.5
H_2N_2	3.5	3.0	4.0	3.0	4.0	3.0	4.0	3.0	3.5
H_2N_3	3.5	4.5	4.0	4.5	4.0	4.5	4.0	4.5	4.0
H_2N_4	3.5	4.5	5.0	4.5	5.0	4.5	4.5	4.5	4.5
H_2N_5	3.0	3.0	3.5	3.0	3.5	3.0	3.5	3.0	3.5
H_2N_6	3.0	3.5	3.5	4.0	3.5	4.0	3.5	4.0	3.5
H_2N_7	3.0	4.0	5.0	4.5	5.0	4.5	4.0	4.5	4.5

Table 46.4: LCC Values Recorded at Different Days After Transplanting in the Two Hybrids as Influenced by Different N Regimes. Mean of 10 observations in *Kharif* 2001

Hybrid/ N-Regime	*Leaf Colour Chart (LCC) Values at Different Days After Transplanting*								
	14	*21*	*28*	*35*	*42*	*49*	*56*	*63*	*70*
H_1N_1	3.5	4.0	5.0	4.5	4.0	4.5	4.5	4.5	4.5
H_1N_2	3.5	3.0	4.0	3.0	4.0	3.0	4.0	3.0	3.5
H_1N_3	3.5	4.5	4.0	4.5	4.0	4.5	4.0	4.5	4.0
H_1N_4	3.5	4.5	5.0	4.5	5.0	4.5	4.5	4.5	4.5
H_1N_5	3.0	3.0	3.5	3.0	4.0	3.0	3.5	3.0	3.5
H_1N_6	3.0	3.5	3.5	4.0	3.5	4.0	3.5	4.0	3.5
H_1N_7	3.0	4.0	5.0	4.5	5.0	4.5	4.0	4.5	4.5
H_1N_8	3.5	4.5	5.0	4.5	5.0	4.5	4.5	4.0	4.5
H_2N_1	3.5	4.0	5.0	4.5	4.0	4.5	4.5	4.0	4.5
H_2N_2	3.5	3.0	4.0	3.0	4.0	3.0	4.0	3.0	3.5
H_2N_3	3.5	4.5	4.0	4.5	4.0	4.5	4.0	4.5	4.0
H_2N_4	3.5	4.5	5.0	4.5	5.0	4.5	4.5	4.5	4.5
H_2N_5	3.0	3.0	3.5	3.0	3.5	3.0	3.5	3.0	3.5
H_2N_6	3.0	3.5	3.5	4.0	3.5	4.0	3.5	4.0	3.5
H_2N_7	3.0	4.0	5.0	4.5	5.0	4.5	4.0	4.5	4.5
H_2N_8	3.5	4.5	5.0	4.5	5.0	4.5	4.5	4.0	4.5

Table 46.5: SPAD Values Recorded at Different Days After Transplanting in the Two Hybrids as Influenced by Different N Regimes. Mean of 10 observations in *Kharif* 2000

Hybrid/	Chlorophyll Meter (SPAD) Values at Different Days After Transplanting								
N-Regime	14	21	28	35	42	49	56	63	70
H_1N_1	38.0	42.5	46.1	43.7	42.3	44.0	43.4	42.9	40.9
H_1N_2	36.5	36.0	38.3	36.0	38.1	35.6	37.7	35.0	36.8
H_1N_3	37.0	41.3	39.6	41.8	40.2	42.1	40.0	42.3	41.0
H_1N_4	39.5	44.0	48.4	45.0	46.5	45.1	44.8	44.0	43.2
H_1N_5	35.1	33.8	37.3	34.0	36.3	34.0	35.5	34.7	36.0
H_1N_6	35.5	39.9	38.5	39.1	37.4	40.2	38.6	41.1	38.8
H_1N_7	37.9	40.2	44.0	41.5	42.7	42.0	40.5	42.5	40.5
H_2N_1	37.2	41.0	44.5	42.3	41.2	42.9	42.1	41.8	39.5
H_2N_2	35.0	35.9	37.3	35.4	37.9	43.1	36.8	34.6	36.0
H_2N_3	36.1	40.0	38.7	40.5	39.1	41.4	39.2	41.1	40.0
H_2N_4	38.4	43.1	46.8	44.3	45.3	44.6	43.5	42.8	42.0
H_2N_5	34.2	32.7	35.6	33.7	35.0	33.8	34.6	34.0	34.0
H_2N_6	34.5	37.8	37.0	40.0	36.3	39.1	37.0	40.1	37.9
H_2N_7	36.4	39.5	42.0	40.8	41.5,	40.2	39.7	41.0	39.0

Table 46.6: SPAD Values Recorded at Different Days After Transplanting in the Two Hybrids as Influenced by Different N Regimes. Mean of 10 observations in *Kharif* 2001

Hybrid/	Chlorophyll Meter (SPAD) Values at Different Days After Transplanting								
N-Regime	14	21	28	35	42	49	56	63	70
H_1N_1	39.1	43.0	47.3	44.0	43.1	45.0	44.0	43.3	41.7
H_1N_2	36.9	35.7	39.9	37.1	39.0	36.0	38.2	35.6	37.2
H_1N_3	38.9	42.0	40.9	42.4	41.3	43.9	40.6	43.5	42.5
H_1N_4	40.0	45.2	49.1	46.7	47.0	46.0	45.0	44.5	44.0
H_1N_5	35.9	34.4	38.4	35.2	37.1	35.0	36.0	35.3	36.4
H_1N_6	36.6	40.3	39.0	40.5	38.3	41.8	39.3	42.2	39.3
H_1N_7	38.0	41.7	45.8	42.3	43.6	43.0	41.0	43.7	41.2
H_1N_8	39.6	44.4	48.1	45.6	45.4	44.4	43.9	43.8	43.1
H_2N_1	39.5	41.5	45.0	43.0	42.3	43.6	42.9	42.5	40.4
H_2N_2	35.9	36.3	38.6	36.0	38.8	36.0	37.5	35.1	36.8
H_2N_3	37.0	40.9	39.4	42.2	40.7	42.3	40.0	42.4	41.6
H_2N_4	39.1	44.6	47.9	45.0	46.2	45.5	44.4	43.6	42.5
H_2N_5	35.0	33.3	36.8	34.6	35.8	34.7	35.0	34.3	35.7
H_2N_6	35.4	38.9	38.0	40.7	37.1	40.0	38.1	41.7	38.4
H_2N_7	37.2	40.0	43.4	41.6	42.6	41.1	40.7	42.0	39.8
H_2N_8	38.2	42.7	46.5	43.9	44.1	43.4	41.8	43.4	41.9

Table 46.7: Correlation Coefficient (r) Between LCC Vs SPAD

Time of Observation (Days after transplanting)	*No. of Observation*	*Kharif 2000*	*Kharif 2001*
14	30	0.53*	0.59**
21	30	0.83**	0.92**
28	30	0.92**	0.94**
35	30	0.93**	0.93**
42	30	0.84**	0.86**
49	30	0.95**	0.95**
56	30	0.89**	0.88**
63	30	0.66**	0.94**
70	30	0.79**	0.82**

Table 46.8: Correlation Coefficient (r) Between LCC Vs Grain Yield

Time of Observation (Days after transplanting)	*No. of Observation*	*Kharif 2000*	*Kharif 2001*
14	30	0.37	0.44
21	30	0.70**	0.79**
28	30	0.73**	0.75**
35	30	0.81**	0.80**
42	30	0.62**	0.67**
49	30	0.81**	0.80**
56	30	0.71**	0.74**
63	30	0.82**	0.75**
70	30	0.79**	0.80**

*: Significant at 5 per cent level (p = 0.05); **: Significant at 1 per cent level (p = 0.01).

References

Black, C.A., 1993. *Soil Fertility Evaluation and Control*. Lewis Publishers, Boca Raton, pp. 171–175.

Greenwood, D.J., Gasta, G., Lemaire, G., Draycott, A., Millard, P. and Neeteson, J.J., 1991. Growth rate and per cent N of field grown crops: Theory and Experiments. *Ann. Bot.*, 67: 181–190.

IRRI, 1996. *Use of Leaf Colour Chart (LCC) for N Management in Rice*. Int. Rice Res. Inst., Manila 1099, Philippines.

Johnkutty, I. and Palaniappan, S.P., 1996. Nitrogen utilization growth and yield of rice as affected by green manuring and timing of isotope and non-isotope N. *Ph.D. Thesis*, Tamil Nadu Agric. Univ., Coimbatore.

Swaminathan, M.S., 1998. Rice and national food and livelihood security. Directorate of Rice Research, Silver Jubilee endowment Lecture, April 16, 1998. Punjab Agricultural University, Ludhiana.

Tanno, F., 1988. Forecasting and diagnosis methods on the plant growth for cultivars, koshihikari and sasanishiki, using multiple parameters. *Jpn. J. Soil Sci. and Plant Nutr.*, 59: 423–428.

Chapter 47

Study of Biodiversity and Water Quality of Kareh River in Samastipur District, North Bihar

Shikha Kumari[1] and B.P. Verma[2]

[1]Department of Home Science, L.N. M.U., Darbhanga

[2]M.K.S. College, Chandauna, Darbhanga, Bihar

ABSTRACT

The river Kareh which courses through Samastipur district is very important from economic point of view. No information is available about the water quality and biodiversity of the river till date. By documenting the water quality and biodiversity of this river we can identify problems prevailing there establish a data for suitable flora and fauna and thereby make maximum beneficial use for the localities. Smt. Shikha Kumari studied the water quality and biodiversity, transparency, temperature, regulation, pH, dissolved oxygen, free CO_2, total alkalinity, nitrate chloride, phosphate, silicate, calcium and magnesium contest of the river. She analyses and suggested that the water quality of the Kareh river is good. The algal community exhibited a high biodiversity. Altogether seventy species of algal taxa were observed, out of which twenty one belonged to cyanophyceae, twenty two belonged to xanthophyceae, five belong to euglenophyceae, and two belonged to rhodo phyceae and two belonged to dianopyceae. All the taxa were not present through the year. The population density of chlorophyceae was much higher than cyanophyceae. In the present investigation he primary peak of blue green algae was found during summer months. Rich biodiversity of algal taxa and the good water and can be exploited to inverse the potential of the fisheries.

Introduction

The survival of our present social system depends upon our ability to develop a harmonious relationship between the industrial and the ecological components. This relationship is dependent upon the development of a "use without abuse". Philosophy by resource developers, industrialist, conservationists, ecologist, politician and others. The Kareh river, which courses through the Samastipur district, is a very important river of North Bihar. By documenting the water quality and biodiversity of this river we can identify problem situations, prevailing there establish priorities and there by maximum beneficial use of the Kareh river.

A number of contributions on the riverine ecosystem are available from India. Bilgrami and Siddiqui (1983). Chakraborty *et al.* (1959), Prasad and Saxena (1980), Prasad and Verma (1988), Saha and Pandit (1990), Singh and Ahmad (1990), Sinha and *et al.* (1996), Khan (1996), Bilgrami and Dutta Munshi (1985).

However, no information is available on the documentation of water quality and biodiversity of Kareh river in Samastipur district for which this research work was undertaken.

Material and Methods

1. Analysis of water quality of river Kareh will be done with the help of APHA (1985).
2. Water samples will be collected from the different sites of the river.
3. Algal community structure of the selected site will be studied periodically.
4. Inter relationship between physico-chemical characteristics and phytoplankton (alga community) will be studied.

Observation

After investigation in the different season of the year the following observation were made.

Table 47.1: Physico-chemical Characteristics of Water of Kareh River

Sl.No.	*Parameter*	*Summer*	*Monsoon*	*Winter*
1.	Temperature	28°C–32°C	29°C–31°C	18°C–19.5°C
2.	Transparency (cm)	28.0–29.0	18.2–20.9	41.2–59.8
3.	pH	7.5–7.9	8.7–8.9	8.1–8.3
4.	Dissolved oxygen	5.4–6.5	5.0–6.1	9.5–10.5
5.	Free CO_2 (mg/l)	1.2	1.3	Nil
6.	Total alkalinity (mg/l)	98.2–102.0	159.0–165.0	190.0–198.0
7.	Nitrate (mg/l)	0.06–0.08	0.2–0.3	0.1
8.	Chloride (mg/l)	12.8	17.5–25.0	17.0–19.0
9.	Phosphate (mg/l)	0.02	0.04	0.03
10.	Silicate (mg/l)	4.9–5.2	7.8–8.9	0.9–1
11.	Calcium (mg/l)	20	25–30	21–27
12.	Magnesium (mg/l)	10–11.3	7.2–8.1	15.20

Seasonal variations in the biodiversity of Kareh river. In the present investigation apparent periodical variation were observed with respect to size as well as the composition of algal spectrum and their population density.

Table 47.2: Seasonal Diversity of Phytoplankton in River Kareh

Sl.No.	*Phytoplankton*	*Summer*	*Monsoon*	*Winter*
Cyanophyceae		+	+	+
1.	Anabaena	+	+	+
2.	Anabaenopsis	–	–	+
3.	Arthrospira	+	–	+
4.	Aphanocapsa	+	–	+
5.	Aphanothece	–	+	+
6.	Chroococcus	+	+	–
7.	Calothrix	–	+	–
8.	Dactylococopsis	–	–	+
9.	Gloeocapsa	+	+	+
10.	Gomphosphaeria	+	+	+
11.	Hapalosiphan	–	+	+
12.	Lyngbya	–	+	+
13.	Nostoc	+	–	–
14.	Merismopedia	+	+	+
15.	Oscillatioria	+	+	+
16.	Pharmidium	+	–	–
17.	Rivularia	+	+	–
18.	Rhabdoderma	+	–	+
19.	Synechococcus	–	–	+
20.	Spirulina	+	+	+
21.	Schizothrix	–	+	+
Chlorophyceae				
1.	Acinastrum	+	+	+
2.	Ankistrodermus	+	–	+
3.	Botryococcus	–	+	–
4.	Chlorella	–	+	+
5.	Closteridium	+	+	+
6.	Coelastrium	+	+	+
7.	Gloeotaenium	–	+	–
8.	Golenkinia	–	+	–
9.	Gonatozygon	+	+	+
10.	Hormidiom	+	+	+
11.	Hydrodictyon	–	–	+
12.	Mesotaenium	–	+	–
13.	Mougeotia	+	+	+
14.	Microspora	–	–	+
15.	Pediastrum	+	+	+
16.	Rhizoclonium	+	+	+

Contd...

Table 47.2–Contd...

Sl.No.	*Phytoplankton*	*Summer*	*Monsoon*	*Winter*
17.	Schizochlamys	−	−	+
18.	Spirogyra	+	+	+
19.	Scnedesmus	+	+	+
20.	Treubaria	−	−	+
21.	Teraspora	+	−	−
22.	Eudorina	+	+	+
Bacillariophyceae				
1.	Achnamthes	+	+	+
2.	Cyclotella	+	−	+
3.	Cymbella	+	+	+
4.	Cymatopleura	−	−	+
5.	Calonies	−	+	−
6.	Diatoma	+	−	−
7.	Diploner's	+	+	+
8.	Hantzschia	+	−	−
9.	Fragilaria	+	+	+
10.	Gomphonema	+	+	+
11.	Gyrosigma	+	+	+
12.	Navicula	+	+	+
13.	Nitzschia	+	+	+
14.	Pinnularia	+	+	+
15.	Melosira	+	+	+
16.	Rhopalodia	+	−	+
17.	Stephanodiscus	−	−	+
18.	Synedra	+	+	+
19.	Surnirella	+	−	−
20.	Tabellaria	+	+	+
Xanthophyceae				
1.	Chlorobotrys	+	+	+
2.	Tribonema	+	−	+
3.	Bumillaria	+	+	+
Euglenophyceae				
1.	Phacus	−	+	−
2.	Euglena	−	+	−
3.	Peronia	−	+	+
4.	Astasia	−	−	+
5.	Trachelomonas	+	−	+
Rhodophyceae				
1.	Lamanea	+	−	+
Dianophyceae				
1.	Ceratium	+	+	−
2.	Peridinium	−	+	+

Discussion

The transparency of river water fluctuated significantly in the different season. The minimum value was in the monsoon and maximum value was in the winter and moderate in the summer season. The temperature in the summer and monsoon almost similar but in winter the temperature was lower than summer.

In the present investigation transparency showed significant, inverse relationship with temperature a low transparency during monsoon may be ascribed to increase in non-algal suspended solid.

The pH of water was found to be in the alkaline range during all the season. In the present investigation the pH of water showed inverse relationship with temperature.

The dissolved oxygen was high in winter in comparison to summer and monsoon. The high value of oxygen in winter showed the good quality of water. Singh and Ahmad (1990) recorded 5.97–9.21 mg/l in river Ganga at Patna. The dissolved oxygen indicated direct relationship with transparency and an inverse relationship with temperature.

The free CO_2 was not detected in winter but it was found in summer and monsoon. Higher amount of free CO_2 during the monsoon months has been attributed to the disturbances of water due to heavy rainfall as reported by Mesfin and Belay (1989). The total alkalinity was highest in winter. Lowest in summer and moderate in monsson. The maximum value of bicarbonate alkalinity was as curtained in winter months as reported by Pahwa and Mehrotra (1966) in river Ganga. The total alkalinity showed universe relationship with temperature and direct relationship with transparency.

The nitrate concentration is high in monsoon. The chloride concentration in the present investigation was found much lower and CO_2 insides, with the report of Pahwa and Mehrotra (1966) Prasad and Singh (1982) and Prasad *et al.* (1988). The river Kareh is not polluted as regard to chloride concentration.

The annual trend in calcium content showed a little variation. It is not too high. That water quality is not polluted that is why then calcium concentration is not high. The magnetism concentration is highest in winter, lower in summer and moderate in monsoon. The concentration of magnesium ions in the present investigation varied from 7.2–20 mg/l which indicated it's unpolluted mature with regard to magnesium.

The phosphate concentration is not very high. It is maximum in the monsoon. The phosphate concentration showed significant correlation with free CO_2 and an inverse relation with pH. In this investigation phosphate concentration is not a problem for good quality of water.

Regarding biodiversity the algal community exhibited a high biodiversity indicating good water quality of the river Kareh. Altogether seventy five (75) algal taxa were observed in the river water during the course of investigation out of which twenty one (21) belonged to cyanophyceae, twenty two (22) belonged to chlorophyceae, twenty one (21) belonged to Bacillaryophyceae, three (3) belonged to Xanthophyceae, five (5) belonged to Euglenophyceae, one (1) belonged to Rhodophyceae, and two (2) belonged to didanophyceae. All the taxa were not present throughout the year. Only sever taxa of cyanophyceae observed throughout the year, Eleven (11) taxa of Chlorophyceae, 12 taxa of bacillariophyceae and two taxa of xanthophyceae. The population density of chlorophyceae was much higher than cyanophyceae. In the present investigation the primary peak of blue green algae was found during summer months.

Conclusions

Rich Biodiversity of algal taxa and the good water quality of Kareh River suggest the good health of the river and can be suggested for fishery.

References

APHA, 1980. *Standard Methods for the Examination of Water and Wastewater*, 15th edition.

Bilgrami, K.S. and Siddiqui, E.N., 1983. Biomonitoring of river Ganges in Mokama Barauni Industrial belt. *Biol. Bull.*, India, 5: 278–281.

Bilgrami, K.S. and Datta Munshi, J.S., 1985. Ecology of the river Ganges. Impact of human activities. Final Technical report, pp. 1–97.

Chakrabarty *et al.*, 1959. A quantitative study of the plankton and the physico-chemical condition of the river Yamuna of Allahabad (1954–55). *Indian J. Fish*, 6(1): 186–203.

Khan, M.A., 1996. Investigation on biomonitoring and ecorestoration measures in selected stretches of the river Ganga and Yamuna. Final Technical Report, pp. 1–64.

Pahwa, D.V. and Mehrotra, S.N., 1966. Observation on the fluctuations in the abundance of plankton in relation to certain hydrological conditions of river Ganga. *Proc. Nat. Acad. Sci., India*, 36: 157–189.

Prasad, B.N. and M. Saxena, 1980. Ecological study of blue algae in river Gomati. *Indian J. Environ. Health*, 22: 151–168.

Prasad, S.S. and Verma, N.N., 1988. Comparative ecology of Gandak and Burhi Gandak, MAB project report.

Saha, L.C. and Pandit, B., 1990. Dynamics of primary productivity between lentic and lotic system in relation to abiotic factors. *J. Indian Bot. Soc.*, 69(3 and 4): 213–217.

Singh, A.K. and Ahmad, S.H., 1990. A comparative study of the phytoplankton of the river Ganga and pond of Patna (Bihar) India. *J. Indian Bot. Soc.*, 69: 153–158.

Sinha, M., *et al.*, 1996. Impact of Farakka barrage on the hydrology and fishery of Hoogly estuary. *Estuaries*, 19(3): 710–722.

Chapter 48

Development of Coatings by Varying Current Density in Electroforming Techniques for High Temperature Applications

S. Ramachandran[1]* and J. Abbas Mohaideen[2]
[1]Assistant Professor, Department of Mechanical Engineering, Sathyabama Deemed University, Old Mamallapuram Road, Jeppiaar Nagar, Chennai – 600 019
[2]Principal, Maamallan Institute of Technology, Sriperumpudur, Kanchipuram (Dist.).

ABSTRACT

Electroforming techniques are used in high temperature and high stress areas like wind tunnel, nozzle, pressure vessels, and turbine blades. In this article this technique is used for surface coating on sample mild steel pipes by varying current densities. The combustion takes place in the combustion chambers and very high temperature fluid passes through the inner passage of pipe and erosion of pipe takes place. The erosion in the pipe changes the velocity of the passing fluid and hence the performance of the engine is affected. For example, the rocket or missile will not attain the target in time in the nozzle or thrust chamber inner surface is eroded due to high temperature fluid. Electroforming coatings are applied on the inner surface of the pipe by various materials and are tested to withstand the elevated temperature so that the erosion at the pipe is avoided.

Keywords: *Electroforming, Current density, Combustion, High temperature.*

---

* Corresponding Author: E-mail: aishram_2006@yahoo.com, 28-A, Karneeshwarar Pagoda Street, Mylapore, Chennai – 4; Phone: 044-24661909, Mobile: 9444001904

Introduction

Electroforming is the production (or) reproduction of articles by electro-deposition. Electroforming technique is used for the production and reproduction of a variety of engineering components, like parts for nuclear and space applications, moulds and dies for plastic, glass and zinc-die cast manufacturers, precision electronic parts, high temperature pressure vessels, turbine blades and nozzles and thrust chambers. Electroforming is the production and reproduction of articles by plating sufficient thickness of metal such as Chromium (or) Nickel over the given mandrel and subsequently the deposit (electroformed deposit) is separated from the mandrel to act as a separate entity. Various structural properties can be controlled during electroforming. The *mandrel* is made to *cathode* of an electroplating circuit and the desired amount of metal then is plated on it (Giggins and Petti, 1951; Giggins, and Petti, 1969; Peter Spiro-Robert Draper Ltd., 1971).

Need for Coating on the Inner Surface of the Pipe by Electroforming Techniques

Electroforming on *pipes through which high temperature is passing* is currently pursued. Coatings on inner surface of the pipes will not be uniformly distributed in other coating techniques or metal working processes. Some nozzles and thrust chambers contain cooling passages on their walls and it is extremely difficult to make coatings by other metal working processes. In other techniques, many components have to be furnace brazed together to form the passages and the tooling costs required to complete the coating is very high and prohibitive. On the other hand, this can be electroformed in one piece conveniently, economically and to exact specification. Since the thickness of the layer electroformed is thin, it is not easily peeled off from the parent metal. Electroformed parts have their own distinct properties and dimensional tolerances are very good (Barkalow *et al.*, 1980; NASA; INCO, 1968; AESF, 1988).

Details of Component

Pipe to be electroformed is shown in Figure 48.1.

Results and Discussion

Pipe Material: Mild steel; *Dimension*: as given in the drawings, *Number of pipes*: 50.

Various types of materials are coated on the inner surface of the pipe using electroforming technique. For each type of coating, the following can be adjusted to get different thickness of coatings.

1. The bath composition
2. The concentration of the electrolyte
3. The pH of electrolyte
4. The current density
5. The forming time

By varying the time duration of electroforming, (or) the concentration of the electrolyte, the thickness of the coating are varied and the data collected are shown in table. By using the table, the different graphs are drawn. The following tables are the effect of forming thickness by varying current densities in Electroforming techniques and are illustrated graphically.

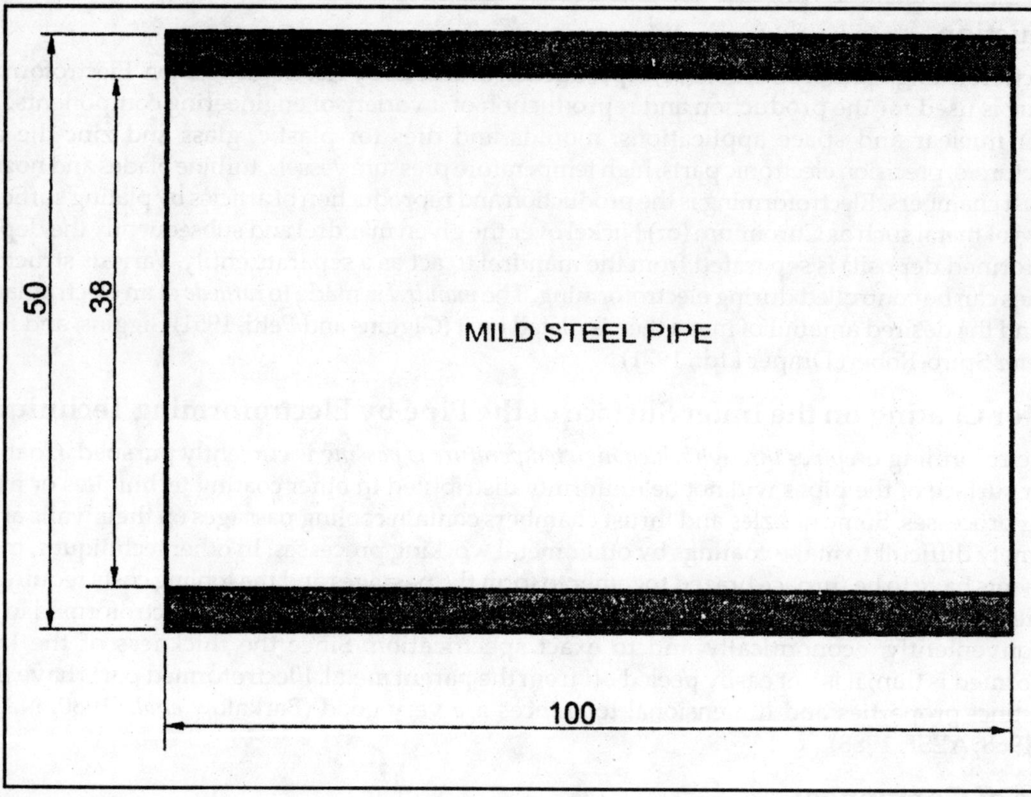

Figure 48.1

Effect of Variation of Current Density on Electroforming

Chromium Forming

The effect of variation of current density on chromium forming is shown in Table 48.1.

Table 48.1

Sl.No.	*Bath Temperature K*	*Rotational Speed of the Cathode (rpm)*	*Forming Time (minute)*	*Current Density (A/dm²)*	*Thickness of the Coating (mμ)*
1.	324	15	30	2	4.32
2.	324	15	30	4	5.60
3.	324	15	30	6	6.76
4.	324	15	30	8	7.24
5.	324	15	30	10	8.38

The effect of variation of current density on chromium forming is drawn graphically in Figure 48.2.

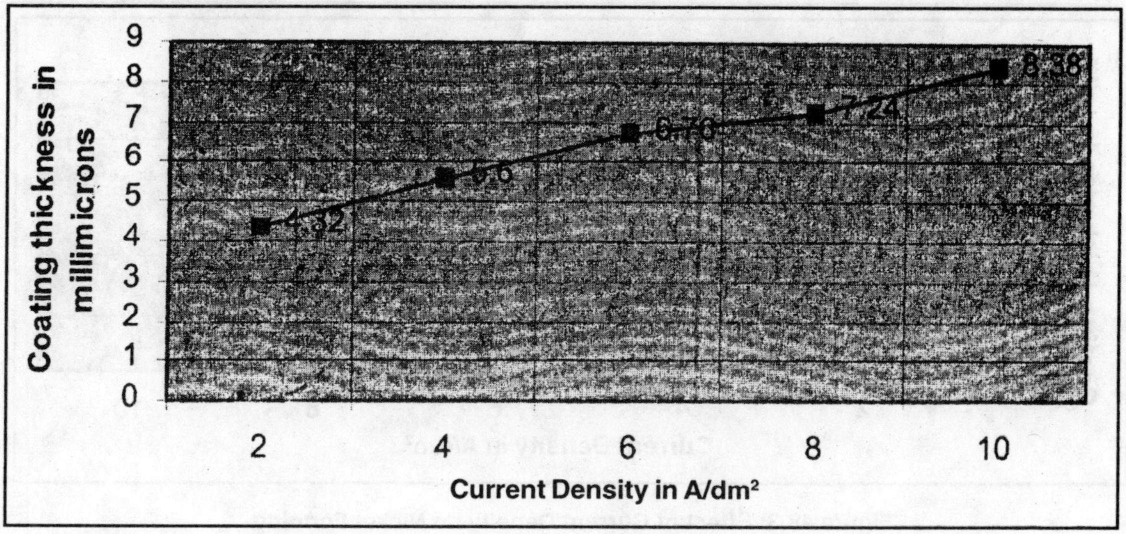

Figure 48.2: Effect of Current Density on Chromium Forming

Nickel Forming

The effect of variation of current density on Nickel forming is shown in Table 48.2.

Table 48.2

Sl.No.	Bath Temperature K	Rotational Speed of the Cathode (rpm)	Forming Time (minute)	Current Density (A/dm^2)	Thickness of the Coating ($m\mu$)
1.	323	15	30	2	5.42
2.	323	15	30	4	6.11
3.	323	15	30	6	8.33
4.	31.3	15	30	8	7.45
5.	323	15	30	10	6.57

The effect of variation of current density on Nickel forming is drawn graphically in Figure 48.3

Ni-Cr Alloy Forming

The effect of variation of current density on Ni-Cr forming is shown in Table 48.3.

Table 48.3

Sl.No.	Bath Temperature K	Rotational Speed of the Cathode (rpm)	Forming Time (minute)	Current Density (A/dm^2)	Thickness of the Coating ($m\mu$)
1.	311	15	30	2	6.42
2.	311	15	30	4	8.67
3.	311	15	30	6	7.37
4.	311	15	30	8	5.97
5.	311	15	30	10	4.31

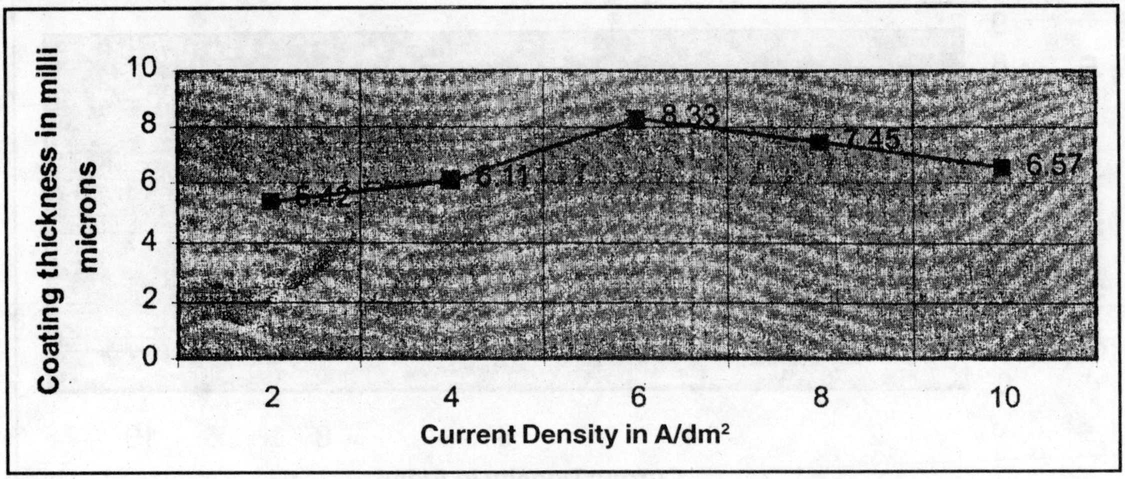

Figure 48.3: Effect of Current Density on Nickel Forming

The effect of variation of current density on Ni-Cr alloy forming is drawn graphically in Figure 48.4.

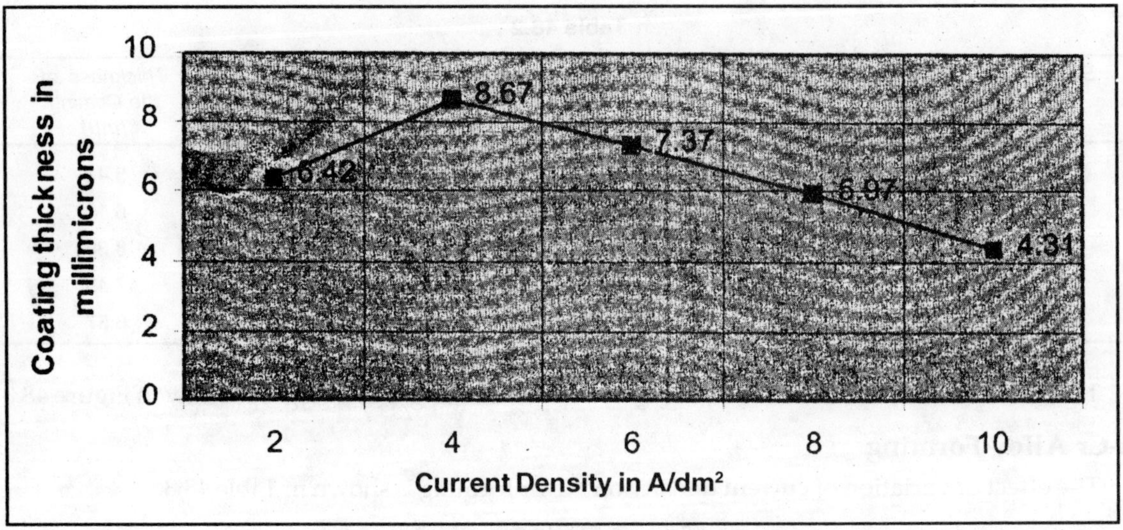

Figure 48.4: Effect of Current Density on Ni-Cr Alloy Forming

Copper Forming

The effect of variation of current density on Copper forming is shown in Table 48.4.

The effect of variation of current density on Copper forming is drawn graphically in Figure 48.5.

Table 48.4

Sl.No.	Bath Temperature K	Rotational Speed of the Cathode (rpm)	Forming Time (minute)	Current Density (A/dm²)	Thickness of the Coating (mµ)
1.	328	15	30	2	4.16
2.	328	15	30	4	5.32
3.	328	15	30	6	6.67
4.	328	15	30	8	7.65
5.	328	15	30	10	6.96

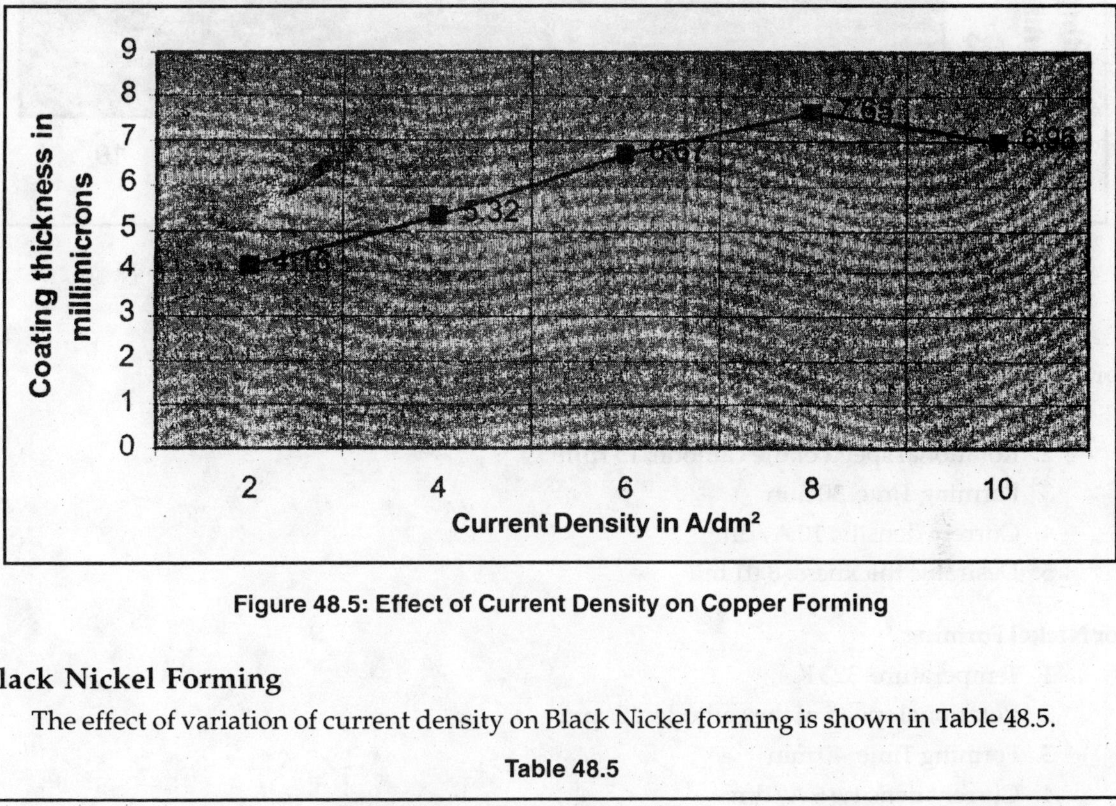

Figure 48.5: Effect of Current Density on Copper Forming

Black Nickel Forming

The effect of variation of current density on Black Nickel forming is shown in Table 48.5.

Table 48.5

Sl.No.	Bath Temperature K	Rotational Speed of the Cathode (rpm)	Forming Time (minute)	Current Density (A/dm²)	Thickness of the Coating (mµ)
1.	327	15	30	2	5.27
2.	327	15	30	4	6.28
3.	327	15	30	6	8.01
4.	327	15	30	8	7.36
5.	327	15	39	10	6.92

The effect of variation of current density on Black Nickel forming is drawn graphically in Figure 48.6.

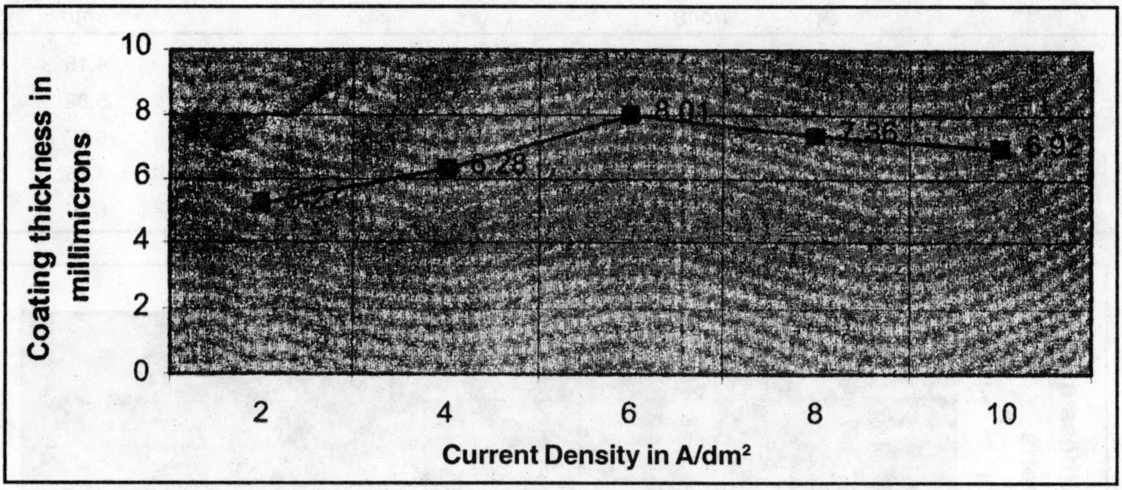

Figure 48.6: Effect of Current Density on Black Nickel Forming

Desirable coatings are formed as given in following:

For Chromium Forming

1. Temperature: 324 K
2. Rotational speed of the cathode: 15 rpm
3. Forming Time: 30 min
4. Current density: 10 A/dm^2
5. Desirable thickness: 8.01 mμ

For Nickel Forming

1. Temperature: 323 K
2. Rotational speed of the cathode: 15 rpm
3. Forming Time: 40 min
4. Current density: 6 A/dm^2
5. Desirable thickness: 8.56 mμ

For Ni-Cr Alloy Forming

1. Temperature: 311 K
2. Rotational speed of the cathode: 15 rpm
3. Forming Time: 30 min
4. Current density: 4 A/dm^2
5. Desirable thickness: 8.68 mμ

For Copper Forming

1. Temperature: 328 K
2. Rotational speed of the cathode: 15 rpm
3. Forming Time: 40 min
4. Current density: 8 A/dm^2
5. Desirable thickness: 8.00 mμ

For Black Nickel Forming

1. Temperature: 327 K
2. Rotational speed of the cathode: 15 rpm
3. Forming Time: 40 min
4. Current density: 6 A/dm^2
5. Desirable thickness: 7.21 m^2

The suitable optimum thickness suggested by electroforming experiment is found to be 8 mμ of thickness. It is not economical when we increase the thickness more than 8 mμ. Also, it saves material consumption, time, money without sacrificing the strength.

After forming on the pipe with various metal/alloying elements and it was examined for high temperature studies for erosion resistance and the data are given in the Table 48.6.

The withstanding temperatures for various forming materials are shown in Figure 48.7.

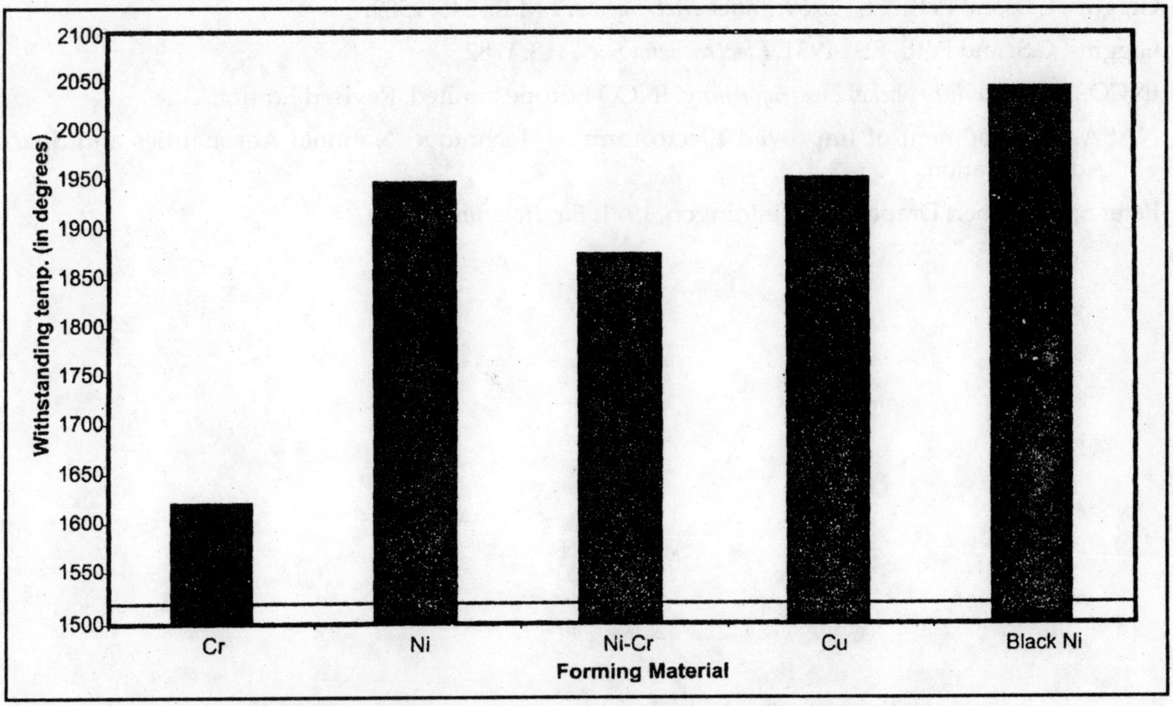

Figure 48.7: High Temperature Studied on Electroformed Nozzle

Table 48.6

Sl.No.	Base Metal (Pipe)	Forming Metal/Alloying Element	Structural Deformation Occurs at °C	
			Prior to Forming the with-standing Stability of Pipe	After Forming the with-standing Stability of Pipe
1.	Mild Steel	Chromium forming	1510°C	1620°C
2.	Mild Steel	Nickel forming	1510°C	1950°C
3.	Mild Steel	Ni-Cr alloy forming	1510°C	1875°C
4.	Mild Steel	Copper forming	1510°C	1952°C
5.	Mild Steel	Black Nickel forming	1.510°C	2045°C

References

AESF Electroforming Symposium–American Electroplaters and Surface Finishers Society, 12644 Research Parkway, Orlando, FL 32826.

AESF, 1989. Paper presented at AESF-Electroforming symposium-1989 at Las Vegas, NY–Mr. Prof. John Paper presented in the international symposium–February 1–5, 1989, Karaikudi B. Electrochem., 4, 12, 1026.

Barkalow, R.H., Goebel, J.A. and Petti, F.S., 1980. *High Temperature Erosion-Corrosion by High Velocity (200 m/s) Particles*. Prepared by Pratt and Whitney Aircraft Group for Electric Power Research Institute, Palo Alto, CA., 94304; EPRI CS-1448, Project 979-4.

Giggins, C.S. and Petti, F.S., 1969. *Trans. Met. Sac.*, A.I.M.E., 245: 2509.

Giggins, C.S. and Petti, F.S., 1981. *Electrochem. Sac.*, 118: 1782.

INCO, 1968. *Guide to Nickel Electroforming*. INCO Europe Limited, Revised Edition.

NASA. Development of Improved Electroforming Technique–National Aeronautics and Space Administration.

Peter Spiro-Robert Draper Ltd., Teddingon, 1971. *Electroforming*.

Chapter 49

Effect of Irrigation and Nitrogen on Phonological Stages, Water Use Efficiency, Oil Yield, Protein Content and Seed Yield of Hybrid Canola (*Brassica napus* L.)

A.S. Sidhu and S.S. Mahal

Department of Agronomy, Agrometeorology and Forestry, PAU, Ludhiana – 141 004

ABSTRACT

A field study on Hybrid Canola (*Brassica napus* L) was carried out during rabi season of 2004 at PAU, Ludhiana to find out the optimum irrigation schedule and rate of nitrogen application. The experiment was conducted in split plot design with 4 replications. Four irrigation schedules *viz.*, based on CPE 40 (I_1), 60 (I_2) and 80 mm (I_3) and three recommended irrigations (I_4) were kept in main plots while five nitrogen levels *viz.*, 0 (N_0), 75 (N_1), 100 (N_2), 125 (N_3) and 150 (N_4) kg ha^{-1} were kept in sub plots. The soil of the experimental field was loamy sand in texture, which was low in organic carbon, available nitrogen and medium in available phosphorus and potassium. The highest seed yield (15.85 q ha^{-1}) was obtained with I_4 irrigation schedule which was significantly superior over I_3 but at par with I_1 and I_2 irrigation schedules. Phenological stages *viz.*, days to 50 per cent flowering and days to 80 per cent siliqua ripening, oil content and protein content were not significantly affected by different irrigation schedules. Highest water use efficiency (35.4 kg ha.^{1} cm^{-1}) was observed in 14 irrigation schedule followed by I_1, I_2 and I_3 irrigation schedules, respectively. Similar trend was observed for water use. N_4 treatment *i.e.* 150 kg N ha^{-1} produced significantly higher seed yield (17.73 q ha^{-1}) as compared to N_0, N_1, N_2 and

N_3 levels of nitrogen. Values of oil yield, protein content and water use efficiency were also highest with N_4 level of nitrogen. Where as oil content was not significantly affected with different nitrogen levels.

Keywords: *Hybrid Canola, CPE, Nitrogen levels, Irrigation Schedule, Water use.*

Introduction

Oilseed crops are next to cereals in production and occupy a premier position in both agricultural and industrial economy of India. They can play an important role in diversification on account of their high yield potential and as a best revenue conservation option. Among the rabi oilseed crops, rapeseed and mustard account for about 28 per cent of total production. In India, total area under rapeseed and mustard cultivation during 2004 was 53.87 lakh ha with a production of 61.98 lakh metric tonnes. Where as, Punjab had an area of 0.52 lakh ha under cultivation with a production of 0.62 lakh metric tonnes (Anonymous 2005). The relative self-sufficiency in oilseed sector has now allowed greater attention towards the quality of oil. This was also necessitated by several reports concerning the undesirability of consuming erucic acid rich brassica oil and seed meal having higher amounts of glucosinolates. The oil quality depends upon nature and proportion of fatty acids. The fatty acids include palmitic, oleic, linoleic, linolenic and erucic acids. The erucic acid content of conventional rapeseed-mustard oil was about 45 per cent. This high level of erucic acid leads to myocardial fibrosis of heart and alters cholesterol metabolism. Similarly, higher glucosinolate in meal in conventional varieties was harmful for animals. So it is the need of the hour to develop canola varieties as they are low in both erucic acid and glucosinolates.

Canola is amphiploid between *Brassica compestris* and *Brassica oleracea,* having 40–45 per cent oil content, 35 per cent crude protein, less than 2 per cent erucic acid and 100–200 micromoles of glucosinolate. Among the various production practices, nitrogen and irrigation are major which play an important role in determining the quality of oil. Canola is an indeterminate plant and such crops can respond well to application of nitrogen. Its response to nitrogen has been reported up to 150 kg ha^{-1} (Gill and Narang 1991, Bilsborrow *et at* 1993). Jackson (2000) at Montana also observed increased seed and oil yield of canola up to 200 Kg N ha^{-1}. Gill and Narang (1993) observed that 3 irrigations significantly increased growth parameters and seed yield of *Brassica napus* on loamy sand soil of Ludhiana. Mahal and Singh (2000) at Ludhiana obtained highest seed yield (15.8 q ha^{-1}) of hybrid gobhi sarson (*Brassica napus*) with two post sowing irrigations applied at 20 and 54 days after sowing. Canola being more succulent, high in quality and having high yield potential, further entails the possibility of different irrigation and nitrogen requirements than the existing strains of gobhi season. Thereafter keeping the above facts in view the present study was carried out to study the effect of management of water and nitrogen on growth and yield of canola.

Materials and Methods

The study was carried out at Punjab Agricultural University, Ludhiana during rabi season of 2004 on loamy sand soil. The field capacity (0.3 bar value) and PWP (15 bar value) of the 120 cm soil profile were 27.54 and 12.07 cm, respectively. The pH and EC values in 0–30 cm soil layer were 8.1 and 0.22 dsm^{-1}, respectively; considered as normal values. The soil was low in organic carbon (0.11 per cent) and available nitrogen (159 kg ha^{-1}), medium in available phosphorus (13.1 kg ha^{-1}) and available potassium (153.1 kg ha^{-1}). The total rainfall during the crop season was 144.2 mm which was above normal. The experiment comprised of four irrigation schedules *viz.*, based on CPE 40 (I_1), 60 (I_2) and 80 mm (I_3) and three recommended (I_4) irrigations in main plot and five nitrogen levels *viz.* 0 (N_0), 75 (N_1), 100 (N_2), 125 (N_3) and 150 kg ha^{-1} (N_4) in subplot in a split plot design. Hybrid canola cultivar Hyola PAC 401 was sown on October 25, 2004 using a seed rate of 3.75 kg ha^{-1} with hand drill at a row spacing of 45 cm.

A common doze of phosphorus (30 kg P_2O_5 ha^{-1}) was drilled at the time of sowing. Nitrogen was applied in the form of urea as per treatments. Half the doze of nitrogen was applied at the time of sowing and remaining half with first irrigation. The crop was thinned 23 days after sowing at 10 cm intra-plant spacing. To control the weeds, one hoeing was given 30 days after sowing. To protect the crop from aphid attacks the crop was sprayed three times, once with thiodon 35EC @ 1000 m/ha^{-1} and twice with Rogor 30 EC @ 1000 m/ha^{-1} in 250 liters of water. The number of irrigations applied in I_1, I_2 and I_3 irrigation schedules were 3, 2 and 1 respectively. Depth of each irrigation was 7.5 cm which was measured with the help of Parshall flume (Parshall, 1950). The crop was harvested manually when siliquae changed their colour from green to yellow. Threshing was also done manually after four days of harvesting. The oil content of seed was determined by nuclear magnetic reasonable spectroscope Newport-analyser (Model MK-IIIA) employing non destructive method of oil estimation in seed.

Results and Discussion

Days to 50 per cent Flowering

Flower opening indicate the beginning of the reproductive phase. Different nitrogen levels significantly affected the days to 50 per cent flowering. The crop which received no nitrogen flowered early as compared to crop with higher doses of nitrogen. The crop in N_4 level took 109 days to flowering as compared to 107, 106, 104 and 100 days in N_3, N_2, N_1 and N_0 levels, respectively (Table 49.1). It can be attributed to more vegetative growth and translocation of assimilates at later stage of crop, which prolonged the vegetative stage of the crop. There was no significant difference in time taken to 50 per cent flowering under different irrigation schedules. These results are in line with the findings of Singh (1995) on *Brassica napus* and Singh (2000) on *Brassica carinata.*

Days to 80 per cent Siliqua Ripening

It is considered as an index of maturity. Different nitrogen levels had a significant influence on days taken to 80 per cent siliqua ripening. Crop which received no nitrogen matured early as compared to crop with higher nitrogen levels. In case of N_0 level, crop took 150 days for maturity as compared to 151, 153, 153 and 156 days under N_1, N_2, N_3 and N_4 levels, respectively (Table 49.1). It is due to more leaf area, succulence of leaves and more chlorophyll content in leaves, because nitrogen is considered integral part of chlorophyll which caused delay in senescence of leaves. There was no significant difference in time taken for maturity under different irrigation schedules. Similar results have been reported by Kumar (1994) on *Brassica juncea* and Singh (1995) on *Brassica napus.*

Seed Yield

Seed yield of crop is the most important criterion for comparing the efficiency of different treatments. Seed yield increased significantly with the increase in levels of nitrogen from N_0 to N_4. Persual of data further evinced that maximum seed yield (17.73 q ha^{-1}) with N_4 level was significantly higher than all other nitrogen levels of N_0, N_1, N_2 and N_3 (Table 49.1). The increase in seed yield over control with N_1, N_2, N_3 and N_4 levels of nitrogen was 13.1, 18.5, 30.9 and 44.1 per cent, respectively. This increase in seed yield might be due to positive effect of nitrogen on plant height, number of primary and secondary branches per plant, leaf area index, dry matter accumulation, number of siliquae per plant and number of seed per siliqua. The higher levels of nitrogen might have increased the supply of assimilates during the vegetative phase to the flowers and further mobilization of these during the seed development phase. These results are in accordance with the findings of Salaria and Dhillon (2003), Jackson (2000), Cheema *et al.* (2001) and Malik *et al.* (2002) on canola (*Brassica napus* L.). The seed yield of hybrid canola was also affected considerably by differential irrigations. I_1, I_4 (3 post sowing irrigations) and I_2 (2 post sowing irrigations) produced significantly higher seed yield than I_3 (1 post sowing irrigation). The increase in seed yield over 13 by I_1, I_2 and I_4 was 16.7, 10.4 and 17.6 per cent, respectively (Table 49.1). The higher seed yield in I_1 and I_4 irrigation schedules might be due to the fact that early irrigation helps in establishment of crop by preventing it from cold injury and better mobilization of nutrients, thus making an effective use of the applied nutrients in crop development. Similar results have been reported by Singh (1997), Gill and Narang (1993) on *Brassica napus*, by Garnayak *et al.* (2000) on Brassica oilseeds and by Singh *et al.* (2001) on *Brassica juncea*. The interaction between irrigation and nitrogen on seed yield was found to be non significant.

Table 49.1: Effect of Irrigation and Nitrogen on Phonological Stages, Water Use Efficiency, Oil Yield, Protein Content and Seed Yield of Hybrid Canola (*Brassica napus* L.)

Treatments	Days to 50% Flowering	Days to 80% Siliquae Ripening	Water Use (cm)	Water Use Efficiency (kg ha^{-1} cm^{-1})	Oil Content (per cent)	Oil Yield (qha^{-1})	Protein Content (per cent)	Seed Yield (qha^{-1})
Nitrogen Levels								
N_0	99.9	149.7	39.7	31.0	40.71	5.01	19.65	12.30
N_1	103.7	150.7	41.4	33.6	40.34	5.63	22.15	13.91
N_2	106.1	153.2	43.1	34.6	40.29	6.02	25.98	14.88
N_3	107.3	152.3	46.4	35.0	40.54	6.54	27.62	16.11
N_4	108.3	155.5	48.6	36.6	39.95	7.08	30.63	17.73
CD (P = 0.05)	4.4	2.8	–	–	NS	0.28	2.12	0.62
Irrigation schedules								
I_1	103.7	152.5	45.8	34.3	40.47	6.38	26.35	15.73
I_2	104.8	152.4	43.9	34.0	40.22	6.00	23.70	14.88
I_3	104.6	151.9	40.9	32.9	40.44	5.45	23.97	13.48
I_4	107.1	152.3	44.7	35.4	40.33	6.40	26.82	15.85
CD (P = 0.05)	NS	NS	–	–	NS	0.59	NS	1.41
Interaction	NS	NS	–	–	NS	NS	NS	NS

Oil Content

The different nitrogen levels did not significantly affected the oil content. Highest oil content (40.7 per cent) was recorded in N_0 level of nitrogen and lowest (39.95 per cent) in N_4 level of nitrogen (Table 49.1). Similarly, oil content was also not significantly affected by different irrigation schedules. Similar results have been reported by Salaria and Dhillon (2003) on canola.

Oil Yield

Oil yield increased significantly with the increase in successive levels of nitrogen from N_0 to N_4 owing to increase in seed yield. The highest oil yield (7.08 q ha^{-1}) was recorded with N_4 which was

41.31, 25.8, 17.6 and 9.2 per cent more over N_0, N_1, N_2 and N_3 levels, respectively (Table 49.1). I_1 and I_4 irrigation schedules significantly increased the oil yield over I_3 but they were statistically at par with I_2 irrigation schedule. Highest oil yield (6.40 q ha^{-1}) was recorded with I_4 irrigation schedule which was 0.4 and 0.95 q ha^{-1} more over I_2 and I_3 irrigation schedules (Table 49.1) respectively. Salaria and Dhillon (2003) and Brar *et al.* (1998) also reported increase in oil yield of *Brassica napus* up to 150 kg N ha^{-1}.

Protein Content

The protein content significantly increased with increase in nitrogen levels from N_0 to N_4 except N_2 and N_3 which were at par (Table 49.1). The maximum protein content of 30.6 per cent was recorded in N_4 level which was 10.98, 8.48, 4.65 and 3.01 per cent higher over N_0, N_1, N_2 and N_3 levels, respectively. There was no significant effect of various irrigation schedules on protein content. However, highest protein content (26.82 per cent) was obtained with I_4 irrigation schedule which was statistically at par with other irrigation schedules. These results are in accordance with the findings of Salaria and Dhillon (2003), Mahal and Singh (2000) in *Brassica napus* and Ali *et al.* (1996) in *Brassica campestris.*

Crop Water Use

The water use increased continuously with the increase in nitrogen levels. The highest seasonal water use (48.6 cm) was obtained with N_4 level which was 8.9, 7.2, 5.5 and 2.2 cm more than N_0, N_1, N_2 and N_3 levels, respectively. Different irrigation schedules also influenced the water use of hybrid canola. Maximum water use (45.8 cm) was recorded with I_1 irrigation schedule which is closely followed by 14 (44.7 cm). Water use in I_2 and I_3 irrigation schedules was 43.9 and 40.9 cm respectively. These results are in line with the findings of Gill and Narang (1991) and Singh (1997) on *Brassica napus* and Bhalerao (2001) on *Brassica juncea.*

Water Use Efficiency

The maximum WUE was obtained with N_4 level of nitrogen (36.6 kg ha^{-1} cm^{-1}). The water use efficiency with N_4 level was 5.6, 3.0, 2.0 and 1.6 kg ha^{-1} cm^{-1} more over N_0, N_1, N_2 and N_3 levels, respectively. Maximum WUE (35.4 kg ha^{-1} cm^{-1}) was observed in I_4 irrigation schedule followed by 34.3, 34 and 32.9 kg ha^{-1} cm^{-1} in I_1, I_2 and I_3 irrigation schedules respectively. Similar results have been reported by Mahal and Singh (2000).

References

Ali, M.H., Zaman, S.M.H. and Altaf Hossain, S.M., 1996. Variation in yield oil and protein content of rapeseed (*Brassica campestris*) in relation to levels of nitrogen, sulphur and plant density. *Indian J. Agron.*, 41: 290–295.

Anonymous, 2005. *Agriculture.* Centre for Monitoring Indian Economy Pvt. Limited, Mumbai, pp. 163.

Bhalerao, P.D., 2001. Effect of irrigation and nitrogen on Indian Mustard (*Brassica juncea*) varieties in Vidharbha region. *Indian J. Agron.*, 46: 727–731.

Bilbsborrow, P.E., Evans, E.J. and Jhao, E.V., 1993. The influence of spring N on yield, yield components and glucosinate content of autumn sown seed rape (*Brassica napus*). *J. Agric. Sci.*, 120: 219–224.

Brar, Z.S., Bal, D.S. and Johl, A.S., 1998. Influence of sowing dates, nitrogen and planting geometry on the performance of gobhi sarson (*Brassica napus* ssp. *Oleifera* var. *annua*). *Indian J. Agron.*, 43: 133–137.

Cheema, M.A., Muhammad, Saleem and Malik, M.A., 2001. Effect of row spacing and nitrogen management of agronomic traits and oil quality of Canola (*Brassica napus* L.). *Pak. J. Agric. Sci.*, 38: 15–18.

Garnayak, L.M., Singh, N.P., Singh, S. and Paikaray, R.K., 2000. Influence of irrigation and nitrogen on growth, yield and nutrient uptake by late-sown Brassica oil seeds. *Indian J. Agron.*, 45: 371–378.

Gill, M.S. and Narang, R.S., 1991. Response of gobhi sarson (*Brassica napus* sub sp *oleifera* var. *annua*) to irrigation and fertilizer nitrogen in Punjab. *Indian J. Agric. Sci.*, 61: 172–177.

Gill, M.S. and Narang, R.S., 1993. Yield analysis in gobhi sarson (*Brassica napus* subsp. *Oleifera* var. *annua*) to irrigation and nitrogen. *Indian J. Agron.*, 38: 257–265.

Jackson, G.D., 2000. Effects of nitrogen and sulfur on canola yield and nutrient uptake. *Agron. J.*, 92: 644–649.

Kumar, S., 1994. Leaf area index relationship with solar radiation interception and yield of raya (*Rrassica juncea* L.) as influenced by plant population and nitrogen levels. *M.Sc. Thesis*, Punjab Agricultural University, Ludhiana, India.

Mahal, S.S. and Singh, K., 2000. Response of hybrid gobhi sarson (*Brassica napus* L.) to different levels of irrigation and nitrogen. *Env. and Eco.*, 18: 722–725.

Malik, M.A., Cheema, M.A. and Saleem, M.F., 2002. Production efficiency of canola (*Rrassica napus* L.) as affected by different NPK levels. *Pak. J. Agric. Sci.*, 39: 278–280.

Parshall, R.L., 1950. Measuring water in irrigation channels with parashall flumes and small weir. *Soil Conserve Surv. U.S. Dept. Agric. Circ.*, 843: 62.

Salaria, A. and Dhillon, A.S., 2003. Effect of Nitrogen and phosphorus on seed yield, oil yield and quality of canola, gobhi sarson (*Brassica napus* L.). *Ann. Bio.*, 19: 195–198.

Singh, G., 2000. Studies on irrigation and nitrogen requirement of late sown Ethiopian mustard (*Brassica carinata* A.Br). *M.Sc. Thesis*, Punjab Agricultural University, Ludhiana, India.

Singh, J., Singh, B. and Yadav, J.S., 2001. Effect of irrigation on Indian mustard (*Brassica juncea*) sown on conserved moisture. *Indian J. Agron.*, 46: 721–726.

Singh, K., 1997. Effect of N and irrigation schedules on growth, yield and quality of hybrid gobhi sarson (*Brassica napus* L). *M.Sc. Thesis*, Punjab Agricultural University, Ludhiana, India.

Singh, P., 1995. Effect of N and spacing on growth, yield and oil content of transplanted gobhi sarson (*Brassica napus* L.). *M.Sc. Thesis*, Punjab Agricultural University, Ludhiana, India.

Chapter 50

Studies on Ecology of the *Phumdis* (Floating Mats) of the Loktak Lake, Manipur

N. Beenakumari Devi[1] and B. Manihar Sharma[2]
[1]Department of Botany, D.M. College of Science, Imphal – 795 001, Manipur
[2]Department of Life Sciences, Manipur University, Canchipur – 795 003, Manipur

ABSTRACT

The ecology of the *phumdis* (floating mats), a unique feature of the Loktak Lake, Manipur (24°25′–24°40′ N latitude and 93°45′–93°55′ E longitude), the largest freshwater lake in the North-Eastern region of India, have been studied with special reference to their morphology and phytosociology of the macrophytic plant species constituting the *phumdis* in 5 study sites.

Vertical core section of the phumdi revealed three distinct vertical zones, *viz.* (*i*) the uppermost root zone, (*ii*) the mat zone, and (*iii*) the lowermost peat zone.

Quadrat method of phytosociological studies showed the presence of 73 plant species contributing to the formation of the phumdis. The detailed study of the frequency, density, abundance as well as the relative values of the plant species were carried out at monthly intervals. The maximum IVI values of the dominant macrophyte species, *viz.*, *Alternanthera philoxeroides*, *Echinocloa stagnina*, *Cyrtococcum accresens* and *Zizania latifolia* in the 5 study sites were estimated.

The importance of the phumdis vis-à-vis the deteriorating ecological condition of the Loktak Lake have been discussed.

Keywords: *Loktak lake, Phumdis, Floating mats, Phytosociology.*

Introduction

Freshwater environment is a natural resource, which has been inseparably associated with mankind since time immemorial. Macrophytes in a freshwater environment play important ecological roles. These vascular aquatic plants accumulate large amounts of nutrients in their tissues that are again released when they die and decompose. The macrophytes are, thus, very important in cycling of minerals in the freshwater ecosystems. They may also be used as indicators of the degree of damage to the ecosystem (Piczynska and Ozimek, 1976).

Floating marshes mainly composed of floating or emergent vascular macrophytes occur in different areas of the world and they represent the natural habitat of many species of plants and animals. *Panicum hemitomon* Schultz floating marshes of Louisiana, the floating *Cyperus* swamps in Africa, Plav on the Danube delta in Europe, floating mats along the Amazon River in Brazil and the Phumdis or floating mats in the Loktak Lake, Manipur are good examples of extensive floating marshes.

The Loktak Lake (24°25′–24°40′ N latitude and 93°45′–93°55′ E longitude) of Manipur is one of the largest freshwater lakes in India. It is located in the Bishnupur District of Manipur, about 40 km to the South of the State capital, Imphal (Figure 50.1). The lake has an inseparable association, since time immemorial, with the people of the State from the socio-cultural, economic and aesthetic view-points. A significant portion of the lake forms the Keibul Lamjao National Park, which is the only natural home of the brow-antlered deer, *Cervus eldi eldi* McClelland. The very existence of the brow-antlered deer living on the floating phumdis as well as a number of economic plants associated with it are threatened as the lake has become grossly polluted and degraded due to human encroachment, water inundation, misuse and overuse of pesticides, insecticides, etc. (Sharma, 2001; Sharma and Devi, 2002).

Considering the importance and uniqueness of the lake biodiversity vis-à-vis the phumdis, the brow-antlered deer, migratory birds and fishes, etc., the IUCN has designated the Loktak as a Ramsar site in 1990.

Assessment of the present condition of the Loktak Lake and finding out measures for conservation of its unique ecosystem is badly needed. In line with this objective, the ecology of the Phumdis has been studied with reference to their morphology, floristic composition and phytosociology.

Materials and Methods

For the present investigations, five different locations in the Loktak Lake were selected as study-sites. They are Kambongphabi (I), Takmu (II), Chinglak Pat (III), Ithing (IV) and Keibul Marnangpat (V). The phumdi morphology was studied by the vertical core section method of O'Neal (1949) and Sasser *et al.* (1991). The different phytosociological characters were studied following the quadtrat method of Curtis (1959) as followed by Mishra (1968). The various phytosociological characters of the different species in all the study sites were investigated at monthly intervals over a period of one year.

Results and Discussion

Vertical sections through the floating mats revealed that the thickness of the mat varied from 50–150 cm in the different study sites, probably corresponding to the variations in the age of formation of the floating mats (Figure 50.2).

Structurally the floating mat is composed of three vertical zones, *viz.*, (1) uppermost root zone, (2) mat zone and (3) pit zone. The uppermost root zone, generally 10–15 cm. in thickness in the different study sites, is composed mostly of fresh roots of the different plants making up the phumdi. The mat zone, located just below the root zone, is a layer of densely interwoven live, dead and decaying roots

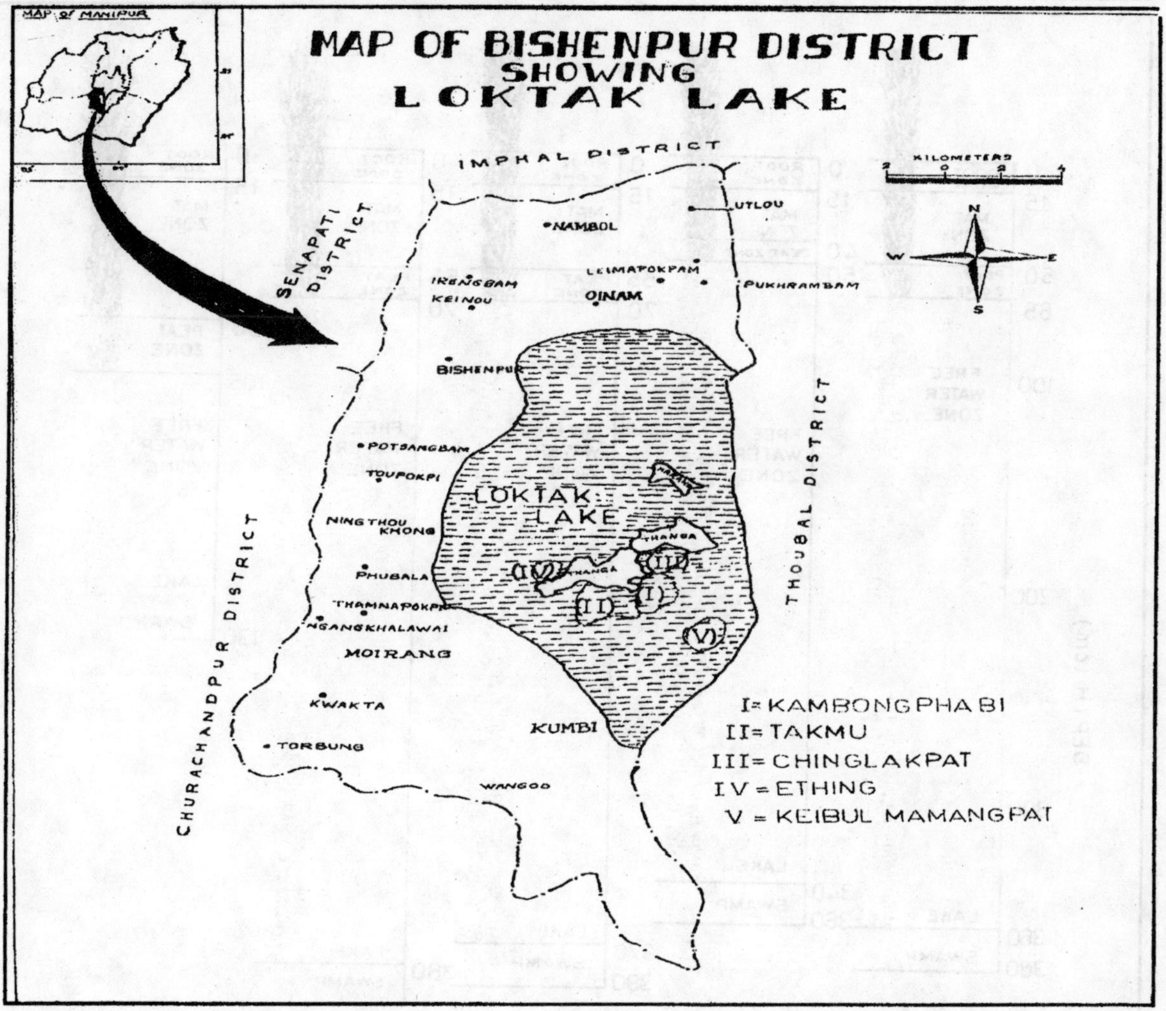

Figure 50.1: Location Map of Loktak Lake in Bishnupur District of Manipur, India

with some litter accumulation in the upper regions. The plant parts in this region still retain their identity. The thickness of this layer varied from 25 cm. in Site 11 to 65 cm in Site V. Below the active root zone is situated a layer of decomposed peat, still intact enough to cling to the overlying roots. The thickness of the peat zone varies from 10 cm Site III to 25 cm in Site V. Below this peat layer is a zone of free water, generally clear in nature and appearance that varies in depth with the lake/marsh water level. Below the free water zone is a layer of organic sludge whose thickness also varies with the water depth. The total distance between the marsh surface and the underlying hard clay surface of the lake bottom varies with the fluctuation of the water level.

The present findings on the structure of the floating mats in Loktak Lake is very similar to the findings of Sasser *et al.* (1991) in case of the *Panicum hemitomon* floating marshes of Louisiana, where the uppermost root zone was 10 cm thick, followed subsequently by a 30 cm mat zone, a 20 cm peat zone, followed by a free water zone and a swamp or sludge zone.

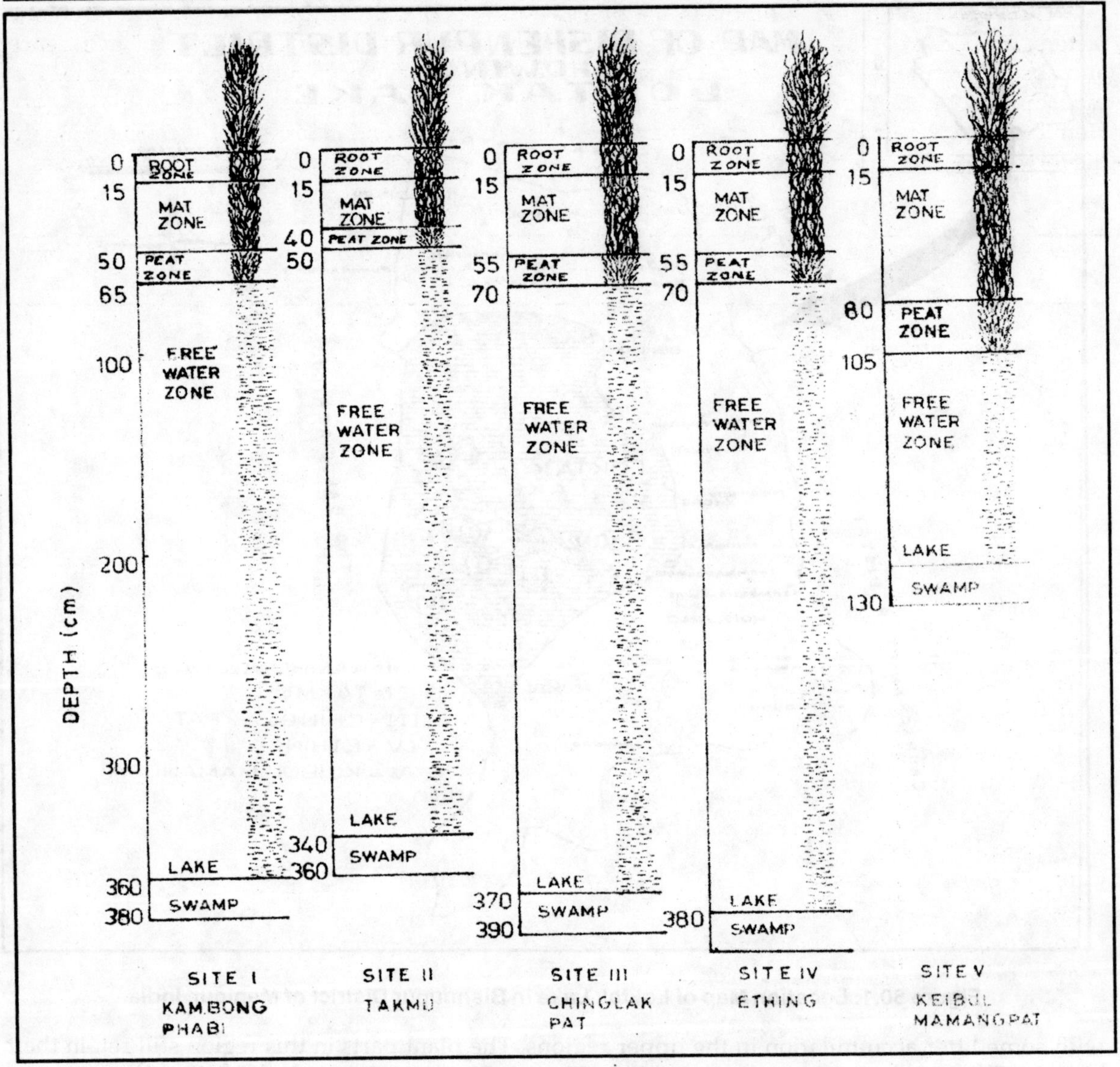

Figure 50.2: Morphology of the Phumdis (Floating Mats) of the Loktak Lake, Manipur

In the Loktak Lake, the phumdis extend over an area of 40 sq km in the Keibul Lamjao National Park, which represents the unique habitat of the brow-antlered deer, *Cervus eldi eldi*. Many terrestrial herbs, shrubs and grasses are found prominently growing in the phumdis, but true trees are absent. Gaudet (1977) reported the occurrences of large areas of floating Cyperus swamps in Africa. Extensive floating masses also occur along the Amazon River in Brazil (Junk, 1970). The Plav is a vast floating mass distributed in the Danube delta in Europe (Pallies, 1915). Small floating marshes are also found distributed in the Netherlands, Hawaii, Arkansas, Florida, Wisconsin and Australia (Cypert, 1972; Van Wirdum, 1979; Wheeles, 1980; Huffman and Lonard, 1983).

The vegetation of the phumdis in the present study is greatly disturbed by biotic and abiotic factors. 73 species of plants were recorded in the phumdis of the five study sites (Table 50.1). Presence and absence of the macrophytic species in the different sites have been furnished in the table. The macrophyte species belong to different life forms like Therophytes (TH), Geophytes (G), Hemicryptophytes (H), chamaephytes (CH), phaneropytes (PH) and Lianas (L). The most dominant life form has been represented by Therophytes followed by Hemicryptophytes and Chamaephytes.

Table 50.1: Floristic Composition of the Phumdis of the Loktak Lake, Manipur

Sl.No.	*Name of Species*	*Sites*					*Life Forms*
		I	*II*	*III*	*IV*	*V*	
1.	*Ageratum conizoides* Linn	+	+	+	+	+	TH
2.	*Alocasia indica* Schott	+	+	+	+	+	G
3.	*Alpinia nigra* Burtt	+	+	+	+	+	G
4.	*Andropogon ascinodes* Clarke	+	+	+	+	+	CH
5.	*Alternanthera philoxeroides* Griseb	+	+	+	+	+	TH
6.	*Arthraxon lanceolatus* (Roxb) Hochst	+	–	+	–	–	CH
7.	*Atylosia ascarabaeoides* (Linn)	+	+	+	+	+	TH
8.	*Bergia capensis* Linn	+	+	+	+	+	H
9.	*Canna indica* Linn	–	–	–	+	–	G
10.	*Carex cruciata* Wahlenb	+	+	–	+	+	TH
11.	*Centella asiatica* Urban	–	–	–	–	+	H
12.	*Cissus assamica* Craib	+	–	+	+	–	L
13.	*Commelina benghalensis* Linn	+	+	+	+	+	H
14.	*Crassocephallum crepidiodes* Mon	+	+	+	+	+	TH
15.	*Cryptocoryne spiralis* Fisch	–	–	–	–	+	G
16.	*Cyperus diffusus* Vahl	+	+	+	+	+	TH
17.	*Cyperus difformis* Linn	+	–	–	–	–	TH
18.	*Cyperus iria* Linn	+	+	+	+	–	TH
19.	*Cyperus michelianus* Linn	+	+	+	+	+	TH
20.	*Cyrtococcum accrescens* Stafp	+	+	–	+	+	TH
21.	*Diplazium porrectum* Wall	+	+	+	+	+	CH
22.	*Dysophylla auricularia* Blume	–	+	+	–	+	CH
23.	*Echinochloa stagnina* (Retz) Beauv	+	+	+	+	+	TH
24.	*Eclipta prostrata* Linn	+	+	+	+	+	TH
25.	*Eleocharis plantaginea* R.Br.	+	–	–	–	+	TH
26.	*Enhydra fluctuans* Lour	+	–	–	–	–	H
27.	*Erianthus munja* Jesw	–	–	–	–	+	CH
28.	*Eriocaulon truncatum* Buch Ham	+	–	–	–	+	TH
29.	*Fagopyrum cymosum* Meissn	+	+	+	+	+	TH
30.	*Forrestia glabrata* Hassk	+	+	+	+	+	H
31.	*Frimbristylis scaberrima* Nees	–	–	+	+	–	TH
32.	*Hedychium coronarium* Koenig	+	+	+	+	+	G
33.	*Hedyotis crataegonum* Spreng	+	+	+	+	+	H

Contd...

Table 50.1–Contd...

Sl.No.	Name of Species	Sites					Life Forms
		I	II	III	IV	V	
34.	*Hewittia sublobata* O. Kuntze	–	–	–	–	–	L
35.	*Hygroryza aristata* Nees	+	–	+	+	–	H
36.	*Impatiens arguta* Hook	+	+	+	+	+	TH
37.	*Impatiens tomentosa* Heyne	+	–	–	–	–	TH
38.	*Ipomoea aquatica* Forsk	+	+	+	+	+	H
39.	*Ipomoea cairica* (Linn) Sweet	+	+	–	–	–	L
40.	*Ipomoea quamoclit* Linn	–	–	+	+	–	L
41.	*Isachne globosa* (Thumb) Kuntze	+	+	+	+	+	TH
42.	*Jussiea repens* Linn	+	+	+	+	+	TH
43.	*Jussiea suffruticosa* Linn	+	+	+	+	+	TH
44.	*Kyllinga triceps* Rottb	+	+	+	+	+	TH
45.	*Laggera flava* Hook	+	–	–	–	+	TH
46.	*Limnophila chinesis* Merr	+	–	–	–	–	H
47.	*Limnophila connata* Pennel	+	–	–	–	–	H
48.	*Limnophila sessiflora*	+	–	–	+	+	H
49.	*Lindernia crustaceae* Muell	+	+	+	–	+	TH
50.	*Mikania micrantha* Kenth	+	+	+	+	+	L
51.	*Myriactis nepalensis* Less	+	+	+	+	+	TH
52.	*Oenanthe javanica* (Bl) D.C.	+	+	+	+	+	TH
53.	*Oryza minuta* Presl	+	+	–	+	–	TH
54.	*Oryza rufipogon* Grift	–	–	+	–	+	H
55.	*Panicum sanguinale* Linn	+	–	–	+	+	H
56.	*Phragmites karka* Trin	+	+	+	+	+	H
57.	*Pilea oxydon* Wedd	+	+	–	+	+	CH
58.	*Pleocnemia trimeni* Bedd	–	–	–	–	+	CH
59.	*Polygonum glabrum* Willd	+	–	+	+	–	TH
60.	*Polygonum pulchrum* Blume	+	+	+	+	+	TH
61.	*Pseudorophis minuta* Pilger	+	+	+	+	+	H
62.	*Rotala baccifera* Koehne	–	–	–	–	+	TH
63.	*Rumex nepalensis* Spreng	–	–	–	–	+	TH
64.	*Saccharum arundinaceum* Retz	+	+	–	–	+	G
65.	*Scirpus supinus* Linn	+	–	–	–	+	TH
66.	*Sclerocarpus africanus* Jacq	–	+	–	–	–	TH
67.	*Senecio wightii* Benth	+	+	+	+	+	TH
68.	*Spiranthes lancea* (Thunb)	–	–	–	–	+	G
69.	*Stephania japonica* Miers	+	+	+	+	+	L
70.	*Thladiantha calarata* Clarke	+	–	–	+	–	L
71.	*Trichosanthes wallichiana* Wight	+	+	+	+	–	L
72.	*Uncaria sessilifructus* Roxb	–	–	–	–	+	PH
73.	*Zizania latifolia* (Griseb) Stapf	+	+	+	+	+	G

Phytosociological studies on the various plant species occurring on the phumdis revealed that seven dominant species, *viz. Alternanthera philoxeroides, Cyrtococcum accrescens, Echinocloa stagnina, Fagopyrum simosum, Mikania micrantha, Oenanthe javanica* and *Zizania latifolia,* occurred throughout the year. The Importance Value Index (IVI) values of these dominant species ranged from 2.91 in *Fagopyrum cymosum* (Site IV) to 185.12 in *Echinocloa stagnina* (Site I) (Table 50.2).

Table 50.2: IVI Value of Dominant Macrophytes of the Phumdis in 5 Study Sites

Sl.No.	Name of Species	Importance Value Index Out of 300 (IVI)				
		Site I	Site II	Site III	Site IV	Site V
1.	Althernanthera philoxeroides	20. 60 (Dec) to 41.92 (Jul)	15.38 (Feb) to 35.86 (Jun)	17.98 (Feb) to 44.75 (Jun)	16.93 (Jan) to 58.38 (Jul)	11.87 (Jan) to 21.58 (Jun)
2.	Cyrtococcum accrescens	18.18 (Jan) to 39.83 (Apr)	61.63 (Apr) to 116.20 (Jun)	16.24 (Apr) to 57.13 (Sep)	11.73 (Jan) to 36.30 (Jul)	6. 69 (Nov) to 55.83 (Mar)
3.	Echinocloa stagnina	84.80 (Dec) to 185.12 (Jun)	82.29 (Oct) to 106.63 (Apr)	76.68 (Dec) to 132.40 (Apr)	69.28 (Dec) to 112.31 (Jun)	72.23 (Mar) to 140.28 (Jun)
4.	Phagopyrum symosum	4.09 (Sep) to 10.23 (Feb)	6.46 (Aug) to 20.09 (Dec)	6.45 (Sep) to 16.65 (Feb)	2.91 (Jul) to 14.59 (Feb)	4.05 (Sep) to 12.59 (Dec)
5.	Mikania micrantha	3.23 (May) to 9.03 (Jan)	3.88 (Jun) to 11.59 (Dec)	2.48 (Sep) to 11.87 (Feb)	3.18 (Aug) to 11.42 (Feb)	2.92 (May) to 10.75 (Dec)
6.	Oenanthe javanica	4.89 (Sep) to 18.53 (Feb)	3.61 (Dec) to 17.26 (Apr)	2.94 (Jul) to 12.65 (Feb)	3.42 (Jul) to 10.50 (Feb)	4.27 (Jun) to 14.98 (Feb)
7.	Zizania latifolia	24.13 (Oct) to 45.08	3.40 (Nov) to 14.50 (Jul)	15.73 (Jan) to 54.09	18.51 (Oct) to 14.74 (Mar)	3.20 (Oct) to 20.54 (Jul)

In Site I, *Echinocloa stagnina* was the most dominant species (IVI value–185.12), *Mikania micrantha* was the species with the least IVI value of 3.23. In Site II, the least IVI value was shown by *Zizania latifolia* (IVI value–3.40), while *Cyrtococcum accrecens* was the most dominant species with an IVI value of 116.20. *Echinocloa stagnina* was the most dominant species in Site III (IVI value–132.40) while *Mikania micrantha* had the lowest IVI value of 2.48. Site IV was also dominated by *Echinocloa stagnina* with an IVI value of 112.31 and the least IVI value of 2.91 was shown by *Fagopyrum cymosum.*

In Site V, the highest IVI was associated with *Echinocloa stagnina* (140.28) and the least value was associated with *Mikania micrantha* (2.92). The Importance Value Index of the phumdis species were found comparable with those in the moist bank community species of Bhaghela Tank of Udaipur, where the values varied from 1.0 in *Boerhaavia repens* to 138.7 in *Polygonum glabrum* (Sankhla and Vyas, 1982). This has been found to be in conformity with the present values of the dominant species, *i.e. E. stagnina.* The present IVI values were found to be comparable with the findings of Handoo and Kaul (1982) in the wetlands of Kashmir. Comparatively high IVI values in the case of macrophytes of different freshwater ecosystems of Manipur were reported by many workers, *viz.*, Devi, Ch. U. and Sharma (1997), Devi, K.I. (1998) and Okram *et al.* (1999).

Generally the high IVI values of the dominant macrophytes were observed during the rainy months while the lower values occur during the winter months. Such high IVI values during the rainy season may be attributed to the result of favourable environmental conditions during these months. The luxuriant growths of the macrophytes of the phumdis contribute significantly to the high productivity of the Loktak Lake. The phumdis also act as a nutrient reservoir for the submerged plants in the lake water. However, significant deterioration in the natural features of the phumdis of the Loktak, including those of the Keibul Lamjao National Park, were occurring due to the action of either natural or artificial agents. This has threatened the existence of the brow-antlered deer and many other animals in the park area. Eutrophication of the lake has also threatened the existence of a number of rare economically important plants like *Alpinia nigra, Hedychium coronarium, Fagopyrum cymosum, Phragmites karka, Spiranthes lancea, Zizania latifolia*, etc. Further investigations on the deterioration in the ecological characteristics of the phumdis of the Loktak Lake will be of great significance in the efforts to conserve this unique ecosystem.

References

Curtis, J.T., 1959. *The Vegetation of Wisconsin*. University of Wisconsin Press, Madson.

Cypert, E., 1972. The origin of the houses Okefenoke Prairies. *Am. Midl. Nat.*, 87: 448–458.

Devi, Ch. Umabati and Sharma, B.M., 1997. Distribution and phytosociology of the macrophytes in the freshwater ecosystems of Canchipur, Imphal. In: *Proc. 8th Manipur Sc. Cong.*, Manipur University, p. 1–11.

Devi, K. Indira, 1998. Ecological studies of the freshwater macrophytes in Uttrapat Lake, Manipur. *Ph.D. Thesis*, Manipur University.

Gaudet, J.J., 1977. Uptake, accumulation and loss of nutrients by *Papyrus* in tropical swamps. *Ecology*, 58: 415–422.

Handoo, J.K. and Kaul, V., 1982. Phytosociological and standing crop studies in wetlands of Kashmir. In: *Wetlands, Ecology and Management*, (Eds.) B. Gopal, R.E. Turner, R.G. Wetzel and D.F. Whigham. National Int. Ecol. and Inter. Scient. Pubs., Jaipur (India), p. 187–195.

Huffman, R. and Lonard, R., 1983. Successional pattern on floating vegetation mats in South West Arkansas *Cyperus* swamps. *Castania*, 48: 73–78.

Misra, R.D., 1968. *Ecology Workbook*. Oxford and IBH Publ. Co., New Delhi.

Okrarn, I.D., Sharma, B.M. and Singh, E.J., 1999. Ecological exploration of the macrophytic vegetation of the Waithou Lake, Manipur, N.E. India, Phytosociological study of the aquatic macrophytes. *Indian J. Environ. And Ecoplan.*, 2(2): 191–199.

O'Neil, T., 1949. The Muskrat in the Louisiana coastal marshes. Louisiana Wildlife and Fisheries Commission, New Orleans.

Pallis, M., 1915. The structure and history of Plav: The floating fen of the delta of the Danube, Linn. *Soc. Bot. J.*, 43: 233–290.

Piczynska, E. and Ozirrek, T., 1976. Ecological significance of macrophytes. *Indian J. Ecol. Environ. Sci.*, 2: 115–128.

Sankhla, S.K. and Vyas, L.N., 1982. Observations on the moist bank community of Baghela Tank, Udaipur (India). pp. 197–206. In: *Wetland, Ecology and Management*, (Eds.) B. Gopal, R.E. Turner, R.G. Wetzel and D.F. Whigham. National Int. Ecol. and Inter. Scient. Pubs., Jaipur (India).

Sharma, B.M., 2001. Impact of Environmental upheaval in Manipur. In: *Constraints in Development of Manipur,* (Eds) Thomas, C., Gopalakrislman, J. and Singh, R.K.R. Regency Publications, New Delhi, pp. 81–82.

Sharma, B.M. and Devi, N.B., 2002. Ecological assessment of macrophytes in Loktak Lake, Manipur: Phytosociology. In: *Ecology of Polluted Waters,* (Ed.) Arvind Kumar. A.P.H. Publishing Corpn., New Delhi, pp. 869–876.

Van Willium, G., 1970. Dynamic aspects of tropic gradients in a mine complex. Committee for hydrological research TNO. The relation between water quality and water quantity in studies of surface water. Proceedings of Technical meeting 35, October 1978, Proceedings and Information No. 25: 66–82

Wheeler, B.D., 1980. Plant communities of rich-fen systems in England and Wales. *J. Ecol.*, 68: 365–395.

Chapter 51

Compatibility Studies of Some Fungicides, Insecticides and Micronutrients in Apple

Dharmesh Gupta[1], V.K. Rana[2], D.D. Sharma[1] and P.L. Sharma[1]
[1]Temperate Horticultural Research Station, Kotkhai, Shimla – 171 202, Himachal Pradesh
[2]Krishi Vigyan Kendra, Kinnaur, Reckong Peo, Himachal Pradesh

ABSTRACT

Compatibility studies of fungicides, insecticides and micronutrients were carried out in apple at Kotkhai, Shimla, Himachal Pradesh. Among 40 combinations of fungicides and insecticides 4 were physically incompatible as there was formation of precipitates. Captan was not compatible with summer spray oil as it caused damage to leaves. Toxicity symptoms were observed on leaves and fruits when zinc sulphate (0. 5 per cent) was sprayed without mixing of lime. Orchex (1 per cent) and profenofos (0.05 per cent) were also not compatible as they caused phytoxicity. Rest all the combinations were compatible.

***Keywords**: Apple, Compatibility, Fungicides, Insecticides, Micronutrients.*

Introduction

Himachal Pradesh has the distinction of being called the "Apple state of India". Apple *Malus domestica* is one of the most important cash crops of Himachal Pradesh and it has lifted the economy of the farmers of the state to a great extent. In the recent past large number of diseases and insect pests have been found to damage apple crop under field conditions. This multiple pest attack on apple usually had additive effects on yield loss every year. Apple is grown under rainfed conditions in Himachal Pradesh and water scarcity is a major problem during the growing seasons. Some times it is not possible to apply fungicides, insecticides and nutrients separately. So keeping in view the scarcity of water, time and to minimize the total number of sprays, it is essential to apply fungicides and insecticides at the same time. In the recent years, the possible pesticidal combination in vogue in apple are: fungicide + fungicide; fungicide + insecticide; fungicide + insecticide + micronutrient etc. The compatibility factor in such cases has been little understood. Compatibility literally means co-existence without undermining efficacy of the other, so that a formulation or a mixture becomes effective in plant disease control.

Combination of materials can cause injury. Combining two emulsifiable concentrates or an emulsifiable concentrate with a wettable powder can lead to compatibility problems. The information on their compatibility as a tank-mix is highly essential to meet such a demand in view of the ease and economy. Though some information is available on the compatibility and toxicity of insecticides and fungicides in Apple (Anonymous, 2002), it is essential to generate similar information on new insecticides and fungicides. In some cases injury results not only when two materials are mixed in the same tank but when one material is applied after another at short interval.

Keeping in view the above problems efforts have been made to study the physical compatibility and phytotoxicity of some new pesticides recommended on apple.

Materials and Methods

Studies were carried out at Kotkhai, Shimla under field conditions on fully bearing apple trees in which 40 combinations of insecticides, fungicides and micronutrients at recommended concentrations were tested (Table 51.1). Sprays were done during the first week of June when the fruit was of medium size. Different combinations of chemicals were sprayed with foot-sprayer only once and observations on physical compatibility were recorded at the time of mixing while phytotoxicity symptoms, if any were noted 2 weeks after spraying. The weather conditions during study period were dry with temperature fluctuating around 30°C.

The experiment was conducted in a randomized block design (RBD) with three/replications.

Results and Discussions

Among different combinations, 4 combinations of fungicides and insecticides were physically incompatible as their mixing led to the formation of white precipitates resulting in unstable mixture. Dodine was not physically compatible with any of the insecticides which could be due to its insolubility in most solvents (Nene and Thapliyal, 1979). Captan (0.3 per cent) when mixed with summer oil resulted in burning of leaves. Similar observations regarding chemical incompatibility have been reported by Sahni (1987) for Captan. Captan applied either before or after oils (within a 2-week period) may cause damage to sensitive varieties. When the micronutrient zinc sulphate (0.5 per cent) was applied alone on apple, it caused black spots on fruits thereby reducing its market value. However, when zinc sulphate (0.5 per cent) was mixed with lime (0.25 per cent), no toxicity symptoms were observed on apple fruits. Combination mixture of Orchex (1 per cent) and Profenofos (0.05 per cent) was also phytotoxic to apple leaves.

The results presented in Table 51.1 shows that except seven combinations (at Sl.No. 24 to 30), all other combinations of fungicides, insecticides and micronutrients were safe to apply on apple.

Table 51.1: Compatibility of Fungicides, Insecticides and Micronutrients in Apple

Sl.No.	Combinations	Physical Compatability/Solubility	Toxicity Symptoms, if any
1.	Orchex 796 (1%) + Chlorpyriphos (0.04%)	Good	–
2.	Orchex 796 (1%) + Thiodan (0.05%)	Good	–
3.	Orchex 796 (1%) Blitox (0.3%)	Good	–
4.	Orchex 796 (1%) + Blitox (0.3%) + Urea (1%)	Good	–
5.	Fenzaquin (0.0025%) + Mancozeb (0.3%)	Good	–
6.	Fenzaquin (0.0025%) + Carbendazim (0.05%)	Good	–
7.	Propergite (0.05%) + Mancozeb (0.3%)	Good	–
8.	Propergite (0.05%) + Carbendazim (0.05%)	Good	–
9.	Orchex 796 (1%) + Captan (0.3%)	Good	–
10.	Combi Product (0.3%) + Fenzaquin (0.0025%)	Good	–
11.	Combi Product (0.3%) + Propergite (0.05%)	Good	–
12.	Hexaconazole (0.05%) + Thiodan (0.05%)	Good	–
13.	Hexaconazole (0.05%) + Chlorpyriphos (0.04%)	Good	–
14.	Hexaconazole (0.05%) + Fenzaquin (0.0025%)	Good	–
15.	Hexaconazole (0.05%) + Propergite (0.05%)	Good	–
16.	Orchex 796 (1%) + Mancozeb (0.3%)	Good	–
17.	Orchex 796 (1%) + Carbendazim (0.05%)	Good	–
18.	Propineb (0.3%) + Fenzaquin (0.0025%)	Good	–
19.	Propineb (0.3%) + Propergite (0.05%)	Good	–
20.	Propineb (0.3%) + Thiodan (0.05%)	Good	–
21.	Propineb (0.3%) + Chlorpyriphos (0.04%)	Good	–
22.	Profenofos (0.05%) + Mancozeb (0.3%)	Good	–
23.	Profenofos (0.05%) + Carbendazim (0.05%)	Good	–
24.	Dodine (0.1%) + Thiodan (0.05%) Precipitation of mixture	–	–
25.	Dodine (0.1%) + Orchex 796 (1%)	–do–	–
26.	Dodine (0.1%) + Chlorpyriphos (0.04%)	–do–	–
27.	Dodine (0.1%) + Lindane (0.05%)	–do–	–
28.	Orchex 796 (1%) + Profenofos (0.05%)	Good	Slight burning and yellowing of leaves
29.	T.S.O. (2%) + Captan (0.3%)	Good	Green bands develop around veins, bronzing of leaves, premature defoliation
30.	Zinc Sulphate (0.5%) alone	Good	Black spots develop on apple fruits
31.	Zinc Sulphate (0.5%) + lime (0.25%)	Good	No toxicity symptoms on fruits
32.	Boric Acid (0.1%) + Chlorpyriphos (0.04%)	Good	–
33.	Boric Acid (0.1%) + T.S.O. (2%)	Good	–

Contd...

Table 51.1–Contd...

Sl.No.	Combinations	Physical Compatability/Solubility	Toxicity Symptoms, if any
34.	Boric Acid (0.1%) + Mancozeb (0.3%)	Good	–
35.	Boric Acid (0.1%) + Carbendazim (0.05%)	Good	–
36.	T.S.O. (2%) + Carbosulfan (0.025%)	Good	–
37.	T.S.O. (2%) + Malalthion (0.05%)	Good	–
38.	Carbosulfan (0.025%) + Mancozeb (0.3%)	Good	–
39.	Carbosulfan (0.025%) + Carbendazim (0.05%)	Good	–
40.	Carbosulfan (0.025%) + Captan (0.3%)	Good	–

–: No Symptoms.

Note: Combi-product consisted of readymade mixture of mancozeb and carbendazim.

The scope of pesticides incompatibility are unlimited in view of developing agrochemical industries not only in India but all over the world. For their effective use, continuous and intensive research has to be carried out under conditions prevalent in India to satisfy the needs of the grower as well as industries. It would be desirable to evaluate possible combinations of the agricultural chemicals consistently, so that the plant pathologists keeps himself/herself ahead in timely as well as appropriate recommendations to the growers.

References

Anonymous, 2002. *Insecticides Fungicides Compatibility Chart*. Published by Department of Horticulture, Navbahar, Shimla-2.

Nene, Y.L. and Thapliyal P.N., 1979. *Fungicides in Plant Disease Control*, 2nd edn. Oxford and IBH Publishing Company, 507 pp.

Sahni, M.L., 1987. Problems in the usage of fungicides particularly in combination with pesticides. *Pesticide Information*, 13(3): 26–29.

Chapter 52

Performance Evaluation of ET_0 Equations for Palleru Sub Basin of River Krishna, India

M.V.S.S. Giridhar[1] and G.K. Viswanadh[2]

[1]Associate Professor, Department of Civil Engineering, MREC, Dulapally (v), Secunderabad – 500 014, India, E-mail: mvssgiri@yahoo.com

[2]Professor, Department of Civil Engineering, J.N.T.U. C.E., JNTU, Hyderabad – 500 072, India, E-mail: gorti_gkv@yahoo.co.in

ABSTRACT

In the present, study twelve existing evapotranspiration equations namely Blaney-Criddle, McCloud, Hargreaves, Turc (1961), Radiation, Priestly-Taylor, FAO-56 Penman-Monteith (PM) methods and Pan evaporation by estimating K_{pan} from different equations *viz.*, Synder, modified Synder, Allen and Pruitt, Cuenca and Orang equations have been used to calculate reference evapotranspiration. FAO-56 PM was used as a standard method to evaluate different ET_0 equations in the present study. The daily meteorological data has been collected from meteorological station, Jagityal, Karimnagar district, A.P. Correlation coefficient (R^2) was developed between reference evapotranspiration estimated from different ET_0 equations with FAO-56 PM method. McCloud method with coefficient Was 1.055, Turc (1961) method, Priestly Taylor method and Orang equation have shown good correlation among all the methods under this study in temperature, radiation, combination and pan evaporation methods respectively. The estimated ET_0 values from Blaney-Criddle method, McCloud method with coefficient W = 1.06, W = 1.065, W = 1.07, Priestly-Taylor method, Hargreaves method, Turc method, Radiation method, Synder equation, Cuenca equation and Allen and Pruitt equation resulted in positive percentage of deviation as 190.34 per cent, 14.83 per cent, 46.55 per cent, 87.07 per cent, 13.40 per cent, 26.68 per cent, 10.63 per cent, 42.69 per cent, 5.09 per cent, 3.12 per cent and 2.60 per cent respectively when compared with FAO-56 Penman-Monteith method. However, McCloud method with coefficient W = 1.05, W = 1.055, Evaporation method, Modified Synder equation and Orang equation reflected negative percentage of deviation as –29.48 per cent, –10.02 per cent, –24.80 per cent, –1.07 per cent and –0.63 per cent respectively when compared with FAO-56 PM method.

Introduction

Doorenbous and Pruitt (1977) defined the ET_0 rate as "the rate of evapotranspiration from an extensive surface of 8 cm to 15 cm tall green grass cover of uniform height, actively growing, completely shading the ground and not short of water." Later, Allen *et al.* (1994) defined the grass reference evapotranspiration as "the rate of evapotranspiration from a hypothetical reference crop with an assumed crop height of 0.12 m, a fixed surface resistance of 70s m^{-1} and an albedo of 0.23 closely resembling the evapotranspiration from an extensive surface of green grass of uniform height actively growing completely shading the ground and with adequate water. ET_0 is a complex phenomenon because it depends on several climatatological factors, such as temperature, humidity, wind speed, radiation, type and growth stage of the crop. ET_0 can be either directly measured using lysimeter or water balance approaches or estimated indirectly using climatalogical data. However, it is difficult to measure ET_0 using lysimeter because it is a time-consuming method and needs precise and carefully planned experiments. The indirect ET_0 estimation methods based on climatalogical data vary from empirical relationships to complex methods such as the Penman Monteith method based on physical processes. These different methods of ET_0 estimation can be grouped into temperature methods (Thornthwaite, SCS Blaney-Criddle, FAO-24 Blaney Criddle, Hargreaves), radiation methods (Turc, Jensen-Haise, Priestley-Taylor, and FAO-24 radiation), combination theory type [Penman Monteith, FAO-24 Penman (c = l), 1982 Kimberly Penman, 1972 Kimberly Penman, FAO-24 corrected Penman] and pan evaporation (FAO-24 pan). Monteith (1965) modified the Penman method by incorporating stomatal resistance term specific to the type of crop in addition to the existing aerodynamic term and formulated the Penman-Monteith evapotranspiration model. The performance of different ET_0 estimation varies with climatic conditions and availability of data, and the data requirements vary from method to method. Furthermore, ET_0 estimation depends upon the quality of the meteorological data. In the present study two equations from temperature theory, three equations from radiation theory, two equations from combination approach and five equations from pan evaporation method have been considered for performance analysis.

Viswanadh *et al.* (2004) developed a computer program in Microsoft Visual C++, which is generalized, to calculate reference evapotranspiration (ET_0) using FAO-56 PM method. This program is based on the FAO-56 Penman-Monteith equation as given by FAO irrigation and Drainage Paper No. 56 (Allen *et al.*, 1988). Giridhar *et al.* (2006) compared different ET_0 equations with FAO-56 PM method at different locations in Andhra Pradesh.

Study Area

The Krishna basin is situated between East longitudes 73°21′ to 81°09′ and North latitudes 13°07′ to 19°25′ in the Deccan plateau covering large areas in the States of Maharashtra, Karnataka and Andhra Pradesh. The Krishna basin extends over an area of 2,58,948 sq.km., which is nearly 8 per cent of the total geographical area of the country. The Palleru (K-11) sub basin lies between latitudes 16°39' and 17°50 North and longitudes 79°17' and 80°09′ East. The Palleru sub basin has a catchment area of 3,263 km², which constitutes 1.3 per cent of the total basin area. The catchment area of the Palleru sub basin entirely lies in the State of Andhra Pradesh. In the present study Palleru sub basin has been considered for the performance of evaluation of different ET_0 equations with the standard FAO-56 PM ET_0 equation.

Methodology

In the present study, two temperature methods namely Blaney Criddle and McCloud methods, three radiation methods, namely Hargreaves, Turc (1961) and FAO-24 Radiation methods, two combination methods, namely Priestly-Taylor method and FAO-56 PM method and five pan evaporation methods namely Synder (1992), Cuenca (1989), Allen and Pruitt (1991), Modified Synder and Orang (1998) methods were considered were considered for performance evaluation with FAO-56 PM method. Their governing equations are listed in equations from 1 to 12. Daily meteorological parameters *viz.,* maximum and minimum temperature, maximum and relative humidity, wind speed, sun shine duration collected for a period of 1989 to 2004 from the meteorological station situated at Jagityal, Karimnagar district, A.P, India was processed and gaps were filled with the averages of previous and subsequent day. Total twelve internationally accepted ET_0 equations were used for the calculation of daily ET_0 out of which two methods are based on temperature, three methods are based on radiation, two methods are based on combination type and five methods are based on Pan evaporation.

In addition to the above mentioned ET_0 equations, an attempt has been made to calculate daily ET_0 values by reducing the coefficient (W) by an interval of 0.005. In McCloud method, ET_0 was calculated by varying the coefficient, *i.e.,* W = 1.07 to W = 1.065, W = 1.06, W = 1.055 and W = 1.05. Direct reading from the Pan was also taken as ET_0 value for that day in the present study. Hence, it is also considered as Evaporation method for further analysis. A program has been developed in VC^{++} to calculate daily ET_0 using different equations. Methodologies on crop water requirements were reviewed at international level to advise on the revision and update of procedures for estimating ET_0 and the panel of experts recommended the adoption of the FAO-56 Penman-Monteith combination method as standard. Hence for the present study, the same method was chosen as the standard method against which performance evaluation of the other methods was carried out. The calculated daily values were converted to monthly and yearly values. The ET_0 values estimated from different methods were compared with the standard method *i.e.,* FAO-56 PM. Performance evaluation was carried out between the values of ET_0 estimated from standard FAO-56 PM and other methods.

$$ET_0 = a_{bc} + b_{bc}\, f \qquad ...1$$

$$ET_p = KW^{(T-32)} \qquad ...2$$

$$ET_0 = 0.0135 \frac{R_s}{\lambda}(T + 17.8) \qquad ...3$$

$$ET_0 = 0.013TM/(TM + 15)\,(23.9\, R_s + 50) \text{ if } RH > 50\% \qquad ...4$$

$$ET_0 = 0.013TM/(TM + 15)\,(23.9\, R_s + 50)\,[1 + (50-RH)/70] \text{ if } RH < 50\%$$

$$ET_0 = c\,(W\, R_s) \qquad ...5$$

$$ET_r = 1.26 \frac{\Delta(R_n - G)}{\Delta + \gamma} \qquad ...6$$

$$ET_0 = \frac{0.408\Delta(R_n - G) + \gamma \frac{900}{T+273} u_2(e_s - e_d)}{\Delta + \gamma(l + 0.34u_2)} \qquad ...7$$

$$K_{pan} = 0.482 + [0.24 \ln(F)] - (0.000376\ U_2) + (0.0045\ RH) \quad ...8$$

$$K_{pan} = 0.475 - (0.24 - 10^{-1}\ U_2) + (0.516 \times 10^{-2}\ RH) + (0.118 \times 10^{-2}\ F) - (0.16 \times 10^{-4}\ RH^2) - (0.101 \times F^2) - (0.8 \times 10^{-8}\ RH^2\ U_2) - (0.1 \times 10^{-7}\ RH^2\ F) \quad ...9$$

$$K_{pan} = 0.108 - 0.00033\ IU + 0.0422 \ln(F) + 0.1434 \ln(H) - 0.000631\ [\ln(F)]^2 \ln(H) \quad ...(10)$$

$$K_{pan} = 0.5321 - 0.00030\ U + 0.0249 \ln(F) + 0.0025\ H \quad ...(11)$$

$$K_{pan} = 0.51206 - 0.00032\ 1U + 0.002889\ H + 0.031886 \ln(F) - 0.000107\ H \ln(F) \quad ...(12)$$

Results and Discussion

Twelve popular ET_0 equations were used for the calculation of daily ET_0 out of which two methods are based on temperature, three methods are based on radiation, two methods are based on combination type and five methods are based on Pan Evaporation. Using these twelve equations, the daily ET_0 values were calculated and sample output data was listed in Table 52.1 for the month of Jan 1989. And from the daily values monthly values were computed and sample data for a period of 2001 to 2003 were shown in Table 52.2. Further, daily ET_0 values were converted into yearly ET_0 values and listed in Table 52.3. Percentage of deviation of ET_0 values by all other methods from FAO-56 PM method were calculated and shown in Figures 52.1 to 52.3. The estimated ET_0 values from Blaney-Criddle method, McCloud method with coefficient W = 1.06, W = 1.065, W = 1.07, Priestly-Taylor method, Hargreaves method, Turc method, Radiation method, Synder equation, Cuenca equation and Allen and Pruitt equation resulted in positive percentage of deviation as 190.34 per cent, 14.83 per cent, 46.55 per cent, 87.07 per cent, 13.40 per cent, 26.68 per cent, 10.63 per cent, 42.69 per cent, 5.09 per cent, 3.12 per cent and 2.60 per cent respectively when compared with FAO-56 Penman-Monteith method. However, McCloud method with coefficient W = 1.05, W = 1.055, Evaporation method, Modified Synder equation and Orang equation reflected negative percentage of deviation as –29.48 per cent, –10.02 per cent, –24.80 per cent, –1.07 per cent and –0.63 per cent respectively when compared with FAO-56 PM method.

ET_0 estimated from McCloud equation with coefficient W = 1.055 has shown –10.02 per cent deviation when compared to FAO-56 Penman Monteith method. This method exhibited least deviation among all the temperature methods studied. Hence for this Palleru sub basin, McCloud equation is suitable with the coefficient of W = 1.055 instead of standard coefficient W = 1.07 as per the literature. Turc (1961) equation gave 10.63 per cent deviation from FAO-56 PM method and it is least among all the radiation methods in the study. Priestly-Taylor equation has shown 13.4 per cent deviation from the standard ET_0 equation. Among all pan evaporation equations Orang equation resulted 0.63 per cent deviation only, from the standard FAO-56 PM method. Correlation coefficient (R^2) has been calculated between the values of ET_0 estimated using FAO-56 PM and other methods on daily and monthly basis and enumerated in Table 52.4.

Conclusions

McCloud equation is found suitable with the coefficient of W = 1.055 among the temperature methods of estimating ET_0. Turc (1961) equation is found suitable among radiation methods of estimating ET_0. Priestly-Taylor equation has shown 13.4 per cent deviation from the standard ET_0 equation and is found as a suitable method in Combination methods for estimating ET_0. Among all pan evaporation equations, Orang equation with 0.63 per cent least deviation has been found to be a suitable method of

Table 52.1: Daily Reference Evapotranspiration Values Calculated Using Different Equations

Date	FAO-56 PM	Blaneny Cridle	Mc-Cloud W = 1.05	Mc-Cloud W = 1.055	Mc-Cloud W = 1.06	Mc-Cloud W = 1.065	Mc-Cloud W = 1.07	Priestly Taylor	Hargreaves	Turc 1961	Radiation Method	Evaporation	Synder	Cuenca	Allen Pruitt	Modified Synder	Orang
1/1/1989	2.89	3.95	1.59	1.89	2.26	2.70	3.22	3.05	3.996	3.564	4.54	2.0	1.88	1.74	1.70	1.66	1.68
1/2/1989	2.98	3.39	1.57	1.88	2.24	2.67	3.18	3.05	3.968	3.543	4.42	3.0	2.91	2.67	2.58	2.55	2.55
1/3/1989	3.28	4.15	1.82	2.20	2.66	3.22	3.89	3.18	4.107	3.631	4.45	4.0	3.80	3.52	3.44	3.36	3.36
1/4/1989	3.12	4.41	1.75	2.11	2.54	3.06	3.68	3.12	3.954	3.517	3.98	3.5	3.29	3.05	2.98	2.91	2.91
1/5/1989	3.04	3.89	1.82	2.21	2.68	3.24	3.91	3.14	3.932	3.492	3.79	3.5	3.33	3.08	3.01	2.94	2.94
1/6/1989	3.01	4.14	1.83	2.22	2.69	3.25	3.93	3.17	3.985	3.533	3.92	3.5	3.52	3.26	3.18	3.11	3.11
1/7/1989	3.37	7.19	1.92	2.34	2.85	3.46	4.21	2.97	3.451	3.105	3.01	3.5	4.23	3.91	3.82	3.73	3.73
1/8/1989	4.35	6.33	2.25	2.79	3.44	4.25	5.24	3.14	3.644	3.228	3.17	3.5	4.23	3.86	3.73	3.73	3.68
1/9/1989	3.20	8.96	1.71	2.06	2.47	2.97	3.57	3.05	3.717	3.332	3.34	3.5	4.30	4.15	4.10	3.95	3.95
1/10/1989	3.15	7.50	1.72	2.08	2.50	3.01	3.61	3.12	3.873	3.454	3.66	3.5	3.07	3.00	3.00	2.85	2.89
1/11/1989	3.19	8.99	1.76	2.13	2.57	3.09	3.72	3.18	3.975	3.532	3.88	3.5	3.08	3.00	3.04	2.89	2.89
1/12/1989	2.93	8.13	1.61	1.92	2.30	2.75	3.28	3.11	3.893	3.48	3.83	3.5	3.32	3.28	3.28	3.12	3.16
1/13/1989	3.09	6.56	1.59	1.90	2.27	2.71	3.24	3.11	3.868	3.461	3.75	3.5	4.09	3.91	3.86	3.73	3.73
1/14/1989	3.52	10.96	1.86	2.25	2.73	3.31	4.01	3.26	4.041	3.575	3.89	3.5	3.06	2.95	2.92	2.81	2.81
1/15/1989	3.33	8.85	1.66	1.99	2.38	2.86	3.42	3.30	4.305	3.803	5.39	3.5	2.43	2.40	2.40	2.28	2.31
1/16/1989	3.44	8.52	1.89	2.30	2.79	3.39	4.11	3.41	4.355	3.818	4.94	3.5	2.87	2.80	2.84	2.70	2.70
1/17/1989	3.46	9.85	1.82	2.20	2.66	3.22	3.89	3.33	4.175	3.685	4.3	3.5	3.74	3.60	3.60	3.47	3.47
1/18/1989	3.39	7.76	1.94	2.36	2.88	3.50	4.26	3.43	4.30	83.83	4.55	3.5	3.16	3.12	3.16	3.00	3.04
1/19/1989	3.57	10.68	2.05	2.52	3.08	3,77	4.61	3.47	4.293	3.859	4.28	3.5	3.59	3.54	3.59	3.40	3.45
1/20/1989	3.74	12.38	2.04	2.49	3.05	3.73	4.55	3.47	4.269	3.735	4.2	3.5	4.46	4.40	4.35	4.18	4.18
1/21/1989	3.28	9.45	1.92	2.34	2.85	3.46	4.21	3.40	4.146	3.651	3.94	3.5	3.61	3.52	3.52	3.34	3.39
1/22/1989	3.27	7.27	1.77	2.14	2.58	3.11	3.75	3.42	4.29	53.784	4.65	3.5	2.32	2.27	2.27	2.16	2.18
1/23/1989	3.21	8.13	1.83	2.22	2.69	3.25	3.93	3.32	3.96	3.513	3.55	3.5	3.24	3.20	3.20	3.04	3.08

Contd...

Table 52.1–Contd...

Date	FAO-56 PM	Blaneny Cridle	Mc-Cloud W = 1.05	Mc-Cloud W = 1.055	Mc-Cloud W = 1.06	Mc-Cloud W = 1.065	Mc-Cloud W = 1.07	Priestly Taylor	Hargreaves	Turc 1961	Radiation Method	Evaporation	Synder	Cuenca	Allen Pruitt	Modified Synder	Orang
1/24/1989	3.37	9.37	1.74	2.10	2.53	3.04	3.66	3.44	4.305	3.958	4.67	3.5	2.57	2.54	2.57	2.48	2.48
1/25/1989	3.55	8.57	1.86	2.25	2.73	3.31	4.01	3.46	4.222	3.718	4.12	3.5	3.00	2.96	3.00	2.85	2.85
1/26/1989	3.95	10.92	2.10	2.58	3.16	3.88	4.75	3.41	3.952	3.481	3.43	3.5	3.44	3.28	3.24	3.12	3.16
1/27/1989	3.56	10.51	1.93	2.35	2.86	3.48	4.23	3.48	4.17	3.669	3.85	3.5	3.32	3.24	3.20	3.08	3.08
1/28/1989	3.74	11.53	1.97	2.40	2.93	3.56	4.34	3.59	4.355	3.81	4.28	3.5	2.96	2.92	2.92	2.78	2.81
1/29/1989	3.59	9.23	2.09	2.57	3.15	3.86	4.72	3.67	4.446	3.868	4.37	3.5	3.81	3.71	3.76	3.57	3.57
1/30/1989	3.58	10.77	1.92	2.34	2.85	3.46	4.21	3.74	4.712	4.154	5.79	3.5	3.95	3.90	3.90	3.75	3.75
1/31/1989	3.74	11.70	1.85	2.24	2.72	3.29	3.98	3.80	4.901	4.254	7.21	3.5	3.95	3.90	3.90	3.75	3.75

Table 52.2: Monthly Reference Evapotranspiration (RET) for the Years 2001, 2002 and 2003

Year	Month	FAO-56 PM	Blaneny Cridle	Mc-Cloud W = 1.05	Mc-Cloud W = 1.055	Mc-Cloud W = 1.06	Mc-Cloud W = 1.065	Mc-Cloud W = 1.07	Priestly Taylor	Hargreaves	Turc 1961	Radiation Method	Evaporation	Synder	Cuenca	Allen Pruitt	Modified Synder	Orang
2001	Jan	88.3	187.3	52.3	62.9	75.7	91.0	109.3	102.1	131.5	119.4	180.3	10.0	83.1	81.0	81.6	77.9	78.4
	Feb	106.9	234.4	56.4	69.0	84.4	103.2	126.0	120.8	153.5	141.3	278.5	23.0	100.5	99.1	100.5	95.7	96.7
	Mar	135.5	297.6	81.5	102.4	128.7	161.5	202.6	164.5	192.3	165.9	265.4	31.0	142.8	140.5	141.4	135.2	136.2
	Apr	147.7	289.2	93.8	120.0	153.4	196.0	250.2	187.1	217.2	188.6	375.5	54.0	149.9	147.3	148.9	142.7	143.8
	May	185.1	789.9	147.0	195.6	259.9	344.8	457.0	213.9	238.0	218.2	301.5	34.0	192.7	197.0	197.8	191.7	193.0
	Jun	126.4	568.7	105.1	135.9	175.4	226.3	291.6	146.9	151.8	126.7	150.7	44.0	141.2	139.8	137.8	132.7	132.9
	Jul	111.3	409.5	86.4	109.2	137.9	173.8	219.0	147.9	151.0	127.4	148.2	18.0	104.1	101.3	99.4	96.0	96.2
	Aug	113.7	305.3	76.8	95.9	119.7	149.3	186.0	130.4	130.8	113.0	141.5	8.0	93.1	88.4	86.1	83.8	83.9
	Sep	117.4	214.8	84.6	107.0	135.3	170.8	215.5	136.9	144.2	121.7	136.8	18.0	99.3	94.9	93.9	90.4	90.9
	Oct	98.4	167.8	59.3	72.3	88.0	107.1	130.3	106.4	112.8	100.9	115.7	19.0	88.6	84.4	83.6	80.6	81.0
	Nov	89.9	197.8	48.9	58.6	70.3	84.2	100.9	87.3	97.3	93.6	90.6	19.0	87.7	86.1	87.3	83.3	84.0
	Dec	77.2	169.2	39.9	46.7	54.7	64.1	74.9	71.1	77.7	79.7	83.4	15.0	79.3	78.2	79.5	75.8	76.6
2002	Jan	92.2	211.5	53.4	64.4	77.6	93.5	112.7	98.0	119.6	107.7	120.9	102.3	85.4	82.9	83.1	79.4	80.0
	Feb	105.8	211.7	66.0	82.1	102.1	126.9	157.6	110.5	125.2	107.8	117.3	103.7	106.6	103.0	102.6	98.4	98.9
	Mar	144.8	298.1	94.2	120.1	153.0	194.7	247.6	163.3	185.0	156.4	204.7	198.1	161.1	157.3	158.0	151.1	152.1
	Apr	168.0	437.3	125.3	164.7	216.3	283.8	372.0	191.6	213.2	183.6	243.3	258.6	197.0	195.5	198.0	189.4	191.0
	May	189.9	1010.3	157.1	210.3	281.3	375.8	501.5	211.0	230.9	208.4	260.6	376.7	250.6	259.1	258.7	249.7	250.8
	Jun	130.7	700.2	131.1	173.4	229.3	303.0	399.9	143,9	145.4	120.0	153.8	238.5	177.8	177.5	174.7	168.4	168.7
	Jul	143.4	829.1	102.5	131.7	169.1	216.9	277.9	155.9	160.0	132.7	153.0	227.4	165.8	167.1	162.5	157.3	156.5
	Aug	107.9	320.0	84.6	106.7	134.4	169.2	212.8	116.5	113.4	98.9	141.3	112.8	97.9	92.8	90.5	88.1	88.1
	Sep	127.9	314.2	90.5	115.2	146.6	186.3	236.6	128.9	132.7	112.4	137.7	132.2	114.1	109.8	108.6	104.7	104.9

Contd...

Table 52.2–Contd...

Year	Month	FAO-56 PM	Blaneny Cridle	Mc-Cloud W = 1.05	Mc-Cloud W = 1.055	Mc-Cloud W = 1.06	Mc-Cloud W = 1.065	Mc-Cloud W = 1.07	Priestly Taylor	Hargreaves	Turc 1961	Radiation Method	Evapo-ration	Synder	Cuenca	Allen Pruitt	Modified Synder	Orang
	Oct	117.6	206.1	88.2	111.8	141.5	178.9	226.1	128.4	141.7	121.0	137.9	130.4	110.8	107.1	107.1	102.6	103.3
	Nov	96.3	200.2	61.1	74.9	91.8	112.3	137.3	106.0	130.8	116.5	142.1	124.6	101.4	99.0	100.2	95.6	96.3
	Dec	89.3	193.1	58.4	71.0	86.3	104.7	126.9	99.0	129.8	115.7	158.7	114.0	92.8	90.8	91.8	87.3	88.0
2003	Jan	88.5	212.3	59.2	72.1	87.9	107.0	130.1	103.4	128.7	115.1	144.0	117.6	96.0	93.9	94.4	90.1	90.8
	Feb	120.5	616.0	74.3	93.4	117.4	147.3	184.7	117.1	134.0	113.8	127.0	145.2	109.0	109.2	107.4	103.2	103.2
	Mar	159.2	1052.8	100.3	128.7	165.0	211.3	270.3	166.2	188.3	167.4	209.9	213.2	142.0	147.4	144.5	139.3	139.2
	Apr	165.8	656.9	128.4	169.5	223.5	294.4	387.4	184.4	203.4	174.3	223.4	280.2	204.2	206.2	206.5	198.0	198.9
	May	205.8	1653.6	171.3	231.4	312.1	420.5	565.8	219.7	240.9	207.9	270.9	377.6	204.1	227.0	218.8	216.4	215.5
	Jun	149.9	728.7	144.7	193.8	259.5	347.4	464.7	176.8	184.2	162.5	176.7	300.0	210.8	212.9	210.8	205.8	206.5
	Jul	111.0	342.9	92.3	117.4	149.1	189.2	239.9	137.8	137.9	116.8	149.7	129.2	112.0	106.6	103.9	101.3	101.3
	Aug	120.6	608.9	86.9	109.9	138.8	175.0	220.5	125.9	124.6	107.3	147.2	98.4	77.3	74.8	71.7	70.4	70.0
	Sep	131.4	619.2	90.2	114.8	145.8	185.1	234.7	125.6	127.8	108.9	136.3	118.8	92.7	89.2	85.4	84.2	83.5
	Oct	115.1	284.8	83.9	105.8	133.2	167.5	210.6	121.4	131.3	112.7	131.0	114.8	101.4	95.8	93.8	91.4	91.3
	Nov	103.6	361.0	62.7	77.0	94.5	116.0	142.2	105.1	127.2	111.2	125.4	113.6	93.6	91.1	90.3	86.6	86.9
	Dec	93.4	471.1	53.6	64.7	78.0	94.1	113.3	94.8	123.8	110.2	161.5	106.0	82.1	81.3	80.5	77.2	77.4

Table 52.3: Yearly Reference Evapotranspiration for the Basin

Year Month	FAO-56 PM	Blaneny Cridle	Mc-Cloud W = 1.05	Mc-Cloud W = 1.055	Mc-Cloud W = 1.06	Mc-Cloud W = 1.065	Mc-Cloud W = 1.07	Priestly Taylor	Hargreaves	Turc 1961	Radiation Method	Evapo-ration	Synder	Cuenca	Allen Pruitt	Modified Synder	Orang
1989	1591	5177	1049	1337	1705	2174	2773	1647	1836	1655	2168	908	1645	1630	1624	1569	1575
1990	1564	4762	1005	1273	1611	2038	2577	1601	1755	1555	1853	1953	1537	1509	1500	1447	1454
1991	1587	4635	1052	1341	1709	2178	2775	1680	1865	1606	1952	1948	1552	1522	1515	1456	1462
1992	1539	4769	1029	1311	1670	2129	2715	1760	2010	1789	2446	2047	1575	1557	1559	1501	1510
1993	1558	4988	1094	1404	1802	2314	2972	1714	1921	1696	2132	2121	1631	1613	1609	1551	1559
1994	1452	3957	1041	1330	1701	2177	2787	1713	1929	1741	2112	395	1628	1580	1570	1527	1535
1995	1516	4019	1116	1433	1838	2359	3028	1691	1879	1626	2109	254	1521	1482	1475	1423	1430
1996	1467	3605	1128	1448	1858	2386	3062	1716	1906	1606	2060	449	1730	1669	1653	1595	1601
1997	1446	3456	1043	1328	1691	2153	2740	1712	1903	1619	1981	324	1622	1566	1551	1496	1502
1998	1401	3680	991	1258	1600	2034	2589	1779	2029	1751	2335	399	1644	1604	1596	1539	1547
1999	1429	4008	1039	1323	1686	2148	2736	1675	1879	1624	2228	314	1463	1433	1431	1374	1380
2000	1402	3756	922	1160	1460	1838	2313	1681	1893	1676	2240	668	1413	1384	1384	1331	1339
2001	1398	3831	932	1175	1483	1872	2363	1615	1798	1596	2268	293	1362	1338	1338	1286	1293
2002	1514	4932	1112	1426	1829	2346	3009	1653	1828	1581	1971	2119	1661	1642	1636	1572	1579
2003	1565	7608	1148	1479	1905	2455	3164	1678	1852	1608	2003	2115	1525	1535	1508	1464	1464
2004	1442	5535	1122	1440	1848	2370	3041	1690	1880	1618	2087	1993	1547	1530	1520	1462	1467

Table 52.4: R^2 Value Between the Values of ET_0 Estimated Using FAO-56 PM and Other Methods for Daily and Monthly

Year Month	FAO-56 PM	Blaneny Cridle	Mc-Cloud W = 1.05	Mc-Cloud W = 1.055	Mc-Cloud W = 1.06	Mc-Cloud W = 1.065	Mc-Cloud W = 1.07	Priestly Taylor	Hargreaves	Turc 1961	Radiation Method	Evapo-ration	Synder	Cuenca	Allen Pruitt	Modified Synder	Orang
R^2 (Daily)		0.64	0.71	0.69	0.67	0.65	0.62	0.83	0.72	0.82	0.26	0.63	0.73	0.72	0.71	0.74	0.76
R^2 (Monthly)		0.56	0.82	0.84	0.75	0.71	0.68	0.87	0.77	0.82	0.53	0.22	0.79	0.77	0.74	0.79	0.81

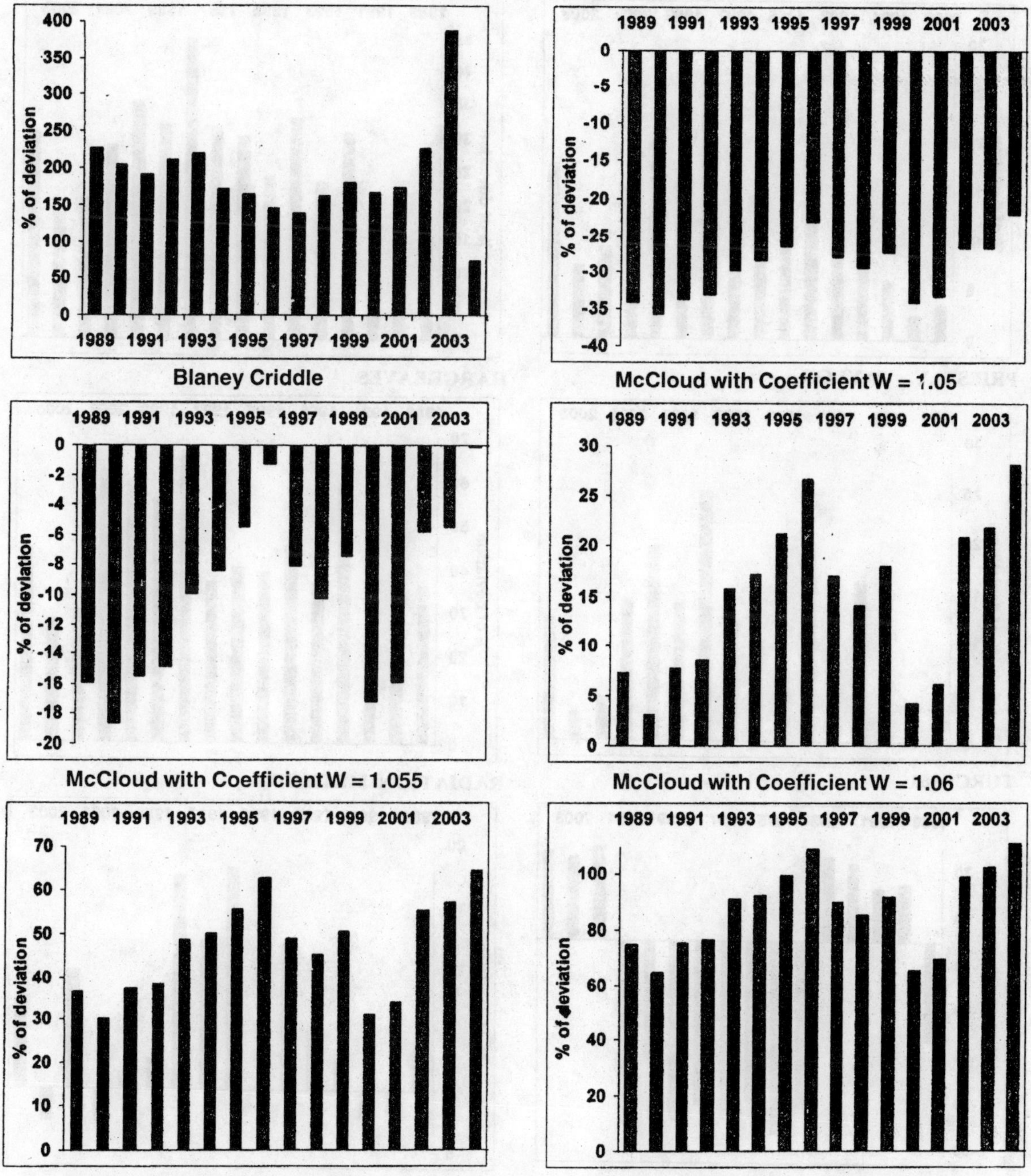

Figure 52.1: Percentage of Deviation of RET Estimated from Different Equations with FAO-56 Penman-Monteith Method

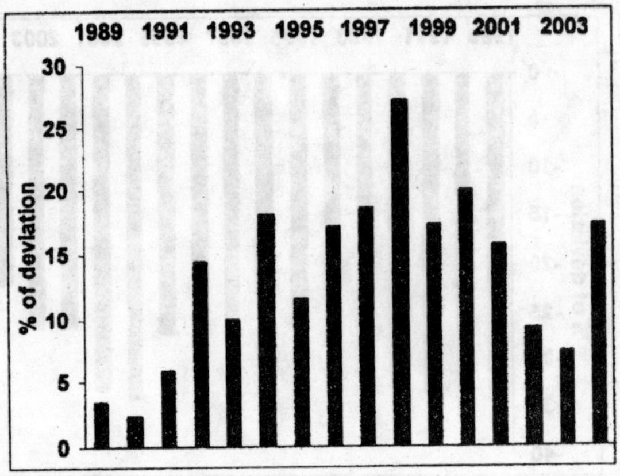

PRIESTLY – TAYLOR

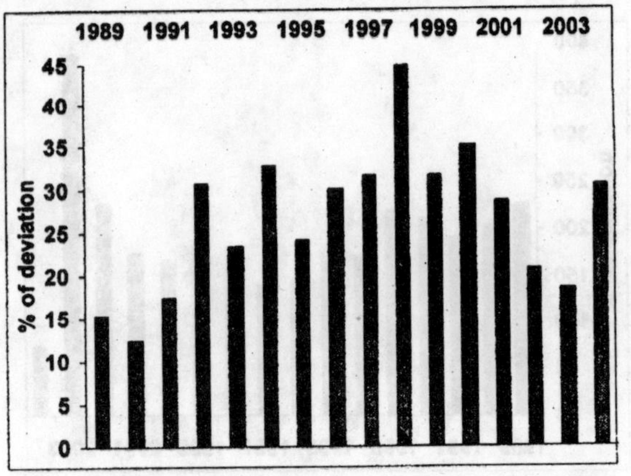

HARGREAVES

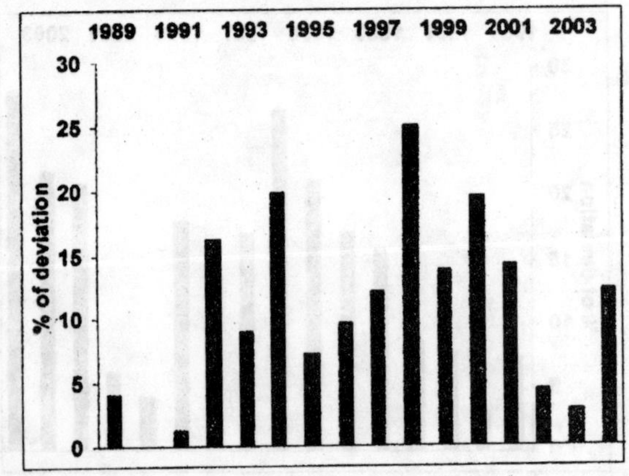

TURC 1961

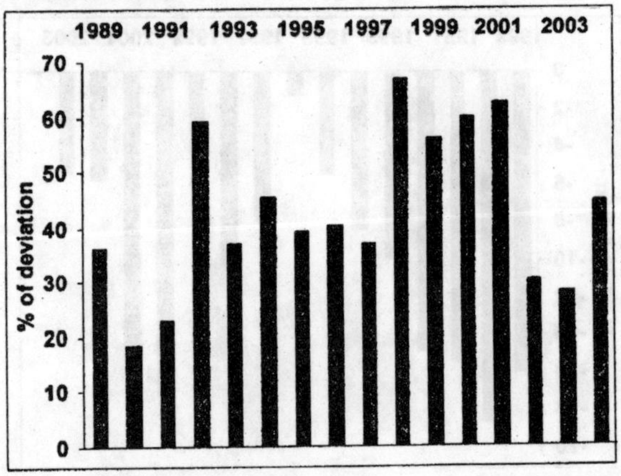

RADIATION METHOD

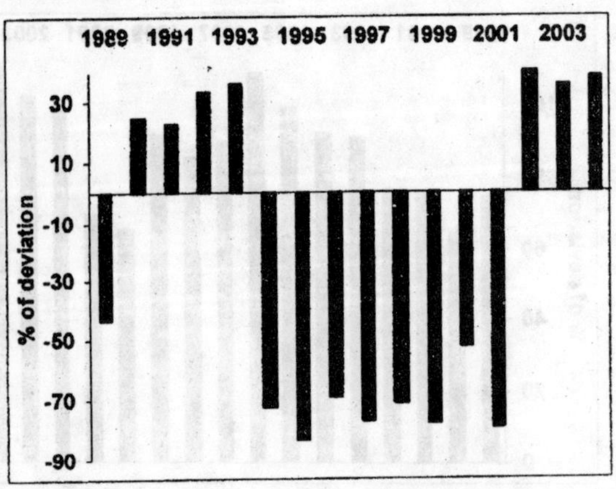

EVAPORATION

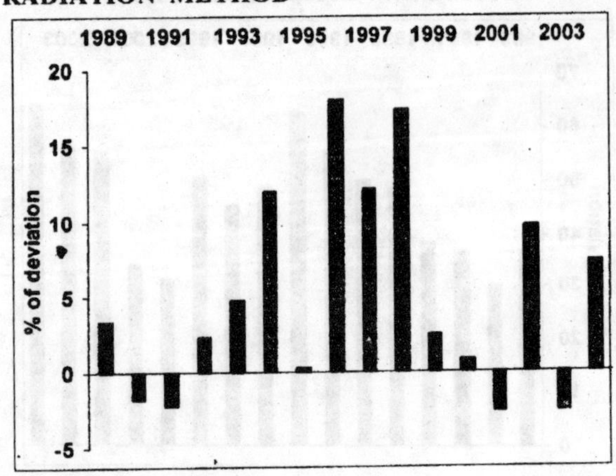

SYNDER (1962)

Figure 52.2: Percentage of Deviation of RET Estimated from Different Equations with FAO-56 Penman-Monteith Method

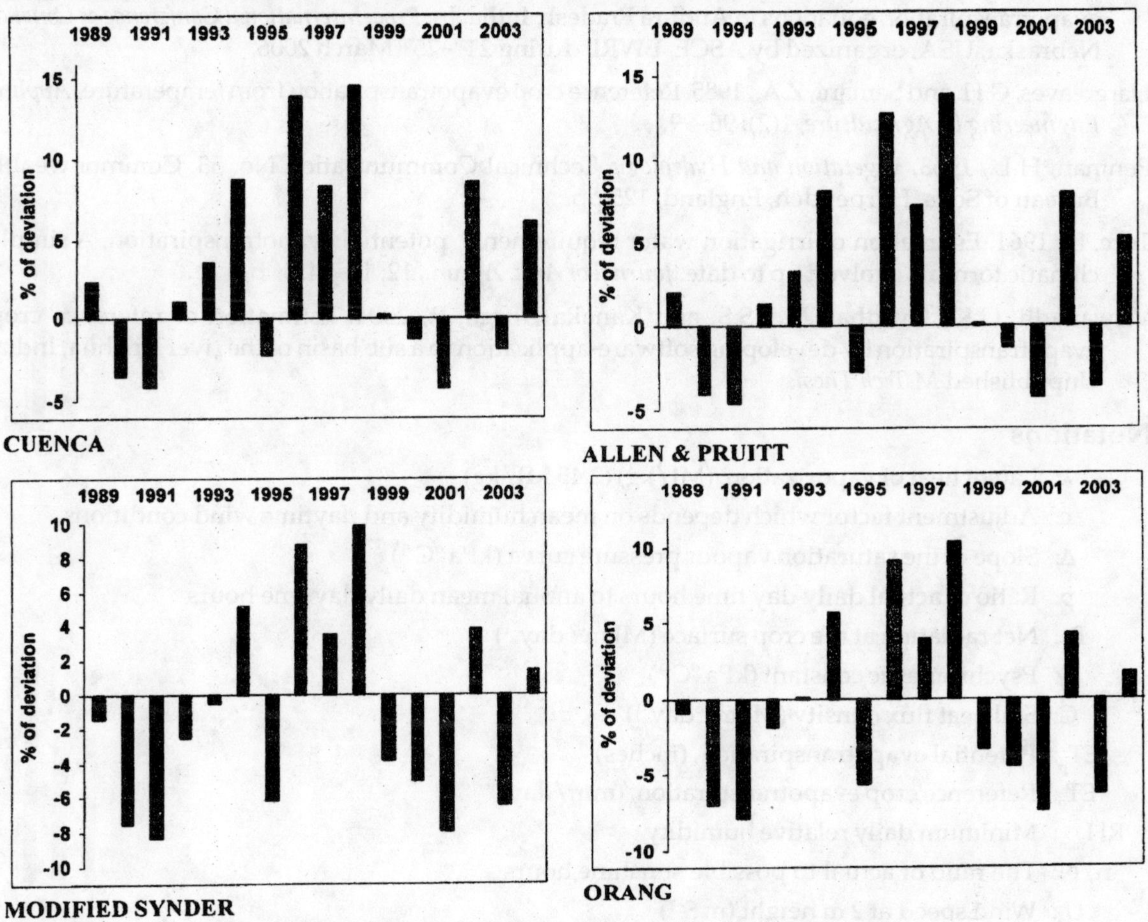

Figure 52.3: Percentage of Deviation of RET Estimated from Different Equations with FAO-56 Penman-Monteith Method

estimating ET_0.

References

Allen, R.G., Smith, M., Pereira, L.S., and Perrier, A., 1994. An update for the calculation of reference evapotranspiration. *ICID Bull.*, 43(2): 35–92.

Allen, R.G., Periera, L.S., Raes, D. and Smith, M., 1998. *Crop Evapotranspiration: Guidelines for Computing Crop Requirements*. Irrigation and Drainage Paper No. 56, FAO, Roffle, Italy, 300 pp.

Blaney, H.F. and Criddle, W.D., 1950. *Determining Water Requirements in Irrigated Areas from Climatological and Irrigation Data*. USDA, Soil Conservation Service Report. SCS-Technical Report-96, 48 pp.

Doorenbos, J. and Pruitt, L.R. *Crop Water Requirements*. Irrigation and Drainage Paper 24, FAO, Rome.

Doorenbos, J. and Pruitt, W.O., 1977. *Guidelines to Crop Water Requirements*. Irrigation and Drainage Paper No. 24 (Revised), FAO, Rome, Italy, 144 pp.

Giridhar, M.V.S.S., Viswanadh, G.K. and Kanaka Durga, P., 2004. Comparison of reference evapotranspiration equations in Andhra Pradesh, India. In: *Proc. International Conference at Omaha*, Nebraska, USA, organized by ASCE, EWRI, during 21st–25th March 2006.

Hargreaves, G.H. and Samani, Z.A., 1985. Reference crop evapotranspiration from temperature. *Applied Engineering in Agriculture*, 1(2): 96–99.

Penman, H.L., 1963. *Vegetation and Hydrology*. Technical Communication No. 53, Commonwealth Bureau of Soils, Harpenden, England, 125 pp.

Turc, L., 1961. Estimation of irrigation water requirements, potential evapotranspiration: A simple climatic formula evolved up to date. *Journal of Ann. Agron.*, 12: 13–14.

Viswanadh, G.K., Giridhar, M.V.S.S. and Kanaka Durga, P., 2004. Estimation of reference crop evapotranspiration by developing software-application to a sub basin of the river Krishna, India. Unpublished *M.Tech Thesis*.

Notations

λ: Latent heat of vaporization, (MJ/kg) (2.45 MJ/kg)

c: Adjustment factor which depends on mean humidity and daytime wind conditions

Δ: Slope of the saturation vapour pressure curve (kPa °C^{-1})

p: Ratio of actual daily day time hours to annual mean daily daytime hours

R_n: Net radiation at the crop surface (MJ m^{-2} day^{-1})

γ: Psychrometric constant (kPa °C^{-1})

G: Soil heat flux density (MJ m^{-2} day^{-1})

ET_p: Potential evapotranspiration, (inches)

ET_0: Reference crop evapotranspiration, (mm/day)

RH_{min}: Minimum daily relative humidity

n/N: The ratio of actual to possible sunshine hours

U_2: Wind speed at 2 m height (m S^{-1})

T: Mean daily air temperature at 2 m height (°C)

e_s: Saturation vapour pressure (kPa)

e_a: Actual vapour pressure (kPa)

R_s: Daily global solar radiation, (KJ/m^2 d^{-1})

U_d: Daytime wind at 2 m height in m/sec

W: Weighting factor which depends on temperature and latitude

RH_{max}: Maximum daily relative humidity

Chapter 53

Synthesis and Pharmacological Screening of 3-(4-Chlorophenyl)-1-phenyl-1H-pyrazole-4-carboxaldehyde and It's Schiff's Bases

Ajit D. Jagtap*, Chandrakant S. Magdum, Shrinivas K. Mohite

Department of Chemistry, Appasaheb Bimale College of Pharmacy, Sangli, South Shivaji Nagar, Sangli-Miraj Road, Sangli – 416 416, Maharashtra

ABSTRACT

3-(4-Chlorophenyl)-1-phenyl-1H-pyrazole-4-carboxaldehyde was obtained from appropriate phenylhydrazone via the Vilsmeier-Haack reaction. This aldehyde was further functionalized employing various substituted anilines to give schiff's bases (III_{a-h}). IR and ^1H-NMR were used for characterization of all new compounds. Antimicrobial, analgesic, antianxiety activities of the corresponding schiff's bases were assessed.

Keywords: Pyrazole, Schiff's bases, Antimicrobial, Analgesic, Antianxiety.

Introduction

Pyrazole containing heterocyclic compounds play an important role in medicinal chemistry. A perusal of literature survey reveals that pyrazoles possesses outstanding biological and

* Corresponding Author: E-mail: jagtapajit@rediffmail.com

pharmacological activities such as antianxiety (Heald *et al.*), antipyretic, analgesic, anti-inflammatory (Reddy *et al.*, 1999) as well as good antibacterial and antifungal properties (Saharia *et al.*, 1974; Sing *et al.*, 1980). The association of pyrazole nucleus with varied biological and pharmacological activities has aroused considerable interest in the synthesis of new series of title compounds (III_{a-h}) and their screening for possible antimicrobial, anxiolytic and analgesic activities. In recent years several methods have been developed for the synthesis of pyrazole-4-carboxaldehyde. Kira *et al.* have reported the formation of pyrazole-4-carboxaldehyde by treating acetophenone phenylhydrazone with DMF/$POCl_3$ complex (Kira *et al.*, 1969). It was two step reactions in which Vilsmeier-Haack reagent was used for cyclization. Compounds were characterized using IR and ^{1}H-NMR. Antimicrobial, analgesic, antianxiety activities of the corresponding schiff's bases were assessed.

Materials and Methods

Melting points were taken on electro thermal digital melting point apparatus and are uncorrected. Single spotted TLC determined completion of reaction. IR and ^{1}H-NMR confirmed structures of all the synthesized compounds.

IR spectra were recorded using KBR disc on a Jasco FTIR-41 0. ^{1}H-NMR spectra were recorded in $CDCl_3$ solution on FTNMR, Varian mercury 300 MHz and proton chemical shift are relative to tetramethylsilane as internal standard. IR Spectra were recorded at FDA Lab., A.B. College of Pharmacy, Sangli, India and ^{1}H-NMR spectra were recorded at Department of Chemistry, University of Pune, India.

Experimental

1. Synthesis of 4-Chloroacetophenone phenylhydrazone (I).
2. Synthesis of 3-(4-chloroacetophenone)-1-phenyl pyrazole-4-carboxaldehyde (II).
3. Synthesis of aldimine derivatives of 3-(4-chlorophenyl)-1-phenylpyrazole-4-carboxaldehyde (III_{a-h}).

Synthesis of 4-Chloroacetophenone Phenylhydrazone (I)

Concentrated acetic acid (1 ml) and phenylhydrazine hydrochloride (0.028 mol) were added to solution of 4-chloroacetophenone (0.0 19 mol) in 60 ml of ethanol. The reaction mixture was then refluxed for 1 hour. The precipitate was filtered and washed with ethanol. After drying in vacuum 86.89 per cent (5.5 g) of product (I) was obtained.

Synthesis of 3-(4-Chloroacetophenone)-1-Phenyl-1H-pyrazole-4-Carboxaldehy (II)

Vilsmeier-Haack reagent was prepared separately cooled Dimethylformamide (35.3 mmol) and $POCl_3$ (35.5 mmol) at 0°C. A solution of (I) in DMF was added dropwise to the Vilsmeier-Haack reagent, which was then warmed at room temperature and refluxed at 60–70°C for four hours. After cooling at room temperature the mixture was neutralized with a cold saturated K_2CO_3 solution. The precipitate was filtered, strongly washed with water and recrystallized from ethanol yielding 78.75 per cent of (II).

M.P.: 138°C; *IR (KBr) cm*$^{-1}$: 3019.18 (Ar C-H_{str}), 1446.97 (Ar C = C_{str}) 1013.12 (Ar C-H_{def}), 1600.8 (C = N_{str}), 1295.72 (C-N_{str}), 1351.4 (Ph-N_{str}), 1093.1 (C-Cl_{str}), 1677 (C = O_{str}), 2829.82 and 2857.47 (C-H_{str} of –CHO).

NMR ($CDCl_3$)*δ ppm*: 7.30–7.42 (m, 4H, $H_{2,3,5,6}$, 3-Phenyl), 7.45–7.61 (m, 3H, $H_{3,4,5}$, 1-Phenyl), 7.79–7.91 (m, 2H, $H_{2,6}$, 1-Phenyl), 8.58 (s, 1H, H_5, Pyrazole), 10.08 (s, 1H, CHO).

General Procedure for the Preparation of Aldimine Derivatives of 3-(4-Chlorophenyl)-1-1Phenyl-1H-pyrazole-4-Carboxaldehyde (III_{a-h})

Compound (II) (0.5 mmol) was dissolved in boiling ethanol and to it 0.2 ml of glacial acetic acid was added. Further (0.5 mmol) substituted anilines were added in it and reaction mixture was refluxed for 7 hours. After cooling reaction mixture at room tempemture the precipitate of the aldimine derivatives were obtained which further recrystallized in suitable solvents and melting points were taken and data recorded in Table 53.1. Reactions were monitored with TLC.

Table 53.1: Characterization Data of All Newly Synthesized Compounds

Compound No.	*Yield*	*M.P.*	*Color*	*R_f Value*
I	86.89 per cent	87°C	Pale Yellow	0.54
II	78.75 per cent	138°C	Pale Yellow	0.203
III_a	75.75 per cent	150°C	Brown	0.60
III_b	58.82 per cent	142°C	Cream	0.78
III_c	69.86 per cent	106°C	Brown	0.81
III_d	87.59 per cent	94°C	Green	0.66
III_e	79.72 per cent	86°C	Brown	0.71
III_f	97.18 per cent	96°C	Orange	0.65
III_g	83.81 per cent	108°C	Yellow	0.83
III_h	90.14 per cent	102°C	Green	0.88

N-{[3-(4-Chlorophenyl)-1-phenyl-1H-pyrazole-4-yl]methylene}-N-(4-methylphenyl amine (III_a)

M.P.: 150°C; *IR (KBr) cm^{-1}*: 3019.18 (Ar $C\text{-}H_{str}$), 1407.7 (Ar $C=C_{str}$), 1012.45 (ArC-H_{def}, 1595.81 ($C=N_{str}$), 1226.51 ($C\text{-}N_{str}$), 1091.5 ($C\text{-}Cl_{str}$).

NMR ($CDCl_3$)δ ppm: 7.341 (d, 2H, $H_{2,6}$, N-Phenyl), 7.095 (d, 2H, $H_{3,5}$, N-Phenyl), 7.208–7.233 (m, 4H, $H_{2,3,5,6'}$ 3-Phenyl), 7.359–7.610 (m, 3H, $H_{3,4,5'}$ 1-Phenyl), 7.654–7.724 (m, 2H, $H_{2,6'}$ 1-Phenyl), 2.209 (s, 3H, methyl grp. of N-Phenyl), 8.312 (s, 1H, H_5, Pyrazole), 8.369 (s, 1H, CH = N).

N-(2-chlorophenyl)-N-{[3-(4-Chlorophenyl)-1-phenyl-1H-pyrazole-4-yl] methylene} amine (III_e)

M.P.: 86°C; *IR (KBr) cm^{-1}*: 3019.18 (Ar $C\text{-}H_{str}$), 1499.38 (Ar $C=C_{str}$), 1596.77 ($C=N_{str}$), 1214.93 ($C\text{-}N_{str}$), 1090.55 ($C\text{-}Cl_{str}$).

NMR ($CDCl_3$)δ ppm: 7.276 (d, 1H, H_6, N-Phenyl), 7.106 (d, 3H, $H_{3,4'5'}$ N-Phenyl), 7.201–7.233 (m, 4H, $H_{2,3,5,6'}$ 3-Phenyl), 7.350–7.491 (m, 3H, $H_{3,4,5'}$ 1-Phenyl), 7.656–7.792 (m, 2H, $H_{2,6'}$ 1-Phenyl), 8.504 (s, 1H, H_5, Pyrazole), 8.979 (s, 1H, CH = N).

N-{[3-(4-Chlorophenyl)-1-phenyl-1H-pyrazole-4-yl]methylene}-N-(2-nitrophenyl)amine (III_f)

M.P.: 96°C; *IR (KBr) cm^{-1}*: 3019.18 (Ar $C\text{-}H_{str}$), 1013.41 (Ar $C\text{-}H_{def}$, 1598.70 ($C=N_{str}$), 1226.51 ($C\text{-}N_{str}$), 1345.11 ($Ph\text{-}N_{str}$), 1092.48 ($C\text{-}Cl_{str}$), 1506.13 (Ar $N=O_{str}$)

NMR ($CDCl_3$)δ ppm: 8.214 (m, 4H, $H_{3,4,6'}$ N-Phenyl), 7.241–7.252 (m, 4H, $H_{2,3,5,6'}$ 3-Phenyl), 7.480–7.752 (m, 5H, $H_{2,3,4,5,6'}$ 1-Phenyl), 8.329 (s, 1 H, H_5, Pyrazole), 8.464 (s, 1H, CH = N).

N-{[3-(4-Chlorophenyl)-1-phenyl-1H-pyrazole-4-yl]methylene}-N-(3-nitrophenyl amine (III_g)

M.P.: 108°C; *IR (KBr) cm^{-1}*: 3019.18 (Ar $C\text{-}H_{str}$), 1013.41 (Ar $C\text{-}H_{def}$), 1598.70 ($C=N_{str}$), 1226.51 ($C\text{-}N_{str}$), 1345.11 ($Ph\text{-}N_{str}$), 1092.48 ($C\text{-}Cl_{str}$), 1506.13 (Ar $N=O_{str}$)

Reaction Scheme

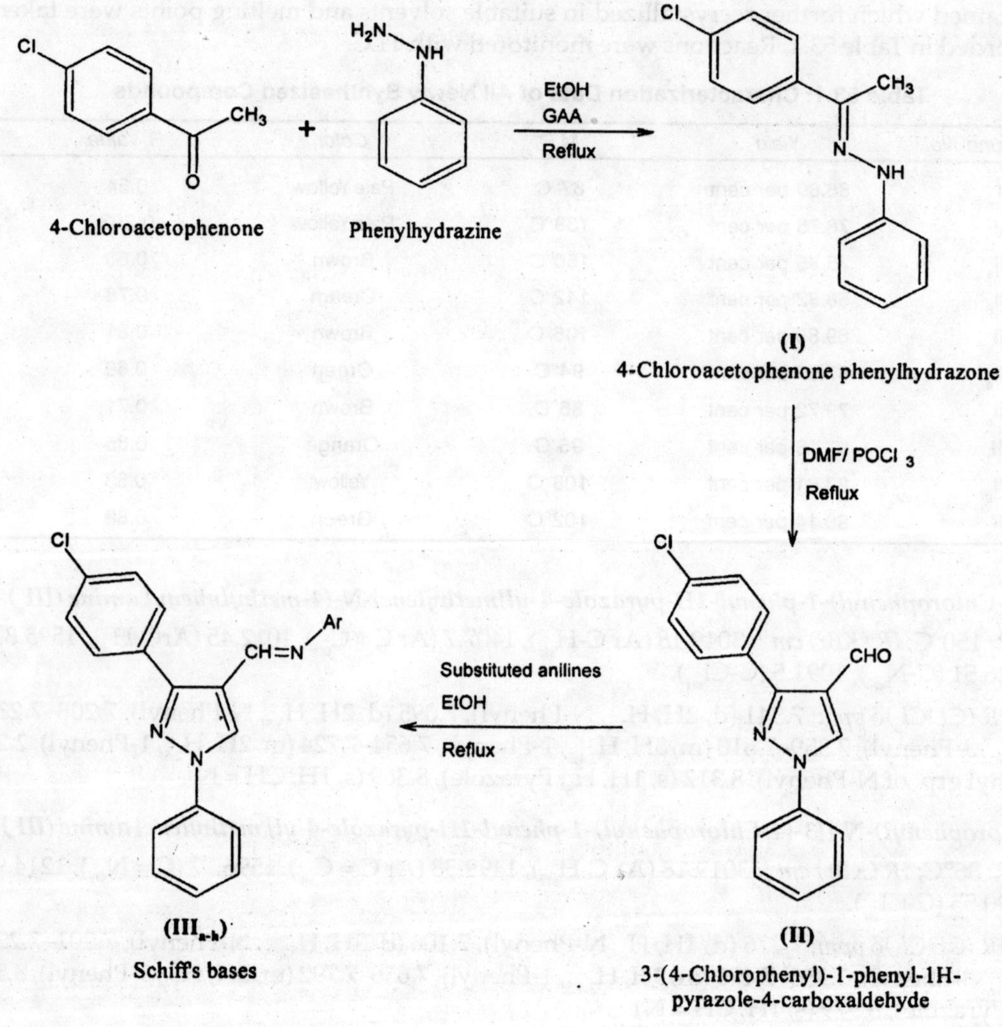

Compound	Ar
III_a	4-Methylphenyl
III_b	2,4-Dimethylphenyl
III_c	2,6-Dimethylphenyl
III_d	4-Methoxyphenyl
III_e	2-Chlorophenyl
III_f	2-Nitrophenyl
III_g	3-Nitrophenyl
III_h	4-Nitrophenyl

NMR ($CDCl_3$)δ ppm: 8.214 (m, 4H, $H_{3,4,6}$, N-Phenyl), 7.241–7.252 (m, 4H, $H_{2,3,5,6}$, 3-Phenyl), 7.480–7.752 (m, 5H, $H_{2,3,4,5,6}$, 1-Phenyl), 8.329 (s, 1H, H_5, Pyrazole), 8.464 (s, 1H, CH = N)

N-{[3-(4-Chlorophenyl)-1-phenyl-1H-pyrazole-4-yl]methylene}-N-(4-nitrophenyl) amine (III_h)

M.P.: 102°C; *IR (KBr) cm^{-1}*: 3019.18 (Ar C-H_{str}), 1013.41 (Ar C-H_{def}), 1598.70 (C = N_{str}), 1226.51 (C-N_{str}), 1345.11 (Ph-N_{str}), 1092.48 (C-Cl_{str}), 1506.13 (ArN = O_{str})

NMR ($CDCl_3$)δ ppm: 8.214 (m, 4H, $H_{3,4,6}$, N-Phenyl), 7.241–7.252 (m, 4H, $_{H2,3,5,6}$, 3-Phenyl), 7.480-7.752 (m, 5H, $H_{2,3,4,5,6}$, 1-Phenyl), 8.329 (s, 1H, H_5, Pyrazole), 8.464 (s, 1H, CH = N).

Pharmacological Screening

Antimicrobial Activity

The filter paper disc method (Robert, 1970; Baller *et al.*, 1960) was performed in nutrient agar for bacterial and saboraud's agar for fungi these agar media were inoculated with 0.5 ml of the 24 hour liquid culture. Filter paper disc (5 mm diameter) saturated with each compound solution (500 µg/ml of DMSO) were placed on indicated agar media the incubation time was 48 hour at 36°C. Amoxicillin (10 mg/ml of DMSO) was used as standard against bacteria and clotrimazole (10 mg/ml of DMSO) was used as standard against fungi. The diameter of inhibition zone in mm were measured and recorded in Tables 53.2 and 53.3.

Table 53.2: Antibacterial Activity of Schiff's Bases

Name of Organism	*Name of Compounds*								*Standard*
	III_a	III_b	III_c	III_d	III_e	III_f	III_g	III_h	
B. substilis	++	+++	+	++	–	–	+++	+++	+++
P. vulgaris	+	–	–	++	–	++	–	–	++
S. aureus	++	+	++	–	–	–	+++	+++	+++
E. coli	–	+	+	–	–	++	++	+	+++
Pseudomonas aeruginosa	–	+	++	–	–	–	+	+	++

Standard: Amoxycillin (10 mg/1 ml of DMSO).

Table 53.3: Antifungal Activity of Schiff's Bases

Name of Organism	*Name of Compounds*								*Standard*
	III_a	III_b	III_c	III_d	III_e	III_f	III_g	III_h	
C. albicans	+	+	–	–	–	–	–	–	+++

Standard: Clotrimazole (10 mg/l ml of DMSO)

Key to symbols:

Highlyactive: +++ (inhibition zone > 18 mm)

Moderately active: ++ (inhibition zone 12–18 mm)

Slightly active: + (inhibition zone 6–12 mm)

Inactive: – (inhibition zone < 6 mm)

Antianxiety Activity

Male Swiss mice (20–25 gm), in a group of six, were used throughout experiments. Mice were housed in plastic cages and kept in 30m² room with controlled 12-hour light/dark cycle, and at constant temperature (22 ± 2°C). As drug is completely soluble in DMSO, DMSO is taken as a control vehicle. The test drug was used in concentration of 10 mg/kg in IP dose.

Elevated Plus Method

The apparatus similar to that described by Pellow *et al.* was used (Pellow *et al.*, 1985). This apparatus was elevated above a table (38 cm above the floor) and consisted of two arms that intersect to form the shape of plus sign. The sides of one arm (30 × 5cm) were closed and the sides of the other arm were opened (30 × 5 × 15 cm). An animals is placed at the intersection of the arms (5 × 5 cm) faced to closed arm or open arm, and the amount of time it chose to spend in each arm was measured. The number of entries and the time spent in the open and closed arms were recorded during a 5-min test period as described (Lister, 1987). The percentage of open arm entries (100 × open/total entries) was calculated for each animals (Pellow *et al.*, 1986). Diazepam (1 mg/kg, I.P.) used as positive control. Observations were recorded in observation (Table 53.4).

Table 53.4: Effect of Schiff's Bases on Mice in Elevated Plus Method

Group Compound	Dose	No. of Entries (n)		Time Spent (Sec.)	
		In Open Arm	In Closed Arm	In Open Arm	In Closed Arm
DMSO (Control)	10 ml/kg	6.9±0.6	8.9±0.4	95.5±1.6	203.1±1.6
Diazepam (Std.)	1 mg/kg	11.8±06	5.7±0.3	156.8±2.2	140.0±3.2
III_a	10 mg/kg	8.8±0.5	8.0±0.5	166.6±2.4	136.5±2.9
III_b	10 mg/kg	7.5±0.4	8.2±0.4	125.6±2.1	176.5±2.8
III_c	10 mg/kg	11.7±0.7	9.4±0.6	143.3±3.7	155.5±1.8
III_d	10 mg/kg	7.2±0.3	6.9±0.4	115.5±2.6	185.6±1.6
III_e	10 mg/kg	10.4±0.6	9.1±0.3	121.8±2.1	168.3±1.7
III_f	10 mg/kg	12.5±0.7	7.6±0.4	168.1±2.8*	130.0±1.9
III_g	10 mg/kg	12.4±0.6	7.0±0.5	180.1±3.2*	127.3±1.4
III_h	10 mg/kg	9.2±0.4	9.3±0.4	147.6±3.2	152.0±3.2

Values are reported as mean ± S.E.M. for the number of mice shown in parentheses.

*: $P < 0.001$ verses control.

(Analysis of variance and Dunnets test as the post-hoc test).

The Light and Dark Exploration Test

This model is based on natural aversion of mice to brightly light places (Crawley, 1981). The apparatus consist of two compartment box, one dark and the other brightly light. A typical apparatus consist of a wooden box (45 × 27 × 27 cm). The box is open topped and dimly illuminated (10W white bulb). Mice were placed individually in the centre of the light compartment and observed for the next 5 minutes for the time spent in the light and dark compartment. Observations were recorded in observation (Table 53.5).

Table 53.5: Effect of Schiff's Bases on Mice in Light and Exploration Test

Group Compound	*Dose*	*No. of entries (n)*	*Time Spent (Sec.)*	
		Light	*Light*	*Dark*
DMSO (Control)	10 ml/kg	3.1±0.4	51.0±1.9	248.0±2.3
Diazepam (Std.)	1 mg/kg	5.1±0.4	137.8±1.6	162.6±2.3
III_a	10 mg/kg	4.1±0.4	105.5±1.9	196.0±1.5
III_b	10 mg/kg	3.7±0.4	69.3±2.7	230.8±4.6
III_c	10 mg/kg	4.1±0.3	110.0±2.8	193.0±2.1
III_d	10 mg/kg	3.5±0.4	75.6±2.7	222.6±3.1
III_e	10 mg/kg	4.1±0.3	95.5±1.4	204.5±2.0
III_f	10 mg/kg	4.5±0.4	115.3±1.5*	182.5±3.2
III_g	10 mg/kg	4.3±0.5	114.3±3.1*	185.6±2.3
III_h	10 mg/kg	3.8±0.3	104.3±3.1	195.6±2.1

Values are reported as mean ± SEM for the number of mice shown in parentheses.

*: P < 0.05 verses control a P < 0.001 verses control.

(Analysis of variance and Dunnets test as the post-hoc test).

Statistical Analysis

All results are presented as mean ± S.E.M. analysis of variance was followed by dunnets test as the post-hoc test. Results were considered significant at $P < 0.01$.

Analgesic Activity

Writhing Test

All the compounds were screened using the method of Ghosh (Ghosh, 1981). Albino mice (20–25 g) of either sex were used for the study. Percentage protection exhibited by the test compounds administered at a dose of 20 mg/kg in DMSO solution IP against the acetic acid (0.6 per cent w/w) induced writhing or stretching syndrome was recorded. Indomethacin (10 mg/kg body weight) was employed as reference standard under similar conditions. Observations were recorded in Table 53.6.

Results and Discussion

The novel Schiff's bases *i.e.* aldimine derivatives of 3-(4-chloroacetophenone)-1-phenyl-1H-pyrazole-4-carboxaldehyde were synthesized and confirmed by their IR, ^{1}H-NMR spectra's and melting points.

All Schiff's bases, except compound III_e, exhibited considerable antibacterial activities against the tested organisms. In some cases their activities are equal to standard substances. Compounds III_g and III_h showed high antibacterial activity against both of the gram positive species (*Bacillus substilis* and *Staphylococcus aureus*) of bacteria used for screening, which is comparable with standard used. However only two compounds *i.e.* compound III_a and III_b showed slight activity against fungal species (*Candida albicans*).

Acute toxicity studies revealed that all the compounds were non-toxic up to the dose as high as 50 mg/kg body weight.

Table 53.6: Analgesic Activity of Compounds

Sl.No.	*Compound*	*Dose (mg/kg)*	*Number of Writhing (±S.E.M.)*	*Per cent Decrease in Writhing*
1.	Control (Saline)	10	30.5±0.5627	–
2.	III_a	20	10.33±0.5578	66.13
3.	III_b	20	9.16±0.8333	69.96
4.	III_c	20	7.66±1.2820	74.88
5.	III_d	20	9.5±0.5627	68.85
6.	III_e	20	11.5±0.4282	62.29
7.	III_f	20	8.5±0.5000	72.13
8.	III_g	20	7.833±0.7032	74.31
9.	III_h	20	8.166±0.3073	73.22
10.	(Indomethcin) Standard	10	1.833±6.0000	93.99

Values are reported as mean ± S.E.M. for the number of mice shown in parentheses.

*: $P < 0.01$ verses control.

(Analysis of variance).

In elevated plus maze method, I.P. administration of Schiff's bases increases number of entries and time spent in open arm. Whereas compound III_g and III_f increases the number of entries in open arm by 79.7 per cent (12.4 ± 1.9) and 81.1 per cent (12.5 ± 2.4) respectively as compared with control DMSO (6.9 ± 0.5). Also compound III_g and III_f showed significant increase in time spent by 89.4 per cent (180.1 ± 12.9) and 75.9 per cent (168.1 ± 12.8) respectively as compared with control DMSO (95 ± 2.1). The results are significant at {$F(9,53) = 4.43$, $P < 0.01$}.

In Light and Dark Exploration method, compound III_g and III_f increases the number of entries in light compartment by 41.9 per cent (4.4 ± 0.2) and 45.10/0 (4.5 ± 0.8) respectively as compared with control DMSO (3.1 ± 0.5). Compound III_g and III_f showed significant increase in time spent in light compartment by 124.1 per cent (114.3 ± 5.4) and 126.2 per cent (115.4 ± 9.6) as compared with control DMSO (51 ± 2.6). Hence compound III_g and compound III_f showed significant anxiolytic activity {$F(9, 53) = 3.24$, $P < 0.01$}.

Compounds III_c, III_f, III_g and III_h showed maximum analgesic activity, while compounds III_a, III_b, III_d and III_e showed minimum activity under the conditions of this experiment.

From the results obtained in pharmacological screening and structures of the compound we can correlate in general that Schiffs' bases (Compound III_c, III_e and III_f) having substitutions on 2nd and 6th position of phenyl ring attached to nitrogen of aldimine group doesn't showed good activity. The Schiffs' bases (Compound III_g and III_h) substituted with Nitro group at position 3rd and 4th of phenyl ring attached to the nitrogen of aldimine group showed very good antimicrobial, antianxiety, analgesic as well as good antitubercular activities. Similarly the Compound III_b prepared from 2,6-dimethylaniline also showed good antimicrobial, antianxiety, analgesic as well as good antitubercular activities.

From this we conclude that substitutions at *ortho* positions of phenyl ring attached to nitrogen of aldimine group in Schiffs' base may affect the activity of the synthesized Schiffs' base.

References

Bailer, A.W., Robert, C.E.R. and Kirky, W.M.M., 1960. *Antibiotica*, 3rd edn. Prentice Hall Inc., New Delhi, p. 574.

Crawley, J.N. and Goodwin, F.K., 1981. Neuropharmacologic specificity of a simple animal model for the behavioural actions of benzodiazepines. *Pharmcol. Biochem. Behav.*, 15: 695–699.

Ghosh, M.N., 1981. *Fundamentals of Experimental Pharmacology*. Scientific Book Agency, Calcutta, p. 153.

Heald, A.F. and Wildonger R.A.L., 1983,98. *US Pat.*, 4, 364, 348; *C.A.*, 1983,98, 89358.

Kira, M.A., Abdel-Rehman, M.O. and Gadalla, K.Z., 1969. Vilsmeier Haack reaction-III cyclization of hydrazones to pyrazoles. *Tetrahedron Lett.*, p. 109.

Lister, R.G., 1987. The effects of repeated doses of ethanol on exploration and its habituation. *Psychopharmacol.*, 92: 78–83.

Pellow, S. and File, S.E., 1986. Anxiolytic anxiogenic drug effects on exploratory activity in an elevated plus maze; a nobel test of anxiety in the rat. *Pharmacol. Biochem. Behav.*, 24: 525–529.

Pellow, S., Chopin, P., File, S.F. and Briley, M., 1985. Validation of open: Closed arm entries in an elevated plus maze as a measure of anxiety in rats. *J. Neurosci. Methods*, 14: 149–167.

Reddy, Y.S.R., Sosamma Mani, T., Rama Sarma, G.V.S. and Suresh, B., 1999. Synthesis and pharmacological activity of substituted pyrazoles. *I. J. Pharma. Sci.*, 61(1): 25–28.

Robert, C., 1970. *Medicinal Microbiology*, 11th edn. ELBS, London, pp. 895 and 901.

Saharia, G.S. and Sharma, H.R., 1974. Studies in Heterocyclic compounds: Part I. Synthesis of 3-(p-Methoxyphenyl)-5-phenyl-4-(substituted sulphonamid benzeneazo) pyrazoles and evaluation of their antibacterial properties. *J. Indian Chem. Soc.*, LI: 351–353.

Sing, C.P. and Ojha, A.C., 198. Synthesis of N-isonicotinyl-3-methyl-4-(substituted hydrazone)-2-pyrazolin-5-one, *J. Indian Chem. Soc.*, LVII(12): 1172.

Chapter 54

Quality Improvement through Second-Order Slope-Rotatable Designs Over All Directions

G.V.S.R. Anjaneyulu[1], P. Venkateswara Rao[2] and V.L. Narasimham[1]

[1]Department of Statistics, Acharya Nagarjuna University, Nagarjuna Nagar, Guntur – 522 510, Andhra Pradesh, India

[2]Christu Jayanthi Jubilee P.G. College, Guntur –522 006, Andhra Pradesh, India

ABSTRACT

Recent years have been an upsurge of interest in the role of statistical ideas and methods in improving the quality and productivity of industrial processes and products. Response surface methodology is, one such idea, useful for analysing problems where several independent variables influence a dependent variable or response. The earlier study of response surface designs mainly emphasized the estimation of absolute response. Estimation of differences in response at different points in the factor space will often be of great importance. If differences at points closed together are involved, estimation of local slope *i.e.,* the rates of change of the response surface is of interest. This problem, estimation of slopes, occurs frequently in practical situations, particularly in fitting second order response surface. This enable the determination of the best operating conditions for the process *i.e.,* the best combination of the levels of the controllable factors which gives the "optimum" value of the second order response function. This would also enable us to determine the best way to control the process. In this article, we made an attempt to study the role of second order slope rotatability to improve the quality of a product. Anjaneyulu *et al.* (1993) introduced embedding in Second-Order Slope-Rotatable Designs and constructed the same using embedding techniques similar to those of Draper (1960), Herzberg (1967), Park (1987) studied the necessary and sufficient conditions for second order slope rotatability over all directions, Anjaneyulu *et al.* (1997) showed that these designs have the variance sum property. In this article an attempt is made to explain the role of embedding in SOSRDOAD for Quality Improvement.

Keywords: *Quality improvement, Response surface methodology, Second order slope rotatability, Second-order-slope-rotatable designs over all directions, Second order rotatable designs, Embedded designs.*

Introduction

Response Surface Methodology is useful for analysing problems where several independent variables influence a dependent variable or response. The RSM enables the determination of the best operating conditions for the process, *i.e.*, the best combination of the levels of the controllable factors which in some sense give the "optimum" value of the response function. The choice of combinations of levels of the independent variables which will enable an experiments to approximate a functional relationship by fitting a duty order polynomial, by the method of least squares. Emphasis was placed on judging a design on the basis of "predition variance". A natural and easily attainable property was that of rotatability, which requires that the variance of a predicted value remain constant at pints that are equidistant from the design center–"stability in prediction variance" (Box *et al.*, 1978).

Data analysis in general are expanding. Japan scientists, who have successfully communicated to us the need for experimental design at the research-and-development level as successful supplement to quality control and process control. The Japanese industrial engineer Genichi Taguchi proposed methodology for product improvement. Here, focus is an "improvement" and not on "optimization" (Kackar, 1985). Taguchi made users aware of the need to include system or process variability as part of the response. As a result the inevitable "nonconstancy of variance is not ignored". Defining "Information" at any point as the reciprocal of the variance at that point, we must ideally then, have experimental designs which provide information contours that would be circles for two variables, spheres or hyperspheres for three or more variables, so that at "constant" distance from the center of the design, the experimenter will gather a constant quantity of information. Such designs are termed as Rotatable Designs. These have the property that whenever the orthogonal axes of the design take any orientation (*i.e.*, of the axes are rotated about the center) of the confidence in a prediction made on the response at any given point will not be altered (Raja and Yayin, 1993).

In many application of RSM, good estimation of the derivatives of the response function may be as important or perhaps more important than estimation of mean response. Certainly, the computation of a stationery point in a second-order analysis depends heavily on the partial derivatives of the estimated response surface function with respect to the design variables. The earlier study of response surface designs mainly emphasized the estimation of absolute response.

Estimation of differences in response at different points in the factor space will often be of great importance. If differences at points close together are involved, estimation of local slope *i.e.*, the rates of change of the response surface are of interest. This problem, estimation of slopes, occurs frequently in practical situations, particularly in fitting second order response surface. Hader and Park (1978) introduced the ideas of slope rotatability and discussed them in the context of central composite designs Park (1978) introduced Second-Order Slope-Rotatability Over All Directions (SROAD) Anjaneyulu *et al.* (1997) established that these designs have the additional property that the sum of the variances of estimates of slopes in all directions at any point is a function of the distance of the point from the design origin.

Draper (1960) gave a method of embedding in Second-Order-Rotatable design of v factors from a know (v-1) factor Second-Order-Rotatable design Herzberg (1967) and Huda (1982) introduced some other methods of embedding in Second-Order-Rotatable design. Hader and Park (1978) studied slope rotatability in central composite designs. Anjaneyulu *et al.* (1993) contracted v factor Second-Order-Slope Rotatable design from known (v-1) factor Second-Order-Slope-Rotatable design using embedding techniques similar to those of Draper (1960) and Herzberg (1967).

Park (1987) studied the necessary and sufficient conditions for second order slope rotatability over all directions. Anjaneyulu *et al.* (1997) showed that these designs have the additional property that the sum of the variances of estimates of slopes in all directions at any point is a function of the distance of the point from the design origin. In this section, we discuss the constructions of v factor Second-Order-Slope-Rotatable design over all direction design from known (v-1) factor Second-Order-Slope-Rotatable design Over All Directions.

Embedding in Second-Order-Slope-Rotatable Designs Over All Directions

Definition of SOSRDOAD

A second order response surface design D is said to be a SOSROAD, if it satisfied the following moment [c.f. Park (1987)].

1. All Odd-Order moments
   i.e. Σx_i, $\Sigma x_i x_j$, Σx_i^3, $\Sigma x_j x_j^2$, $\Sigma x_i^2 x_j$, $\Sigma x_i x_j^3$, $\Sigma x_i x_j x_k x_l$, etc., are become zero, $i \neq j \neq k \neq 1$.
2. Σx_i^2, are equal for all i.
3. Σx_i^4, are equal for all i
4. $\Sigma x_i^2 x_j^2$, are equal for all $i \neq j$ ......(1)

Variance-sum SOSRDOAD

Anjaneyulu *et al.* (1997) established that SOSRDOAD has the additional interesting property that the sum of the variances of estimates of slopes in all axial directions at any pint is a function of the distance of the point from the design origin.

i.e. $$\sum_{i=1}^{v} V\left[\frac{\partial \hat{V}}{dx_i}\right] = f(p^2), \text{ where } p^2 = \sum_{i=1}^{v} x^2_{ui}$$

They also observed that any symmetric second order response surface design is a SOSRDOAD. Hence every SOSRDOAD has the additional property that the trace of the dispersion matrix of estimates of slopes in all directions has a type of variance-sum second order slope rotatability property. Hence every SOSRDOAD is a variance-sum SOSRD on axial directions and conversely.

Embedded designs may be particularly useful when the experimenter feels the necessity to introduce additional factor(s) in the known experiment without discarding the original results. Very particular in agriculture field experiments, quality improvement programmes these are highly useful. With this technique repeating the same experiment for high number of factors does not arise. Hence time and cost of expenditure will be saved. In view of this practical importance many authors developed several kinds of embedded designs.

The second order rotatable designs which are a subset of second order slope rotatable design over all directions are restrictive. At the same time second order slope rotatable design over axial direction is also highly restrictive when compared SOSRDOAD. So it is desirable to have SOSRDSOAD for all practical purposes. Hence an attempt is made to construct SOSRDOAD in this article.

Draper (1960) presented a method for construction a SORD in v dimensions from a known SORD of (v-1) dimensions. Draper's method mainly be applied to some restrictive classes of given designs. In this, introducing the se of vth factor design pints is not in the hands of the experimenter. Herzberg (1967) presented another method which works in all cases and which does not need to discard the results of the experiments that may have been performed in accordance with the (v-1) dimensional design. In many situations, it may be necessary to extend a (v-1) dimensional Second-Order-Rotatable design into v-dimensional SORD, for some $1 \in [1, \ldots\ldots(k-1)]$. Draper's method and Herzberg's methods of embedding could be applied successively many times. For large values of v and L, the design derived by the successive application of Herzberg's method is likely to require more experiments. Huda (1982) suggested another method of embedding and it works if a suitable balanced incomplete block design with a small number of blocks exists.

Anjaneyulu *et al.* (1993) complicated v factors Second-Order-Slope-Rotatable design over axial directions from know (v-1) factor Second-Order-Slope-Rotatable design over axial direction using embedding techniques similar to those of Draper (1960) and Herzberg (1967). They found that even in the case of construction of SOSRD over axial direction Herzberg method of embedding techniques always succeeds. However Draper's method requires less number of design pints, even though its success is conditional.

Construction of Embedded SOSRDOAD

If the experimenter feels to improve the quality of a product by adding one or more desirable factor (s) to the already existing SOSRDOAD experimental set-up keeping the earlier results as it is, he can feel free to choose any factor or any number of factors as per the customer's specifications. If it is an agriculture experiment or it can be a food experiment, the desired factors can be added to make the considered (v-1) or (v-1) SORDOARD as v dimensional SOSRDOAD, which satisfied the moment conditions given in (2.1). Unlike in other methods of embedding, here, in the construction of embedded SOSRD(OAD) no extra selective or restrictive point sets needed except the required and suitable factor (s). The total number of design pints in the embedded SOSRDOAD are considerably less when compared embedded SORD or embedded SOSRD over axial direction. The embedding procedure is also quite simple. The procedure is as follows.

Consider (v-1) or (v-1) dimensional ASOSRDOAD. Add the vth factor or 1 factors to the considered design, in such way that which satisfies the moment conditions given in equation (1). Therefore v-dimensional design is a SOSRDOAD. Suppose the considered design is SORD in lower dimension but the experimenter requires SOSRDOAD of higher dimensions without discarding the earlier model and results, then apply embedding techniques proposed by various authors, which suits to the model, the embedded higher order SORD, itself is the desired SOSRDOAD of higher dimensions.

Similarly the considered design is SOSRD over axial directions. Here also the experimenter feels not to loose the earlier model of the experiment and not to discard the results, but feels to have a SOSRDOAD, then it can be achieved by variance-sum property introduced by Anjaneyulu *et al.* (1997). For this append the vth factor by embedding techniques given in Anjaneyulu *et al.* (1993) and the sum of the variances of the slopes in all axial directions at any point is a function of the distance of the point from the design origin, which is the v factor SOSRDOAD. Embedding in SOSRDOAD is always less restrictive and unconditional. In this, always, the total number of design pints are less.

References

Anjaneyulu, G.V.S.R., Dattatreya Rao, A.V. and Narasimham, V.L.,1993. Embedding in second-order-slope-rotatable designs. *Reports of Statistical Application Research, JUCE*, 40(4): 1–11.

Anjaneylu, G.V.S.R., Varma, D.N. and Narasimham, V.L., 1997. A note in second-order-slope-rotatable designs over all directions. *Communications in Statistics: Theory and Methods*, 26(6): 1477– 1479.

Box, G.E.P., Hunter, W.G., and Hunter J.S., 1978. *Statistics for Experiments*. Wiley, New York.

Box, George and Jones, Stephen, 1992. Designing Products that are robust to the environment. *Total Quality Management*, 3(3) 265–283.

Draper, N.R., 1960a. Second-order-rotatable-designs in four or more dimensions. *Ann. Math. Stat.*, 31: 23–33.

Draper, N.R. and Ying, L.H., 1994. A note on slope rotatability over all directions. *Journal of Statistical Planning and Inference*, 41: 113–119.

Hader, R.J. and Park, S.H., 1978. Slope rotatable central composite designs. *Technometrics*, 20: 413–417.

Herzberg, A.M., 1967. A method for the construction of second order rotatable designs in k-dimensions. *Ann. Math. Stat.*, 38: 177–180.

Huda, S., 1982. A method for constructing second-order-rotatable desig. *Cal. Stat. Assn. Bull.*, 30: 139–144.

Kacker, R.N., 1985. Off-line quality control, parameter design and the Taguchi method. *Journal of Quality Technology*, 17(4): 176–209.

Park, S.H., 1987. A class of multifactor designs for estimating the slope of response surfaces. *Technometrics*, 29(4): 449–453.

Raja, A. and Yayin, I., 1993. Application of response surface methodology for the maximization of concora crush resistance of paper board. *Quality Engineering*, 6(1): 1–20.

Victor Babu, B.Re. and Narasimham, V.L., 1991. Construction of second-order-slope-rotatable designs through balanced incomplete block designs. *Coffin. Stat. Theory and Methods*, 20(8): 2467-2478.

Chapter 55

FT-IR Spectroscopic Analysis of Some Archaeological Pottery Shreds Recently Excavated at Adichanallur in Tamil Nadu

G. Velraj and M. Vigneswari
Department of Physics, Periyar University, Salem – 636 011, Tamil Nadu

ABSTRACT

The Archaeological potteries excavated recently in Tamil Nadu from the archaeological site at Adichanallur in Tuticorin District, were analysed using FT-IR spectra in the mid range 4000–400 cm^{-1} to identify the mineral composition of the clay of the potteries. The clay minerals were identified from the tentative vibrational assignments or the absorptions observed in the spectra of all the samples of interest for the present study.

Keywords: FT-IR, Archaeological pottery, Clay minerals.

Introduction

The importance of the clay and its widespread use by ancient men arises from its possibility to mould moistured clay into useful shapes. The more important aspect is the shape which is being retained on drying and firing afterwards. On drying clays they have become harder and the volume shrinks. The decrease in volume is approximately proportional to the water lost. Further shrinkage takes place on firing together with other physical and chemical changes and then the fired clay products resulted in increased strength (Jack and Gillot, 1968).

Potteries are made up of clay materials, which include primary, secondary accessory minerals and amorphous materials. Clay minerals are made up of chains of SiO_4 tetrahedral and AlO_6 octahedral, which are the basic structural units of silicates. The silicons are tetrahedrally surrounded by oxygen atoms and hydroxyl groups to from a hexagonal network which is repeated infinitely. The tetrahedral are arranged, so that their apexes point in the same direction and their bases are in the same plane. On the other hand the oxygen and the hydroxyls on the free apexes are coordinated to Al ions forming octahedral sheets. Accessory minerals generally are the various iron oxides and oxyhydroxides. The are often dispersed as finely divided particles or as surface coatings. Minute quantities of these minerals can influence the morphological and mineralogical potteries of the clay minerals (Halavay *et al.*, 1977).

Potteries are made up of baked clay minerals (Keller and Dicket, 1949; Venkatachalapathy *et al.*, 2001). The clay minerals commonly found in the pottery samples and their phase transformation, during firing have been studied by using a number of techniques. The structural composition of the clay in the normal and its transformed state is also identified from the changes observed in the characteristic absorption band frequencies.

Materials and Methods

The Infrared spectra oft he pottery samples were recorded in the mid region 4000–400 cm^{-1} using Perkin-Elmer Spectrometer with the ±4 cm^{-1} resolution. The sample prepared using KBr pellet technique. In this spectrometer the Be-Ne Laser source is used. The spectrometer recorded the frequency values to the second decimal place accuracy. Each sample is mixed with spectra grade KBr in powder form and finely grounded using Agate mortar, at the ratio 1 : 20 by weight. The samples of 2 mg mixed with 40 mg of spectra grade KBr. The pellets were prepared at a pressure of 8 atmospheres (Russell, 1987). The FT-IR spectra were recorded for each sample in the set mode after attaining consistency in the appearance of the spectra on the monitor on the consecutive runs.

The FT-IR spectra of the archaeological samples collected from different depths were recorded in the received state is shown in Figure 55.1. Except the lower most spectrum, remaining spectra were displaced vertically. The position of the characteristic absorption bands with the tentative assignments and visually estimated relative intensities are given in Table 55.1

Results and Discussion

The clay minerals present in the pottery can be characterized by the general formula of the type (Ramasamy *et al.*, 1991; Prasad, 1965). $O_6Si_4\ [O_4(OH)_2]\ [(AlFe)^{3+}{}_{4-2n}\ X_n{}^{2+}]\ (OH)_{6-4n}$. Where n-represent the degree of order. In the case of well ordered Kaolinite n = 0. For large value of n (*i.e.*) n = 1 the structure becomes more disordered resulting in Illite structure.

The changes in the structure of the clay are occurred due to the presence of Fe, Mg and other metal substituents in the octahedral lattice. Basically typical clay mineral has the following composition. Al_2O_3 (19.7 per cent); CaO (11.8 per cent); Fe_2O_3 (7.3 per cent) and MgO (2.4 per cent) with traces of TiO_2, MnO and NiO.

The Infrared spectra were recorded for the five samples named as ADR1, ADR2, ADR3, ADR4 and ADRS. They were collected from the different depths at Adichanallur. It is well known that the infrared absorption bands in the region 3800–3600 cm^{-1} is the spectra of clay minerals are due to the stretching vibrations of the free hydroxyl groups present in them (Ramasamy *et al.*, 1991). In this study the sample ADR1 show the weak and medium absorption bands at 3802 cm^{-1} and 3749 cm^{-1} respectively, it may be due to the presence of stretching vibrations of free hydroxyl groups. The band at 3440 cm^{-1}

along with weak band at 1640 cm^{-1} due to adsorbed water (Sankaran and Ramasamy, 2000). All the five samples ADR1 to ADR5 show very strong absorption bands in the range 3423–3443 cm^{-1} along with weak and medium absorption bands in the range 1640–1646 cm^{-1}, which is may be due to adsorbed water. The absorption bands at 1420 cm^{-1}, 1470 cm^{-1} and 1483 cm^{-1} are analogous to carbonate groups fined in the clay sample (Eissa *et al.*, 1979). Except ADR2 all the other samples show absorption band around at 1457 cm^{-1}, which is attributed to the carbonate groups present in them. The presence of carbonate is shown by the weak band at 1400 cm^{-1} (Ghosh, 1978). The samples ADR1, ADR3 and ADR4 show a very weak band around 1397 cm^{-1} indicates the presence of carbonates. Ghosh (1978) had indicated that the absorption band observed around 1040 cm^{-1}is due to the position of silicate band.

Table 55.1 FT-IR Adsorption Frequencies (cm^{-1}) and Relative Intensities of Archaeological Pottery Samples with Tentative Vibrational Assignment

ADR1	*ADR2*	*ADR3*	*ADR4*	*ADR5*	*Tentative Vibrational Assignment*
3802.78 W	–	–	–	–	O-H Streching
3749.90 M	–	–	–	–	O-H Stretching
3443.27 VS	3430.03 VS	3423.60 VS	3426.95 VS	3442.48 S	O-H Stretching adsorbed water
1646.90 M	1642.52 W	1646.43 M	1640.70 M	1645.74 W	O-H Stretching adsorbed water
1457.86 VW	–	1457.40 VW	1457.86 VW	1457.67 VW	Carbonate groups
1397.02 VW	–	1396.76 VW	1397.52 VW	–	Carbonates
1034.94 VS	1032.12 VS	1035.64 VS	1035.57 VS	1035.52 VS	Si-O stretching
777.59 W	777.97 W.	778.08 W	778.08 W	778.87 W	Si-O Quartz
–	–	–	–	694.82 VW	Quartz
639.64 W	–	642.07 W	–	–	Mn-O (Manganese Oxide)
585.45 W	–	–	–	–	Fe-O Fe_3O_4 (Magnetite)
540.22 W	537.52 M	540.93 W	538.98 M	540.94 M	Fe-O Fe_2O_3 (Hematite)
466.75 M	467.69 S	–	468.68 S	467.98 M	Fe-O Fe_3O_4 (Magnetite)
–	–	422.60 S	–	–	Si-O Mixed deformation

VW: Very weak; W: Weak; M: Medium; VS: Very strong; S: Strong.

The very strong absorption band in all the samples, in the region 1032–1035 cm^{-1} is due to Si-O stretching of silicate band (Grienier Wronowa *et al.*, 1999) have identified the absorption bands at 778

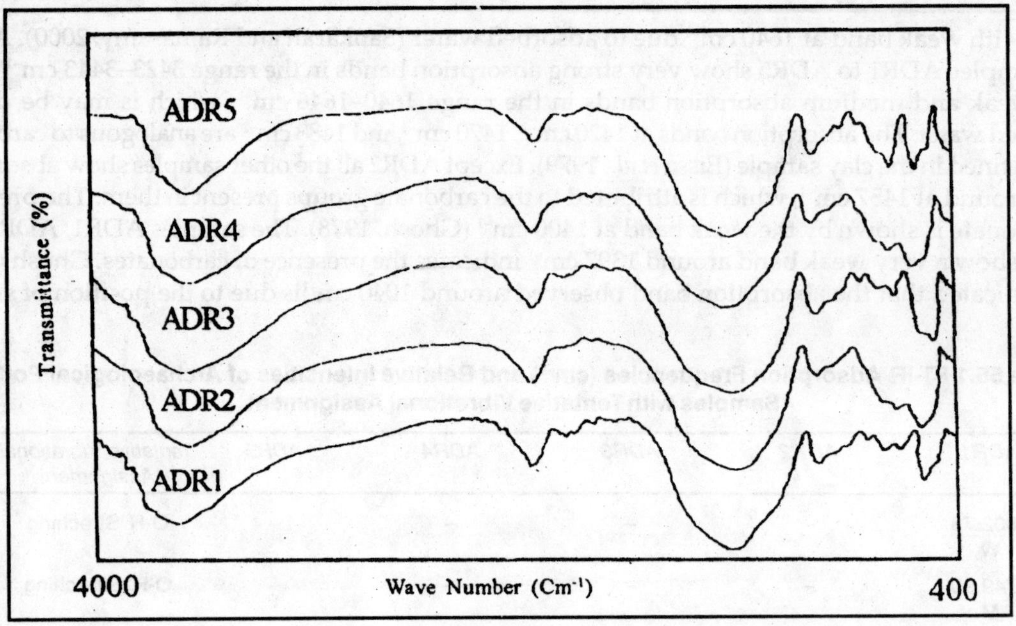

Figure 55.1: FT-IR Spectra of Pottery Shreds Excavated in Adichanallur in Received State

cm^{-1} and 779 cm^{-1} using IR technique on finding clay materials and referred to crystalline quartz. The samples taken for the present study shows the presence of secondary mineral quartz as indicated by the weak absorption band centered around 777 cm^{-1}. According to Elsaas and Oliver (1978) and Farmer and Russell (1968), the presence of the sharp band at 694 cm^{-1} indicates the size of the quartz particles as thin and in the case of thick particles, this band would be shifted to 689 cm^{-1}. In the pottery sample ADR5 the band appears at 694 cm^{-1} is may be due to the presence of quartz and indicate that it may be thin particles. A small shoulder of the band centered at 644 cm^{-1} is due to manganese oxide (Barilaro *et al.*, 2005).

The samples ADR1 and ADR3 show the weak absorption bands at 639 cm^{-1} and 642 cm^{-1} respectively, may be due to the presence of manganese oxide. The broad absorption bands appearing at 580 cm^{-1} and at 540 cm^{-1} attributed to the presence of FeO of Fe_3O_4 (magnetite) and FeO of (Fe_2O_3) (hematite) respectively (Russell, 1987; McDevitt and Baun, 1964; Barilaro *et al.*, 2005). The weak band at 585 cm^{-1} in the sample ADR1 indicates the presence of magnetite (Fe_3O_4). The absorption band around 540 cm^{-1} in all the samples indicates the presence of Hematite (Fe_2O_3). A band at 470 cm^{-1} is characterized by the presence of magnetite (Fe_3O_4) (Ramaswamy *et al.*, 1991). In this study all the samples show strong and medium absorption bands around 467 cm^{-1} may be due to presence of magnetite (Fe_3O_4). The strong absorption band around 430 cm^{-1} and 425 cm^{-1} are attributed to mixed deformation vibrations of Si-O (Ramaswamy and Kamalakkannan, 1987). The sample ADR2 has the strong absorption band at 422 cm^{-1} which is due to the mixed vibrations of Si-O.

Acknowledgement

The authors are grateful to Dr. V. Krishnakumar, Head of the Department, Department of Physics, Periyar University and thankful to K. Nagalakshmi, Nehru Memorial College, Puthanampatti for recording the FT-IR spectra of the samples. We acknowledge with thanks to Thiru. S. Mathivanan, Conservation Assistant, Archaeological Survey of India, Salem Branch, for providing the archaeological pottery samples for the present study.

References

Barilaro, D., Barone, G., Crupi, V., Donato, M.G., Majolino, D., Messina, G. and Ponterio, R., 2005. *Journal of Molecular Structure*, 744–747, 827–831.

Eissa, N.A., Sallam, H.A., Sanad, A.M. and Mira, A.F., 1979. *Indian Journal of Pure and Applied Physics*, 17: 731.

Elsaas, P. and Olivier, D., 1978. *Clay Minerals*, 13: 299.

Farmer, V.C. and Russell, J.D., 1968. *Spectrochemica Acta*, 20: 1149.

Ghosh, S.N., 1978. *Journal of Material Science*, 13: 1677.

Griener-Wronowa, E., Paluszkiewiez, C. and Stoch, L., 1999. *Journal of Molecular Structure*, 511–512: 199–204.

Hlavay, J., Jonas, K., Elak, S. and Inczedy, J., 1977. *Clays and Clay Minerals*, 25: 451.

Jack, E. Gillot, 1968. *Clay in Engineering Geology*. Elsevier Publishing Company, New York, p. 183–185.

Keller, W.D. and Picket, E.E., 1949. *Am Minerals*, 34: 855.

McDevitt, N.T. and Baun, W.L., 1964. *Spectrochemica Acta*, 20: 799.

Prasad, J., 1965. *Trans Indian Ceram Soc.*, 24(3): 78.

Ramasamy, K., Duraisamy, D. and Venkatachalapathy, R., 1991. *Proc. of the 1st International Colloq on Role of Chemical in Archaeology*, (Eds.) Ganorkar M.C., Rama Rao, N., Hyderabad, India, pp. 47–54.

Ramasamy, K. and Kamalakkannan, N., 1987. *Indian Journal of Pure and Applied Physics*, 20: 284–286.

Russell, J.D., 1987. *A Handbook of Determinative Methods in Clay Mineralogy*, (Ed.) Wilson M.J. Blackie and Son Ltd., 1st Edn., p. 163.

Sankaran, S. and Ramasamy, K., 2000. *Asian Journal of Physics*, 9(2): 362.

Ramasamy, K. and Kamalakkannan, N., 1987. *Indian Journal of Pure and Applied Physics*, 20: 284–286.

Venkatachalapathy, R., Manokaran, C., Sridharan, T. and Basilraj, C.N., 2001. *Indian Journal of Pure and Applied Physics*, 40: 207.

Chapter 56

Phase Equilibria Studies for Multicomponent Systems

V.A. Shinde, S.J. Raut, S.V. Gadekar, K.S. Kulkarni and S.N. Kaul
Department of Chemical Engineering,
Bharati Vidyapeeth Deemed University, Pune – 411 043, Maharashtra, India

ABSTRACT

Separation processes in process industries depend mainly on the vapour liquid equilibrium data for the components present. Whereas binary and ternary systems are studied analyzed and reported extensively in technical literature. Data for multi-components exceeding four components are hardly reported. However, industrially relevant systems contains large no of components often exceeding four. There are large no of examples in industry wherein, a right split of feed of multi-components into distillate and heavies or distribution of minor components in light and heavy phases or selective separation of product is essential for the process feasibility. Lack of correct knowledge about V LE of multi-components can make the process non viable. In this study an analysis of industrially important multi-components V LE is presented using simple and known principles.

Keywords: *Flash vaporization, VLE, Enthalpy, Feed split.*

Introduction

The importance of phase equilibria studies with special reference to process design has not diminished even after the advent of computer and information technology era. It is true that there is tremendous emphasis on creation of data banks for thermodynamic, thermochemical data during the last five decades. It is also true that chemical solution theories and group contribution models have been undergoing refinements and mathematical rigors in the last five decades leading us to believe that a priori designs can be made instead of waiting for experiments. Harold R Null, Engineering fellow (Monsanto, USA) and an expert in process design made these remarks in his book (Harold, 1970).

However, it is clearly seen that when stakes are high in the form of process licensing of specialty chemicals, phase equilibria data are generated experimentally rather than by simulation for use in process design. In the case of carbonylation of methanol to acetic acid using one of the several homogenous catalysis routes, preserving and recycling the catalyst components of the process is regarded as a proprietary intellectual property by all the Carbonylation practioners.

In this article authors attempt to analyse and improve predictability of the multi-component, high-pressure vapour liquid equilibrium data for the čarbonylation processes from various angles. The simplest approach is to treat the system as non-ideal mixture of components in a flash equipment; and carry out process design.

Theoretical Analysis

Adiabatic flash operation is an important unit operation in any carbonylation process. Acetic Acid process scheme consists of the components as reported in Table 56.1. It is available at a total pressure of 70 bar (absolute) and temperature 235°C and allowed to flash adiabatically to new lower pressure *viz.*, 10 bar, for the purpose of this study. We aim to determine the fraction of the mixture that will be in the vapour phase at equilibrium and the composition of the liquid and vapour phases. The solution of this flash problem combines a material balance with the phase-equilibrium calculations and enthalpy balance.

Basis : 1000 gms of Feed Mixture

P = 70 bar, T = 235°C = 508.15°k

Table 57.1: Feed Composition

Component	*Formula*	*Wt in gms*	*Mol. Wt.*	*No of Moles*	*Mole Fraction*
Methyl Iodide	CH_3I	40.0	141.9	0.2818	0.01545
Methanol	CH_3OH	80.0	32.0	2.5	0.1371
Methyl Acetate	$CH_3CO_2CH_3$	120.0	74.0	1.6216	0.0889
Acetic acid	CH_3COOH	730.0	60.0	12.167	0.667
Water	H_2O	30.0	18.0	1.67	0.0916
Total		**1000.00**		**18.24**	**1.000**

Computational Technique

Following are the equations which are used for this study. Considerable number of iterations are carried out to find out the results which are in terms of vapour pressure, component balance and enthalpy balance.

Individual component vapour pressure (P_i^*) is calculated by using Antoine type eq^n

$$Log_{10}P_i^* = A+B/T+Clog_{10}T+DT+ET^2 \quad T = \text{temperature } ^\circ k \qquad \ldots(1)$$

Composition of vapour phase is calculated by:

$$y_i = P_i^* x_i / P \qquad ...(2)$$

Overall mole balance

$$F = L + V \qquad ...(3)$$

Individual component mole balance

$$Fx_{fi} = Lx_i + Vy_i \qquad ...(4)$$

Enthalpy of feed Q_F, liquid Q_L, vapour Q_V

$$Q_F = Q_L + Q_V \qquad ...(5)$$

$$Q_L = m \int_{T_1}^{T_2} Cp_{mix}\, dT \qquad ...(6)$$

where,

$$Cp_{mix} = \sum_{i=1}^{n} x_i Cp_i \qquad ...(7)$$

$$Q_F = m \int_{T_1}^{T_2} \sum_{i=1}^{n} y_i Cp_i(l)\, dT + m \sum_{i=1}^{n} y_i \Delta H_{vap} + m \int_{T_1}^{T_2} \sum_{i=1}^{n} y_i Cp_i(g)\, dT \qquad ...(8)$$

For Heat capacity, Heat of vapourisation, Vapour Pressure (b).

Result

Table 56.2: Individual Component Vapour Pressure Data at Five Different Relevant Temperature

Component	*P_i^* bar*				
	T = 150°C	*T = 160°C*	*T = 170°C*	*T = 183°C*	*T = 235°C*
Methyl Iodide	14.375262	17.187575	20.410099	25.292498	55.475533
Methanol	13.895354	17.5537	22.202919	28.867883	75.44093
Methyl Acetate	11.568797	14.046279	16.190087	21.235539	47.40662
Acetic Acid	2.4809718	3.1929193	4.05972	5.4564717	15.26205
Water	4.751411	6.1688726	7.9046042	10.71613	30.582927

Basis: 1 Mole of feed mixture.

Enthalpy of feed (Q_F) = 520183.12 joule Refer equations (6) and (7).

Case 1

Reduced pressure P = 10 bar, T = 186°C = 459.15 °K

P_i^* is calculated from Antoine equation (1)

Table 56.3: T-x-y Results

Component	P_i*Bar	x_i	y_i
Mehtyl Iodide	26.541882	0.015	0.0398128
Methanol	30.688692	0.1	0.3068869
Methyl Acetate	22.343983	0.055	0.1228919
Acetic Acid	5.8273432	0.75	0.4370507
Water	11.464731	0.08	0.0917178
			Σy_i = 0.9983601

We assume Feed split into 80 per cent liquid and 20 per cent vapour

Moles of liquid = 14.592 Moles of Vapour = 3.648

For individual component balance refer equation (3) and (4).

Table 56.4: Feed Split and Component Mole Balance

Component	*L Moles*	*V Moles*	*(L + V) Moles*	*F Moles*
Mehtyl Iodide	0.21888	0.145237	0.364117	0.281808
Methanol	1.4592	1.1195234	2.5787234	2.500704
Methyl Acetate	0.80256	0.4483096	1.2508696	1.621536
Acetic Acid	10.944	1.594361	12.538361	12.16608
Water	1.16736	0.3345865	1.5019465	1.4592
Total	**14.592**	**3.6420175**	**18.234018**	**18.24**

Enthalpy of liquid Q_L.

Basis: 1 Mole of liquid

Enthalpy of Liquid (Q_L) = 309.62871 joules Refer equations (6) and (7).

Enthalpy of vapour Q_V

Basis: 1 mole of gas

Enthalpy of vapour (Q_V) = 209.51918 K joule Refer equation (8)

$Q_L + Q_V$ = 519.14789 K joule

Table 56.5: Enthalpy Balance

Pressure P Bar	*Temperature °C*	*Enthalpy (K Joule)*			
		Q_F	Q_L	Q_V	$Q_L + Q_V$
10	186	520.18312	309.62871	209.51918	519.14789

Discussion of Results

Results of an adiabatic flash operation of a practically and industrially relevant system are reported based on the data provided. Acetic acid system taken as a case study in this report is representative of a large number of industrially relevant systems of similar nature and complexity. The numerical results generated and reported in this report are not validated by experiment. But this statement is not to be stretched to mean that the data are incorrect. The fact is that, these data and several similar ones in petroleum, petrochemical industry wherein a large number of volatile components are handled in stage-wise separation apparatus, are closely guarded proprietary knowledge of several foreign process licensors.

Conclusion

This exercise should be considered as a case study in analyzing the steps involved in developing reliable data for process design. It should also be viewed as a supplement to the experiment and other predictive methods. We find that an exercise like this helps in Planning an experiment/experiments with better insight. Developing better control strategy for the experimental programme and finally. Develop and cautiously use both chemical and analytical tools for experiments keeping in minds accuracy requirements.

Nomenclature

Symbol	Description	Units
F:	Molar or mass feed per unit time	gm moles/sec gms/sec
V:	Vapour flow in moles or mass per unit time	gm moles/sec gms/sec
L:	Liquid flow in moles or mass per unit time	gm moles/sec gms/sec
X_{fi}:	Bulk composition of mixture	
x_i:	Liquid phase mole fraction of component i	
y_i:	Vapour phase mole fraction of component i	
P_i^*:	Vapour pressure of pure component i	bar
P_i:	Partial pressure of pure component i	bar
P:	Total pressure	bar
T:	Absolute temperature	°K
V_i:	Volume of pure component i	cc
R:	Universal gas constant cc	bar/mole °K
Q_F:	Enthalpy of feed per mole	J/mole
Q_L:	Enthalpy of liquid per mole	J/mole
Q_V:	Enthalpy of vapour per mole	J/mole
m:	No of moles	
A,B,C,D,E,n:	Regression coefficients for chemical compound	
Cpi(l):	Heat capacity of liquid component i	J/mole °K
Cpi(g):	Heat capacity of gaseous component i	J/mole °K
ΔHvap:	Enthalpy of vapourization	J/mole °K
Tc:	Critical temperature	°K

Acknowledgement

Authors are thankful to Dr. A.R. Bhalerao Principal B.V.D.U.C.O.E Pune. Dr. S.N. Kaul, Dr. A.K. Dey for their support during this work.

References

Bhatt, B.I. and S.M. Vora, 194. *Stichiometry*, ed. Tata McGraw Hill Publishing Company Ltd., New Delhi.

Gmehling, J. and U. Onken, 1977. *Vapour Liquid Equilibrium Data Collection.* Chemistry Data Series (DECHEMA), Vol. 1, Part 1.

Gmehling, J., U. Onken, and W. Arit, 1981. *Vapour Liquid Equilibrium Data Collection.* Chemistry Data Series (DECHEMA), Vol. 1, Part 1a.

Kirk-Othmer, 1992. *Encyclopedia of Chemical Technology*, 4th ed. Wiley Interscience.

Null, Harold R., 1970. *Phase Equilibrium in Process Design.* John Wiley and Sons.

Ortega, Juan and Pedro Susial, 1993. Vapour-liquid equilibria behaviour of methyl esters and propan-2-o1 at 74.66, 101.32 and 127.99 Kpa. *J Chem. Eng. Japan*, 26(3): 259–265.

Raal, J.D.and A.L. Mullbauer, 1994. Measurement of high pressure vapour liquid equilibria. In: *Development in Chemical Engineering and Mineral Processing*, 2(2/3): 69–103.

Viktorov, A.I., N.A. Smirnover and A.G. Moracheviskii, 1982. Calculation of liquid vapour equilibrium in ternary partially miscible system with the aid of NRTL model. *J. Appl. Chem., USSR*, 55: 949–952.

Vinadevi, Patil V. Ph.*D. Thesis*, M.E. (Chemical), Shivaji University, Kolhapur.

Yaws, Carl L. *Chemical Properties Handbook.* McGraw Hill Book Company.

Zoller, Joseph, R. *Acetic Acid and its Derivatives.* Eastman Chemical Company.

Chapter 57

Comparison of Toxicity Between Plant Extracts and Malathion Against Two Beetles

Sameer Mishra and Ranjana Saxena
Pest and Parasite Research Lab-II, Department of Zoology, Bareilly College, Bareilly – 243 005, U.P.

ABSTRACT

Comparative effect of toxicity of Malathion and two most effective plant extracts* *i.e.* Red Kaner leaves (for *Tribolium castanellm*) and Garlic bulbs (for *Oryzaephilus surinamensis*) was tested by "Dry Eilm Technique" in the prior case and through "Topical Application Method" in the latter one. The result were more significant with Malathion for both the pests, however with the plant extracts also, the recorded values for mortality were quite considerable.

***Keywords**: Malathion, Plant extracts, T. castaneum, O. surinamensis, Toxicity.*

Introduction

Due to the restless efforts of different Agriculturists and scientists the problem of shortage of foods products get decreased to a great extent. As a result now, India is a second largest wheat producer in the World, follows China. Several insects-pest cause severe damage at field and storage level and plays a drastic role in reducing the amount of different food products. To combat the 'effect

* *Most effective extracts were taken, out of total 11 tested plant extract against both the pests.

of these harmful creatures, synthetic pesticides have widely been used since long, but due to their several harmful effects like pollution, untargeted toxicity (Makanjuola, 1989) etc. the pest control scenario get changed greatly and plant products are preferred these days because of their innate biodegradability and non hazardous value (Saxena and Saxena, 1991).

On one hand, these plant product are ecofriendly and safe, but on the other this is also true that these products may not be able to suppress the pest population solely. Keeping this view in mind here an attempt has been made to study the comparison of toxicity between a known insecticide 'Malathion' and two plant extracts, which were proved effective against the experimental pests.

Materials and Methods

To make a comparison of toxicity between Malathion and most effective plant extract proved against *T. castaneum* and *O. surinamensis*, the Dry Film Technique (Rani and Jamil, 1989) and Topical Application Method were applied respectively. The tested concentration for the plant extract were 0.5, 1.0, 2.0 and 5.0 per cent whereas, for Malathion the concentrations were 0.03125, 0.0625, 0.125 and 0.25 per cent. 10 adults of about one week old of both the genus were taken for experiment. Mortality counts were made after each 24 hours interval for 3 days. One the basis of mortality, the Corrected Values (Abbott's, 1925) Probit Kill and LC_{50} values were calculated.

Results and Discussion

To make a comparison of toxicity between the most effective plant extracts proved against *Tribolium castaneum* and *Oryzaephilus surinamensis,* a known insecticide, "Malathion" was taken and following observation were made.

Effect on *Tribolium castaneum*

In this case the leaf extract of Yellow Kaner was found to be the most effective by showing the LC_{50} values as 0.40 per cent and relative toxicity as 0.03075 per cent, after an exposure period of 24 hours. Here with plant extracts, total four concentrations *i.e.* 5.0, 2.0, 1.0 and 0.5 per cent were taken and the mortality per cent of the pest was recorded as 90 per cent, 83.3 per cent, 80 per cent and 76.6 per cent respectively, whereas, with Malathion the used concentrations were very low in comparison to the plant extract and they were 0.25, 0.125, 0.065 and 0.03125 per cent, here the rate of mortality was 96.6 per cent 93.3 per cent, 86.6 per cent and 83.3 per cent respectively and the LC_{50} value was 0.0123 per cent (Table 57.1).

Table 57.1: A Comparison of Most Effective Plant Extract (Yellow Kaner Leaves) with Malathion, in Case of *Tribolium castaneum*

Sl.No.	*Treatment*	*Concs. Used (Per cent)*	*Average Per cent Mortality*	*Corrected Values*	*Per cent Probit Kill*	*LC_{50} Values*	*Relative Toxicity*
1.	Yellow Kaner	0.5	76.6	74.0	5.64		
	Leaves	1.0	80.0	77.7	5.77		
		2.0	83.3	81.4	5.84	0.40	0.03075
		5.0	90.0	88.8	6.23		
2.	Malathion	0.03125	83.3	81.4	5.88		
		0.0625	86.6	85.1	6.04	0.0123	–
		0.125	93.3	92.5	6.48		
		0.25	96.6	96.2	6.75		

Effect on *Oryzaephilus surinamensis*

In case of *O. surinamensis*, the best result was obtained with Garlic bulb extract in which the LC_{50} value and relative toxicity were 0.35 and 0.03857 per cent respectively, however, at the individual doses of 5.0, 2.0, 1.0 and 0.5 per cent total 86.0, 80.0, 73.3 and 66.6 per cent pest showed mortality. With "Malathion", the mortality per cent was recorded as 100 per cent, 96.6 per cent, 93.3 per cent and 76.6 per cent at the respective doses of 0.25, 0.125, 0.0625 and 0.03125. In this case the LC_{50} value was found to be 0.0135 per cent (Table 57.2).

Table 57.2: A Comparison of Most Effective Plant Extract (Garlic Bulbs) with Malathion in Case of *Oryzaephilus surinamensis*

Sl.No.	*Treatment*	*Concs. Used (Per cent)*	*Average Per cent Mortality*	*Corrected Values*	*Per cent Probit Kill*	*LC_{50} Values*	*Relative Toxicity*
1.	Garlic bulb	0.5	66.6	628	5.28		
		1.0	73.3	70.3	5.61		
		2.0	80.0	77.7	5.74	0.35	0.03857
		5.0	86.6	85.1	6.04		
2.	Malathion	0.03125	76.6	74.0	5.64		
		0.0625	93.3	92.5	6.84	0.0135	–
		0.125	96.6	96.2	675		
		0.25	100.0	100	8.09		

Here, it was found that Malathion was more effective in causing the lethal effect in the pests, as even at the lowest does the result were significant. Effect of Malathion has widely been studied on various pests of stores. Like the present study, Shelby (1979) evaluated the level of malathion needed to produce mortality in *T. castaneum* and found that 848.5μ malathion/gm. body weight of adult was the optimum dose. Similarly Yadav *et al.* (1983) tested the effect of seven organophosphorous insecticide against several stored product beetles, there was some variability in LC_{50} values of the present study and the values given by the workers. Like the present study, the effective control of *O. surinamensis* with different chemicals has also been reported by Barson (1991), Sontakke (1993) reported the field efficacy of some insecticide alone and in combination with neem oil against some insect/pest and their predators in rice, and found that insecticide alone prove more effective (like the present study) than in combination. As in the present study, Ahmed *et al.* (2001) compared the toxicity of two insecticide with Neem Extract against *T. castaneum* and found the better results with insecticide.

The overall inference is drawn that no doubt chemical insecticides proved better than plant extracts, but due to their several demerits one has to think twice before using them and the statement *"Chemical Cart Can Run Away With The Biological Horse"* is strongly supported and recommended.

References

Abbott, W.S., 1925. A method of computing the effectiveness of insecticides. *Journal of Economic Entomology*, 18(1): 265–267.

Ahmed, I., Ali, R.R., Tabassum, R., Azmi, M.A., Naqvi, S.N.H. and Khan, F., 2001. Toxicity determinations of two pyrethroids Karate (Cyhalothrin) and deltamethrin as compared to Neem extract on *T. castaneum*. *Journal of Experimental Zoology*, 4(1): 169–173.

Barson, G., 1991. Laboratory assessment of the residual toxicity of commercial formulations of insecticides to adult *O. surinamensis* exposed for short time intervals. *Journal of Stored Product Research*, 21(4): 205–211.

Makanjuola, W.A., 1989. Evaluation of extract of neem for the control of some stored product pests. *Journal of Stored Product Research*, 25(4): 231–237.

Rani, P.U. and Jamil, L., 1989. Effect of water hyacinth leaf extract on mortality, growth and metamorphosis of certain pest of stored products. *Insect Science and its Application*, 10(3): 327–332.

Saxena, A. and Saxena, R.C., 1991. A new plant extract to suppress the population of malaria vector *A. culcifacies*. *Geobios New Reports*, 10: 154–156.

Shelby, C.E., 1979. Levels of Malathion needed to produce a normal distribution of mortality in *T. castaneum*. *Journal of Georgia Entomological Society*, 34(4): 305–311.

Sontakke, B.K., 1993. Field efficacy of insecticide alone and in combination with neem oil against insect plant and their predators in rice. *Indian Journal of Entomology*, 55(3): 260–266.

Taulkdar, F.A. and Howse, P.E., 1995. Evaluation of *A. polystachya* as a source of repellent, antifeedants, toxicants and protectants in storage against *T. castaneum*. *Journal of Stored Product Research*, 31(1): 55–61.

Yadav, T.D., Singh, S., Khanna, S.C. and Pawar, C.S., 1983. Toxicity of dusts of Organophosphorous insecticide against stored product beetles. *Indian Journal of Entomology*, 45(3): 247–253.

Chapter 58

On a New Species of the Genus *Agriocnemis* Selys (Order : Odonata) from India

K.P. Shinde and T.V. Sathe
Department of Zoology, Shivaji University, Kolhapur – 416 004, India

ABSTRACT

A Damsel fly, *Agriocnemis kolhapurensis* sp. nov. (Odonata : Coenagriidae) was described for the first time from India. Male 36.00 mm long, abdomen 30.00 mm long, superior 1.3 mm long and 0.9 mm broad, inferior top black and hairy. *A. kolhapurensis* biological control agent of paddy pests.

Keywords: *Agriocnemis kolhapurensis sp. nov., Description, Biological pest control agent.*

Introduction

Damsel flies are important biocontrol agents of Insect pests. Hence, present study was objected to described damsel fly recorded from paddy agroecosystem of Kolhapur, India. The present species belongs to the Genus *Agriocnemis.* This Genus I was erected by Selys in 1 869. The Genus is distributed through out the tropical zones of the old world from Africa to Australia and Ocenea. However, this genus is supposed to be of the oriental region since the bulk of the species are recorded from Southern Asia. Ten species of this genus have been reported from India (Fraser, 1933). In past Fraser (1933, 1934, 1936), Prasad (1995, 1996), Lahiri (1987), Prasad and Varshney (1995), Prasad and Kulkarni (2001, 2002), Kulkarni and Prasad (2002) etc. have worked on taxonomy of Odonata from India.

Results

Male

Body 36.00 mm long, 3.85 m broad excluding anal appendages; Head 2.00 mm long; 3.5 mm broad; Abdomen 30.00 mm long, 2.00 mm broad; Thorax 4.0 mm long, 3.00 mm broad; fare wing 21.00 mm long, 4.5 mm broad; Hind wing 22.00 mm long, 4.5 mm broad; superior 1.3 mm long and 0.9 mm broad.

Head

2.00 mm long, 3.85 mm broad; compound eyes large, circular, brownish above, yellowish blue at sides and below, occular distance 1.8 mm, interoccular distance 1.75 mm, occiput brownish with black spots; three ocellie in triangle reddish brown, small, shining, tubercles; ocellar region spiny, long spines at bases of compound eyes, frons yellowish with faint green tinch; clypeus half circular anteriorly and small blunt strip posteriorly, yellowish with faint green colour; genae yellowish with three dark yellow blue spots; Labrum yellowish blue with no black spots, non-metalic; mandibles brown, bases dark brown to black; Labium whitish yellow.

Antenna

6.42 mm long, 0.6 mm broad, brownish, 7 segmented, scape, yellowish, 0.02 mm long; pedicel reddish brown, 1.2 mm long, 0.8 mm broad; flagellum brownish, 5.2 mm long; 1st segment 1.6 mm long, 0.6 mm broad, 2nd segment 1.00 mm long, 0.3 mm broad, 3rd segment 0.9 mm long, 0.2 mm broad; 4th segment 0.9 mm long 0.15 mm broad; 5th segment 0.8 mm long, 0.1 mm broad.

Flagellar Formula

L/W 1/2 = 0.79, 2/2 = 1.00, 3/2 = 1.35, 4/2 = 1.80, 5/2 = 2.40, A=1.46.

Thorax

4.00 mm long, 3.00 mm broad, brownish black on dorsum, two black stripes parallel on lateral side running obliguely, one is broad on prothorax associated with white stripe. Three light green blue stripes on lateral side running parallel to each other.

Legs

15.00 mm long, 1.2 mm broad, yellow with black spines. Coxa 2.8 mm long yellow; trochanters 1.3 mm long yellow; Femur 6.00 mm long, internally yellowish, outer side black, 12 black spines from inner side; tibia yellow from inner side, black from outer side, 7.00 mm long, with 18 black spines; tarsus yellow, 2.00 mm long, two segmented, 1st segment 1.00 mm long, 2nd segment 1.00 mm long; claw 0.75 mm long, black spines arranged from inner side in two rows.

Wings: Hyaline

Fore wing 21.00 mm long, 4.50 mm broad; hind wing 22.00 mm long, 4.5 mm broad, fore wings with 8 postnodal nervures; both wings transparent with black veins, pterostigma brown 1.2 mm long, 0.55 mm broad, roughly rectangular, similar in hind and fore wings.

Abdomen

30.00 mm long, 2.00 mm broad, 1st segment with transverse stripes black, white, black colour, 3.5 mm long, 2.00 mm broad; 2nd segment 4.20 mm long, 1.8 mm broad, laterally dense hairs. Abdomen darker blue; terminal segments bluish with black dorsal marking; black dorsal marking on segment 2,

3, 4, 5, 6, 7, 9 and 10, blue eye spots on segment second. 9th abdominal segment 0.4 mm long, 0.47 mm board; 10th abdominal segment 2.6 mm long, 2.00 mm broad, greenish blue, Anal appendages brownish yellow, not black entirely. Inferior and superior of same colour. Superior 1.3 mm long, 0.9 mm broad; interier black and 0.9 mm long, 0.5 mm broad, hairy at tip, not acutely angular at its lateral tips but some what curved inwardly. Superior hairy, terminal lobs not pointed, superior shorter than length of 10th segment no ridges and furrows on the border of 9th segment, 10th segment with black *damru* shaped patch dorsally, arms of which not reaching the lateral borders of the segment.

Female: Unknown

Preys: Jassids, *Sogatella furcifera* (Harvath), *Nephoteltix virescens*, Pyralid moth, *Scirophaga incertulus* (WLK).

Holotype

Male, Kolhapur, Maharashtra, India. Coll. 20-VII-2005, Kolhapur, T.V. Sathe; antenna, legs, wings, abdominal tergites on slide labelled as above.

Paratype

Male 4, Coll. K.P. Shinde from June to August same data as above.

Distribution Record

Kolhapur (1) 12-VII-2005, Amboli Ghats (1) 15-VII-2005, Koyananagar (1) 15-VII-2006, Patan (1) 15-VII-2006.

Discussion

Agriocnemis kolhapurensis sp. nov. runs close to *A. aborensis* (Laidlaw) in Fraserls key (1933) by having (1) Labrum non-metalic and (2) Fore wings with post nodal nervures. However, it differs from the above species by having following characters.

1. Superior shorter than length of 10th segment.
2. Labrum yellowish blue with no black spots on it.
3. Pterostigma proportion.
4. Terminal structure of inferior tip black and hairy.
5. Abdominal length 30.00 mm.
6. Flageller formula:
   L/W 1/2 = 0.79, 2/2 = 1.00, 3/2 = 1.35, 4/2 = 1.80, 5/2 = 2.40, A = 1.46.

Acknowledgement

Authors are thankful to Shivaji University, Kolhapur for providing facilities.

References

Fraser, F.C., 1933, 1934 and 1936. *Fauna of British India including Ceylon and Burma*. Odonata, 1: XIII + 423 pp., 2: XXIII + 398 pp., 3: XI + 461 pp., Taylor and Francis, London.

Fraser, F.C., 1957. A classification of the order Odonata. *Rec. Zool. Soc.*, N.S.W. Sydney, 133 pp.

Kulkarni, P.P. and Prasad, M., 2002. Insecta: Odonata. *Rec. Zool. Surv.*, India: Wetland Ecosystem Series No. 3: Fauna of Ujani, p. 91–104.

Lahiri, A.R., 1987. Studies on Odonata fauna of Megalaya. *Rec. Zool. Surv.*, India Misc. Publ. Paper, 99: 402.

Prasad, M., 1995. *On a Collection of Odonata from Goa, India*, 2(12): 7–8.

Prasad, M., 1996. An account of the Odonata of Maharashtra State, India. *Rec. Zool. Surv. of India*, 95(3–14): 305–327.

Prasad, M., 1999. Faunal diversity in India: Odonata. *Rec. Zool. Surv. of India*, Calcutta, p. 171–178.

Prasad, M. and Varshney, R.K., 1995. A checklist of India including data on larval studies. *Oriental Ins.*, 29: 385–428.

Prasad, M. and P.P. Kulkarni, 2001. Insecta: Odonata. *Rec. Zool. Surv. of India.* Fauna of Conservation Area Series, 11: Fauna or Nilgiri Biosphere Reserve, p. 73–83.

Prasad, M. and P.P. Kulkarni, 2002. Insecta: Odonata. *Rec. Zool. Surv. of India.* Fauna of Eravikulam National Park, Conservation Area Series, 13: 7–9.

Silsby, J., 2001. *Dragonflies of the World.* National History Museum and Plymbridge Distributors Ltd., p. 1–124.

Chapter 59

Variability and Heritability in Castor (*Ricinus communis* L.)

S. Rajamani
Scientist (Plant Breeding), Regional Agricultural Research Station, Lam – 522 034, Andhra Pradesh

ABSTRACT

Seven genotypes were evaluated for genetic variation studies for six characters *viz.*, plant height, days to 50 per cent flowering, nodes up to primary spike, spike length, number of nodes up to primary spike, number of spikes and yield. The variance analysis revealed that the high genetic variability present in the quantitative characters studied. Yield, days to 50 per cent flowering and plant height has high GCV, PCV, heritability and genetic advance reveals prevalence of additive gene action and simple selection is effective for improvement of these characters.

Keywords: *Castor, Genetic variance, Heritability, Genetic advance.*

Introduction

The two basic requirements of plant breeding are presence of genetic variation and exploitation of variation through selection. Presence of genetic variation in a character is a must for any crop improvement in that character. The nature and magnitude of genetic variability in the population is of most valuable for executing efficient breeding programme to improve the yield potential of the genotypes. Genotypic coefficient of variation in addition with other heritable estimates are useful for improvement in genotypic values over the population. Knowledge on heritability with genetic advance will be of much value for selection. Hence, considering these aspects genetic variability studies were carried with diverse castor genotypes.

Materials and Methods

Field experiment was conducted with 7 castor genotypes were sown in randomized block deign with three replications at agricultural research station, Darsi. The data was recorded on characters *viz.*, plant height, days to 50 per cent flowering, nodes up to primary spike, spike length (cm), number of spikes and yield (Kg/ha). Data was subjected to statistical analysis and various genetic parameters such as GCV, PCV, heritability and genetic advance were estimated as per Johnson *et al.* (1995) and Hanson (1963).

Results and Discussion

Analysis of variance revealed that highly significant differences among the genotypes for seed yield and other characters shows considerable genetic variation in the genotypes (Table 59.1). Study of genetic parameters (Table 59.2) revealed that plant height, yield, days to 50 per cent flowering and spike length (cm) had high GCV and PCV values indicating that these characters are under the influence of genetic control Moderate GCV and PCV values were observed for spike length and number of nodes up to primary spike.

Table 59.1: Analysis of Variance for Seed Yield and Yield Attributing Characters

Sl.No.	*Character*	*Man Sum of Squares*	
		Replications	*Treatments*
1.	Plant height	7.64	82.18*
2.	No. of nodes up to primary spike	0.20	2.12*
3.	Spike length (cm)	4.08	28.22*
4.	Days to 50 per cent flowering	1.71	112.63*
5.	No. of spikes	0.038	4.513*
6.	Yield (kg/ha)	1225	181753*

Table 59.2: Mean, Range, GCV, PCV, Heritability, Genetic Advance and Genetic Advance as Per cent Over Mean for Seed Yield and Yield Attributes

Sl.No.	*Character*	*Mean*	*Range*	*GCV*	*PCV*	*Heritability*	*Genetic Advances*	*GA as Per cent Over Mean*
1.	Plant height	50.50	40.80–60.40	26.08	29.18	0.89	27.02	53.51
2.	No. of nodes up to primary spike	14.08	12.20–15.20	10.75	11.57	0.92	3.08	21.94
3.	Spike length (cm)	24.73	24.20–33.40	17.58	21.35	0.81	9.01	36.43
4.	Days to 50 per cent flowering	50.42	42–60	37.00	37.66	0.98	38.33	76.03
5.	No. of spikes	5.23	3.30–7.80	13.85	15.29	0.90	1.48	28.35
6.	Yield (kg/ha)	1564.28	1050–1920	1900.7	2008.75	0.94	60846.52	3890.44

The phenotypic coefficient of variation was higher than GCV indicating that presence of environmental influence in the expression of these characters studied. Since variability estimates are influenced by environment the use of genetic material and their utility is limited unless we study the genetic advance in association with heritability.

High heritability and low genetic advance was observed in number of spikes, number of primary spikes and spike length indicating that there is prevalence of non additive gene action. The high heritability is being exhibited due to favourable influence of environment rather than genotype. Selection for such traits may not be rewarding. Moderate heritability and high genetic advance was observed in plant height indicating that the character is governed by additive gene action. The low heritability is being exhibited due to high environmental efforts. Selection may be effective in such cases. The characters yield and days to 50 per cent flowering shows high heritability and high genetic advance indicates that most likely that the heritability is due to additive gene effects and selection may be effective.

From these discussions, it can be concluded that high GCV, PCV, heritability and genetic advance as per cent over mean were observed for yield, days to 50 per cent flowering and plant height indicating that prevalence of additive gene action in the control of these characters and simple selection may be effective to improve these characters. These results are in accordance with Reddy *et al.* (1999).

References

Hanson, W.O., 1963. *Heritability in Statistical Genetics and Plant Breeding*. (Eds.) Hanson, W.O. and Robinson, H.F., 1982. National Academy of Science and National Research Council, Washington D.C.

Johnson, H.W., Robinson, H.F. and Comstock, R.E., 1955. Estimates of genetic and environmental variability in soybean. *Agronomy Journal*, 47: 314–318.

Reddy, P.R., Sankar, G.R.M., Vanaja, M., Rao, C.M., Venkateswarulu, S. and Eastin, J.D., 1999. Heritability and genetic advance in castor (*Ricinus communis*) over varied soils and dates of sowing. *Indian Journal of Agriculture Sciences*, 69(2): 371–375.

Chapter 60

Synthesis and Biological Evaluation of 2,5-Disubstituted 1,3,4-Oxadiazoles

S.K. Mohite*, C.S. Magdum, L.V. Sonawane, A.T. Dalvi, S.S. Patil
Department of Chemistry, Appasaheb Bimale College of Pharmacy, Sangli, South Shivaji Nagar, Sangli-Miraj Road, Sangli – 416 416, Maharashtra

ABSTRACT

In the present investigation, few 1,3,4-oxadiazoles derivatives were synthesized as potential antimicrobial agents. The results of the biological activities revealed that the compounds (IV a_1–a_3), (IV b_1–b_3) and (IV c_1–c_4) exhibited both antibacterial activities against *K. pneumonia, B. substilis, P. vulgaris, S. aureus, E. coli, S. typhi* and *P. aeruginosa* as well as antifungal activity against *C. albicans.* Structures of the compounds were confirmed by IR and HI-NMR.

Keywords: *1,3,4-oxadiazole, Antibacterial, Antifungal.*

Introduction

Oxadiazoles are very important class of the compounds because of the immense medicinal importance. Many of the oxadiazole derivatives have been patented for their valuable pharmacological and biological properties. The compound possessing oxadiazole moiety acts as an analgesics (Umio *et al.*, 1969), hypoglycemic, bactericidal (Saikawa and Maeda, 1970), antipyretid (Wilder, 1964), and local anesthetics (Bhargav and Nair, 1957; Thomas, 1974). They are also known for their potential anti-inflammatory, diuretic, antiemetic, and anticonvulsant activity. They also act as hypnotic, sedative

* Corresponding Author: E-mail: mohite_shrinivas@rediffmail.com; Phone: +919850000946.

and antidiarrheal agents. In some cases they act as insecticides (Ram and Pandey, 1974) and show fungicidal activity. Some halogenated alkyl derivatives of 1,3,4-oxadiazole are used in photosensitive compositions. Anti tubercular and leprostatic activity of 1,3,4-oxadiazole has been described. Biphenyl or napthyl substituted derivatives of 1,3,4-oxadiazole have been used as scintillators. In view of these facts some of substituted 1,3,4-oxadiazole compounds (IV a_1–a_3) (IV b_1–b_3) and (IV c_1–c_4).

Completion of reaction was determined single spotted TLC and structures of the compounds were confirmed by IR, H[1]-NMR. Compounds were screened for antibacterial activity against *K. pneumonia, B. substilis, P. vulgaris, S. aureus, E. coli, S. typhi* and *P. aeroginosa* and for antifungal activity against *C. albicans.*

Material and Methods

Melting points were taken on Electro thermal digital melting point apparatus and are uncorrected. Completion of reaction was determined single spotted TLC. Structures of the compounds were confirmed by Jasco FTIR-410 and Varian mercury 300 MHz, H[1]-NMR.

Experimental

Synthesis of substituted 1,3,4-oxadiazole derivatives was carried out in steps such as:

Preparation of Hydrazides of Substituted Phenoxy Acetic Acid

Ethyl ester of substituted phenoxy acetic acid (2) (0.01 mol) (o-methyl, m-methyl, p-methyl) and hydrazine hydrate (0.02 mol) were refluxed in absolute ethanol. The mixture was concentrated, cooled and poured in ice. The solid thus separated out was filtered, dried and recrystallized to give hydrazides (I a-c).

Preparation of N-chloroacetyl-substituted Phenoxy Acetic Acid Hydrazide

Substituted phenoxy acetic acid hydrazides (I a-c) (0.01 mol) (o-methyl, m-methyl, p-methyl) was placed in round bottom flask fitted with reflux condenser and guard tube. To this solution chloroacetyl chloride (0.01 mol) was added and mixture was refluxed for 3-8 hrs. The solvent was evaporated under reduced pressure residue obtained was recrystallized to give N-acetyl-substituted Phenoxy Acetic Acid Hydrazides (II a-c)

Preparation of 2-(chloromethyl)-5-substituted Phenoxymethyl-1,3,4-oxadiazole

N-chloroacetyl-substituted Phenoxy Acetic Acid Hydrazide (II a-c) (0.1 mol) was dissolved in DMF/$POCl_3$ and mixture was refluxed. Excess of solvent was removed under pressure. The corresponding 2-(chloromethyl)-5-(n-methylphenoxy) methyl-1,3,4-oxadiazoles (III a-c) residue was recrystallized.

Preparation of Substituted 1,3,4-oxadiazole Derivatives (IV a1-2, b1-3, c1-4)

The separately (0.002 mol) of 2-(chloromethyl)-5′-(2-methylphenoxy) methyl-1,3,4-oxadiazole (III a), 2-(chloromethyl)-5-(3-methylphenoxy) methyl-1,3,4-oxadiazole (III b) (0.002 mol) and 2-(chloromethyl) 5-(4-methylphenoxy) methyl-1,3,4-oxadiazole (III c) (0.002 mol) was mixed with substituted amines (o-chloroaniline, m-chloroaniline, p-chloroaniline) and 25 ml ethanol was refluxed.

The solvent was then evaporated under reduced pressure and the crude product was gave corresponding N-substituted phenyl-N-({5-[(2-methylphenoxy) methyl]-1,3,4-oxadiazole-2-yl} methyl) amine (IV a_1-a_3), N-substituted phenyl-N-({5-[(3-methylphenoxy) methyl]-1,3,4-oxadiazole-2-yl} methyl) amine (IV $b_{1-}b_3$) and N-substituted phenyl-N-({5-[(2-methylphenoxy) methyl]-1,3,4-oxadiazole-2-yl} methyl) amine (IV $c_{1-}c_4$).

N-(2-ch)oropheny)-N-({5-[(2-methylphenoxy) methy))-1,3,4-oxadiazole-2-yl} methyl) Amine (IVa$_1$)

MP156-158°C, IR (KBrpallet) cm^{-1} 3463 (N-Hstr), 3150 (ArC-Hstr), 1650, 1670 (ArC = Cstr), 1495, 1456 (C = Nstr), 749 (C-H def),

H^1NMR ($CDCl_3$) d (ppm) 7.2 (s,4HC_6H_4), 5.2 (s, 1H, NH), 2.9 (s, 1H, NH), 1.7 (s, 2H, CH_2)

N-(3-chlorophenyl)-N-({5-[(2-methylphenoxy) methyl]-1,3,4-oxadiazole-2-yl} methyl) Amine (IV a$_2$)

MP-171°C, IR (KBr pallet) cm^{-1} 13230 (N-Hstr), 1650, 1575 (ArC = Cstr), 1667, 1150 (C = Nstr), 706 (C-Hdef),

H^1NMR (CDC13) d (ppm) 6.9 (s,4H,C_6H_4), 7.2 (s,4H,C_6H_4), 2.9 (s,1H,NH), 1.3 (s,2H,CH_2).

N-(4-chloropheny)-N-({5-[(2-methylphenoxy) methyl]-1,3,4-oxadiazole-2-yl} methyl) Amine (IV a$_3$)

MP-189°C, IR (KBr pallet) cm^{-1} 3332 (N-Hstr), 2799 (ArC-Hstr), 1650, 1540 (ArC = Cstr), 1638 (C = Ostr), 1451, 1399 (C = N str), 771 (C-H def), 660 (C-Cl str),

H^1NMR ($CDCl_3$) d (ppm) 6.9 (s,4H,C_6H_4), 7.2 (s,4H,C_6H_4), 2.9 (s,1H,NH), 1.3 (s,2H,CH_2).

N-(2-chlorophenyl)-N-({5-[(3-methylphenoxy) methyl]-1,3,4-oxadiazole-2-yl} methyl) Amine (IV b$_1$)

MP-155°C, IR (KBr pallet) cm^{-1} 3350 (N-Hstr), 1690, 1599 (ArC = Cstr), 1470 (C = Nstr), 710 (C-Hdef).

N-(3-chlorophenyl)-N-({5-[(3-methylphenoxy) methyl]-1,3,4-oxadiazole-2-yl} methyl) Amine (IV b$_2$)

MP-191°C, IR (KBr pallet) cm^{-1} 3431 (N-Hstr), 2913 (ArC-Hstr), 1670, 1621 (ArC = Cstr), 1456, 1360 (C = Nstr), 829,860 (C-Hdef),

H^1NMR ($CDCl_3$) d (ppm) 7.3 (s,4HC_6H_4), 3.5 (s, 1H,NH), 2.2 (s,2H,CH_2), 1.4 (s,2H,CH_2).

N-(4-chlorophenyl)-N-({5-[(3-methylphenoxy) methyl]-1,3,4-oxadiazole-2-yl} methyl) Amine (IV b$_3$)

MP-176°C, IR (KBr pallet) cm^{-1} 3445 (N-Hstr), 1690, 1615 (ArC = Cstr), 1455, 1408 (C = Nstr), 828, 886 (C-Hdef)

N-(2-chlorophenyl}-N-({5-[(4-methylphenoxy) methyl]-1,3,4-oxadiazole-2-yl} methyl) Amine (IV c$_1$)

MP-189°C, IR (KBr pallet) cm^{-1} 3467 (N-Hstr), 3150 (ArC-Hstr), 1596, 1500 (ArC = Cstr), 1469, 1382 (C = Nstr),

H^1NMR ($CDCl_3$) d (ppm) 6.5-7.5 (s,4HC_6H_4), 7.2 (s,4HC_6H_4), 1.9 (s,1H,NH), 1.2 (s,2H,CH_2)

N-(3-chlorophenyl)-N-({5-[(4-methylphenoxy) methyl]-1,3,4-oxadiazole-2-yl} methyl) Amine (IV c$_2$)

MP-216°C, IR (KBr pallet) cm^{-1} 3331 (N-Hstr), 3050 (ArC-Hstr), 1650, 1580 (ArC = Cstr), 1480 (C = Nstr).

N-(4-chlorophenyl)-N-({5-[(4-methylphenoxy) methyl)-1,3,4-oxadiazole-2-yl} methyl) Amine (IV c$_3$)

MP-192°C, IR (KBr pallet) cm^{-1} 3209 (N-Hstr), 1650, 1600 (ArC = Cstr), 1490 (C = Nstr), 771 (C-Hdef),

H^1NMR ($CPCl_3$) d (ppm) 7-8 (s,4H,C_6H_4), 2.7 (s, 1H,NH), 2.5(s,2H,CH_2).

N,Ndiethyl-N-([5-1(4-methylphenoxy)methyl]-1,3,4-oxadiazole-2-yl Amine (IV c_4)

MP-152°C, IR (KBr pallet) cm^{-1} 3179 (N-Hstr), 2980 (ArC-Hstr), 1690, 1610 (ArC = Cstr), 1450, 1400 (C = Nstr)

Biological Activity

All the synthesized compounds (IV a_1–a_3), (IV b_1–b_3) and (IV c_1–c_4) were screened for their *in vitro* antibacterial activity against *K. pneumonia, B. substilis, P. vulgaris, S. aureus, E. coli, S. typhi* and *P. aeruginosa* as well as antifungal activity against *C. albicans* by disc diffusion method using DMF as a solvent. The compounds including standard amoxycillin and fluconazole were used as a standard for antibacterial and antifungal at a concentration of (250 μg/0.1 ml) respectively. The results were mentioned by measuring zone of inhibition in Tables 60.1 and 60.2.

Table 60.1: Zone of Inhibition of Substituted 1,3,4-oxadiazoles Derivatives for Different Bacteria

Name of Organism	*Zone of Inhibition in mm*										
	IV a_1	*IV a_2*	*IV a_3*	*IV b_1*	*IV b_2*	*IV b_3*	*IV c_1*	*IV c_2*	*IV c_3*	*IV c_4*	*Std.*
K. pneumoniae	++	+	++	+++	+	++	++	+	+	++	+++
B. substilis	++	+	+	+	+++	++	+	++	++	+++	++
P. vulgaris	+++	–	+	–	++	+++	++	+	+	++	+++
S. aureus	++	++	++	++	+	++	–	+	–	+	++
E. coli	+	+	–	+	+	–	–	–	–	+	++
S. typhi	++	+	+	++	–	+++	+	++	+	++	+++
P. aeruginosa	–	–	–	–	–	–	++	–	++	++	+++

Standard–Amoxycillin.

Zone of inhibition in mm

+: 8–4; ++: 15–20; +++: 20–25

Table 60.2: Zone of Inhibition of Substituted 1,3,4-oxadiazoles Derivatives for Fungal Activity

Name of Organism	*Zone of Inhibition in mm*										
	IV a_1	*IV a_2*	*IV a_3*	*IV b_1*	*IV b_2*	*IV b_3*	*IV c_1*	*IV c_2*	*IV c_3*	*IV c_4*	*Std.*
C. albicans	+	++	–	+++	++	++	+	++	++	+++	+++

Standard: Fluconazole.

Zone of inhibition in mm

+: 6–13; ++: 14–19; +++: 19–23.

Results and Discussion

Compound no. IV a_1, IV a_3, IV b_3, IV c_3 and IV c_4 had showed moderate antibacterial activity as compared to that of standard *i.e.* amoxycillin and all compounds excluding IV a_3 showed mild to good antifungal activity against C. *albicans* as compared to that of standard *i.e.* fluconazole.

References

Bhargav, N. and Nair, M.G.R., 1957. Electrochemical studies on some pyrazole, oxadiazole and thiadiazole derivatives. *J. Indian Chem. Soc.*, 34: 42.

Crushshank, R. *et al.*, 1998. *Medical Microbiology*. Churchill Livingstone, 2, pp. 190.

Hyman, M., Jr., 1977. *U.S. Patent*, 4017738; *Chem. Abs.*, 87: 30914.

Ram, V.J. and Pandey, H.N., 1974. *Chem. Abs.*, 80: 14880.

Saikawa, Isamu and Maeda, Toyyo, 1970. *Japan Patent*, 7008, 862.

Thomas, J., 1974. *Gen. Offen*; 2403357; *Chem. Abs.*, 81: 136153.

Umio, Suminori, Karyone, Kazuo, and Kishimoto, Teiji, 1969. *Japan Patent*, 7006, 2313.

Weilder, Smith E., 1964. U.S. Patent, 3, 127, 410; *Chem. Abs.*, 61: 3118.

Chapter 61

Carbamate Pesticide Methomyl (Lannate) Induced Chromosomal Aberrations and Micronucleus Test on the Freshwater Cichlid *Oreochromis mossambicus* (Peters)

J.J. Arockia Rita and M.C. John Milton
P.G. and Research Department of Advanced Zoology and Biotechnology, Loyola College, Chennai – 600 034

ABSTRACT

Oreochromis mossambicus revealed 22 pairs of chromosomes. Out of 22 pairs observed, the first pair being large and subtelocentic designated as 'Marker pair' which could be easily identified in all the well spread compliments the second pair was also subtelocentric but was small, while the rest were acrocentric. When the fishes were exposed to pesticide methomyl for a period of 15, 30 and 45 days chromosomal aberrations like ploidy, endoreduplications, chromotid break, chromatid gap, centromeric deletion, fusions, ring chromosome, acentric fragments, sticky plates, clumping and partial sticky plates were observed. No micronucleus was observed in control tissues and it was observed at 24hrs exposure and declined to the maximum at 45 days sublethal exposure.

Keywords: *Oreochromis mossambicus, Methomyl (Lannate), Chromosomal aberrations and micronucleus.*

Introduction

Out of 20,000 odd species of fishes described by taxonomists, the cytological informations of about only 1,300 species of inland and marine fishes were reported (Ojima, 1981). A preliminary study on the chromosomes of *Tilapia mossambica* from solid tissues such as gills of male and female specimen was conducted by Natarajan and Subramanyan (1968). The chromosome number from both sexes was observed as 2n = 44. Youssef *et al.* (1989) studied cellular DNA content and chromosome number in 4 species of *Tilapia* using gill, spleen and gonad. Franck *et al.*, 1992, studied the genetic variability in a family of satellite DNAs from Tilapia. An improved technique for chromosome karyotyping from embryonic and soft tissues of *Tilapia* and Salmonids were reported by Hussain and McAndrew (1994). Since most of the chemicals and pesticides used terrestrially end up finally in aquatic systems, it is the fishes which are affected and should be monitored for these subtle but genetically damaging changes. Cytological end points like Sister Chromatic Exchange (SCE), Chromosomal aberrations (CA) and Micronucleus (MN) formation can be used in very sensitive genetic assays for detecting environmental mutagens and genotoxic pollutants even at subtoxic levels (Kligerman, 1982).

In India *Oreochromis mossambicus* has been advocated to be a better model for mutagenicity testing (Manna, 1984). The maximum study has been carried out in *Tilapia* by Manna and his collaborators since this species satisfy most of the requirements as experimental model suggested by Kligerman (1982). A comparative study on the mitotic indices of *Oreochromis mossambicus* in natural and polluted habitats was carried out by Ouseph *et al.* (1994). Genetic toxicity assays have been enriched by large number of methods for characterising the genotoxcity of chemicals and other environmental pollutants.

However, micronucleus test is considered to be the best short term screening test over the other chromosome aberration and sister chromatic exchange analysis for many reasons. Scoring of micronuclei may indicate chromosome aberrations and genotoxic effects through clastogenic and spindle apparatus damage (Manna *et al.*, 1985). Micronucleus test, which utilise fish erythrocytes was termed as "Piscine erythrocytic micronucleus test" Micronucleus test in the peripheral erythrocytes of the exotic fish *Oreochromis mossambicus* was studied by Manna *et al.*, 1985. Micronucleus test in *Etrophus suratensis* was carried out by Muruganandam, 1993.

Materials and Methods

Fishes were exposed to sublethal concentrations of Methomyl for a period of 15, 30 and 45 days. The sublethal concentrations of 0.03, 0.045 and 0.09 ppm were responsible for 50 per cent mortality of the fish at 96 hrs. Chromosome plates were made from the gill tissues of the fish as per the procedures of Kligerman (1982). The procedure of Manna *et al.* (1985) was followed for micronucleus test.

Results

The gross karyotypical analysis of *Oreochromis mossambicus* revealed 22 pairs of chromosomes. Out of 22 pairs observed, the first pair being large and subtelocentic designated as 'Marker pair' which could be easily identified in all the well spread compliments the second pair was also subtelocentric but was small, while the rest were acrocentric. The chromosomal aberrations observed were ploidy, endoreduplications. Structural aberrations include, chromatid break, chromatid gap, centromeric deletion, fusions, ring chromosome, acentric fragments, sticky plates, clumping and partial sticky plates (Plates 61.1 and 61.2). A general increase in the percentage of chromosomal aberrations were observed at 24 h and a gradual decrease was observed from 48 h to 96 h (Table 61.1). The number of chromosomal aberrations were more in 15 days exposure and there was a slight decrease in 30 and 45 days exposure (Table 61.2).

Table 61.1: Frequency of Chromosomal Aberrations Induced by Methomyl in the Gill Cells of *Oreochromis mossambicus* Exposed for 24, 48, 72 and 96 hrs

Sl.No	Treatment (hrs)	Metaphase Studied	AF	CB	ED	CG	CF	PL	SP	DC	PS	Total no. of Aberrations	Total no. of Aberrated Metaphase	Per cent Aberration	t Value
1.	Control	100	01	02	00	01	0	0	0	0	0	04	05	04	
2.	24	100	08	09	02	07	03	05	06	02	03	45	31	45	3.40
3.	48	100	05	04	00	04	02	03	03	01	01	23	28	23	2.25
4.	72	100	03	02	01	04	01	01	03	01	02	18	20	18	1.64
5.	96	100	03	03	00	03	01	01	01	00	01	13	17	13	1.73

AF: Acentric Fragment; CB: Chromatid Break; ED: Eendoreduplication; CG: Chromatid Gap; CF: Centromeric Fusion; PL: Ploidy; SP: Sticky Plate; DC: Dicentric Chromosome; PS: Partial Stickyplate.

Table 61.2: Frequency of Chromosomal Aberrations Induced by Methomyl in the Gill Cells of *Oreochromis mossambicus* Exposed for 15, 30 and 45 Days

Sl.No	Treatment (hrs)	Metaphase Studied	AF	CB	ED	CG	CF	PL	SP	DC	PS	Total no. of Aberrations	Total no. of Aberrated Metaphase	Per cent Aberration	t Value
1.	Control	100	02	02	0	01	0	01	01	0	0	07	09	07	0
2.	15	100	10	12	04	06	03	04	03	02	01	45	41	45	3.40
3.	30	100	08	08	03	04	02	05	06	01	0	37	38	37	2.25
4.	45	100	06	07	02	02	01	03	04	01	02	25	27	25	1.64

AF: Acentric Fragment; CB: Chromatid Break; ED: Eendoreduplication; CG: Chromatid Gap; CF: Centromeric Fusion; PL: Ploidy; SP: Sticky Plate; DC: Dicentric Chromosome; PS: Partial Stickyplate.

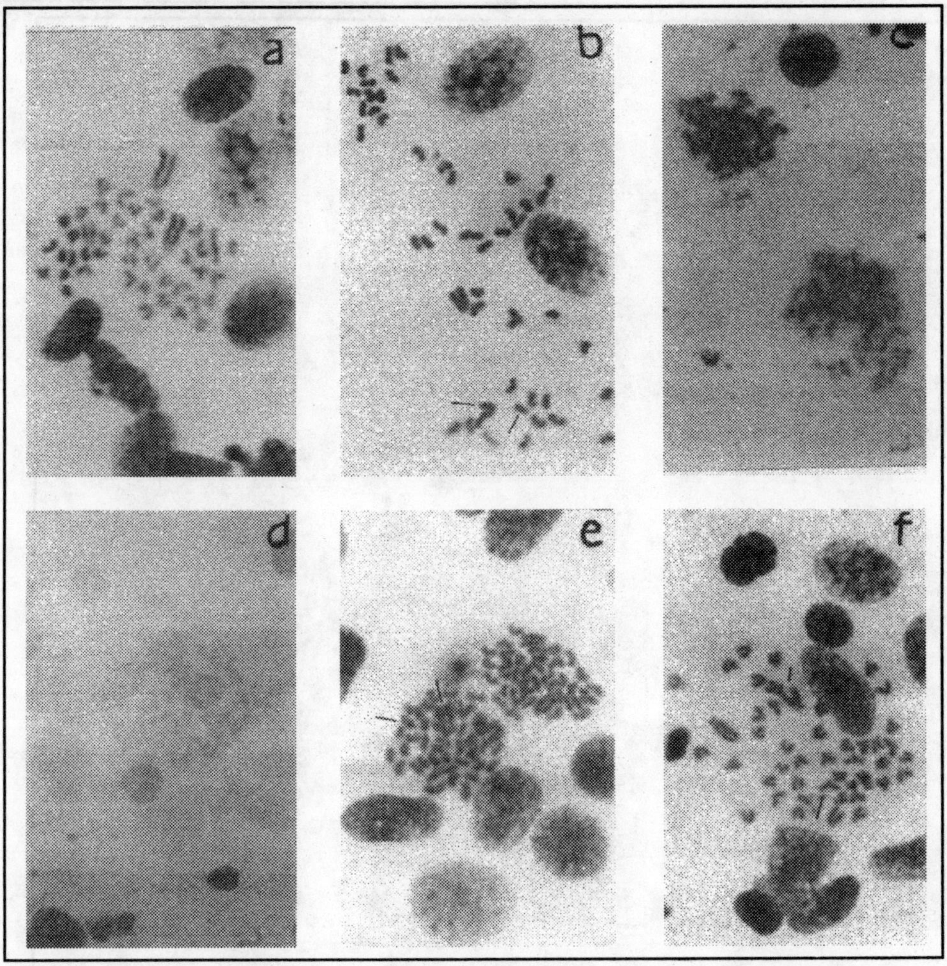

Plate 61.1: Methomyl (Lannate) Induced in Chromosomal Aberrations of *Oreochromis mossambicus* (Peters)

(a) Normal plate; (b) Break in the marker pair and acentric fragment; (c) Sticky plate; (d) Endoreduplication; (e) Break and fusion; (f) Break in the marker pair.

In control group of fishes, absolutely no micronucleus was observed. In pesticide treated fishes at 24 h, the number of micronucleus was observed to be 12 per cent and the number gradually decreased at 96 h (Plate 61.3). When exposed to sublethal concentrations, the micronucleus number decreased to a minimum at 45 days at highest sublethal concentration (Table 61.3). The frequency of micronucleus was more in 15 days exposure than 30 and 45 days exposure (Table 61.4).

Discussion

In *Oreochromis mossambicus* the tissues of juvenile and growing fishes served as best materials for the cytogenetic studies. The gill epithelium gave good results than the kidney tissue. The diploid number of chromosome observed in the present study is in conformity with the earlier reports of

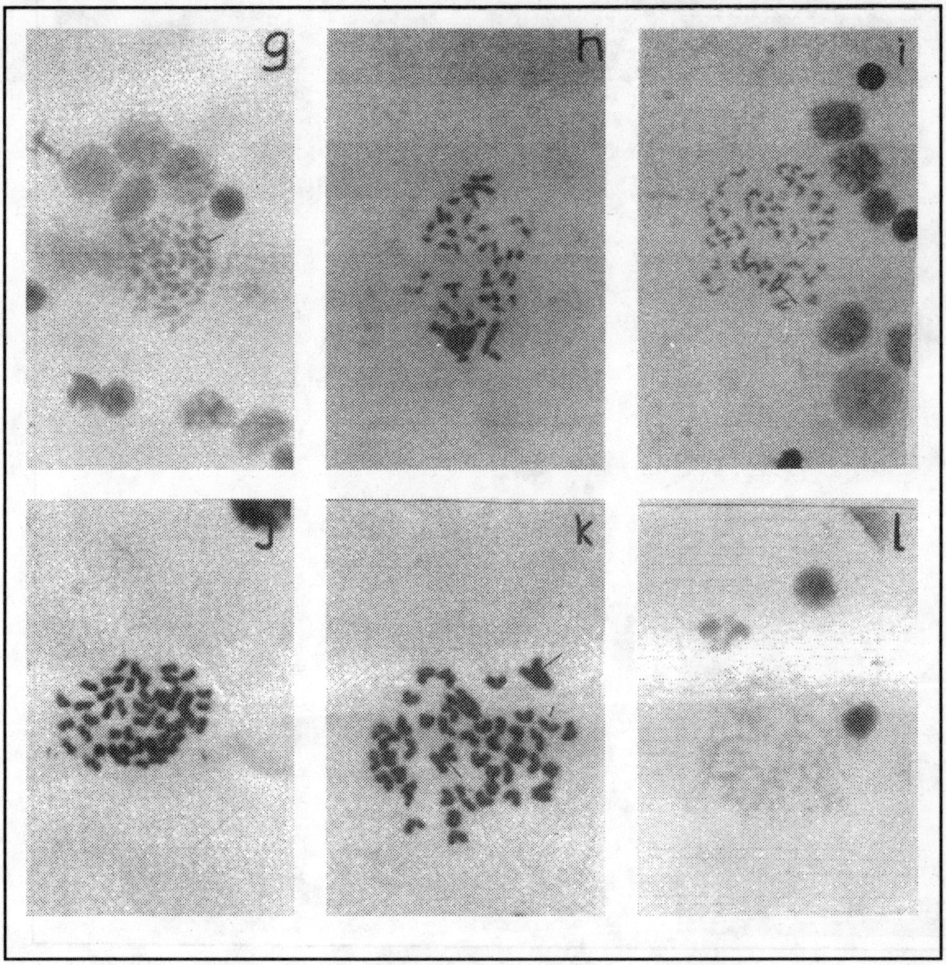

Plate 61.2: Methomyl (Lannate) Induced in Chromosomal Aberrations of *Oreochromis mossambicus* (Peters)

(g) Ring chromosome; (h) Rod like chromosomes; (i) Dicentric chromosomes; (j) Short and stumpy chromosomes; (k) Centomeric fusion, constriction in the marker arm, gap (l) Polyploidy

Manna *et al.*, 1985. The types of aberrations observed in the present study were chromatid breaks, polyploidy, acentric fragments, centromeric fusion, stickiness, subchromatid breaks and such other aberrations. These observations were in accordance with the findings of Rishi and Grewal (1993 and 1995). Such chromosomal aberrations in the tissues of the fishes due to pesticide pollution has been reported several times (Manna, 1990). Evaluation of the genotoxic effects of pesticides is of great importance since their extensive use could be deleterious to the ecosystem in general and mankind in particular. The problem of this kind has received much attention in recent past (Mohammed *et al.*, 1990).

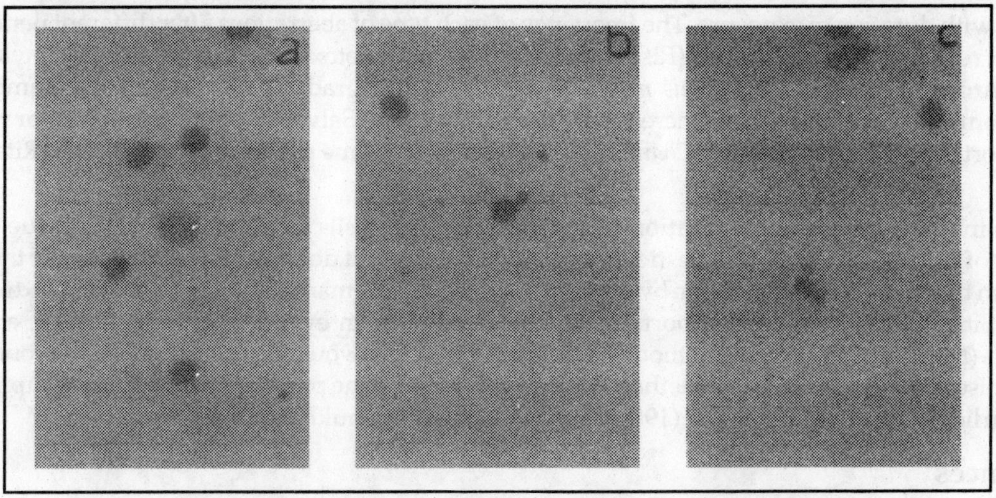

Plate 61.3: Methomyl (Lannate) Induced in Chromosomal Aberrations of *Oreochromis mossambicus* (Peters)

(a) Normal RBCs; (b) Micronucleus; (c) Trinucleated cell.

Table 61.3: Frequency of Micronucleus Induced by Methomyl in *Oreochromis mossambicus* Exposed for 24, 48, 72 and 96 hrs

Sl.No.	*Treatment (hrs)*	*No. of Cells Oserved*	*ANM*	*PC*	*SC*	*Total No. of Micro-nucleus*	*Per cent of Micro-nucleus*	*t Value*
1.	Control	10,000	0	0	0	0	0	0
2.	24	10,000	05	03	04	12	12	3.40
3.	48	10,000	04	04	02	10	10	2.25
4.	72	10,000	02	01	05	08	08	1.64
5.	96	10,000	02	01	01	04	04	1.73

Table 61.4: Frequency of Micronucleus Induced by Methomyl in *Oreochomis mossambicus* Exposed for 15, 30 and 45 Days

Sl.No.	*Treatment (hrs)*	*No. of Cells Oserved*	*ANM*	*PC*	*SC*	*Total No. of Micro-nucleus*	*Per cent of Micro-nucleus*	*t Value*
1.	Control	10,000	0	0	0	0	0	0
2.	15	10,000	07	05	05	17	17	3.40
3.	30	10,000	05	06	04	15	15	2.25
4.	45	10,000	03	03	03	09	09	1.64

In the present study majority of the control cells included metaphases with 2n = 44. But the irregular cells of treated tissue have metaphase (4n) and polyploidy. The results revealed that tetraploidy were the highest degree of ploidy reached in the juvenile fishes. In general, the aberrations tend to decrease with duration of exposure. The frequency of each type of aberrations after different periods of treatment revealed a regular decline (Rishi and Sunita, 1995). Genotoxic effect of pyrethroid fenvalerate on the chromosomes of *Oreochromis mossambicus* revealed a gradual decrease in the number of chromosomal aberrations with the increase in duration (Rita and Selvanayagam, 1998). Similar results were reported in the feshwater fish *Channa punctatus* using the same pyrethroid fenvalerate (Rita *et al.*, 1998).

This may be due to the elimination of aberrated or dead cells from the cell cycle or due to the recovery of defects caused by the pesticide by the various inbuilt detoxifying mechanism of the fish. Studies on the micronuclei formation by genotoxic agents in non-mammalian vertebrates is extremely limited. Since fish is the most important biological monitor in evaluating the genotoxic aquatic pollutants (Manna, 1983), a quick method of detection of the latter would be useful and the micronucleus test promises as one such possible method (Manna *et al.*, 1985). The present investigation is supported by the earlier works of Manna *et al.* (1985) and Manna and Sadhukhan (1992).

References

Franck, J.P.C., J.M. Wright and B.J. McAndrew, 1992. Cadmium induced metabolic changes in Zooplankton. *Ind. J. Env. Prot.*, 12(10): 727–732.

Hussain, M.G and McAndrew, 1994. An improved technique for chromosome karyotyping from embryonic and sift tissue of *Tilapia* and Salmonids. *Asian. Fish. Sci.*, 7: 187–190.

Kligerman, A.D., 1982. The use of cytogenetics to study genotoxic agents in fishes. In: *Cytogenetic Assays of Environmental Mutagens*, (Ed.) T.C. Hsu. Allenheld, Osmum and Co., N.J., p. 161–181.

Manna, G.K and A. Sadhukan, 1992. Mutagenic potential of anisole (Methyl Phenyl Ether) on *Oreochromis mossambicus. J. Inland Fish. Soc.*, India, 24(1): 40–49.

Manna, G.K., 1983. Cytogenitic studies on fishes and amphibian. *General Research in India*, ICAR Publ., p. 244–273.

Manna, G.K., 1984. Progress in fish cytogenetics. *Nucleus*, 27(3): 203–231.

Manna, G.K., 1990. Convenor's Address. In: *National Symposium on Environmental Impacts on Animals and Aquaculture*, (Eds.) G.K. Manna and B.B. Jana, p. 11–15.

Manna, G.K., G. Banerjee and S. Gupta, 1985. Micronuclei test of different genotoxic agents in the peripheral erythrocytes of the exotic fish *Saratherodon mossambicus. Nucleus*, 28: 176–179.

Mohammed, R., V. Vasudev and N.B. Krishnamurthy, 1990. Response of *Drosophila melanogaster* to a carbamate pesticide furadan. In: *Trends in Ecotoxicology*, p. 179–184.

Muruganandam, M., 1993. Genotoxic studies in *Etroplus suratensis* (Bloch) using micronuclei test. *Dissertation* submitted for the degree of Master of Science (Mariculture), Cochin University of Science and Technology, Cochin, p. 30–35.

Natarajan, R. and K. Subramanyan, 1968. A preliminary study on the chromosomes of *Tilapia mossambica* (Peters). *Curr. Sci.*, 37(9): 262–263.

Ojima, Y., 1981. *Report of the Biological Laboratory*, Faculty of Science, Kwansei Gakuin Univ., 4: 1–137.

Ouseph, A., D. Sudarsanam, P. Gandheeswari and T. Ambrose, 1994. Influence of physical and chemical parameters of river Cooum, Madras on the mitotic index of *Oreochromis mossambicus. Proc. Sys. Pest. Resid. Env. Poll.*, p. 97–102.

Rishi, K.K. and S. Grewal, 1993. Chromosome aberration test for the insecticide Dichlorvos on fish chromosomes. *Mut. Res.*, 344: 1–4.

Rishi, K.K and S. Grewal, 1995. Cytogenic screening of effect of a mosquito larvicide on kidney cell chromosomes of *Channa punctatus*. Abst in English. In: *All India Congress on Cytology and genetics*, pp. 87.

Rita, A.J.J and M. Selvanayagam, 1998. Genotoxic effect of fenvelarate on the chromosomes of fish *Oreochromis mossambicus* (Peters). *Poll. Res.*, 17(2): 119–122.

Rita, A.J.J., M. Valarmathy and M. Selvanayagam, 1998. Chromosomal aberration test for the pesticide fenvelarate on the fish *Channa puctatus* (Bloch). *Proc. Acad. Environ. Biol.*, 7(2): 167–171.

Youssef, M.K., E.A. Bad, H. Badaw, S. El Deed, 1989. Cellular DNA content and chromosome number in 4 sps of *Tilapia. Alexandria, Egypt*, p. 81–847.

Chapter 62

Behavioural Responses of Male Lesser Bandicoot Rat *Bandicota bengalensis* (Gray and Hardwicke) Towards Cyclic and Pregnant Female Urine

Gurpreet Kaur and V.R. Parshad
Department of Zoology and Fisheries,
Punjab Agricultural University, Ludhiana – 141 004, India

ABSTRACT

Adult males (6–7 months old) littermates of the lesser bandicoot rat, *Bandicota bengalensis* were exposed to urine odour of cyclic and pregnant female rats in different choice combinations. The male rats were exposed to distilled water versus urine of cyclic and pregnant female rats, was attracted to urine odour than to distilled water, which was maximum to oestrous urine odour than to that of other stages and of pregnant female urine. The male rats after sniffing the urine from a distance of about 30 cm licked the urine and then applied it to 1 × 3 cm size area on either sides off the rump extending toward thigh. Body marking was immediately followed by self grooming around the mouth and face descending ventrum and anogenital area and tail. The male rats were also exposed to urine of different stages of cyclic female *B. bengalensis.* The male rats did not exhibit any distinct preference between any two urine odours of the cyclic rats, except for the increased response to urine of oestrous and metoestrous compared to that of the proestrous rats. The male rats showed significantly more number of visits/3 mins to proestrous than to oestrous and metoestrous female urine but visits to dioestrous were more than to proestrous urine. The male rats when exposed to cyclic and pregnant female urine simultaneously. The number of visits/3 mins were more towards pregnant female urine as compared to four stages of oestrous cycle but time spent/visit was more with oestrous urine similarly body marking and self grooming acts were more with oestrous urine as compared to other stages. The male rats were exposed to each stage of oestrous cycle in comparison to pregnant female urine. The male rats showed more number of visits, licking time/visit to pregnant female urine but number of visits, sniffing time/visit, grooming time/visit, total time spent/visit were more towards oestrous urine as compared to pregnant female urine.

***Keywords:** Urine, Bandicota bengalensis, Oestrous, Cyclicity, Pregnancy.*

Introduction

Urine, the principal medium by which metabolic waste is eliminated from an organism, functions centrally in the lives of many species of rodents as a source of chemical signals. The urine odours of mammals contain information about the species, sex, age (Osada *et al.*, 2003), dominance status (Desjardens *et al.*, 1973), reproductive status (Vandenbergh, 1983) and individual identity of each animal (Isles *et al.*, 2002).

Rodent urinary semiochemicals may have major effects on the reproductive physiology of conspecifics (Primer pheromone effects) and they may have more immediate behavioural consequences. Male and female attractants have been noted in urines of the opposite sex while the presence of female mouse urine will, for example, increase social investigation within established groups ofmale mice (Davies and Bellamy, 1974).

The lesser bandicoot rat, *Bandicota bengalensis* which is the most predominant agricultural pest in South Asia (Parshad, 1999) breeds throughout the year with large litters and maintains its large population (Parshad *et al.*, 1989). The lesser bandicoot rat is a nonseasonal, polyoestrous mammal. The duration of oestrous cycle in it was 4–5 days (Sahu and Maiti, 1978) The rats exhibited strong discrimination between urine and water and this discrimination was effected by oestrous cycle of the female (Shumake *et al.*, 1971). Preliminary studies on the effect of oestrous cyclicity on urine attraction were available on *Tatera indica* (Idris and Prakash, 1983) *Mus musculus* (Shumake *et al.*, 1971; Dixon and Mackintosh, 1975) and *Rattus norvegicus* (Gao, 1991). However, no information is available on effects of oestrous cyclicity on urine attraction in *Bandicota bengalensis.* This paper describes the behavioural responses of male lesser bandicoot rat *Bandicota bengalensis* towards cyclic and pregnant female urine.

Materials and Methods

Behavioural responses of adult male rats to urine of cyclic and pregnant female rats were determined by exposing males to urine samples of females in different stages of oestrous cycle and to urine of pregnant females.

Grouping of Rats

In this experiment, 5 laboratory bred adult males 6–7 months old littermates were used as recipients and 8–10 months old adult females derived from different colonies were used as donors of urine. All were individually housed in cages.

Oestrous Cyclicity

Vaginal smears of female rats were taken with 0.9 per cent saline solution and stages of the oestrous cycle were determined by the method of Long and Evans (1922).

Proestrous was characterized by the presence of small, round nucleated cells. Oestrous by cornified cells, Metoestrous by cornified cells and leucocytes and Dioestrous by Epithelial cells and leucocytes in the vaginal smear.

Collection of Urine

For the collection of urine samples, the sheets of plastic were placed under each of the donors (Gao, 1991). Rat urinated on the plastic sheet, samples were collected with a clean glass pipette and stored for 15 days in sterilized vials at –20°C and thawed immediately before use.

Experimental Design

All recipient male rats were individually housed from day 30 of their age onwards in experimental cages. Each cage consisted of a set of two laboratory cages (A and B) joined together through their one of the passage so that the rat had free access to each cage A and B.

The door connecting the cages had a perforated shutter. Food and water were provided *ad libitum* in cage A and cage B served as the open area. Small glass cavities for test samples of urine and or water were placed at 6 cm distance from each other. The odour testing session was started when the male rats were 6 months old. During the odour testing session no other cage/animal were placed in the experimental room except for the observer for whom the rats had developed familiarity during daily care and feeding. Odour testing experiment was carried out between 0930 to 1030 hrs under normal temperature and day light conditions. The perforated barrier between the cages was closed 10 mins before the experiment with rat separated in cage A.

The main hypothesis of this experiment was to determine the ability of male rats to detect variations in urine odour due to oestrous cyclicity and pregnancy of female donors. Therefore, males were exposed to urine odour of cyclic and pregnant rats in different choice combinations as given below:

Different Combinations of Distilled Water v/s Urine of Cyclic and Pregnant Females to which Male Recipients were Exposed for Odour Testing

Distilled water v/s proestrous urine

Distilled water v/s oestrous urine

Distilled water v/s metoestrous urine

Distilled water v/s dioestrous urine

Distilled water v/s pregnant female urine

Different Combinations of Urine of Cyclic and Pregnant Rats to which the Male Recipients were Exposed during Odour Testing

Proestrous v/s Oestrous

Proestrous v/s Metoestrous

Proestrous v/s Dioestrous

Oestrous v/s Metoestrous

Oestrous v/s Dioestrous

Proestrous v/s Pregnant

Oestrous v/s Pregnant

Metoestrous v/s Pregnant

Dioestrous v/s Pregnant

Simultaneous Exposure of Urine of Cyclic and Pregnant Rats to Male Recipients

To determine the discriminatory ability of male rats among urine odours of 4 stages of oestrous cycle and pregnant rats, their samples were exposed simultaneously in the test cage system, so that each male rat had access to urine samples of the following stages simultaneously.

Proestrous – Oestrous – Metoestrous – Dioestrous – Pregnant.

The 3 min urine odour test began 10 mins after the glass cavities were placed into the test cage when the perforated barrier between the cages were removed. This method was earlier adopted by Ferkin (1999). Criterion for investigation of urine odour was that the rat licked or sniffed a stimulus urine and/or its nose directed towards the urine odour. The distance between the two glass cavities was about 6 cm. The location of the glass cavities was rotated daily. The experimenter wore disposable latex gloves to minimize human scent transfer while handling glass cavities.

Observations

When the glass cavities containing urine were placed in the cage the observations were taken for 3 mins period. The frequency of behavioural acts was noted with the help of a stopwatch. The following observations were recorded:

First Response

First directed visit to the water/urine samples.

Time Spent

The total time and time spent per visit to each sample in glass cavities.

Number of Visits

Number of visits/3 mins to each urine sample.

Sniffing

Sniffing of sample as revealed by movements of the snout directed towards urine.

Licking

Licking of sample by tongue by the rat.

Self Grooming

Behavioural act involving cephalocaudal progression of rhythmic movements of the paws around the mouth and face descending to ventrum and anogenital, area and tail.

Statistical Analysis

The time spent/visit, number of visit, sniffing/visit, licking/visit and self grooming/visit by male rats were analyzed by a nonparametric test that is Wilcoxon paired-sample test based on ranks (Zar, 1974).

The differences between the attraction of male rat to urine of different stages of oestrous cycle and pregnant female were analyzed by Friedman's analysis of variance by ranks (Zar, 1974).

Results and Discussion

Urine of Cyclic and Pregnant Rats in Choice to Water

The lesser bandicoot rat exhibited strong discrimination between urine and water and this discrimination was effected by oestrous cycle of the female. The number of visits, time spent/visit, sniffing, licking/visit and grooming/visit were significantly more towards the female urine as compared to distilled water (Table 62.1). During the odour test the male littermate rats (which are genetically similar) after sniffing the urine from a distance of about 30 cm licked the urine and applied it to 1 × 3 cm size area on either sides of the rump extending toward thigh. The marked area became wet as a result of the application of urine possibly mixed with some saliva. Body marking with urine on the rump and thighs was immediately followed by vigorous self grooming involving cephalocaudal progression of rhythmic movements of the paws around the mouth and face descending to ventrum and the anogenital area and tail.

Relationship of oestrous cycle and reproduction is well known in rodents. The active period of oestrous cycle is oestrous which is the period of sexual desire of the female. During proestrous the animal prepares for heat, during metoestrous and oestrous heat subsides and dioestrous is the period of quiescence (inactive stage) (Greenwald, 1978). The lesser bandicoot rat is a non seasonal, polyestrous mammal. The duration of oestrous cycle in it was 4–5 days (Sahu and Maiti, 1978).

The male *B. bengalensis* was attracted to urine odour than to distilled water as evident by first response (66.7 to 86.7 per cent) which was maximum to oestrous urine odour than to that of other stages and of pregnant rats. The rats showed significantly more sniffing acts towards female urine than to distilled water as determined by Wilcoxon's signed rank test ($P < 0.05$) and the frequency of sniffing acts in choice to water was significantly more ($\chi^2 = 68.933$ $0.01 > P < 0.001$) towards oestrous urine as compared to urine of proestrous, metoestrous, dioestrous and pregnant females urine as determined by Friedman's analysis of variance based on ranks. However, the duration of licking and self grooming acts was significantly more with dioestrous urine ($\chi^2 = 0.005 > P < 0.001$) as compared to oestrous, proestrous, metoestrous and pregnant female urine but the number of visits were significantly more towards oestrous urine ($\chi^2 = 0.01 > P < 0.001$) as compared to other stages of oestrous cycle.

The time spent/visit was significantly more towards metoestrous urine ($\chi^2 = 0.005 > P < 0.001$) as compared to that other stages in choice to water. This observation indicated that the rats visited and spent considerable time with odour of inactive rats.

The results of present studies on *B. bengalensis* can be compared with that of Gao (1991) in which oestrous urine was also most attractive to the male rats, but was not significantly different from urine from other stages of oestrous cycle ($P > 0.05$). In contrast in *B. bengalensis* variations in the attraction towards urine of different stages of oestrous stages and pregnant rats were statistically significant ($\chi^2 = 0.005 > P < 0.001$). The time spent was significantly more ($\chi^2 = 0.005 > P < 0.001$) towards metoestrous urine as compared to other stages. According to Idris and Prakash (1983) the frequency of marking in *Tatera indica* increased significantly in the vicinity of oestrous ($P < 0.01$) and dioestrous urine ($P < 0.01$). Similarly in *B. bengalensis* the frequency of licking and self grooming acts followed by body marking was significantly more with dioestrous urine and oestrous urine in choice to distilled water compared to other stages of oestrous cycle and pregnant female urine.

The attraction of the male *B. bengalensis* to urine odour as affected by cyclicity of female donors may be due to certain volatile compounds were released at different stages of the oestrous cycle.

Table 62.1: Differential Behavioural Responses of Male *B. bengalensis* Toward Female Urine

Treatment Av/s B (n = 15)	*Number (%) of Rats Showing First Response*		*Number of Visits*		*Sniffing Time (Secs)/Visit*		*Licking Time/ Visit*		*Body Marking and Self Grooming/Visit (Secs) (Secs)*		*Time Spent/Visit (Secs)*	
	A	*B*	*A*	*B*	*A*	*B*	*A*	*B*	*A*	*B*	*A*	*B*
Distilled water v/s proestrous Urine	4 (26.66)	11 (73.33)	1.2±0.17	4.2±0.59*	4.06±0.80	5.00±1.46*	0	2.73±0.77	0	3.12±1.06	4.06±0.80	10.88±0.94*
Distilled water v/s oestrous urine	2 (13.33)	13 (86.66)	1.13±0.25	7.4±1.13*	3.36±1.06	1.84±0.52*	0	2.69±0.44	0	5.34±0.79	3.36±1.06	9.87±1.13*
Distilled water v/s metoestrous urine	4 (26.66)	11 (73.33)	1.2±0.17	7.06±0.67*	3.76±0.73	1.36±0.62*	0	2.86±0.36	0	8.14±1.65	3.76±0.73	12.37±1.38*
Distilled water v/s dioestrous urine	5 (33.33)	10 (66.66)	1.26±0.11	5.2±0.77*	5.3±0.96	2.01±0.80	0	4.85±1.20	0	5.31±1.42	5.3±0.963	12.19±1.86*
Distilled water v/s pregnant female urine	3 (20)	12 (80)	1.2±0.17	5.66±0.58*	6.21±1.16	3.941±1.34*	0	2.23±0.52	0	3.68±0.82	6.21±1.16	9.83±0.73*
			$\chi^2 = 68.933$ df = 11		$\chi^2 = 8.16$ df = 11		$\chi^2 = 44.673$ df = 11		$\chi^2 = 34.169$ df = 11		$\chi^2 = 51.07$ df = 11	
			$0.01 > P < 0.001$		$0.001 < P < 0.005$		$0.005 > P < 0.001$		$0.005 > P < 0.001$		$0.005 > P < 0.001$	

*: Significant differences between A and B by Wilcoxon paired sample test ($P < 0.05$).

χ^2 Significance of variations among all treatments by Friedman's analysis of variance.

Comparison Between Two Stages of Oestrous Cycle

The male bandicoots did not exhibit any distinct preference between any two urine odours of the cyclic rats as revealed by non-significant differences in the first response of males in different tests except for the increased response to urine of oestrous and metoestrous compared to that of the proestrous rats. Analysis of other behavioural responses of male rats in choice tests using urine of two stages of oestrous by Wilcoxon paired sample test based on ranks revealed significant differences in the response of male rats to female urine from different stages of the oestrous cycle (Table 62.2). The male rats showed significantly more number of visits/3 mins to proestrous than to oestrous ($\chi^2 = 0.005 > P < 0.50$) and metoestrous female urine ($\chi^2 = 0.005 > P < 0.50$) but visits to dioestrous were more ($\chi^2 = 0.005 > P < 0.50$) than to proestrous urine. Similarly compared to oestrous and dioestrous urine the males showed more visits to metoestrous urine ($\chi^2 = 0.005 < P < 0.50$) but compared to dioestrous the males preferred to visit oestrous urine ($\chi^2 = 0.005 > P < 0.50$).

Time spent/visit to the urine sample had no correlation with number of visits in different choice odour tests except between proestrous versus oestrous and metoestrous female urine.

The sniffing time/visit (secs) was significantly more to proestrous urine than metoestrous and dioestrous urine and in comparison to proestrous and metoestrous it was more to oestrous urine ($\chi^2 = 0.005 > P < 0.50$). The sniffing time/visit was significantly more towards dioestrous than oestrous urine. The licking time/visit (secs) was significantly more towards proestrous urine compared to oestrous and metoestrous urine and was significantly less than to dioestrous urine ($\chi^2 = 0.005 > P < 0.50$). Licking time/visit was significantly less towards oestrous than metoestrous and dioestrous urine. As compared to metoestrous urine the licking time/visit was significantly more to dioestrous urine ($\chi^2 = 0.005 > P < 0.50$).

The comparison among the stages were analyzed by Friedman's two way analysis of variance on the basis of ranks. The number of visits/3 mins, total time spent/visit (secs), licking time/visit and sniffing time/visit (secs) during their choice visits to different combinations of urine 2 different stages of oestrous cycle were statistically significant (Table 62.2).

The number of body marking and self grooming acts were significantly ($P < 0.25$) more towards proestrous urine except as compared to dioestrous urine similarly the time spent/act was significantly more towards proestrous urine except dioestrous urine. The time spent/act was significantly more towards metoestrous as compared to oestrous and, dioestrous urine, but significantly more towards oestrous as compared to dioestrous urine. The difference in time spent/act (secs) among all the stages was analyzed by Friedman's two way analysis of variance on the basis of ranks. The time spent/act was statistically significant ($\chi^2 = 14.776$ $P < 0.25$ df = 11).

Comparison of Responses of Male Towards Urine of Four Stages of Oestrous and Pregnant Rats

The male bandicoot rats did not show distinct discrimination among urine odours of four stages of oestrous cycle and pregnant rats as revealed by non-significant differences in the first response except to urine of pregnant rats. The number of visits/3 mins were significantly ($\chi^2 = 3.93$ $P < 0.50$) more towards pregnant female urine as compared to four stages of oestrous cycle. The time spent/visit was significantly ($\chi^2 = 2.63$ $P < 0.75$) more with oestrous urine. The sniffing time/visit was significantly ($\chi^2 = 6.62$ $P < 0.25$) more with metoestrous urine but licking time/visit was significantly ($\chi^2 = 4.82$ $P < 0.50$) more with dioestrous urine (Table 62.3). The body marking and self grooming time/visit was significantly more ($\chi^2 = 6.69$ $P < 0.25$) with oestrous urine as compared to other stages (Kaur, 2005).

Table 62.2: Comparison of Behavioural Responses of Male Towards Urine of Different Stages of Cyclic Female *B. bengalensis*

Treatment A v/s B (n = 15)	*Number (%) of Rats Showing First Response*		*Number of Visits/ 3 min*		*Sniffing Time (Secs)/Visit*		*Licking Time/ Visit*		*Body Marking and Self Grooming/Visit (Secs) (Secs)*		*Time Spent/Visit (Secs)*	
	A	*B*	*A*	*B*	*A*	*B*	*A*	*B*	*A*	*B*	*A*	*B*
Proestrous v/s oestrous	6 (40)	9 (60)	6.8±1.13	5.93±0.97	0.93±0.48	2.01±1.20*	4.04±1.48	2.03±0.38	5.21±1.01	4.95±0.79*	10.18±1.3.7	8.99±1.01
Proestrous v/s metoestrous	6 (40)	9 (60)	6.33±0.94	4.73±0.73*	1.92±0.81	1.32±0.45*	3.47±0.78	2.26±0.39*	4.61±0.83	3.71±1.20*	10.00±0.65	7.29±1.19*
Proestrous v/s dioestrous	7 (46.66)	8 (53.33)	3.06±0.50	6.4±0.84	2.33±0.94	0.73±0.36*	1.72±0.37	2.94±0.41*	3.19±0.74	6.31±1.42*	7.24±0.82	9.98±1.43*
Oestrous v/s metoestrous	7 (46.66)	8 (53.33)	5.06±0.75	5.33±0.48*	2.76±1.02	1.27±0.59*	1.59±0.31	2.61±0.91*	4.52±1.05	6.52±1.07*	8.87±0.93	10.4±1.10*
Oestrous v/s dioestrous	8 (53.33)	7 (46.66)	5.86±1.03	4.33±0.65*	2.26±0.85	23.01±0.76*	.03±0.36	1.99±0.79*	5.25±1.42	3.07±0.92*	9.54±1.32	8.07±1.00*
Metoestrous v/s dioestrous	8 (53.33)	7 (46.66)	5.2±0.61	4.86±0.85*	1.99±0.65	1.60±0.72*	1.83±0.25	1.81±0.31*	5.25±0.72	5.09±1.18*	9.07±0.55	8.5±1.17*
			χ^2 = 15.660 (P < 0.25) df = 11		χ^2 = 10.97 (P < 0.50) df = 11		χ^2 = 15.624 (P < 0.25) df = 11		χ^2 =14.776 (P < 0.25) df = 11		χ^2 = 19.9886 (P < 0.05) df = 11	

*: Significant differences between A and B by Wilcoxon paired sample test (P < 0.05).

χ^2 Significance of variations among all treatments by Friedman's analysis of variance.

Table 62.3: Behavioural Responses of Male *B. bengalensis* Exposed to Cyclic and Pregnant Female Urine Simultaneously

Treatment (n = 15)	*Number of Rats Showing First Response*	*Number of Visits/3 mins*	*Sniffing Time/ Visit (Secs)*	*Licking Time/ Visit (Secs)*	*Body Marking and Self Grooming Time/ Visit (Secs)*	*Total Time Spent/Visit (Secs)*
Proestrous	3 (20)	2.53±0.58	1.96±0.77	1.72±0.51	7.31±1.24	
Oestrous	3 (20)	2.66±0.54	1.50±0.45	1.81±0.42	4.38±0.96	7.69±1.02
Metestrous	2 (13.33)	2.4±0.37	2.40±1.03	1.13±0.33	3.02±0.93	6.55±1.09
Diestrous	2 (13.33)	3.53±0.68	0.73±0.47	0.73±0.47	3.99±0.80	7.32±1.02
Pregnant female urine	5 (33.33)	4.13±0.85	1.04±0.49	1.48±0.34	3.73±0.89	6.25±1.09
		χ^2 = 3.93 (0.75 > P < 0.50) df = 4	χ^2 = 6.62 (0.50 > P < 0.25) df = 4	χ^2 = 4.82 (0.75 > P < 0.50) df = 4	χ^2 = 6.69 (0. 50 > P < 0.25) df = 4	χ^2 = 2.63 (0.90 > P < 0.75) df = 4

χ^2 Significantly different by Friedman's analysis of variance by ranks.

Though the bandicoots could not immediately discriminate 5 urine odours, but other variations in other parameters like the number of visits, time spent per visit, licking, body marking and grooming clearly showed the effect of cyclicity and pregnancy of female rats on attraction of males. This may be due to ovarian steroids which are released at different stages of oestrous cycle. Late during metoestrous through early dioestrous plasma levels of oestradiol begin to rise. The increase continues through dioestrous and early pro estrous to reach peak values and plateau by mid proestrous. Also the peripheral plasma levels of testosterone and androstenedione are similar to those for oestradiol (Dupon and Kim, 1973). Possibly different volatile compounds are excreted in urine during different stages of oestrous cycle which may explain variations in the responses of males to urine of different stages. The volatile compounds which were earlier identified in mice during oestrous stage were 2-hexanone, 2-heptanone, trans-5-heptene-2-one and trans-4-heptene-2-one (Andreolini *et al.*, 1987). Statistically significant alternations in the excretion of urinary volatiles were correlated with the biologically characterized periods.

Self grooming of male *B. bengalensis* may allow individuals to enhance short-range communication with conspecifics. By self grooming and body marking bandicoots may increase the active space that surrounds their bodies within which a conspecifics can detect them as observed earlier in prairie voles (Ferkin *et al.*, 2001).

Comparison of Responses of Male Toward Urine of Stages of Oestrous Cycle and Pregnant Rat

Present studies revealed no clear difference between the first response of male *B. bengalensis* toward proestrous and pregnant female urine odour, but attraction toward urine odour of oestrous, metoestrous and dioestrous in choice to that pregnant rats was evident (Table 62.4).

The male rats showed significantly more number of visits, licking time/visit to pregnant rat urine than to proestrous urine. However, sniffing time/visit, grooming time/visit and total time spent/ visit toward proestrous were significantly ($P < 0.05$) more than to the pregnant female urine. The number of

Table 62.4: Comparison of Behavioural Responses of Male *B. bengalensis* Towards Urine of a Stage of Oestrous Cycle and Pregnant Female

Treatment A v/s B (n = 15)	Number (%) of Rats Showing First Response		Number of Visits/ 3 min		Sniffing Time (Secs)/Visit		Licking Time/ Visit (Secs)		Body Marking and Self Grooming/Visit (Secs)		Time Spent/Visit (Secs)	
	A	B	A	B	A	B	A	B	A	B	A	B
Proestrous v/s pregnant	7 (46.66)	8 (53.33)	4.53±0.65	5.0±0.63*	2.31±0.75	1.60±0.32*	1.91±0.65	3.11±1.05*	4.98±1.31	3.98±0.67*	9.2±1.51	8.69±1.17*
Oestrous v/s pregnant	10 (66.66)	5 (33.33)	6.00±0.88	5.40±0.56	1.58±0.64	1.31±0.51*	2.04±0.32	2.42±0.48*	5.30±0.85	4.41±0.78*	8.92±0.76	8.14±0.87*
Metestrous v/s pregnant	10 (66.66)	5 (33.33)	3.46±0.55	4.86±0.78*	3.90±1.38	2.27±0.72*	3.31±1.09	2.03±0.51*	4.54±1.15	4.95±1.15*	11.76±1.15*	9.25±1.12*
Dioestrous v/s pregnant	9 (60)	6 (40)	4.6±0.57	4.13±0.66*	1.95±0.78	2.54±0.85*	2.43±0.43	2.16±0.52*	6.22±1.29	4.99±1.41*	10.62±1.28	9.69±1.27*
			$\chi^2 = 0.80$ 0.25 < P < 0.50 df = 1		$\chi^2 = -6500^{NS}$		$\chi^2 = -21.21^{NS}$		$\chi^2 = -6.25^{NS}$		$\chi^2 = -16.83^{NS}$	

*: Significantly differences between by Wilcoxon signed rank test (P < 0.05).

χ^2 Significance of variations among all treatments by Friedman's analysis of variance by ranks.

visits, sniffing time/visit, grooming time/ visit, total time spent/visit were significantly ($P < 0.05$) more towards oestrous urine as compared to pregnant female urine. However, licking time/visit was significantly more ($P < 0.05$) towards pregnant female urine as compared to oestrous urine. The sniffing time/visit, licking time/visit and time spent/visit were significantly more ($P < 0.05$) towards metoestrous urine as compared to pregnant female urine, but number of visits/3 mins were significantly more towards pregnant female urine as compared to metoestrous urine. The number of visits/3 mins, sniffing time/visit were significantly more towards pregnant female urine as compared to dioestrous urine, but licking time/visit, grooming time/visit and time spent/visit were significantly more towards dioestrous urine as compared to pregnant female urine.

Comparison of the behavioural responses of male rats among 4 paired combinations of female urine samples by Friedman's two way analysis showed non-significant differences in sniffing time/visit, licking time/visit, grooming time/visit, time spent/visit but significant ($\chi^2 = 0.80$ $0.25 < P < 0.50$) differences occurred in number of visits/3 mins. These results clearly showed that the male rats responded differently to paired presentations of urine samples.

Behavioural differences in male bandicoot rats in response to urine samples of cyclic and pregnant rats observed during the present study may be due to in stage related differences in chemistry of urine of female rats (Jemiolo *et al.*, 1987). The body marking and self grooming act by male *B. bengalensis* may be associated with courtship or mating as male prairie voles self groomed at higher rates when they were exposed to odours of reproductively active females and postpartum oestrous females (Witt *et al.*, 1990).

Self grooming and body marking acts by male *B. bengalensis* with pregnant female urine may be associated with major histocompatiability complex type that influence choice behaviour as observed in mice (Beauchamp *et al.*, 2001). Sexually inexperienced male mice spent more time near pregnant females that carried fetuses most genetically different from the males themselves and foetal major histocompatiability complex influences choice behaviour via maternal secretions/excretions.

Acknowledgements

Authors are thankful to Dr. M.H. Ferkin, Department of Ecology and Organismal Biology, University of Memphis, Memphis, TN 38152, USA., Dr. Lucille Lumley, 503 Robert Grant Avenue, Silver Spring, MD 20910-7500 USA and Dr. R.E. Brown, Department of Psychology, Dalhousie University, Halifax, Nova Scotia, Canada B3H 4J1 for providing literature.

References

Andreolini, F., Jemiolo, B. and Novotny, M., 1987. Dynamics of excretion of urinary chemosignals in the house mouse (*Mus musculus*) during the natural estrous cycle. *Experientia*, 43: 998–1002.

Beauchamp, G.K., Curran, M. and Yamazaki, K., 2001. MHC-mediated fetal odour types expressed by pregnant females influence associative behaviour. *Anim. Behav.*, 60: 289–295.

Davies, V.J. and Bellamy, D., 1974. Effects of female urine on social investigation in male mice. *Anim. Behav.*, 22: 239–241.

Desjadens, C., Maruniak, J.A. and Bronson, F.H., 1973. Social rank in house mice: Differentiation by ultra-violet visualization of urinary marking patterns. *Science*, 182: 939–941.

Dixon, A.K. and Mackintosh, J.H., 1975. The relationship between the physiological condition of female mice and the effects of their urine on the social behaviour of adult males. *Anim. Behav.*, 23: 513–520.

Dupon, C. and Kim, M.H., 1973. Peripheral plasma levels of testosterone, androstenedione and oestradiol during the rat estrous cycle. *J. Endocrinol.*, 59: 653–654.

Ferkin, M.H., 1999. Attractiveness of opposite-sex odour and responses to it vary with age and sex in meadow voles (*Microtus pennsylvanicus*). *J. Chem. Ecol.*, 25: 757–769.

Ferkin, M.H., Leonard, S.T., Heath, L.A. and Mino, C.G.P., 2001. Self grooming as a, tactic used by Prarie voles *Microtus ochrogaster* to enhance sexual communication. *Ethology*, 107: 939–949.

Gao, Y., 1991. Behavioural responses of rats to the smell of urine from conspecifics. *Anim. Behav.*, 42: 506–508.

Greenwald, G.S., 1978. Follicular activity in mammalian ovary. In: *The Vertebrate Ovary: Comparative Biology and Evolution*, (Ed.) Jones, R.E. Plenum Press, New York, pp. 639–689.

Idris, M.D. and Prakash, I., 1983. Response of male Indian gerbil *Tatera indica* to urine from females. *Ind. J. Exp. Biol.*, 21: 97–98.

Isles, A.R., Baum, M.J., Ma, D., Szeto A., Keverne, E.B. and Allen, N.D., 2002. A possible role for imprinted genes in inbreeding avoidance and dispersal from the natal area in mice. *Proceedings of the Royal Society. Proc. R. Soc. Lond. B.*

Jemiolo, B., Andreolini, F., Wiesler, D. and Novotny, M., 1987. Variations in mouse (*Mus musculus*) urinary volatiles during different periods of pregnancy and lactation. *J. Chem. Ecol.*, 13: 1941–1956.

Kaur, G., 2005. Urinary signals in the lesser bandicoot rat *Bandicota bengalensis* (Gray) and their application for management of its population. *Ph. D Dissertation*, Punjab Agricultural University, Ludhiana, India, pp. 1–132.

Long, J.A. and Evans, H.M., 1922. The estrous cycle of the rat and its associated phenomena. *Mem. Univ. Calif.*, 6: 1–148.

Osada, K., Yamzaki, K., Curran, M., Bard, J., Smith, B.P.C. and Beauchamp, G.K., 2003. The scent of age. *Proc. R. Soc. London*, 270: 929–933.

Parshad, V.R., 1999. Rodent control in India. In: *Pest. Mgmt. Rev.*, 4: 97–126.

Parshad, V.R., Kaur, P. and Guraya, S.S., 1989. Reproductive cycles of mammals: Rodentia. In: *Reproductive Cycles of Indian Vertebrates*, (Eds.) Saidapur, S.K. Allied Publishers Limited, New Delhi, pp. 347–408.

Sahu, A. and Maiti, B.R., 1978. Oestrous cycle of the bandicoot rat: A rodent pest. *Zool. J. Linn. Soc.*, 63: 309–314.

Shumake, S.A., Thompson, R.D. and Candill, C.J., 1971. Taste preference behaviour of laboratory versus wild Norway rats. *J. Comp. Physiol. Soc.*, 77: 489–494.

Vandenbergh, J.G., 1983. Social factors controlling puberty in the female mouse. In: *Hormones and Behaviour in Higher Vertebrates*, (Eds.) Balthazart, J., Prove, E.and Gilles, R. Springer-Verlag, Berlin Heidelberg.

Witt, D.M., Carter, C.S., Chayer, R. and Adams, K., 1990. Patterns of behaviour during postpartum estrous in prairie voles. *Anim. Behav.*, 39: 528–534.

Zar, J.H., 1974. *Biostatistical Analysis.* Prentice Hall, Inc. Englewood Cliffs, N.J.

Chapter 63

Ethnobotany and Biodiversity Conservation in the Niger Delta, Nigeria

Rufus M. Ubom

Department of Botany, University of Uyo, P.M.B. 1017, Uyo, Nigeria

ABSTRACT

A complimentary ethnobotanical uses of forest and homestead garden plant species in the Niger Delta are presented. A total of 339 plant species were identified in the forests and homestead gardens belonging to 88 families; all the species have at least, one reported use. Species with more uses are *Elaeis guineensis* (12), *Raphia hookeri* (12), *Cocas nucifera* (11), *Irvingia gabonensis* (11), *Hevea brasiliensis* (9), *Lonchocarpus cyanescens* (9), *Pterocarpus santalinaides* (9) and *Dacryodes edulis* (8). Ethnobotanical uses with more species include medicine (178/52.5 per cent), fuel (107/31.6 per cent), food/condiment (199/29.4 per cent), timber (72/21.2 per cent), fodder/feed (70/20.6 per cent), commercial (58/17.1 per cent), furniture (55/16.2 per cent), fibre/cordage (55/16.2 per cent), hometool (45/13.2 per cent) and fence (38/11.2 per cent). The presence of commercial species (fruits, flavouring, beverages, thickeners, spices, timber) in the forests allows for the discussion of the importance of the lowland rainforests of Niger Delta and the suggestion that modern management technologies, if applied, could lead to cost effective conservation methods and sustainable production of animal food source, building materials for houses, economic and social benefits of the rural dwellers.

Keywords: *Ethnobotany, Biodiversity, Conservation, Niger Delta.*

Introduction

Biodiversity is a concept that involves different facets of biological variety (Peters, 1991) including inter alia, taxonomic richness, genetic differences in each taxon, communities, ecosystems, landscapes within which organisms inhabit, and the indigenous knowledge of nature possessed by the natives living on the land. Biodiversity is thus a prized variety of interwoven moral, aesthetic, utilitarian and economic reasons. The role of biodiversity for agriculture and medicine is often underestimated. Plant breeding often turns to wild relatives and increased genetic materials from them (Spellberg, 1995). The biodiversity of the Niger Delta is intimately associated with the habitat diversity, each supporting diverse and distinctive plants and animals within their niches, some endemic to the region.

The Niger Delta region of Nigeria is particularly important for its oil exploitation. In addition, its vast plants and animal resources have also been immensely exploited. Niger Delta occupies 37700 km^2 of the 923700 km^2 land area of Nigeria (Figure 6.1). It is situated in the southern most part of Nigeria, and bordered to the south by Atlantic Ocean. It comprises eight states of Nigeria: Akwa Ibom, Bayelsa, Cross-River, Delta, Edo, Imo, Ondo and Rivers. The area is essentially a lowland belt about 30m below sea level. The area supports about 19.21 million people and various densities of livestock. These exert pressure on the resources in addition to oil exploitation by numerous multinational companies. Plants and animal species of economic importance are threatened or endangered. Endemic species of ecological importance are also threatened and need be conserved.

Rural communities, in this region, depend on their interactions with plants for commercial values and thus exploit, albeit, sustainably their environment, including forests. Exploitation of these forests (can) jeopardize certain species existence or even lead them to extinction. Commercial values of the plants are the direct and indirect economic, aesthetic and ethical values. Emphasis on direct commercial value of forest products is placed on a small number of timber species and a short-term profit derived from felled portions of forests for agricultural purposes. Multi-use exploitation and conservation of forest resources, as practiced by these rural communities, have not been adopted as a viable alternative to modern production schemes (Clay, 1998). From ecological perspective rural communities can achieve production in harmony with nature's laws that their production capacity tends to maintain the renewing capacity of the ecosystems, albeit, forests. From this standpoint, the sum of empirical knowledge that the people in the Niger Delta region posses about plants, soils, climate and ecophysiographic units serving their production strategic becomes important in unciphering and understanding traditional economic rationality. Each method of production reveals a strategy of appropriation from nature, a mix of technologies, knowledge about the components of the ecosystems and its interrelations. These depend upon forest compositions, species abundance, appreciation and usage of the species by the native. Due to exploration and exploitation of crude oil much of the area has been polluted and the needed medicinal plants and other useful non-timber species have been going extinct. Little attention has been paid to this seemingly unnoticed but important degradation and elimination of species much averred by the indigenous people of the Niger Delta. Thus, this study attempts to assess the values of forests and their species vis-à-vis the diversity of uses the plants species have been put to and marketed or traded by the natives. An attempt has also been made in eliciting and proffering conservation method(s) the natives should practice in order to protect and preserve these plants for posterity.

Materials and Methods

Study Sites

The Niger Delta, one of the world's largest wetlands in southern Nigeria, is a vast flood plain (over 2000 km^2) built up by the accumulation of sedimentary materials washed down the Niger and Benue Rivers (Pangely, 1994). It comprises four ecological zones; coastal barrier islands, mangrove swamp forests, freshwater swamp forests are low lame rainforests. The later two forests are the most

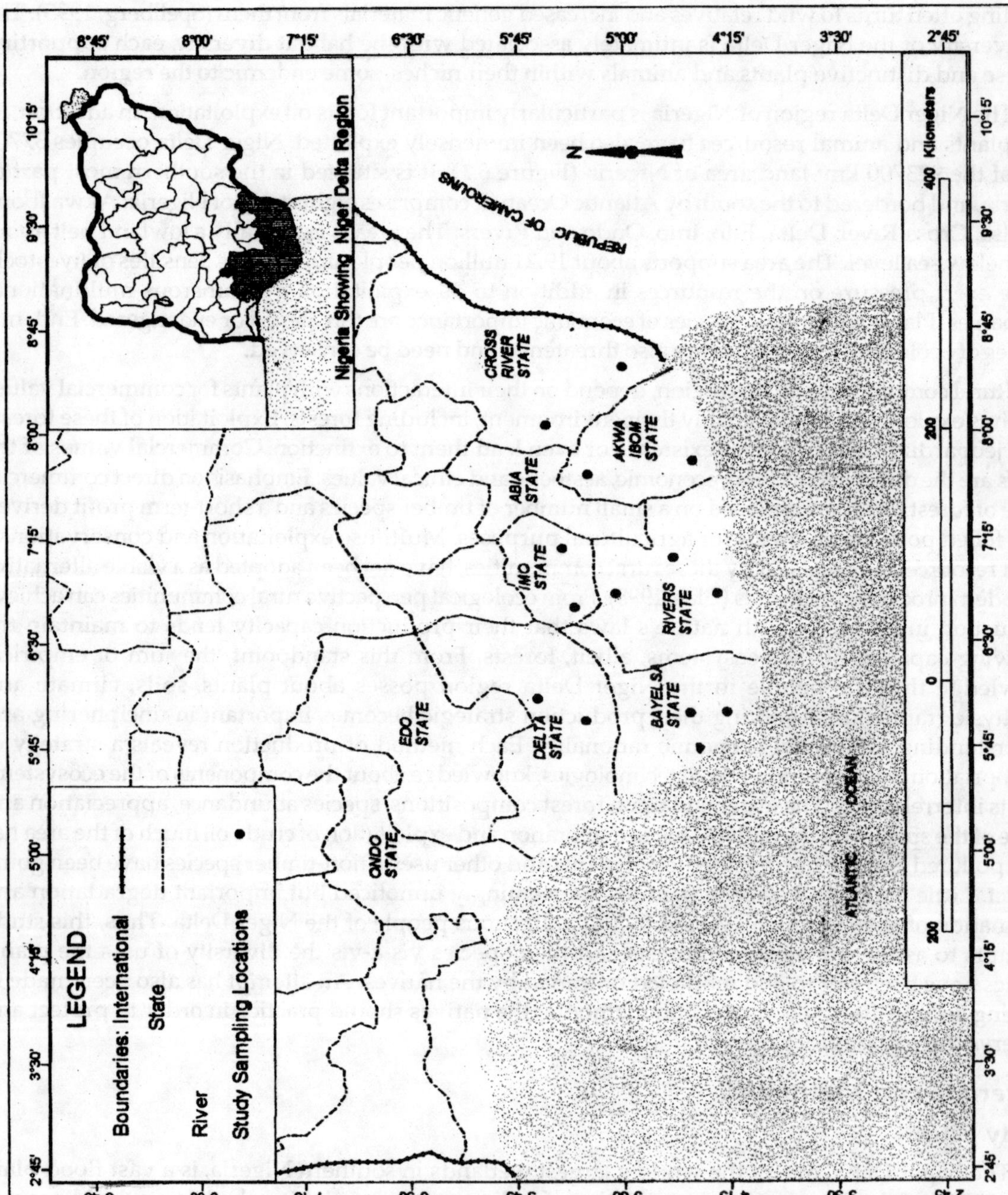

Figure 63.1: Niger Delta Region Showing Study Sites

extensive ecological zones in the region. However, anthropogenic activities (timber and fuel wood removal, cultivation, capital project deployment, oil exploration and exploitation) are fast degrading these forests and their biodiversity.

The region is located between latitudes 4º00'N and 7º'30'N and longitudes 4º30'E and 9º15'E. The area experiences two main season: wet and dry seasons. The wet season is about 7 months in duration from April to October. The dry season lasts about 5 months beginning from November and ending in March. The mean temperature for the area is 25ºC; annual rainfall 4921 mm and relative humility 85 per cent. The main crops are cassava, *Manihot esculentus* Krantz; Cocoyam, *Colocasia antiquorum* Schott, yam, *Dioscorea* sp. and oil palm, *Elaeis guineensis.*

Six out of eight states in the region were selected for the study: Akwa-Ibom, Bayelsa, Cross River, Delta, Imo and Rivers. Selection was based on previous knowledge, experience and cumulative work (unpublished) done in these areas by the author. Thirteen study sites (Figure 63.1) were chosen for the study: two in each state except Rivers with three sites.

Sampling and Analysis

Field study was conducted for four years (1999, 2000, 2002, 2004) for both wet and dry seasons. The vegetation (freshwater swamp and lowland rainforests) was sampled systematically within 10 m × 10 m Quadrats along established transects, and randomly throughout the sampling areas. Interviews with different informants were also conducted as to the type of species found in the areas, the use they are being put to and the methods used in conserving the species. The Shannon-Weiner species diversity was used to assess the species quantitatively (Sokal and Rohlf, 1995) according to the formula:

$$H^1 = -\sum_{i=1}^{s} pi \ln pi$$

where,

- H^1: Shannon-Weiner index
- S: Number of species
- pi: Proportion of individuals or abundance of the ith species expressed as a proportion of the total number of individuals of all species.
- ln: Log base.

Endangered and endemic species were recorded; identification of species was accomplished by the assistance of Forestry Institute of Nigeria (FRIN) herbarium and texts (Hutchinson and Dalziel, 1954–72; Akobundu and Agyakwa, 1987). Classification of ethnobotanical uses of species involved interviews, complimented by literature (Sofowora, 1982; Kokwara, 1995; Etukudo, 2000, 2003). Diversity of uses was calculated using Shannon-Weiner index ($H^1 = -\Sigma$ pi log pi) *where,* I: uses and pi: proportion of species for the ith use (Rico-Gray *et al.*, 1991). Sorensen index (S.I = 2a/+btc) was used to calculate species similarity between sites; a: number of species common to sites compared, b: number of species in site 1 and c: number of species in site 2.

Results and Discussion

Plant Diversity

Three hundred and thirty nine plant species were encountered in the forests and homestead gardens in the study area (Tables 63.1, 63.2). Of these, 206 species were recorded for the forests and 79 species for the homestead gardens. All the 339 species belong to 88 families (44 families for forest, 44 for gardens) and 266 genera (205 for forests, 61 for gardens). Fabaccae has the highest number of species (45), followed by Euphorbiaceae (25), Rubiaceae (12) Apocynaceae (11), Asteraceae (11) and Cecropiaceae (10). The species diversity index between the forest and the homestend gardens is 4.89. Species similarity index among the forest is low (47.78 per cent) where 147 species are common to all forests. Similarity index between the homestead gardens is high (49.5 per cent) as 76 species are common to the gardens.

Table 63.1: List of Plant Species Encountered in the Forests of Study Areas. Nomenclature follows Hutchinson and Dalziel (1972) and Akobundu and Agyakwa (1987).

Species	*Habit*	*Family*	*Ethnobotanical Uses*	*Plant Parts Used*
Abrus precatorius Linn	Climber	Fabaceae	17, 21, 25	Seeds
Abutilon mauritianum (Jacq.) Medic	Shrub	Malvaceae	7, 17	Seeds, bark, leaves
Acanthus montamus (Nees) T. Anders	Herb	Acanthaceae	17, 19	Leaves, whole plant
Achyranthes aspera Linn	Herb	Amaranthaceae	10, 17	Leaves, whole plant
Acioabarteri (Hook f. ex Oliv) Engl.	Shrub	Rosaeceae	1, 2, 3, 5	Stems, fruits
**Aframomum melegueta* K. Schum.	Herb	Zingiberaceae	7, 17, 19, 21, 23, 24	Stems, seeds
**A. sceptrum* (Oliv. and Hanb) K. Schum.	Herb	Zingiberaceae	7, 17, 19, 21, 24	Stem, seeds
**A. strobilaceym* (Sm.) Hepper	Herb	Zingiberaceae	7, 17, 19, 21, 24	Stem, seeds, leaves
Ageratum conizoides Linn.	Herb	Asteraceae	5, 17, 18	Leaves, flowers
Albizia zygia (DC.) J.F. Machr.	Tree	Fabaceae	9, 10, 12, 13, 15, 26, 29	Leaves, stems
Albizia adianthifolia (Schum) W.F. Wight	Tree	Fabaceae	9, 11, 17, 27	Leaves, stem, bark
Alchornea cordifolia (Schum. and Thonn.) Mull. Arg.	Shrub	Euphorbiaceae	3, 6, 7, 9, 11, 30	Stem, bark, leaves, shoots
A. laxiflora (Benth.) Pax and K. Hoffin	Shrub	Euphorbiaceae	3, 6, 7, 11, 17	Stem, bark, leaves
Allanblackia floribunda Oliv.	Tree	Clusiaceae	2, 4, 9, 11, 12, 27, 29	Stems, leaves, fruits
Alstonia boonei De Wild	Tree	Apocynaceae	2, 4, 11, 12, 14, 15, 27	Stems, leaves, bark
A. congensis Engl.	Tree	Apocynaceae	2, 4, 11, 12, 14, 15, 27	Stems, leaves, bark
Alternanthera bettzickiana (Regel) Voss	Herb	Amaranthaceae	9, 19	Leaves, shoots
Alternanthera pungens H.B.K.	Herb	Amaranthaceae	10, 17, 18, 19	Leaves, shoots, stems
A. sesslis (Linn.) R.Br. ex Roth	Herb	Amaranthaceae	10, 17	Leaves, shoots, stems
Anchomanes difformis (Bl.) Engl.	Herb	Araceae	1, 10, 17, 21	Stems tubers, leaves, fruits
**Ancistrophyllum secundiflorum* (P. Beauv.) Wendl.	Shrub	Palmaceae	4, 7, 14, 15, 30	Stems, shoots
Aningeria robusta (A. Chev.) Aubrev. and Pellegr.	Tree	Sapotaceae	4, 11, 12, 15, 27, 29	Bark, stem
Anthocleista vogelii Planch.	Tree	Loganiaceae	4, 9, 11, 12, 17, 27, 29, 30	Stems, bark, leaves
A. dyalonensis A. Chev.	Tree	Loganiaceae	2, 8, 11, 15, 17, 30	Leaves, stems
**Anthostema aubryamum* Baill.	Tree	Euphorbiaceae	11, 15, 17, 27	Bark, leaves, stem
Antiaris africana Engl.	Tree	Moraceae	7, 11, 12, 27, 29	Bark, stems, leaves
Antidesma vogelianum Mull. Arg.	Tree	Euphorbiaceae	2, 12, 15, 27	Stems
***Anthonatha macrophylla* P. Beauv.	Shrub	Fabaceae	6, 11, 13	Stems

Contd...

Table 63.1–Contd...

Species	*Habit*	*Family*	*Ethnobotanical Uses*	*Plant Parts Used*
Araliopsis soyauxii Engl.	Tree	Rutaceae	3, 11, 15, 27	Stems, roots
Argemone mexicana Linn.	Herb	Papaveraceae	17	Whole plant
Aristolochia albida Duchatre	Herb	Aristolochiaceae	1, 11, 12	Stems
Aspilia africana (Pers.) C.D. Adams	Herb	Asteraceae	5, 9, 17, 30	Leaves, shoots, flowers
Asystacia gangetica (Linn.) T. Anders	Herb	Acanthaceae	9, 17, 30	Leaves
Bacopa creanata (P. Beauv.) Hepper	Herb	Scrophulariaceae	10, 11, 17	Leaves, stem
Bambusa vulgaris Schrad. ex Wendel	Shrub	Poaceae	2, 6, 9, 12, 14	Stem, shoots, leaves
Baphia nitida Lodd.	Shrub	Fabaceae	3, 5, 6, 7, 9, 21, 30	Leaves, stems, bark
Barteria nigritiana Hook. f.	Shrub	Passifloraceae	6, 91 10, 11	Stems, leaves
Begonia microcarpa Warb	Herb	Begoniaceae	17, 21	Leaves, roots
Berlinia grandiflora (Vahl.) Hutch and Dalz.	Tree	Fabaceae	2, 6, 11, 13	Stems
Bixa orellana Linn.	Shrub	Bixaceae	5, 17, 19, 22, 30	Fruits, stems, leaves, whole plant
Blighia sapida Konig	Tree	Sapindaceae	17, 27	Leaves, seeds, stems
Boerhavia erecta Linn.	Herb	Nyctaginaceae	9, 17	Leaves, roots, shoots
Bombax buonopozense P. Beau V.	Tree	Bombacaceae	4, 7, 11, 12, 14, 15	Stems, fruits, leaves
Bosqueia angolensis Ficalho	Tree	Moraceae	11, 12, 14, 17	Stems
Brachystegia eurycoma Harms	Tree	Fabaceae	9, 10, 11, 27	Leaves, stems, seeds
Bridelia grandis Pierre ex Hutch	Tree	Euphorbiaceae	4, 12, 27	Stems
B. micrantha (Hochst.) Baill.	Tree	Euphorbiaceae	4, 12, 14, 27	Stems
Bryophyllum pinnatum (Lan.) Okon	Herb	Crassutaceae	17	Leaves, whole plant
Calamus deerratus Mann and Wendl.	Climber	Palmaceae	4, 7, 14, 15	Stems
Calopogonium mucunoides Desv.	Creeper	Fabaceae	9	Leaves, fruits
Canarium schweinfurthii Linn.	Tree	Burseraceae	4, 10, 11, 17, 24, 29	Stems, fruits, bark
Carpolobia lutea G. Don	Shrub	Polygalaceae	3, 15	Stems
Cassytha filiformis Linn.	Climber	Lauraceae	17	Whole plant
Cathormion altissimum (Hook. f.) Hutch and Dandy	Tree	Fabaceae	11, 12, 27	Stems
Cedrella odorata Linn	Tree	Meliaceae	11, 12, 15, 27	Stems
Ceiba pentandra (L.) Gaertn.	Tree	Bombacaceae	4, 7, 11, 13, 17, 19, 27	Stems, fruits, roots
Celtis zenkeri Engl.	Tree	Ulmaceaee	2, 4, 5, 6, 12	Stems
Centrosoma pubescens Benth.	Creeper	Fabaceae	9	Leaves, fruits
Ceratophylluma demersum Linn.	Herb	Ceratophyllaceae	17	Whole plant
Chromolaena odorata (L.) R.M. King and Robinson	Herb	Asteraceae	17, 30	Leaves, shoots
Chlysobalamus ellipticus Soland. e Sabine	Shrub	Rosaceae	2, 14, 17	Bark, stems
Cissus quadrangularis Linn.	Climber	Ampelidaceae	10, 17	Leaves, stems, fruits
Clappertonia ficifolia (Willdo) Decne	Shrub	Tiliaceae	7, 17	Bark, leaves, flowers
Cleistopholis patens (Benth.) Engl. and Diels	Tree	Annonaceae	2, 11, 27, 29	Stems, leaves

Contd...

Table 63.1–Contd...

Species	Habit	Family	Ethnobotanical Uses	Plant Parts Used
Cleome viscosa Linn	Herb	Capparidaceae	9, 17	Leaves, fruits
Clerodendrum splendens G. Don.	Climbing, Shrub	Verbenaceae	5, 17, 19	Leaves, fruits, whole plant
Clitoria ternatea Linn.	Climber	Fabaceae	17, 18	Seeds, roots
Cnestis ferruginea DC.	Shrub	Connaraceae	5, 17	Leaves, fruits, roots
Coelocaryon preusii Warb.	Tree	Myristicaceae	3, 5, 6, 11, 15	Stems, leaves, bark, roots
Coix lachrymal-jobi Linn.	Herb	Poaceae	6, 7, 9	Whole plant
Combretum hispidum Laws	Shrub	Combretaceae	17, 19	Roots
C. racemosum P. Beauv.	Shrub	Combretaceae	17,	Twigs, shoots
***C. zenkeri* Engl. and Diels	Shrub	Combretaceae	5, 7, 17	Leaves, stems
Commelina africana Linn.	Herb	Commelinaceae	17, 19	Whole plant
C. diffusa Burm. f.	Herb	Commelinaceae	17, 19	Whole plant
C. erecta Linn.	Herb	Commelinaceae	17, 19	Whole plant
***Costus afer* Ker-Gawl	Herb	Costaceae	3, 9, 17, 24	Stems, shoots, roots
Costus lucanusianus J. Braun and K. Schum.	Herb	Costaceae	7, 9, 12	Stems, leaves, roots
***C. schlechteri* Winkler	Herb	Costaceae	9, 17, 24	Stems, roots
Crassocephalum togoense C.D. Adams	Herb	Asteraceae	9, 17	Leaves, stems
Crassocephalum crepidinoides S. Moore	Herb	Asteraceae	10, 17	Leaves
Crotalaria retusa Linn.	Herb	Fabaceae	7, 19	Whole plant, bark
Crudia klainei Pierre ex De Wild	Tree	Fabaceae	2, 11	Stems
Cuscuta adstralis R.Br.	Twinner	Cuscutaceae	1, 17	Whole plant
Cynometra vogelii Hook. f	Herb	Fabaceae	11, 29	Stems
Cyperus difformis Linn	Herb	Cypereceae	10, 17	Stems, leaves
Cyperus esculentus Linn	Herb	Cypereceae	7, 10, 17	Tubers, stems, leaves
Dalbergia latifolia Roxb.	Tree	Fabaceae	12, 15, 29	Stems
Daniellia ogea (Harnes) Rolfe ex Holl.	Tree	Fabaceae	4, 91 11, 12, 15, 27	Stems, leaves
Datura stramonium Linn.	Herb	Solanaceae	17, 24, 28	Leaves, seeds, fruit, flowers
Desmodium turtuosum (Sw.) DC.	Herb	Fabaceae	9, 17, 29	Whole plant, roots
Dialium guineense Wild	Shrub	Fabaceae	3, 5, 7, 9, 10, 11	Leaves, stems, bark
Diodia sarmentosa Sw.	Herb	Rubiaceae	17	Leaves
**Diospyros alboflavescens* (Gurke) F. White	Tree	Ebenaceae	4, 8, 11, 12, 15, 27, 29	Stems, leaves
Diplazium sammatii (Kuhn) C. Chr.	Herb	Athyriaceae	10	Shoots
**Dracaena arborea* Link	Tree	Liliaceae	6, 19, 21	Leaves, whole plant
**D. mannii* Bak.	Tree	Uliaceae	6, 19, 21	Leaves, whole plant
Drypetes floribunda (Mull. Arg.) Hutch.	Shrub	Euphorbiaceae	1, 28	Fruits, leaves
Emilia coccinea (Sims) G. Don	Herb	Asteraceae	9, 17	Leaves, shoots

Contd...

Table 63.1–Contd...

Species	*Habit*	*Family*	*Ethnobotanical Uses*	*Plant Parts Used*
E. sonchifolia (Linn.) DC.	Herb	Asteraceae	9, 17	Leaves, shoots
Erythrina senegalensis DC.	Tree	Fabaceae	6, 27	Stems
Erythrophelm ivorense A. Chev.	Tree	Fabaceae	11, 27, 29	Leaves, stems
Euphorbia heterophylla Linn.	Herb	Euphorbiaceae	13, 17, 19	Leaves, stems, whole plant
Fagara macrophylla Engl.	Tree	Rutaceae	11, 17, 27	Stems, leaves, roots
F. zanthoxyloides Lam	Tree	Rutaceae	3, 6, 11, 17, 27	Stems, leaves, roots
Ficus exasperata Vahl.	Tree	Moraceae	6, 9, 13, 16, 17	Stems, fruits, shoots, leaves
Ficus lepreuri Miq.	Tree	Moraceae	13, 16, 17	Stems, fruits, shoots, leaves
Ambristylis ferruginea (Linn.) Vahl.	Herb	Cyperaceae	17	Fruits
Fimbristylis littoralis Linn.	Herb	Cyperaceae	17	Fruits
Fleurya aestuans (Linn.) Gaudex Miq.	Herb	Urticaceae	17, 28	Leaves, flowers
F. ovalifolia (Schum. and Thonn.) Dandy	Herb	Urticaceae	17	Fruits, roots
Fuirena umbellate Rottb.	Herb	Cyperaceae	17	Fruits
Funtumia elastica (Preuss.) Stapf.	Tree	Apocynaceae	13, 16	Stems, leaves, bark
***Garcinia mannii* Oliv.	Tree	Clusiaceae	3, 13	Stems
Gloriosa superba Linn.	Herb	Liliaceae	17	Whole plant
Glyphaea brevis (Spreng.) Monachino	Shrub	Tiliaceae	3, 5, 6, 7, 10, 11, 17	Stems, leaves, bark
Guarea cedrata (A. Chev.) Pellegr.	Tree	Meliaceae	2, 8, 11, 27, 29	Leaves, stems
G. thompsonii Sprague and Hutch.	Tree	Meliaceae	2, 8, 11, 27, 29	Leaves, stems
Hallea stipulosa (DC) J.F. Leroy	Tree	Rubiaceae	4, 11, 12, 15, 17, 27, 29	Stems, roots, bark
Halopegia azurea K. Schum	Herb	Marantaceae	7, 30	Leaves, stems
Hallea ciliata Aub and Pellegr.	Tree	Rubiaceae	4, 11, 12, 15, 17, 27	Stems, roots, bark
Hannoa klaineana Pierre and Engl.	Tree	Simaroubaceae	2, 4, 6, 11, 12, 27, 29	Leaves, stems
Harungana madagascariensis Lam. ep Poir	Shrub	Hypericaceae	5, 9, 11, 13, 17	Leaves, stems, bark
Heteranthera callifolia Rchb. ex Kunth	Herb	Pontederiaceae	17	Leaves, roots
Heterotis rotundifolia (Sm.) Jac-fet	Herb	Melastomataceae	17	Leaves, fruits, flowers
Hewittia sublobata Linn.	Shrub	Convolvulaceae	17, 19	Flowers, whole plant
**Hibiscus sabdarifa* Linn.	Shrub	Malvaceae	7, 10, 23	Bark, leaves, fruits, flowers
***Hippocratea pallens* Planch. ex Oliv.	Climbing	Hippocrateaceae	11, 17, 25	Fruits, roots shrub
Homalium letestui Pellegr.	Tree	Samydaceae	3, 6, 9	Leaves, stems
Hylodendron gabunense Taub.	Tree	Fabaceae	2, 6, 11, 27	Stems
cacina trichantha Oliv.	Straggling shrub	Icacinaceae	17	Tubers
**Ipomoea aquatica* Forsk.	Twinner	Convolvulaceae	9, 10, 19	Leaves, stems, whole plant
I. involucrata P. Beauv.	Creeper	Convolvulaceae	7, 17	Stems, leaves

Contd...

Table 63.1–Contd...

Species	Habit	Family	Ethnobotanical Uses	Plant Parts Used
Irvingia gabonensis Baill.	Tree	Irvingiaceae	9, 10, 11, 18, 23	Fruits, seeds, stems
Justicia flava (Forsk.) Vahl.	Herb	Acanthaceae	17	Whole plant
Khaya grandifoliola C. DC.	Tree	Meliaceae	2, 4, 8, 11, 12, 15, 27, 29	Leaves, stems
K. ivorensis A. Chev.	Tree	Meliaceae	2, 4, 11, 12, 15, 27, 29	Leaves, stems
Klainedoxa gabonensis Pierre ex Engl.	Tree	Irvingiaceae	2, 4, 11, 12, 15, 27, 29	Stems, leaves
Landolphia heudelotll A. DC.	Shrub	Apocynaceae	6, 11, 13, 15, 16	Stems, leaves
L. owariensis P. Beauv.	Liane, shrub	Apocynaceae	6, 11, 13, 15, 16	Stems, leaves
Lannea acida A. Rich	Tree	Anacardiaceae	10, 11, 17	Leaves, stems, fruits
**Leea guineensis* G. Don	Shrub	Leeaceae	17, 19	Leaves, roots, bark
**Lonchocarpus cyanescens* Benth (Schum. and Thonn.)	Shrub	Fabaceae	2, 3, 5, 6, 9, 10, 11, 27, 29	Leaves, stem, roots, seeds
L. griffoneanus (Baill.) Dunn.	Shrub	Fabaceae	1, 2, 6, 11, 21, 27, 29	Leaves, stems, seeds
Lophira alata Barks ex Gaertn. f.	Tree	Ochnaceae	2, 4, 6, 8, 11, 12, 15, 27	Stems
Loranthus aphyllus Sprangue	Parasitic Shrub	Loranthaceae	17	Leaves, fruits
Lovol trichilioides Harms	Tree	Meliaceae	2, 4, 8, 11, 12, 14, 15, 27, 29	Leaves, stems
Ludwigia decurrens Walt.	Herb	Onagraceae	17	Leaves, flowers
Ludwlgia hyssopifolia (G. Don.) Exell	Herb	Onagraceae	17	Leaves, flowers
**Macaranga barteri* Muell. Arg.	Tree	Euphorbiaceae	9, 11, 13, 14, 25	Leaves, stems, fruits, roots
**Maesobotrya barteri* (Baill.) Hutch	Shrub	Euphorbiaceae	3, 9, 10, 11, 14	Stems, leaves, fruits
Mallotus oppositifolius (Geisel) Mull. Arg.	Shrub	Euphorbiaceae	5, 9, 11, 17	Leaves, stems, roots
Manihot glaziovii Mull. Arg.	Shrub	Euphorbiaceae	16, 19, 28	Leaves, tubers
Manniophyton fulvum Mull. Arg.	Shrub	Euphorbiaceae	9	Shoots
Mansonia altissima A. Chev.	Tree	Sterculiaceae	4, 11, 12, 27	Stems
Massularia acuminata (G. Don.) Bull. ex Hoyle	Shrub	Rubiaceae	3, 6, 17	Stems
Maranthochloa cuspidata (Rosc.) Milne-Redh.	Herb	Marantaceae	2, 7, 30	Leaves, stems
Melanthera scandens (Schum. and Thonn.) Roberty	Herb	Asteraceae	17	Leaves, roots
Melastomastrum capitatum (Vahl.) A. and R. Fern	Herb	Melastomataceae	17	Leaves, fruits, flowers
Melochia corchorifolia Linn.	Herb	Sterculiaceae	7, 10, 17	Bark, leaves, roots
**Microdesmis puberula* Hook. f. ex Planch.	Shrub	Euphorbiaceae	3, 9, 10	Stems, shoots, leaves
Milicia excelsa (Welw.) C.C. Berg.	Tree	Melaiceae	4, 11, 12, 21, 27	Stems

Contd...

Table 63.1–Contd...

Species	*Habit*	*Family*	*Ethnobotanical Uses*	*Plant Parts Used*
Millettia thonningii (Schum. and Thonn.) Bak	Shrub	Fabaceae	6, 8, 10, 11, 17, 29	Leaves, stems, roots
Mimosa pigra Linn.	Herb	Fabaceae	19, 27	Leaves, whole plant
Mimosa peduca Linn.	Herb	Fabaceae	19, 27	Whole plant
MImosa invisa Mart.	Herb	Fabaceae	4, 11, 12, 17, 27, 29	Leaves, whole plant
Momordica charantia Linn.	Climber	Cucurbitaceae	10, 17	Fruits, leaves
Momodora myristica (Gaertn) Dunal	Climber	Annonaceae	19, 23, 27	Leaves, fruits, whole plant
Musanga cecropioides R.Br.	Tree	Moraceae	9, 11, 16, 17, 30	Stems, leaves, bark, roots
***Mussaenda isertiana* DC	Climbing shrub	Rubiaceae	19, 27	Leaves, whole plant
Myrianthus arboreus P. Beauv.	Tree	Moraceae	2, 11, 12, 15, 17, 27	Stems, leaves, bark, roots
**Napoleona vogelii* Hook. and Planch.	Tree	Lecythidaceae	1, 3, 11, 30	Stems, leaves
Nauclea diderrichii (De Wild. and Th. Dur.) Merrill	Tree	Rubiaceae	6, 17, 27	Stems, fruits
Nauclea latifolia Sm.	Strangling shrub	Rubiaceae	2, 3, 5, 13, 27	Leaves, stems, roots, bark
Nauclea vanderguchtii (De Wild) Petit.	Shrub	Rubiaceae	2, 3, 5, 13	Stems, roots
**Newbouldia laevis* (P. Beauv.) Seemann ex Bureau	Tree	Bignoniaceae	6, 11, 19, 21	Stems, shoots
**Oxystigma mannii* (Baill.) Harms	Tree	Fabaceae	2, 4, 11, 12, 15, 27, 29	Leaves, stem
Pachystela brevipes (Bak.) Baillon ex Engl.	Tree	Sapotaceae	2, 4, 8, 11, 12, 15, 27, 29	Stems
Palisota hirsuta (Thunb.) K. Schum.	Herb	Commelinaceae	9, 17	Leaves, stems
**Pandanus candelabrum* Beauv.	Tree	Pandanaceae	4, 6, 7, 12, 14, 21	Stems
Parkia bicolour A. Chev.	Tree	Fabaceae	4, 9, 10	Leaves, fruits, stems
Parinari excelsa Sabine	Shrub	Chrysobalanaceae	4, 11, 12	Stems
Paspalum vaginatum Sw.	Herb	Poaceae	9	Shoots
Paullinia pinnata Linn.	Climber	Sapindaceae	3, 17	Leaves, stem, roots
**Pentachlethra macrophylla* Benth.	Tree	Fabaceae	4, 9, 10, 11, 18, 21, 23	Leaves, stem, seeds
Pentadesma butryracea Sabine	Tree	Clusiaceae	13, 14, 17	Bark, stems
Peperomia pellucida (L) H.B. and K.	Herb	Piperaceae	17, 28	Whole plant, leaves
Phyllanthus amarus Schum and Thonn.	Herb	Euphorbiaceae	16, 17	Leaves, stem, fruits, flowers
Phyllantus muelleriamus (O. Ktz) Exell	Herb	Euphorbiaceae	17	Flowers, fruits
Physalis angulata Linn.	Herb	Solanaceae	17	Leaves
Piper umbellatum Linn.	Climber	Piperaceae	10, 17, 21	Leaves, roots, whole plant

Contd...

Table 63.1–Contd...

Species	Habit	Family	Ethnobotanical Uses	Plant Parts Used
Plukenetia conophora Mull. Arg.	Climbing shrub	Euphorbiaceae	10, 11, 12, 14, 17, 18	Leaves, seeds, fruits, stems
Poga oleosa Pierre	Tree	Rhizophoraceae	4, 10, 11, 12, 27, 29	Leaves, stems
**Psychotria nigerica* Hepper	Shrub	Rubiaceae	17, 24	Leaves
Pterocarpus mildbraedii Harms	Tree	Fabaceae	5, 6, 9, 11, 15, 17, 29	Leaves, bark, stem
Pterokcarpus santalinoides L'Her Ex DC.	Tree	Fabaceae	5, 6, 9, 10, 12, 14, 15, 17, 26	Bark, leaves, stems
P. osun Craib	Tree	Fabaceae	5, 6, 9, 11, 15, 17, 29	Leaves, bark, stems
Pterygota macrocarpa K. Schum	Tree	Sterculiaceae	12, 27	Stems
Pupalia lappacea (Linn.) Juss.	Herb	Amaranthaceae	17	Shoots
Pycanthus angolensis (Welw.) Warb.	Tree	Myristicaceae	4, 11, 12, 15, 21, 27	Stems, twigs
Rauvolfia vomitoria Afzel.	Shrub	Apocynaceae	9, 11, 16, 17, 28	Leaves, stems, roots
***Raphia vinifera* P. Beauv.	Tree	Palmaceae	2, 7, 11, 25	Leaves, fruits
Ricinodendron heudelotti (Baill) Pierre ex Pax.	Tree	Euphorbiaceae	9, 10, 17, 18, 27	Bark, stems, seeds, leaves
Rothmania hispida (K. Schum.) Fager	Shrub	Rubiaceae	5, 17, 19	Leaves, fruits, whole plant
Rhynchospora corynbosa (Linn.) Britt.	Herb	Cyperaceae	17	Leaves, fruits
***Sacoglottis gabonensis* (Baill.) Urb.	Tree	Humiriaceae	11, 12, 23, 25, 29	Bark, leaves, stems
**Sarcophyrynium brachystachys* (Benth.) K. Schum.	Herb	Marantaceae	17, 21	Leaves, whole plant
Schrankia leptocarpa DC	Straggling herb	Fabaceae	19	Whole plant
Scleria naumanniana Boeck	Herb	Cyperacese	17	Fruits
S. vorrucosa Wild.	Herb	Cyperaceae	17	Fruits
Scoparia dulcis Linn.	Shrub	Scrophulariaceae	17	Whole plant
Selaginella nyosurus (Sw) Alston	Creeper	Selaginellaceae	17, 21	Whole plant
S. scandens (P. Beauv.) Spring	Creeper	Selaginellaceae	17, 21	Whole plant
Setaria megaphylla (Steud.) Dur. and Schinz.	Herb	Poaceae	21, 30	Leaves, roots
Senna alata (Linn.) Roxb.	Shrub	Fabaceae	5, 11, 17, 19	Leaves, roots, stems
Senna hirsuta (Linn.) Irwin and Barneby	Shrub	Fabaceae	17	Whole plant
Senna obtusifolia (Linn.) Irwin and Barneby	Shrub	Fabaceae	10, 17, 25	Whole plant
Senna occidentalis (Linn.) Link	Herb	Fabaceae	19	Whole plant
Senna podocarpa (Guill and Perr.) Lock	Shrub	Fabaceae	19	Bark
S. rhombifolia Linn.	Herb	Malvaceae	7, 17	Bark, leaves, stem, roots
Smilax anceps Willd.	Climbing herb	Smilacaceae	7, 17	Roots, tubers, whole plant
Spigelia anthelma Linn.	Herb	Loganiaceae	28	Whole plant
Spilenthes filicaulis (Schum and Thonn.) C.D. Adam	Creeping herb	Asteraceae	17	Flowers, whole plant

Contd...

Table 63.1–Contd...

Species	*Habit*	*Family*	*Ethnobotanical Uses*	*Plant Parts Used*
Spondiathus preussii Engl.	Tree	Euphorbiaceae	2, 11, 28	Leaves, stems
Stachytarpheta cayenensis (L.C. Rich) Schau.	Shrub	Verbanaceae	17	Whole plant
S. indica (Linn.) Vahl.	Shrub	Verbanaceae	17	Whole plant
Staudtia stipitata Warb	Tree	Myristicaceae	2, 4, 11, 12, 16, 27	Stems, bark
Saterculia tragacantha Lindi	Tree	Sterculiaceae	7, 10, 13, 15, 17	Bark, leaves, stems
Strophanthus sarmentosus DC	Woody climber	Apocynaceae	17, 19	Bark, leaves, seeds
Symphonia globulifera Linn. f.	Tree	Clusiaceae	2, 3, 4, 11, 12, 15, 17, 27	Stems, leaves, roots
***Synsepallum dulcificum* (Schum. and Thonn.) Daniell	Shrub	Sapotaceae	4, 10, 11, 25	Seeds, leaves, stem
Terminalia ivorensis A. Chev.	Tree	Combretaceae	12, 15, 17, 27	Bark, stems
**Tetracera podotricha* Gilg	Shrub	Dilleniaceae	11, 17	Stems, bark
**Tetrapleura tetraptera* (Schum. and Thonn.) Taub.	Tree	Fabaceae	4, 11, 15, 17, 23, 27	Stem, fruits, bark
Thalia welwitschii Ridl	Herb	Marantaceae	19, 30	Leaves, whole plant
Thaumatococcus danielii (Benn.) Benth	Herb	Marantaceae	10, 25, 28, 30	Fruits, leaves
Thevetia peruviana Juss.	Shrub	Apocynaceae	6, 17, 28	Bark, stem, leaves, roots
Treculia africana Decne	Tree	Moraceae	10, 12, 17, 18, 21, 27	Fruits, seeds, stems
Trichilia heudelotii Planch ex Oliv.	Tree, Shrub	Meliaceae	11,12, 27	Stems
Trloax procumbens Linn.	Herb	Asteraceae	9, 17	Leaves, roots
Tristemma incompletum R. Br.	Herb	Melastomataceae	10	Fruits
Triumpheta cordifolia A. Rich	Herb	Tiliaceae	7, 30	Leaves, bark
Uapaca staudtii Pax.	Tree	Euphorbiaceae	2, 4, 11, 17, 27, 30	Stem, leaves, roots
U. togoensis Pax	Tree	Euphorbiaceae	2, 4, 11, 17, 27, 30	Stem, leaves, roots
Urena lobata Linn.	Shrub	Malvaceae	7, 9, 17	Bark, roots, seeds
**Uvaria chamae* P. Beauv	Shrub	Annonaceae	11, 17, 28	Stems, fruits, roots
**U. cristata* R.Br.	Shrub	Annonaceae	11, 17, 28	Stems, fruits, roots
Voacanga africana Stapf.	Tree	Apocynaceae	9, 11, 16, 28	Shoots, stem, leaves
Vossia cuspidata (Roxb.) Griff.	Herb	Poaceae	2, 7, 9	Stems, leaves
Waltheria indica Linn.	Herb	Steruliaceae	7, 21	Bark, whole plant
***Xylopia aethiopica* (Dunal) A. Rich.	Tree	Annonaceae	2, 4, 10, 17, 23, 27	Stems, fruits
***Xylopia villosa* Chipp.	Tree	Annonaceae	14, 15, 17, 27, 29	Stems, seeds

*: Endangered species; **: Endemic species.

Ethnobotanical Uses: 1: Bait; 2: Building; 3: Toothbrush; 4: Commercial; 5: Dyes; 6: Fence; 7: Fibre/Cordage; 8: Field-tool; 9: Fodder, Feed; 10: Food/Condiment; 11: Fuel; 12: Furniture; 13: Gum/Resins; 14: Handicrafts; 15: Home tool; 16: Latex; 17: Medicine; 18: Oil, Oranmental; 20: Plaiting/Weaving; 21: Religious/Cultural; 22: Soap; 23: Spice/Flavouring; 24: Stimulant; 25: Sweetners; 27: Timber; 28: Toxic; 29: Work tool; 30: Wraps/Packing.

Table 63.2: List of Plant Species Present in the Homestead Gardens in the Study Areas. Nomenclature and ethnobotanical uses of the species follow those in Table 63.1.

Species	*Habit*	*Family*	*Ethnobotanical Uses*	*Plant Parts Used*
Abelmoschus esculentus (Linn.) Moech	Herb	Malvaceae	7, 10, 17, 23	Bark, leaves seeds
Amaranthus hybridus Linn.	Herb	Amaranthaceae	7, 10	Bark, leaves
Ananas comosus (Linn.) Merril	Herb	Bromeliaceae	7, 10, 17	Fruits, stems
**Ananas muricata* Linn.	Shrub	Annonaceae	10, 14	Fruits, stems
Arachis hypogea Linn.	Herb	Fabaceae	9, 10, 18, 22, 23	Fruits
**Artocarpus communis* Forest	Tree	Moraceae	10, 12, 13, 14, 30	Fruits, stems, leaves
***A. heterophylla* Lam.	Tree	Moraceae	10, 12, 13, 14, 30	Fruits, stems, leaves
Azadirachta indica A. Juss	Tree	Meliaceae	3, 11, 17, 19	Bark, leaves, stems
Caesalpina pulcherima Swartz	Tree	Fabaceae	11, 12, 19, 27	Flowers, stems
Capsicum annum Linn.	Shrub	Solanaceae	4, 10, 17, 23, 24, 28	Fruits, seeds
C. frutescens Linn.	Shrub	Solanaceae	4, 10, 17, 23, 24, 28	Fruits, seeds
Carica papaya Linn.	Herb	Caricaceae	10, 13, 17, 26	Bark, stems, leaves
**Chrysophyllum albidum* G.Don.	Tree	Sapotaceae	10, 11, 13, 16, 27	Bark, fruits, stems
**Citrullus colosynthis* Schrad	Climber	Cucurbitaceae	10, 17	Fruits, roots
C. vulgaris Linn.	Climber	Cucurbitaceae	10, 17	Fruits, roots
Citrus aurantifolia (Christm.) Swing	Shrub	Rutaceae	17, 24	Fruits
C. limon (L.) Burm. f.	Shrub	Rutaceae	10, 17, 24	Fruits
C. reticulata Blanco	Shrub	Rutaceae	10, 17, 22, 24	Fruits
C. sinensis (Linn.) Osbeck	Shrub	Rutaceae	10, 17, 22, 24	Fruits
Cocos nucifera Linn.	Tree	Palmaceae	2, 4, 7, 10, 14, 15, 17, 18, 22, 24, 27	Fruits, seeds, leaves, whole plant
***Cola argentea* Mast	Tree	Stercuriaceae	2, 4, 11, 24, 27	Fruits, stems
C. nitida (Vent) Schott and Endl.	Tree	Sterculiaceae	2, 4, 11, 12, 24, 27	Fruits, stems
***Cofocasia antiquorum* Schott	Herb	Araceae	9, 10, 30	Leaves, corns
C. esculenta (Linn.) Schott.	Herb	Araceae	9, 10, 19, 2.3, 30	Leaves, corns, whole plant
Corchorus olitorius Linn.	Herb	Tiliaceae	7, 10	Bark, leaves
Coula edulis Baill.	Tree	Olacaceae	3, 8, 12, 17, 24, 27	Fruits, stems
Crescentia cujete Linn.	Tree	Bignoniaceae	11, 15, 19, 30	Fruits, stems, whole plant
Cucumis melo Linn.	Herb	Cucurbitaceae	9, 10, 18, 23	Fruits, seeds
Cucumis sativus Linn.	Herb	Cucurbitaceae	10	Fruits, seeds
***Curcubita moschata* Linn.	Herb	Cucurbitaceae	9, 10, 15, 30	Fruits, leaves
Cymbopogon citratus (DC) Stapf.	Herb	Poaceae	17, 24	Leaves, whole plant
***Dacryodes edulis* (G. Don.) H.J. Lam.	Tree	Burseraceae	3,9, 10, 11, 12, 13, 15, 18	Bark, fruits, leaves stems
***Dennettia tripetala* Bak. f.	Shrub	Annonaceae	10, 11, 23, 24	Fruits, seeds, stems, leaves

Contd...

Table 63.2–Contd...

Species	*Habit*	*Family*	*Ethnobotanical Uses*	*Plant Parts Used*
Dioscorea alata Linn.	Climber	Dioscoreaceae	7, 10	Stems, tubers
D. bulbifera Linn.	Climber	Dioscoreaceae	7, 10	Stems, tubers
D. cayenensis Lam	Climber	Dioscoreaceae	7, 10	Stems, tubers
D. dumentorum (Kunth) Pax.	Climber	Dioscoreaceae	7, 9, 10	Stems, tubers
D. esculenta (Lour.) Burkm	Climber	Dioscoreaceae	7, 10	Stems, tubers
**D. rotundata* Poir	Climber	Diosoreaceae	7, 10, 23	Stems, tubers
Elaeis guineensus Jacq	Tree	Palmaceae	2, 7, 9, 11, 12, 14, 15, 17, 18, 20, 22, 23	Fruits, seeds, leaves, bark, stems
Eugenia owariensis P. Beauv.	Tree	Myrtaceae	10, 11, 15	Fruits, stems
Garcinia kola Heckel	Tree	Clusiaceae	3, 8, 11, 12, 17, 24	Fruits, seeds, stems
Gmelina arborea Roxb	Tree	Verbanaceae	4, 7, 11, 15, 29	Stems
***Gnetum africannum* Welw.	Herb	Gnetaceae	4, 7, 10	Leaves, stems
***Gongronema lafiolium* Benth.	Climber	Asclepiadaceae	3, 7, 10, 21, 23	Leaves, stems, bark
***Heinsia crinata* (Afzel) G. Tayl.	Shrub	Rubiaceae	3, 4, 10, 15, 17	Leaves, stems
Hevea braziliensis (Kunth) Muell. Arg.	Tree	Euphorbiaceae	2, 4, 6, 9, 11, 13, 14, 16, 18	Bark, fruits, stems, leaves, seed
***Hyptis suaveolens* Poit	Herb	Lamiaceae	10, 17, 23, 24, 25	Leaves, fruits
**Irvingia gabonensis* (Aubru-Lecomte ex O'Rorke) Baill	Tree	Irvingiaceae	2, 4, 8, 9, 10, 11, 12, 14, 15, 23, 27	Fruits, seeds, stems
L. grandifolia (Engl.) Engl.	Tree	Irvingiaceae	2, 4, 8, 10, 11, 12, 14, 27	Fruits, seeds, stems
***Lasianthera africana* P. Beauv.	Shrub	Icacinaceae	3, 6, 10, 17	Leaves, stems
Lycopersicon esculentum Mill.	Herb	Solanaceae	4, 10	Fruits, seeds
Mangifera indica Linn.	Tree	Anacardiaceae	4, 10, 11, 12, 13, 17, 27, 30	Bark, fruits, stems, leaves
Manihot esculentus Krantz	Shrub	Euphorbiaceae	1, 4, 9, 10, 13, 16, 19	Leaves, stems, tubers, whole plant
Mucuna urens Adans.	Herb	Fabaceae	7, 10, 23	Seeds, stems
Musa paradisiacal Linn.	Herb	Musaceae	7, 10, 17, 22, 30	Leaves, stems, fruits, whole plant
M. paradisiaca subsp. *sapientum* Linn.	Herb	Musaceae	7, 10, 17, 22, 30	Leaves, stems, fruits, whole plants
***Ocimum gratissimum* Linn.	Shrub	Lamiaceae	9, 10, 17, 23, 24	Leaves, fruits
Persea Americana Mill	Tree	Lauraceae	5, 9, 10, 11, 18	Leaves, fruits, stems
Phaseolus lunatus Linn.	Herb	Fabaceae	10, 17	Seeds, leaves, pods
***Piper guineensis* Schurn and Thonn.	Climber	Piperceae	10, 17, 23, 24	Leaves, seeds
Plumeria alba Linn.	Tree	Apocynaceae	16, 17, 19	Bark, stems, whole plant
P. rubra Linn.	Tree	Apocynaceae	16, 17, 19	Bark, stems, whole plant
Psidium guajava Linn.	Shrub	Myrtaceae	3, 9, 10, 11, 15, 17, 26	Fruits, bark, leaves

Contd...

Table 63.2–Contd...

Species	*Habit*	*Family*	*Ethnobotanical Uses*	*Plant Parts Used*
***Raphia hookeri* Mann. and Wendl.	Tree	Palmaceae	2, 4, 6, 7, 11, 12, 13, 14, 16, 20, 24, 28	Fruits, leaves, stems, seed, whole plant
Ricinus communis Linn.	Shrub	Euphorbiaceae	4, 7, 17, 18, 22	Fruits, seeds, stems, bark, leaves
Saccharium officinarum Linn.	Herb	Poaceae	4, 7, 10, 25	Stems
**Sesamum indicam* Linn.	Herb	Pedaliaceae	10, 17, 18	Leaves, fruits, seeds
Solanum macrocarpum Linn.	Shrub	Solanaceae	9, 10	Fruits
**Spondias mombin* Linn.	Tree	Anacardiaceae	6, 9, 10, 12, 27	Fruits, stems
Talinum triangulare (Jacq.) Willd.	Herb	Portulaceaceae	9, 10, 17	Leaves, whole plant
***Telfairia occidentailis* Hook f.	Climber	Cucurbitaceae	9, 10, 17, 23	Fruits, leaves seeds, roots
Terminalia catappa Linn.	Tree	Combretaceae	8, 9, 10, 17, 18, 19, 22, 26, 27	Fruits, seeds, stems, bark, leaves
Theobroma cacao Linn.	Tree	Sterculiaceae	11, 17, 24	Fruits, stems, seeds
Trema guineensis (Schunn. & Thonn.) Ficalho	Tree	Ulmaceae	6, 11, 17	Stems, roots
Vernonia amygdalina Del.	Shrub	Asteraceae	3, 10, 17, 23	Leaves, stems, roots
**Xanthosoma sagittifolium* Schott	Herb	Araceae	10, 23, 30	Leaves, corms, whole plants
Zea mays Linn.	Herb	Poaceae	4, 7, 9, 10, 17, 30	Fruits, stems
Zingiber officinale Rosc.	Herb	Zingiberaceae	4, 17, 18, 23, 24	Stems, rhizomes

*: Endangered species; **: Endemic species.

Diversity of Uses of Plant Species

All plant species have at one reported use. Two hundred and fifty nine forest species and eighty homestead garden species are used for at least a single purpose (Tables 63.1, 63.2). Species with more uses include *Elaeis guineensis* (12), *Raphia hookeri* (12), *Cocos nucifera* (11), *Irvingiagabonensis* (11), *Hevea brasiliensis* (9), *Lonchocarpus cyanescens* (9), *Lovoa trichilioides* (9), *Pterocarpus santalinoides* (9), *Dacryodes edulis* (8), *Khaya grandifoliola* (8), *Lophira alata* (8), *Anthocleista vogelii* (8), *Pachystela brevipes* (8), and *Symphonia globulifera* (8). Ethnobotaincal use with more species include Medicine (178/52.5 per cent), Fuel (107/31.6 per cent), Food/condiment (100/29.4 per cent), Timber (72/21.2 per cent), Food/feed (70/20.6 per cent), Commercial (58/17.1 per cent), Furniture (55/16.2 per cent), Fibre/cordage (55/16.2 per cent), Homestool (45/13.2 per cent) and Fence (38/11.2 per cent). These figures are inappropriate for medicinal and fuel plants as almost every species is used in healing a disease or is used as fuel wood. Diversity index for species ethnobotanical use is high, 7.9 for forest species and 5.0 for homestead garden species. This is indicative of the multiple uses the species are put to.

There is an indication that people from rural areas within the study areas use more forest and homestead garden species than people in towns as the former places are more distant and isolated villages. The surplus forest produce is traded locally. The people in better-communicated towns buy and sell their produce in well-organized markets in and around the towns. Thus, the use of forest and homestead garden produce in the Niger Delta communities is affected by communication systems such as roads and transport.

Economic Importance of the Forests Plants

The type, quality and quantity of resources from the Niger Delta forests have diverse uses (Etukudo, 2000). More important are the timber species (*Antiaris africana, Albizia zygia, Brachystegia enrycoma, Ceiba pentandra, Diospyros alboflavescens, Erythrophleum ivorense, Funtumia elastica, Hallea stipulosa, Hannoa klaineana, Hylodendron gabunense, Khaya grandifoliola, Lannea welwitschii, Lovoa trichilioides, Nauclea diderrichii, Oxystigma mannii, Staudtia stipitata, Tlipochiton scleroxylon* and *Uapaca togoensis*); the fruit species (*Canalium schweinfulrhii, Dialium guineense, Irvingia gabonensis, Maesobotrya balteri, Microdesmis peberula, Momordica charantia, Pentaclethra macrophylla, Synsepallum dulcifilum,* and *Tetracarpidium conophorum*), the spices, flavouring and thickners (*Aframomum melegueta, Irvingia gabonensis, Piper guineense, Tetrapleura tetraptera, Xylopia aethiopica*), the beverages (*Cocos nucifera, Elaeis guineensis, Ananas comosus, Annona muricata, Sacoglottis gabonensis, Synsepalum dulcificum*) and the medical species (Table 63.3). The homestead gardens contain a great number of these species (Table 63.2).

The Forests and Animal Diets

The forests are natural habitats of various animal species, some endemic to Nigeria (*Cercopithecus sclateri, Cercocebus torquatus, Cercopithecus mona, C. niditans*). Some of the animals are important bush meat, source most appreciated by the people of the Niger Delta. The most important venison of these forests are *Cephalophus* sp. and *Thryonomys swinderianus.* The population sizes of these animals have been greatly reduced due to indiscriminate poaching, habitat fragmentation and loss to various forms of anthropogenic activities including oil exploration and exploitation. The plant species used by *cephalophus* sp. and *T. swinderianus.* as food sources include those cultivated in homestead gardens (*e.g. Colocasia* sp., *Corchorus olitorius, Curcubita moschata, Dioscorea dumentorium, Manihot esculentus, Solanum macrocarpum, Telfairia occidentalis, Zea mays*) as well as the fruits of *Elaeis guineensis, Irvingia gabonensis, Persea americana* and *Psidium guajava;* and those found in the forests: *Alchornea cordifolia, Allanblackia floribunda, Centrosoma pubescens, Dialium guineense, Lonchocarpus cyanescens, Macaranga balteri, Maesobotrya balteri, Manniophyton fulvum, Microdesmis puberula, Pterocarpus* sp. and *Voacanga africana.* The monkeys (*Cercopithecus sp., Cercocebus torquatus*) feed on the above fruits species.

The Forests as Sources of Building Materials

The traditional house in the Niger Delta is built from the materials obtained from the forests. This is obvious in the more isolated villages. Materials for the buildings come from species such as *Acioa barteri, Alstonia* sp., *Bambusa vulgaris, Betfinia grandiflora, Cocos nucifera, Elaeis guineensis, Klainedoxa gabonensis, Myrianthus arboreus, Uapaca* sp. and *Raphia hookeri* among others. These traditional houses are fast disappearing and are being replaced by modern ones because over-exploitation of the forests had brought about scarcity of these building materials, the regenerating forests do not contain materials old enough to be used in building; it takes more time to search for and obtain available and needed materials; and since the advent of oil exploration and exploitation some of the rural dwellers who are gainfully employed want to have same standards of living as city dwellers (Rico-Gray *et al.*, 1991). This has resulted in dwindling economic values of these natural materials and their turnover rates. It is suggested that rather than abandoning (the use of) these materials, modern ideas should help improve the production, quality and sustainability of these natural materials, which will enhance their cost benefit status and high turnover rates. These could be used for alternative or different purposes.

Table 63.3: Some Uncultivated Plant Species Identified for Medical Use in the Niger Delta Region, Nigeria

Species	*Family*	*Action*	*Usage*
Abrusprecatorius Linn.	Malvavceae	Nervous disorders treatment	Infusion of seeds
Abutilon mauritianum (Jacq.) Medic	Malvaceae	Alleviates pectoral pains	Mucilage of leaves
Achyranthes aspera L.	Amaranthaceae	Diuretic, anti-pneumonia	Infusion, decoction of whole plant
Aframonum melegueta K. Schum.	Zingiberaceae	Bechic, expectorant, mucolytic	Fresh seeds
Aframomonum sceptrum (Oliv. and Hanb) K. Schum.	Zingiberaceae	Anti-flatulent, calms nervous stomach spasms	Fresh fruits, essence
Aframomum strobilaceum (Sm) Hepper	Zingiberaceae	Anthelmintic	Root decoction
Ageratum conizoides Linn.	Asteraceae	Wound healer, antiseptic	Poultice leaves
Alchornea laxiflora (Benth) Pax. and K. Hoffin	Euphorbiaceae	Gonorrhoea treatment	Leaves and roots decoction
Alternanthera pungens H.B.K	Amaranthaceae	Febrifuge	Infusion of leaves
Althernanthera sessilis (Linn.) R.Br.ex Roth	Amaranthaceae	Febrifuge, snake bite	Infusion, decoction of stems and leaves
Anthonatha macrophylla P. Beauv	Fabaceae	Wound healer	Compress with sap of bark
Argemone mexicana Linn.	Papaveraceae	Skin mycosis, scabies	Lotions, latex
Aspilia africana (Pers.) C.D. Adams	Asteraceae	Alleviates asthma	Leaf decoction
Asystasiagangetica (Linn.) T. Anders	Acanthaceae	Anti-rheumatic, analgesic	Leaf juice and lotion
Bryophyllum pinnatum (Lam.) Oken	Crassulaceae	Coughs, boils treatment	Leaf juice, liniment
Canarium schweinfurthii Linn	Burseraceae	Cough, piles treatment	Bark decoction
Chromolaena odorata (L.) R.M. King and Robinson	Asteraceae	Hepatitis, jaundice healing	Infusion of leaves, decoction of whole plant
Cleome viscosa Linn.	Capparidaceae	Ear troubles treatment	Juice, lotion of leaves, whole plant
Cnestls ferruginea D.C.	Connaraceae	Enema for dysentery, diarrhea,	Root decoction and juice sinusitis, gonorrhea
Combretum hispidium Laws	Combretaceae	Jaundice treatment	Raw root chewed
Combretum zenkeri Engl and Die's	Combretaceae	Depurative, diuretic, sudorific	Leaf extract, decoction
Commelina africana Linn.	Commelinaceae	Laxative, beneficial in leprosy	Infusion, decoction of whole plant
Commelina diffusa Burm f.	Commelinaceae	Soothes skin itching; wound	Poultice plant, compress healer, abscess
Costus afer Ker-Gawl	Zingiberaceae	Rheumatism relief	Leaf decoction
Costus schlechteri Winkler	Zingiberaceae	Wound healer, earaches	Plant extract, decoction

Contd...

Table 63.3–Contd...

Species	*Family*	*Action*	*Usage*
Coula edulis Baill.	Olacaceae	Anaemia, dysentery alleviation	Powered bark decoction
Datura stramonium Linn.	Solanaceae	Anti-rheumatic, asthmatic earache	Poultice of leaves with oil, flower juice
Desmodium turtuosium (Slo.) DC	Fabaceae	Soothing, anti inflammatory in	Leaf extract in enema haemorrhioids
Diodia scandens Sw.	Rubiaceae	Asthma treatment	Leaf extract as emetic
Emilia coccinea (Sims) G. Don.	Asteraceae	Antitussive, calms coughs	Infusion of plant
Emilia sonchifolia (Linn.) DC	Asteraceae	Antitussive, bechic, calms	Infusion of plant tonsillitis
Eramomastox speciosa (Hockst.) Cufod	Acanthaceae	Treatment of diabetes	Leaf juice, lotion
Euphorbia heterophylla Linn.	Euphorbiaceae	Laxative	Decoction of leaves
Euphorbia hirta Linn.	Euphorbiaceae	Sickle cell ailment treatment	Decoction of leaves, roots
Fagara zanthoxyloides Lam.	Rutaceae	Treatment of sickle cell ailment	Decoction, infusion of leaves, bark
Ficus exasperata Vahl	Moraceae	Stops haemorrhage, eye sore	Leaf juice
Aeurya aesatuans (Linn.) Graudex Miq	Urticaceae	Assuages asthma	Air-dried leaves
Fleurya ovalifolia (Schum. and Thorn.) Dandy	Urticaceae	Enema for epilepsy	Leaf decoction
Gloriosa superba Linn.	Liliaceae	Heals leprosy	Root decoction
Glyphaea brevis (Spreug.) Monachino	Tiliaceae	Antiseptic in venereal discharges	Decoction of leaves
Hallea stipulosa (DC) J.F. Lersy	Rubiaceae	Prevents stomach spasms	Boiled root chewed
Heterotis rotundifolia (Sm) Jac-fet	Melastomataceae	Treatment of conjunctivitis	Leaf lotion
Hippocratea pallens Planch. ex Olive	Hippocrateaceae	Calms stomach ache	Fruit pulp
Icacina tirichantha Oliv	Icacinaceae	Treatment of mumps	Tuber juice, lotion
Ipomoea involucrata P. Beauv.	Convolvulaceae	Laxative, analgesic	Leaf infusion
Irvingia gabonensis Bail.	Irvingiaceae	Useful for malaria	Bark decoction
Justicia schimperi (Hochst) Dandy	Acanthaceae	Enema for umbilical pains in	Leaf decoction, infusion children
Loranthes asphyllus Sprauge	Loranthaceae	Antidiabetic	Leaf decoction
Mallotus oppositifolius (Geisel) Mull. Arg.	Euphorbiaceae	Stops bleeding	Leaf juice
Melanthera scandens (Schum and Thonn.) Roberty	Asteraceae	Wound healer, stops bleeding	Leaf juice
Melochia cochorifolia Linn.	Sterculiaceae	Prevents bronchial spasms	Root lotion
Milletia thonningii (Schum. and Thonn.) Bak.	Fabaceae	Prevents abortion, miscarriage in goats	Raw leaves, young branches

Contd...

Table 63.3–Contd...

Species	*Family*	*Action*	*Usage*
Momordica charantia Linn.	Cucurbitaceae	Useful in piles, jaundice, as vermiguge	Fruits, leaves
Musanga cecropioides R.Br.	Cecropiaceae	Indigenous medicine for toothache	Bark infusion
Myrianthus arboreus P. Beauv.	Moraceae	Heals heamaturia	Roots liquid
Nauclea diderrichii (De Wild and Dur.) Merril	Rubiaceae	Enema for indigestion	Bark decoction
Palisota hirsuta (Thunb.) K. Schum	Commelinaceae	Treatment of gonorrhoea	Pounded root decoction with lime juice
Peperoma pellucida (L.) H.B.K.	Piperaceae	Heals earache	Juice of leaf
Phyllanthus amarus Schum and Thonn.	Euphorbiaceae	Eliminates fungi, heals ringworm, scabbies	Poultices of leaves
Physalis angulata Linn.	Solanaceae	Stomachic invigorator	Leaf infusion, decoction
Piper guineense Schum and Thonn.	Piperaceae	Antirheumatic, carminative, insecticidal	Pulverized seeds and fruits
Platostoma africanum P. Beauv.	Lamiaceae	Insecticidal	Pulverized leaves
Pupalia lappacea (Linn.) Juss.	Amaranthaceae	Antitussive, eases cough	Leaf decoction
Rauvolfia vomitoria Afzel	Apocynaceae	Reduces blood pressure	Root decoction, alkaloids
Sida acuta Burm. f.	Malvaceae	Astringent, useful in urinary	Root decoction diseases
Stachytarpheta cayennensis L.C. (Rich) Schau	Verbenaceae	Demulcent, soothes inflammation	Poultice of leaves
Stachytapheta indica (Linn.) Vahl.	Verbenaceae	Indigenous treatment for malaria	Root infusion, decoction
Symphonia globullrera Linn. f.	Clusiaceae	Stomachic invigorator	Bark decoction
Tetrapleura tetraptera (Schum. and Thonn.) Taub.	Fabaceae	Emetic and purgative	Bark decoction
Tridax procumbens Linn.	Asteraceae	Anti-haemorrhage	Plant juice, poultice
Uapaca staudtii Pax	Euphorbiaceae	Useful in cough, antitussive	Raw fruit
Uapaca topensis Pax.	Euphorbiaceae	Useful for whooping cough, antibiotic	Plant and root sap
Urena lobata Linn.	Malvaceae	Anti-rheumatic	Poultice of roots
Uvaria chamae P. Beauv.	Annonaceae	Useful in and stops nose bleeding, diarrhoea	Roots decoction, infusion
Uvaria cristata R.Br.	Annonaceae	Treatment of waist pains	Root decoction
Xylopia aethiopica (Dunal) A. Rich	Annonaceae	Carminative, treatment of kidney and bladder afflictions	Fruit infusion, decoction

Homestead Gardens Produce

A variety of produce is obtained from the homestead gardens: fruits, flavouring, oils, ornamentals, spices, vegetables and thickeners in quantities that are sold or traded locally. Their commercial values can be enhanced if production is co-ordinated and harnessed effectively from the different villages to towns where they are in high demand. These can advance the per capita income of the producers and in turn improve their living standards.

Medical Plants from the Forests

The Niger Delta forests habour a large number of plant species known to be of importance in the medical and pharmaceutical industries. *Alchomea laxinora, Cnestis feniginea, Fagara zanthoxyloides, Icacina trichantha, Piper guineense* and *Uvaria cristata* are widely known and used as species of medicinal utility. *Aspilia africana* contains thiarubine–A in its root, a red oil that kms parasites, viruses, fungi and bacteria. *Smilax kraussiana* contains saponinic glycosides, resin and essential oil that promote elimination of urea, uric acid and other organic wastes as well as decreasing blood cholesterol level. *Strophanthus sarmentosus* is a source of drug called cortisone for the treatment of rheumatoid arthritis. Vernonioside-B_1, is a constituent of *Vernonia amygdalina* which has an antitumoral and antibacterial properties: It is used in treating schistosomiasis. The list of some plants occurring in the Niger Delta forest with their medicinal values is presented in Table 63.3. The information contained therein is based on indigenous knowledge of the local people. The usage of the plants has been, in the main, through extracts or in raw form for ages. However, there is need for scientific analyses of these plants for proven efficacies by biochemists, pharmacognosists, pharmacologists and pharmaceutical industries.

Endemic and Endangered Plants Species

A number of plants species in the Niger Delta forest are endemic and endangered (Tables 63.1 and 63.2). Endemism and a number of sub-species presence in the study area are evidenced in phylogenetic plasticity and the dynamics of ecological and environmental perturbations (Khan *et al.*, 2003).

The loss of plant biodiversity in the Niger Delta is precipitated by anthropogenic factors of over-extraction, unsustainable agriculture and forestry practices, pollution, land use changes, urbanization and probably global climate change. These culminate in biodiversity depletion and diminished productivity. Over extraction target species are those used as fuel wood, in medicine, in building, as timber and as fodder. Observations show that due to extensive use of plant species for these purpose, a large number of plants are endangered.

Conservation Methods

An important element in the relationship between ethnobotany and conservation is the detailed knowledge of the components of biodiversity possessed by the natives over the years and the dynamics of the ecosystem. Through observations, assessments and experimentations, these natives have successfully adopted their production methods towards the characteristics of these biological resources. The possession of this biodiversity-related knowledge should be preserved through capacity building in order to compliment scientific knowledge and provide information on biological system. This will also make it possible for the socio-cultural, economic and religious significance of many plants to be transferred from generation to generation. This requires education strategies that will raise the level of awareness of the natives in the conservation of biodiversity.

The indiscriminate over-exploitation of plants that impinges on food security and health care should be avoided and halted. No wild plant species should be endangered through international trade. There should be the promotion of *in situ* and *ex situ* cultivation of the endemic, medicinal and more obvious economic species. Conservation of the plants should also take the form of systematic protection through breeding. The protected areas in form of groves and nature reserves in the region should be preserved. All these could be achieved through collaborative cooperation of governments, Non-governmental organisations (NGO's), community-based organisations (CBO's), the natives, schools, colleges and other higher institutions.

As biodiversity involves global environmental change, species extirpation or extinction and changes in societal values (Khan *et al.*, 2003), ethnobotany offers an effective approach to plant biodiversity conservation, since this provides a wealth of data on timber and non-timber products which can be rationally exploited. Thus, commodities can be extracted from plant communities with minimal environmental damage and hence provide incentive for conservation. Conservation for diversity and stability, aesthetic, ethical, scientific values and utilitarian purposes is rooted in ethnobotany, which is the most cost-effect method to be practiced.

Conclusion

The economic importance of the tropical lowland rainforests rural dwellers in Niger Delta is emphasized. Though the natural environments of these forests have been modified by activities of man (arable farming, animal and oil exploitation, capital development, forestry production), the establishment of management systems that are flexible, adaptive and experimental would allow for sustainable exploitation while conserving them (Sayer *et al.*, 2000). These management systems should involve immediate economic benefits by integrating components (*e.g.* agroforestry, agrosilviculture) that will promote and complement long-term economic benefits. In addition, enforcement of forests laws, improvement of forest policy and adoption of rules and regulations should be in practice (Li *et al.*, 1999). When these strategies are combined with other vegetation studies and effectively applied, they would provide sustainable means of utilizing the natural resources of these forests and probably bring about compatibility of conservation of nature.

Acknowledgements

The author thanks the Akwa Ibom State Environmental Protection Agency, Uyo; Universal, Scientific and Industrial Company, Uyo and the Prodec-Fugro, Port Harcourt for their financial supports.

References

Akobundu, I.O. and Agyakwa, C.W., 1987. *Guide to West African Weeds.* I.I.T.A. Ibadan, 522 p.

Clay, J.W., 1988. *Indigenous Peoples and Tropical Forests.* Cultural Survival, Inc. Cambridge, M.A.

Etukudo, I., 2000. *Forests: Our Divine Treasure.* Dorand Publishers, Uyo, 194 p.

Etukudo, I., 2003. *Ethnobotany: Conventional and Traditional Uses of Plants.* The Verdict Press, Uyo, 191 p.

Hutchinson, J. and Dalziel, J.M., 1954–72. *Flora of West Tropical Africa,* (Revised by Keay, R.W.J. and Hepper, F.N). Crown Agents for Overseas Governments, London.

Khan, T.I., Dular, A.K. and Solomon, D.M., 2003. Biodiversity conservation in the Thar desert with emphasis on endemic and medicinal plants. *The Environmentalis,* 23: 137–144.

Kokwaro, J.O., 1995. Ethnobotany in Africa. In: *Ethnobotany: Evolution of a Discipline,* (Eds.) Schultes, R.E. and Von Reis, S. Chapman and Hall, London, pp 216–225.

Li, C., Koskela, J. and Luukkan, O., 1999. Protective forest system in China: Current status, problems and perspective. *Ambio*, 28(4): 341–345.

Pangely, R., 1994. *International River Basin Organisation in Sub Saharan Africa Technical paper* 250, World Bank, Washington DC.

Peters, R.H., 1991. *A Critique for Ecology.* Cambridge University Press, New York.

Rico-Gray, V., Chemas, A. and Mandujano, S., 1991. Uses of tropical deciduous forest species by the Yacatecan Maya. *Agroforestry System*, 14: 149–161.

Sayer, J., Ishwaran, N., Thorsell, J. and Sigaty, T., 2000. Tropical forest biodiversity and the world heritage convention. *Ambio*, 29(6): 302–309.

Soforowa, A., 1982. *Medicinal Plants and Traditional Medicine in Africa.* John Wiley and Sons, Chichester.

Sokal, R.R. and Rohlf, F.J., 1995. *Biometry.* Colt Freeman and Company, New York, 887 p.

Spellerberg, I.F., 1995. Conserving biological diversity. In: *Conservation Biology*, (Ed.) Spellerberg, I.F. Longman, Harlow, pp. 25–35.

Chapter 64

Growth Parameters, Biomass Production and Nutrient Uptake by Greengram as Influenced by Potash and Zinc Levels

S.V. Patil, A.V. Solanke, T.K. Tiwari and B.J. Shete
Department of Agronomy, Mahatma Phule Krishi Vidyapeeth, Rahuri – 413 722, India

ABSTRACT

A field experiment was conducted during 2003–2004 at Rahuri, to study the effect of potash and zinc on greengram. Both potash and zinc affected the seed yield and yield attributes significantly. Application of potash @25 kg ha^{-1} and zinc @15 kg ha^{-1} were found optimum to obtain significantly more number of primary branches per plant, dry matter production and yield.

Application of potash @37.5 kg ha^{-1} and zinc @15 kg ha^{-1} recorded significantly more uptake of zinc by greengram crop.

Keywords: *Greengram, Zinc, Potash, Yield and Nutrient uptake.*

Introduction

Besides being an indispensable component of vegetarian diet, greengram plays a vital role in sustaining the long-term productivity of our soil resources. Among production inputs, fertilizer application plays a key role in enhancing the productivity levels. Pulse crops are able to fix atmospheric nitrogen; consequently the major part of N requirements is met by N fixation. Phosphorus fertilization also being given importance now a day and its recommendations for pulse crops is made in Maharashtra state. Potassium application is almost neglected resulting in imbalanced nutrients and lower crop yields. Zinc has also gained significant importance in Maharashtra agriculture due to its widespread deficiency. A significant response of applied K has been reported by Duke *et al.* (1980) and that of zinc sulphate by Singh *et al.* (2001). However, the study on combined effect of potash and zinc simultaneously has not largely been studied particularly in Maharashtra. The present experiment was therefore planned to study the effect of varying levels of potash and zinc on yield and agro-physiological parameters of green gram in clayey soils of western Maharashtra.

Materials and Methods

The investigation was carried out at experimental research farm of Mahatma Phule Krishi Vidyapeeth, Rahuri (M.S.), for rainy season of 2003–2004 in clayey soil. It had 178 kg ha^{-1} available nitrogen, 16.8 kg ha^{-1} available phosphorus, 448 kg ha^{-1} available potassium and 0.42 mg kg^{-1} zinc. The study was carried out in factorial randomized block design. The treatments consisted of 5 levels of potash (0, 12.5, 25.0, 37.5 and 50.0 kg ha^{-1} and 2 levels of zinc (0 and 15 kg ha^{-1}), thus forming 10 treatment combinations replicated 3 times (Table 64.1). Muriate of potash and zinc sulphate were the source of potash and zinc, respectively. Full amount of potash and zinc according to treatments was applied at planting below the seed. Nitrogen @ 25 kg ha^{-1} and phosphorus @ 50 kg ha^{-1} were also applied basal at planting below the seed. FYM @ 5t ha^{-1} was applied to all the plots during land preparation. Cultivar 'Vaibhav' of greengram was planted on 24th July whereas final picking of pods was done on 12th October. Concentration of Zn in plants in acid digest was estimated using DTPA method (Lindsay and Norvell, 1978) using atomic absorption spectrometer (AAS-l). The crop received total precipitation of 193 mm (14 rainy days) during the crop season. Other agronomical practices were performed as per the recommendations of crop treating the field homogeneous.

Table 64.1: Effect of Potash and Zinc Sulphate on Morphological Parameters, Yield and Nutrient Uptake of Green Gram

Treatment	*Plant Height (cm)*		*Primary Branches per Plant*	*Dry Matter per Plant (g)*	*Yield (q ha^{-1})*		*Nutrient Uptake: Major Nutrients (kg ha^{-1})*			*Micro-nutrient (ppm)*
	45 DAS	*Harvest*		*At Harvest*	*Grain*	*Straw*	*N*	*P*	*K*	*Zn*
Potash levels (kg ha^{-1})										
K_0 : 0	33.08	46.71	2.81	9.43	7.49	11.65	187.50	15.83	458.82	2.52
K_1 : 12.5	33.63	48.09	2.95	10.20	8.07	12.13	187.91	16.03	459.92	2.92
K_2 : 25.0	36.42	51.85	3.28	13.68	8.98	13.54	188.13	16.14	464.15	5.92
K_3 : 37.5	37.73	53.73	3.48	14.04	9.23	14.23	189.00	16.23	467.63	6.22
K_4 : 50.0	37.30	52.88	3.38	13.70	9.10	13.74	189.40	16.61	467.68	6.72
S.Em. ±	0.88	1.11	0.10	0.28	0.17	0.27	3.88	0.33	11.93	0.18
CD at 5 per cent	2.63	3.31	0.30	0.83	0.51	0.80	NS	NS	NS	0.55
Zinc Sulphate (kg ha^{-1})										
Z_0 : 0	34.04	48.36	3.04	11.22	8.12	12.47	188.26	16.10	463.55	0.44
Z_1 : 15	37.22	52.94	3.32	13.20	9.03	13.64	189.11	16.23	463.73	9.28
S.Em. ±	0.56	0.70	0.06	0.17	0.11	0.17	2.45	0.21	7.50	0.11
CD at 5 per cent	1.66	2.09	0.19	0.52	0.32	0.50	NS	NS	NS	0.35

Results and Discussion

Effect of Potash

Potash influenced the plant height, number of primary branches per plant and dry matter per plant significantly and application of potash @ 37.5 kg ha^{-1} registered significantly more plant height (53.73 cm), number of primary branches per plant (3.48) and dry matter per plant (14.04 g) than rest of the treatments except 25 and 50 kg K_2O ha^{-1} which were at par with it (Table 64.1). The significant response of potash on grain and straw yield of greengram was restricted only up to 37.5 kg ha^{-1} but application of 25 kg K_2O ha^{-1} yielded statistically identical seed and straw yield of greengram. The restricted response of potash in the present study on yield and yield attributes may be due to higher potassium status of soil. Response of potash on greengram upto 20 kg ha^{-1} has also been reported by Saxena *et al.* (1996).

N, P and K uptake of greengram was not influenced significantly due to incremental levels of potash from 0 to 50 kg ha^{-1}, but the linear increase in uptake of these nutrients was observed with increased levels of potash. Application of potash @ 50 kg ha^{-1} recorded significantly more uptake of zinc than rest of the potash levels except 37.5 kg K_2O ha^{-1}.

Effect of Zinc

Significant effect of zinc was noticed on the plant height, number of primary branches per plant and dry matter per plant. Application of 15 kg Zn ha^{-1} recorded significantly the highest plant height (52.94), number of primary branches per plant (3.32) and dry matter per plant (13.20 g) (Table 641.). Higher dry matter might be due to conducive influence of zinc in improving the metabolic processes by Zn (Price *et al.*, 1972 and Sarkar *et al.*, 1998). Malewar *et al.* (1979) also reported the increase in dry matter yield of greengram due to Zinc application. The response of green gram crop to the application of 15 kg Zn ha^{-1} was positive and the seed and straw yield obtained in these plots were significantly more than the control without zinc application. The beneficial effects of zinc may be attributed to the marginal content of available Zn in the soil and its role in metabolic activities mainly in protein synthesis in plants as well as in N fixation. The results confirm the findings of Singh *et al.* (2001).

N, P and K uptake of greengram was not influenced significantly with application of zinc @ 15 kg ha^{-1}. Application of zinc @ 15 kg ha^{-1} recorded significantly more uptake of zinc than control without zinc application.

References

Duke, S.H., Collings, M. and Solenlaske, R.M., 1980. Effect of K-fertilization on N-fixation, node enzyme of nitrogen metabolism in alfalfa. *Crop Sci.*, 20: 213–219.

Lindsay, W.L. and Norvell, W.A., 1978. Development of DTPA soil test for Zn, Fe, Mn and Cu. *Proc. Soil Sci. Soc. America J.*, 42: 421–425.

Malewar, G.U., Budhewar, L.M., and Iadhav, N.S., 1979. Response of greengram to zinc application. *Indian J. Agron.*, 24: 164–165.

Price, C.A., Clark, H.E. and Funkhouser, 1972. Functions of micronutrients in plants. In: *Micronutrients in Agriculture* (Eds.) Morvedi, J.J., Giordano, P.M. and Lindsay, W.L. Soil Sci. Soc. America, Inc. Madison, Wisconsin, pp. 1731–1742.

Sarkar, R.K., Bhattacharya, B., Chakraborty, A. and Bala, B., 1998. Growth and productivity of green gram (*Phaseolus radiatus*) in relation to micronutrients application on rice fallow land. *Indian J. Agron.*, 43: 555–559.

Saxena, K.K., Varma, H.R. and Saxena, H.K., 1996. Effect of phosphorus and potassium on greengram (*Phaseolus radiatus*). *Indian J. Agron.*, 41: 81–87.

Singh, R.C., Singh, Mehar, Kumar, Rakesh, Singh, M. and Kumar, R., 2001. Response of mungbean [*Vigna radiata* (L.) Wilczek] to phosphorus application with and without FYM and $ZnSO_4$. *Haryana Agric. Univ. J. Res.*, 31: 89–92.

Chapter 65

Utilization of Waste Sludge from Paper Mill and Pressmud from Sugar Mill for the Culture of a Compost Earthworm *Perionyx excavatus* (Perrier)

S. Chandran and R. Ramalingam
Vermitechnology Research Unit, Department of Zoology, Annamalai University, Annamalainagar – 608 002, India

ABSTRACT

Paper mill sludge is one of the largest under utilized organic waste and causes environmental pollution. Ten newly emerged hatchlings of *Perionyx excavatus* were cultured under laboratory conditions with 60–75 per cent moisture in plastic trays containing the following treatments: T_1: PM alone; T_2: PMS + PM (1 : 3); T_3: PMS + PM (1 : 1); T_4: PMS + PM (3 : 1). Once in 15 days upto 90 days the weight of earthworms, worm growth rate and number of hatchlings were determined and analysed statistically. On the basis of the results the efficiency of four treatments to support the growth and reproduction of *P. excavatus* could be ranked in the following order: PMS + PM 1 : 3 ratio (T_2) > PM alone (T_1) > PMS + PM 1 : 1 ratio (T_3) > PMS + PM 3 : 1 ratio (T_4). The higher growth rate, weight gain, and hatchlings production observed in T_2 (PMS + PM 1 : 3 ratio) could be due to rich nutrients contribution by PM and PMS. On the other hand, the lesser growth rate, weight gain and hatchlings production recorded in T_4 (PMS + PM 3 : 1) could be due to addition of 75 per cent PMS in the feed combination, since wider C : N ratio and lesser quantity of NPK were reported in PMS. So paper mill sludge in combination with pressmud (1 : 3 or 1 : 1 ratio) can be used for vermiculture and production of vermicompost.

***Keywords:** Paper mill sludge, Pressumud, Perionyx excavatus, growth, reproduction.*

Introduction

The paper mill is the greatest consumer of water and also generate industrial solid waste sludge. The paper mills treat their waste by activated sludge process. The excess sludge obtained is usually either dehydrated to cake and land filled or incinerated. Both methods lead to loss of organic resources. The overloading of sludge on land leads to environmental pollution. The high organic matter content 44 per cent low trace metal and organic pollutants in paper mill sludge (PMS) suggest that it may provide a valuable resource for soil amendment (Bellamy *et al.*, 1995). In sugar factories. during sugarcane juice clarification, pressmud (PM) is produced as a by product about 4–5 per cent of the cane weight. The production of pressmud amounts to 3 million tons annually in India. Though pressmud contains higher levels of nutrients its wax content (8.15 per cent) inhibits its nutritional value for the plant (Thopate *et al.*, 1997). Further, its decomposition under dry natural condition required 5–7 months. One of the author's (Ramalingam, 1997) earlier studies have established that PM not only serving as an ideal medium for vermiculture but also vermicomposted into good quality organic manure besides acting as a booster for biodegradation of other wastes/residues.

At present biodegradation of organic wastes using earthworms is gaining momentum in India. Previous studies had shown that though PMS can be used as a feed source by epigeic earthworms, it could not be used alone as feeding media. PMS need to be mixed with other nitrogen rich organic wastes in order to provide nutrients and inoclum of microorganisms (Butt, 1993; Elvira *et al.*, 1995). *Perionyx excavatus* an Indian epigeic (surface dwelling) earthworm has been effectively used in India for vermicomposting of organic wastes (Ismail, 1997). The survey of literature indicated that data on growth and reproduction of compost worms cultured in varieties of organic wastes are available in plenty. But information about the growth and reproduction of worms cultured in paper mill sludge is scandy. Hence the present work aims to investigate the influence of PMS, pressmud mixture on the growth and reproduction of *Perionyx excavatus*.

Materials and Methods

Paper mill sludge (PMS) in semi-solid state was collected from a near by paper mill, sun dried and pulverized. Pressmud (PM) was obtained from a nearby sugar factory, cured by mixing with water and repeated air drying to remove heat and odour. On the basis of trial studies, using PMS and PM the following four feed combinations were prepared on weight basis *viz.*, T_1: PM alone; T_2: PMS + PM 1 : 3 ratio; T_3: PMS + PM 1 : 1 ratio; T_4: PMS + PM 1 : 3 ratio. In each treatment one kg of substrate was prepared using water (60–70 per cent moisture) in circular plastic troughs (35 cm dia × 15 cm depth).

After 15 days of initial decomposition, 10 newly emerged hatchlings of *P. excavatus* were collected from the stock cultured and introduced in T_1–T_4 troughs after recording their individual initial weight. The plastic troughs were covered by nylon nets. For each treatments three replicates were maintained. Once in 15 days upto 90 days weight of the earthworms were recorded. Every time hatchlings and adults were counted and weighted separately and the adults were reintroduced in the respective plastic troughs, but the hatchlings were discarded. Once in a month the newly prepared (in specific proportions) feed mixtures were added and surface layer of vermicast was removed from the experimental troughs and discarded after checking cocoons and hatchlings. The worm growth rate for specific periods was calculated using the formula by Mazantseva (1982). The statistical significance of the data were tested by Student's 't' test.

Results and Discussion

The growth and reproductive efficiency of *Perionyx excavatus* in PM and PMS + PM feed combinations were determined by its biomass (weight gain), growth rate and the rate of multiplication. Generally the body weight of *P. excavatus* increased continuously (but in different magnitude) up to 90 days in all the four treatments. The results indicated that *P.excavatus* gained lesser weight in T_3 and T_4 and more weight in T_2 for all the periods compared to control (T_1) (Table 64.1). The efficiency of four treatments to support the growth of *P. excavatus* could be ranked in the following order: PMS + PM 1 : 3 (T_2) > PM alone (T_1) > PMS + PM 1 : 1 (T_3) > PMS + PM 3 : 1 (T_4). BID values indicated that in T_2 the biomass of *P. excavatus* increased significantly ($P < 0.05$) over control (T_1). On the other hand in T_3 the biomass decreased insignificantly ($P > 0.05$) (Barring 45th and 60th day values). While in T_4 (75 per cent PMS) the biomass decreased significantly ($P < 0.05$) over control (Table 65.1). Comparison of growth rate values indicated that on 45th day *P. excavatus* recorded maximum mean growth rate of 6.40 mg/ worm/day^{-1} in T_2 followed by 6.0 mg/worm/day^{-1} in T_1; 4.13 mg/worm/ day^{-1} in T_3 and minimum growth rate of 3.47 mg/worm/day^{-1} in T_4 (Table 65.1). Further, the worms showed higher growth rate upto 45 days thereafter the growth rate declined upto 90 days in all the treatments (T_1–T_4). *P. excavatus* produced maximum number of hatchlings in T_2 (1062) followed by T_1 (906); T_3 (507) and less number in T_4 (261) (Table 65.1).

The epigeic earthworms are highly sensitive to the quality and availability of feed for their growth and reproduction. Different types of organic wastes originated from different sources have been used for vermicomposting (Neuhauser *et al.*, 1980; Ramalingam and Thilagar, 2003; Tripathi and Bhardwaj, 2004). But all the wastes are not either readily acceptable by the worms nor effectively support the growth and reproduction. The quality and quantity of feed, various physico-chemical parameters influenced earthworms growth and reproduction (Reinecke and Hallatt, 1989). A close observation of results indicated that among the three feed combinations of PMS + PM (T_2 1 : 3; T_3 1 : 1 and T_4 3 : 1 ratio), comparatively in 1 : 3 ratio (T_2) *Perionyx excavatus* recorded higher biomass, showed higher growth rate and produced more number of hatchlings (Table 65.1). The earlier investigators proved that the earthworms prefers food with higher quantity fungus and calcium (Cooke and Luxton, 1980), higher content of sugar and nitrogen (Edwards and Bohlen, 1996). It is already reported that PM containing higher quantities of nitrogen N, P, K, Ca, Mg, S, Fe (Ramalingam, 1997; Ramalingam and Ranganathan, 2001) and PMS containing higher quantities of organic matter (44 per cent), organic carbon (32 per cent) (Piearce *et al.*, 2003). Hence, it can be concluded that 1 : 3 ratio of PMS + PM feed mixture (T_2) could provide all the nutrients for the worms growth and reproduction and become comparatively nutrious food for *P. excavatus*. Further, the treatment T_3 PMS + PM (1 : 1) was equivalent to T_1 (control).

The treatment T_4 (PM + PM 3 : 1) was inferior to T_2 and control T_1 (Table 64.1). This could be due to addition of higher quantity of PMS (75 per cent) in T_4 which contained meagre concentrations of N (0.4 per cent), P (0.08 per cent) and K (0.08 per cent) (Bellamy *et al.*, 1995) since availability adequate nitrogen is essential for growth of worms (Edwards and Bohlen, 1996). Further it could be also due to very wider C : N ratio (85 : 1) of PMS with which the worms were not able to grow. This fact find support from the report of Neuhauser *et al.* (1980), who observed that *Eisenia fetida* accept the food between CN ratio 15 : 1 to 35 : 1. In nutshell the underutilized, organic matter rich papermill sludge in balanced combination with pressmud (PMS + PM 1 : 3 or 1 : 1) could be use as a culture medium to raise compost earthworms and production of organic fertilizer 'vermicompost' is an eco-friendly way.

Table 65.1: Growth in Terms of Biomass Gain by *Perionyx excavatus* Cultured in Pressmud (Control) and Paper Mill Sludge + Pressmud (1 : 3, 1 : 1, 3 : 1 Ratio) Feed Mixture

Age of Worms (Days)	T_1 Pressmud Alone (PM) (Control)			T_2 PMS + PM 1 : 3 (25% PMS)				T_3 PMS + PM 1 : 1 (50% PMS)				T_4 PMS + PM 3 : 1 (75% PMS)			
	WB (mg)	Growth rate (mg/Worm/ Day^{-1}	HP (No)	WB (mg)	Growth rate (mg/Worm/ Day^{-1}	BID (%)	HP (No)	WB (mg)	Growth rate (mg/Worm/ Day^{-1}	BID (%)	HP (No)	WB (mg)	Growth rate (mg/Worm Day^{-1}	BID (%)	HP (No)
1	4.87±0.044		–	4.85±0.029	–	–	–	4.97±0.017	–	–	–	4.85±0.058	–	–	–
15	34±0.058	1.94±0.021	–	39.3±0.467	2.30±0.028	+15.59*	–	34.5±0.233	1.97±0.006	+1.47NS	–	34.2±0.120	1.96±0.012	+0.59NS	–
30	71±0.706	2.5±0.053	–	21±1.86	3.47±0.172	+28.17*	–	70±1.45	2.37±0.158	–1.41NS	–	63±1.00	1.92±0.073	–11.27*	–
45	161±0.577	6.0±0.068	–	187±0.333	6.40±0.108	+16.15*	–	132±0.741	4.13±0.133	–18.01*	–1	15±0.392	3.47±0.043	–28.57*	–
60	182±0.896	1.4±0.067	95±2.60	203±1.40	1.07±0.037	+11.54*	97±6.81	177±0.683	3.0±0.078	–2.76*	53±4.30	166±1.53	3.40±0.027	–8.79*	26±2.33
75	193±1.33	0.73±0.006	456±24.0	214±0.333	0.73±0.076	+10.88*	451±17.1	191±0.857	0.93±0.025	–1.04NS	197±7.60	162±1.20	0.27±0.023	–16.02*	113±4.63
90	198±0.881	0.33±0.050	355±1.75	217±0.437	0.20±0.023	+9.6*	514±10.4	195±2.19	0.27±0.019	–1.51NS	257±3.80	164±0.33	0.13±0.059	–17.17*	122±5.50
THP			906				1062				507				261

Values are mean of 30 observations ± S.E.

*: Significant at 5 per cent level ($P < 0.05$)

THP: Total hatchlings production; BID: Biomass increase or decrease over control; HP: Hatchlings production; NS: Not Significant ($P > 0.05$); WB: Worm biomass.

Acknowledgement

The authors thank the authorities of Annamalai University and the Head of Department of Zoology for providing necessary research facilities.

References

Bellamy, K.L., Chong, C. and Cline, R.A., 1995. Paper sludge utilization in agriculture and container nursery culture. *J. Environ. Qual.*, 24: 1074–1082.

Butt, K.R., 1993. Utilization of solid papermill sludge and spent brewery yeast as a feed for soil-dwelling earthworms. *Bioresource Technology*, 44: 105–107.

Cooke, A. and Luxton, M., 1980. Effect of microbes on food selection by *Lumbricus terrestris. Rev. Ecol. Biol. Sol.*, 17: 365–370.

Edwards, C.A. and Bohlen, P.J., 1996. *Biological and Ecology of Earthworms*, 3rd Edn. Chapman and Hall, London.

Elvira, C. Dominguez, J., Sampetro, L. and Mato, S., 1995. Vermicomposting from the paper pulp industry. *Bio Cycle*, 36(6): 62–63.

Ismail, S.A., 1997. *Vermicology: The Biology of Earthworms*. Orient Longman Ltd. Chennai, India.

Mazantseva, G.P., 1982. Growth patterns in the earthworm *Nicodrilus caliginosus* (Oligochaeta : Lumbricadae) during the first year of life. *Pedobiologia*, 23: 272–276.

Neuhauser, E.F., Kaplan, D.L., Malecki, M.R. and Hartenstein, 1980. Materials supportive of weight gain by the earthworm *Eisenia fetida* in waste conversion systems. *Agrie Wastes*, 2: 43–60.

Piearce, T.G., Budd, T., Hayhol, J.M., Sleep, D. and Clasper, P.J., 2003. Earthworms of a land restoration site treated with paper mill sludge. *Pedobiologia*, 47: 792–795.

Ramalingam, R., 1997. Studies on the life cycle, growth and population dynamics of *Lampito mauritii* (Kingberg) and *Eudrilus eugeniae* (Kingberg) cultured in different organic wastes and analysis of nutrients and microbes of vermicompost. *Ph.D. thesis*, Annamalai University, India.

Ramalingam, R. and Ranganathan, L.S., 2001. Vermicomposting of pressmud boosts, nutrient quality of compost. *Ecol. Env. and Cons.*, 7: 297–299.

Ramalingam, R. and Thilagar, M., 2003. Bio-conversion of agro waste sugarcane trash using an Indian epigeic earthworm *Perionyx excavatus* (perrier). In: *Dimensions of Environmental Threats*, (Ed.) Arvind Kumar. Daya Publishing House, Delhi, p. 203–208.

Reinecke, A.J. and Hallatt, L., 1989. Growth and Cocoon production of *Perionyx excavatus* (Oligochaeta). *Biol. Fertil. Soils*, 8: 303–306.

Thopate, A.M., Hapase, P.R., Jadhau, S.B., 1997. Sugarcane by products: An important source for vermicompost production. In: *Seminar on Recycling of Biomass and Industrial By-products of Sugarcane*, TNAU, India, pp. 49–53.

Tripathi, G. and Bhardwaj, P., 2004. Decomposition of kitchen waste amended with cow manure using an epigeic species (*Eisenia fetida*) and an anecic species *Lampito mauritii. Bioresource Technology*, 92: 215–218.

Chapter 66

Combined Effect of Copper Coating and Extended Expansion of SI Engine

V. Krishnamoorthy[1]* and S.Chandrasekaran[2]

[1]Assistant Professor, Department of Mechanical Engineering, Sathyabama Deemed University, Old Mahabalipuram Road, Chennai – 600 119, Tamil Nadu, India

[2]Professor and Head, Department Automobile Engineering, Anna University, MIT Campus Chennai –600 048, Tamil Nadu, India

ABSTRACT

The concept of Extended Expansion (EE) in IC engines though very old is gaining importance nowadays due to the scope it possess in areas of fuel economy and engine emission. In this paper, a single cylinder engine with variable ER/CR ratio was analysed for its performance when copper coating is done on the surface of cylinder head piston combination. Copper an element, which can promote pre-flame reaction was coated on the surface of cylinder head and piston, ER/CR ratio was varied with LIVCT and the load was regulated with LIVCT-VCR combination. Due to the catalytic effect of copper coating it was found that for the copper coated EE engine, the brake thermal efficiency was improved by 4 per cent compared to standard engine operating at same conditions. The SFC of the copper coated engine was reduced by 0.8 *kg/kw-hr* compared to that of standard engine. It was found that for ER/CR ratio equal to 1.5, the copper coated EE engine was giving best performance in terms of load and thermal efficiency. It was also seen that the lean misfire limit of the engine was extended up to 0.75.

***Keywords**: Extended Expansion (EE), ER/CR, SFC, Efficiency, Lean misfire limit.*

---

* E-mail: krishna153@hotmail.com.

Introduction

With the increasing trend of usage of petroleum products, it becomes necessary to use fuel in efficient manner in internal combustion engines. One such attempt to achieve fuel economy was extended expansion concept which dates back to 19th century. James Atkinson (Heywood, 1988) introduced the concept of higher expansion to compression ratio (ER/CR), so that the advantage of extended expansion in terms of higher engine work output and reduced blow down losses can be achieved. When expansion ratio is more than the compression ratio it becomes extended expansion engine (Eakin *et al.*).

There were lot of research work being carried out in the past in EE engine (Ma and Rajabu, 1988), however the effect of copper coating on EE engine was not tried yet. In this work performance comparison between an EE engine and copper coated EE engine has been done. Experiments were conducted on a modified single cylinder SI engine with variable ER/CR ratio. Initially performance tests were conducted with EE engine for different ER/CR values. Then copper was coated on the engine cylinder head-piston combination of EE engine. Performance test were conducted on the copper coated EE engine and results were compared with that of EE engine. The basic idea of using catalytic material copper (Beyerlein and Wojcicki, 1988) is to improve the pre-flame reaction so that we can achieve lean burn combustion.

Engine Modification

Initially a water cooled single cylinder diesel engine with specification shown in Table 66.1 was converted into spark ignited gasoline engine by altering the fuel injection system and providing electronic ignition system.

Table 66.1: Specification of Diesel Engine Before Modification

Rated power	5 bhp
Rated speed (constant)	159° rpm
Type of cooling	Water cooling
Compression ratio	16.5 : 1
No. of cylinders	1
ER/CR	1

Further to achieve extended expansion, we have to delay the inlet valve closing time. Hence in the standard camshaft only the inlet valve cam was modified leaving the exhaust valve cam untouched.

The amount of delay in closing the inlet valve decides the ER/CR ratio for extended expansion. Figure 66.1 shows the valve-timing diagram for standard engine and for modified SI engine to operate on extended expansion with ER/CR ratio is 1.5 and 2.

To achieve an ER/CR ratio equal to 1.5 we have to delay the inlet valve closing by 67° *i.e.* inlet valve should close 102° aBDC. Similarly to achieve an ER/CR ratio of 2 we have to delay the inlet valve closing by 83° *i.e.* inlet valve should close 118° aBDC. Hence the Figure 66.1 clearly shows the inlet valve closing time for corresponding ER/CR ratios. Other modification was done on piston crown. It was machined on top to have required peak pressure.

By using plasma coating technique, the surface of the cylinder head and piston combination was coated with copper up to a depth of 100 microns.

Results and Discussion

Performance tests conducted with standard engine, EE engine and copper coated EE engine for different CR values clearly indicate the improvement in efficiency for copper coated EE engine.

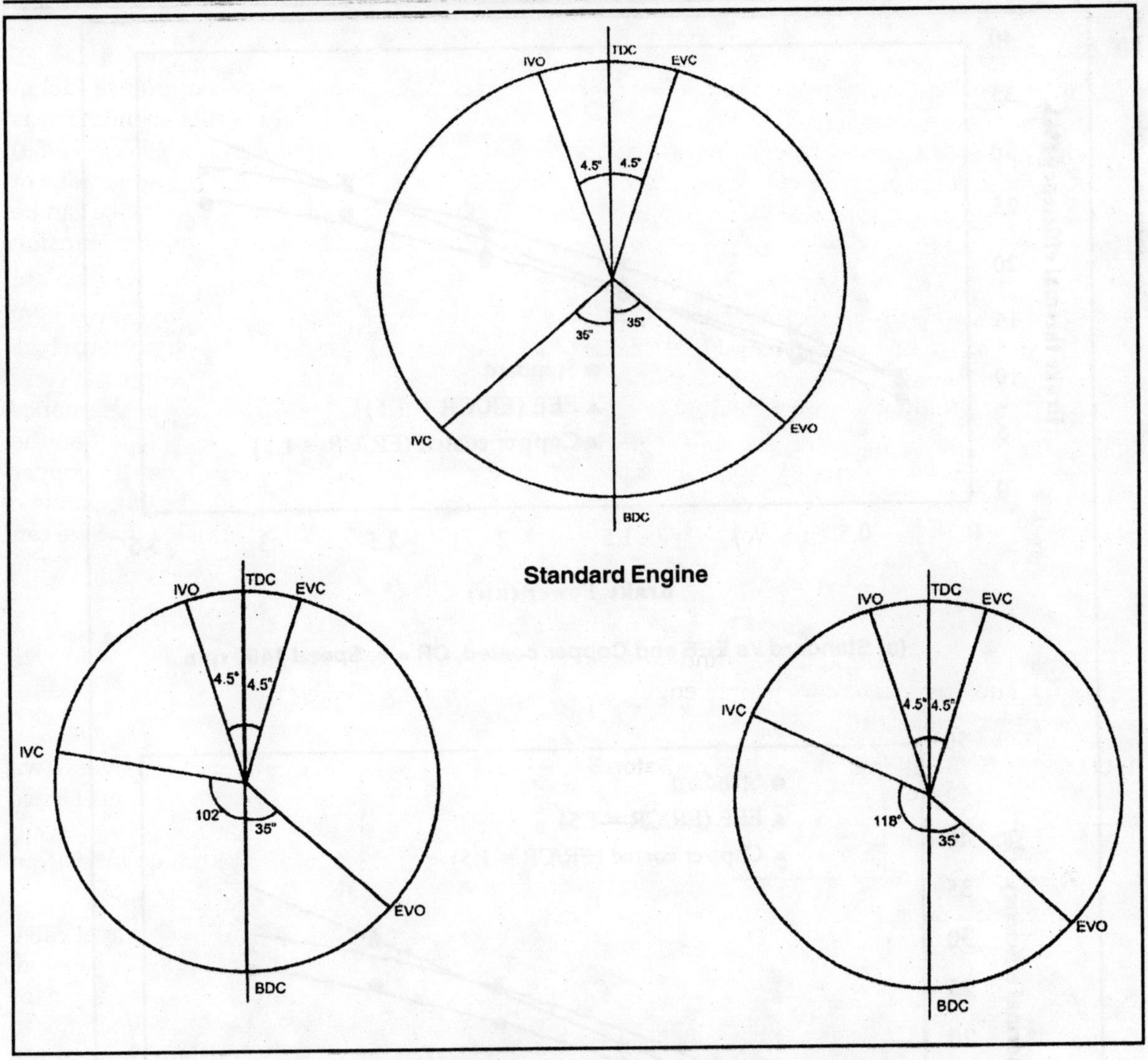

Figure 66.1: Valve Timing for Standard Engine and EE Engine

Effect of Copper Coating on Engine Efficiency

Copper coating was done on cylinder head and on piston crown for the EE engine with optimized value of ER/CR ratio equal to 1.5. Due to catalytic effect of copper the performance test done on copper coated EE engine showed remarkable effect on brake thermal efficiency. Figures 66.2 and 66.3 shows the improvement in efficiencies for modified engines over standard engine.

It is observed that compared to standard engine, copper coated engine with ER/CR equal to 1.5 was having better thermal efficiency. The Figure 66.2 indicates the improvement in efficiency of modified engine when CR is 8 as compared to the Standard Engine (ER/CR = 1) and EE engine (ER/CR = 1.5). The above figure clearly indicate that at speed 1400 rpm and 1500 rpm, the part load efficiency of the modified engine is up by 4 per cent. The copper coated EE engine with ER/CR= 1.5 is better than EE engine and Standard Engine.

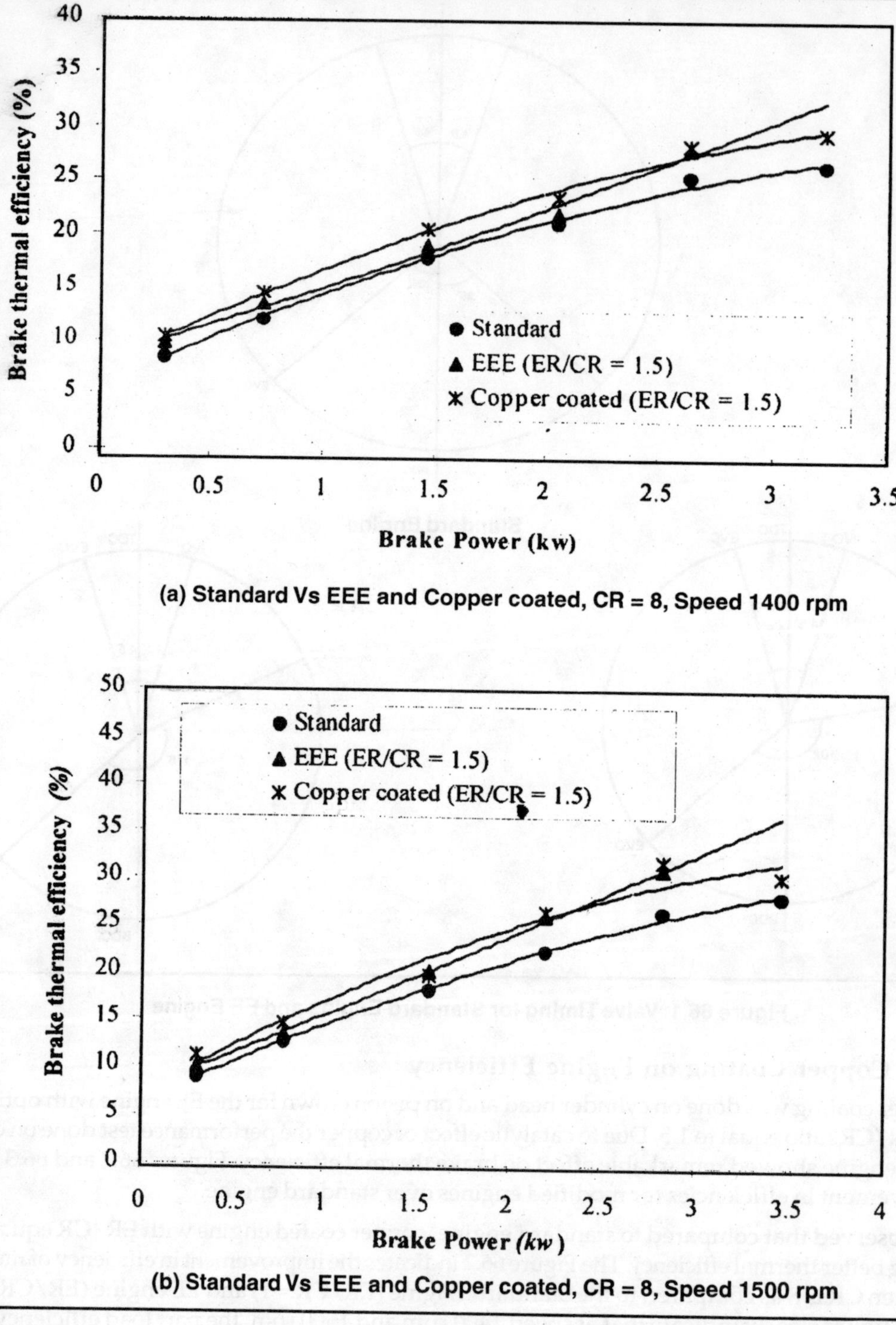

(a) Standard Vs EEE and Copper coated, CR = 8, Speed 1400 rpm

(b) Standard Vs EEE and Copper coated, CR = 8, Speed 1500 rpm

Figure 66.2: Effect of Copper Coating on Engine Efficiency

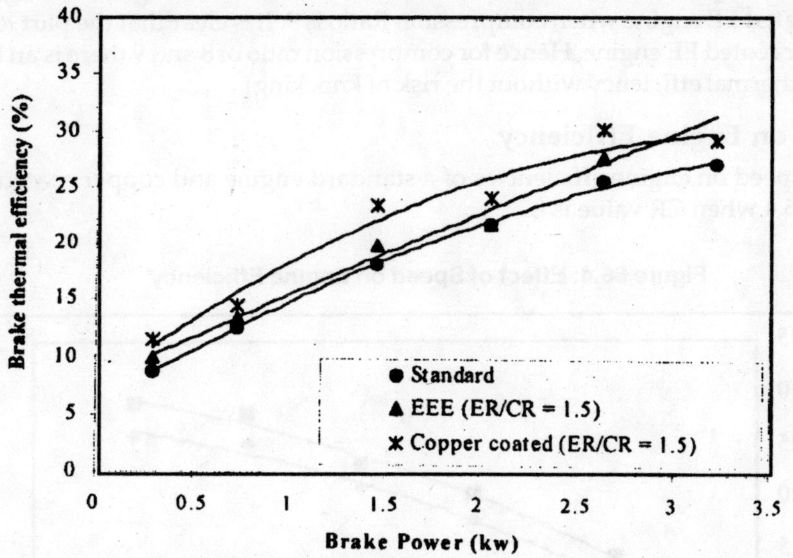

(a) Standard Vs EEE and Copper coated, CR = 9, Speed 1400 rpm

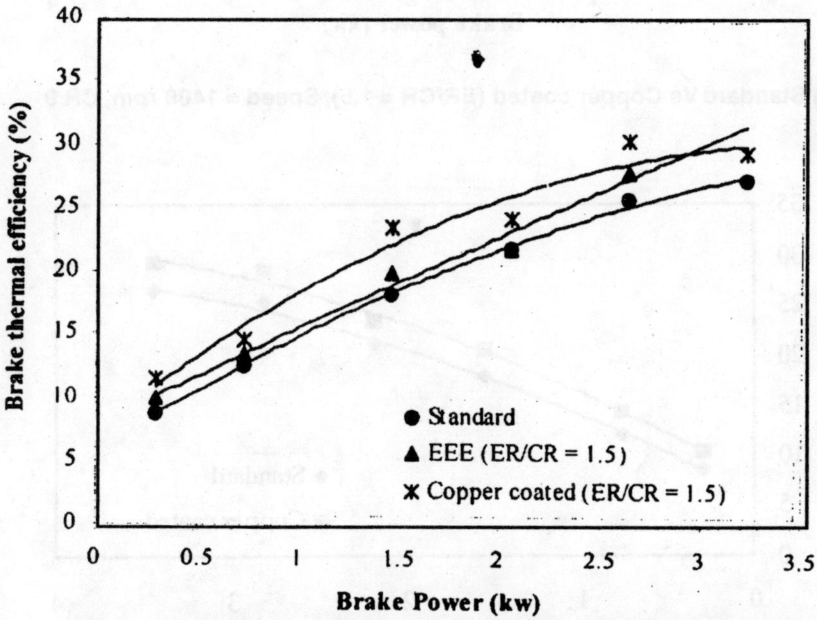

(b) Standard Vs EEE and Copper coated, CR = 9, Speed 1500 rpm

Figure 66.3: Effect of Copper Coating on Engine Efficiency

The Figure 66.3 compares the Brake thermal efficiency of standard engine, EE engine (ER/CR = 1.5) and copper coated EE engine when compression Ratio is 9. It is clear that the part load efficiency is higher for copper coated EE engine. Hence for compression ratio of 8 and 9 there is an improvement in part load brake thermal efficiency without the risk of knocking).

Effect of Speed on Engine Efficiency

The effect of speed on engine efficiencies of a standard engine and copper coated EE engine is shown in Figure 66.4 when CR value is 8.

Figure 66.4: Effect of Speed on Engine Efficiency

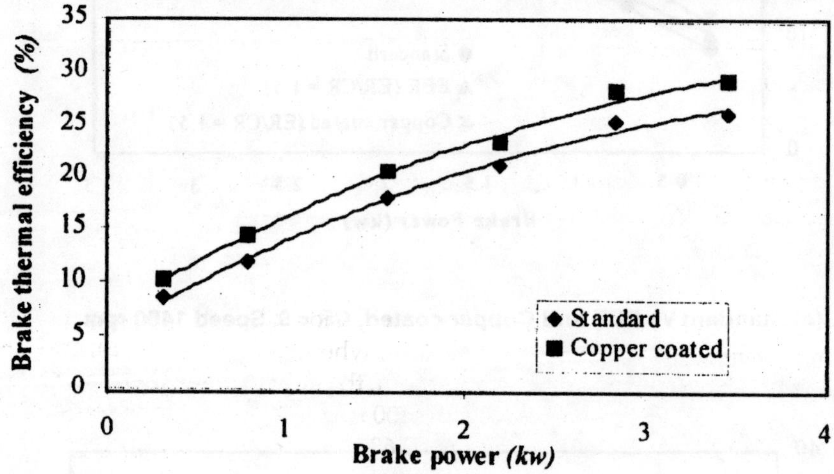

(a) Standard Vs Copper coated (ER/CR = 1.5), Speed = 1400 rpm, CR 9

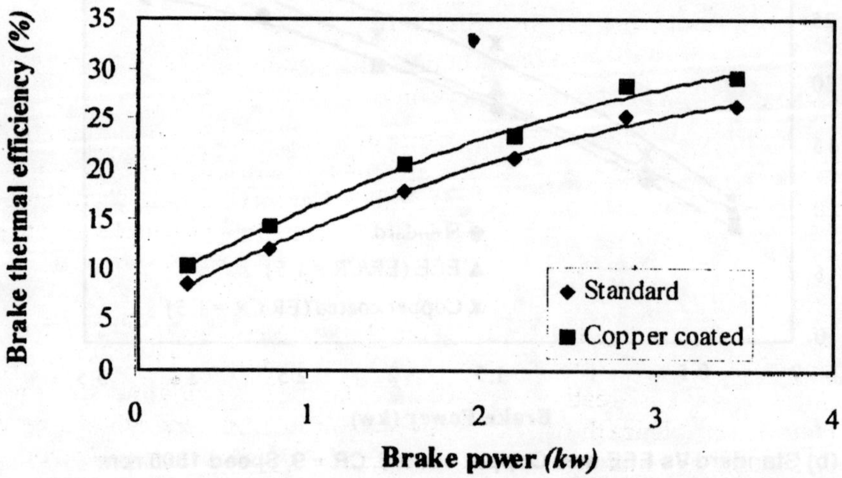

(b) Standard vs Copper coated (ER/CR = 1.5), Speed = 1500 rpm, CR 8

Contd...

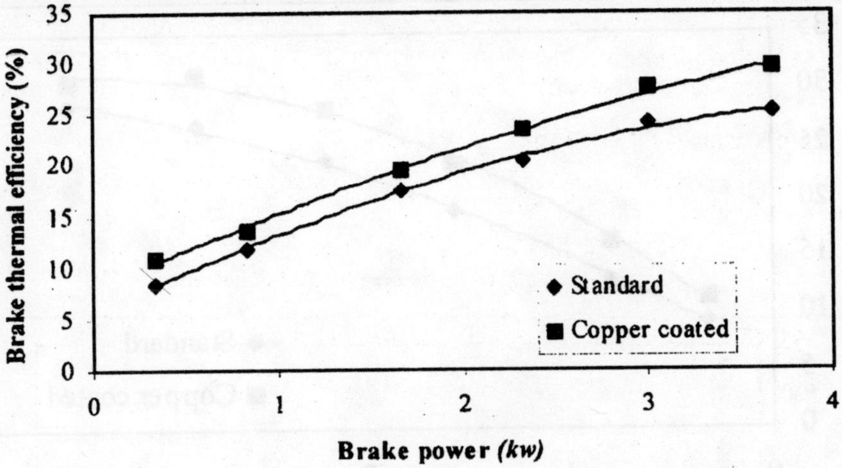

(c) Standard vs Copper Coated (ER/CR = 1.5), Speed = 1600 rpm, CR 8

The above figure clearly indicates that, for the compression ratio of 8, the copper coated EE engine has better part load efficiency. For apart load of 2.63 kw when speed is 1400 rpm, and 2.826 kw when speed is 1500 rpm and 3.01 kw when speed is 1600 rpm, the gain in brake thermal efficiency is more. The increase in efficiency is 3 per cent for a speed of 1400 rpm, 5.4 per cent for a speed of 1500 rpm and 3.3 per cent for a speed of 1600 rpm for part loads of 2.63 kw, 2.826 and 3.01 kw respectively. It is clear that speed of 1500 rpm would enhance the efficiency at part loads more effectively than 1400 and 1600 rpm. Similarly the Figure 66.5 shows the effect of speed on efficiency when CR value is 9.

For a speed of 1600 rpm the increase in part load efficiency is 6.1 per cent at part load of 2.34 kw. For 1500 rpm the increase is 4.5 per cent for a wide range of part load compared to 1400 and 1600 rpm

Effect of Equivalence Ratio on Engine Efficiency

As performance tests were conducted on finding the lean misfire limit for copper coated EE engine, the effect of equivalence ratio on efficiency could be analyzed.

Figures 66.6 shows clearly the onset of lean misfire limit and its effect on engine efficiency. It can be easily ascertained from the figures that, as the lean limit is reached the efficiency drops suddenly owing to low brake power output from the engine. It is quite clear that the thermal efficiency is more for copper coated EE engine than standard engine and also the lean limit is extended up to 0.76.

Conclusion

The above analysis explains lucidly the advantage of copper coating on cylinder head and piston of EE engine. Not only the brake thermal efficiency and SFC are improved for the EE engine, also the lean limit of the air fuel mixture is greatly extended to 0.75 for the EE engine implying the significance and advantages of lean burn. Hence the idea of copper coating on EE engine cylinder head and piston combination proves to be an advantageous trend towards fuel economy in internal combustion engines.

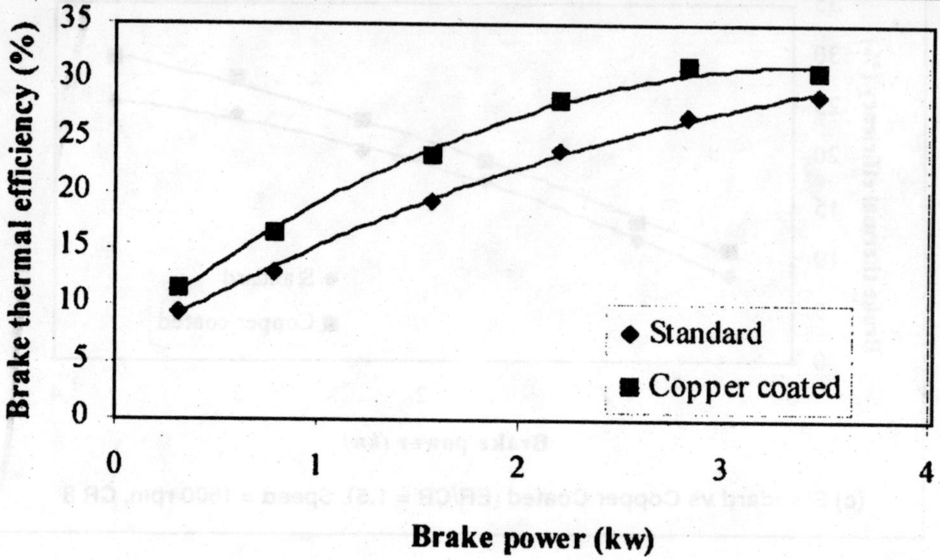

(a) Standard Vs Copper coated (ER/CR = 1.5), Speed = 1500 rpm, CR 9

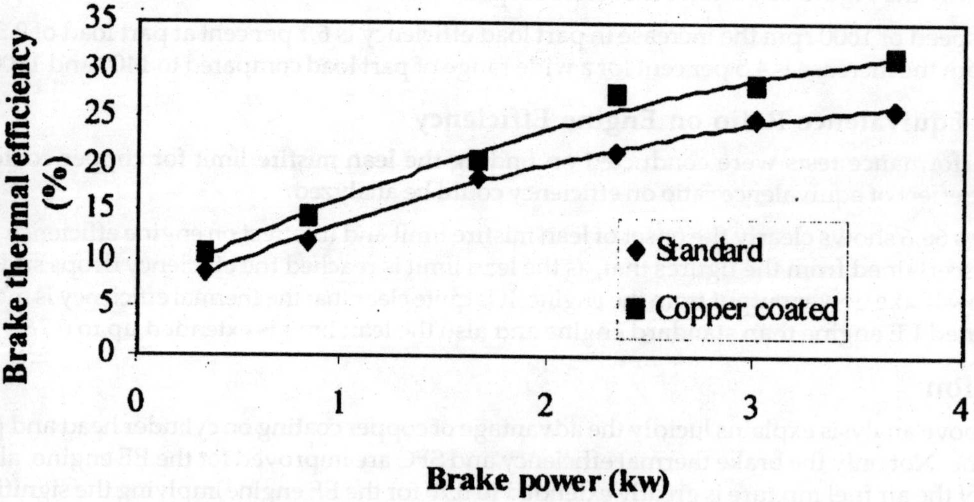

(b) Standard vs Copper coated (ER/CR = 1.5), Speed = 1600 rpm, CR 8

Figure 66.5: Effect of Speed on Engine Efficiency

Figure 66.7: Effect of Equivalence Ratio on Engine Efficiency

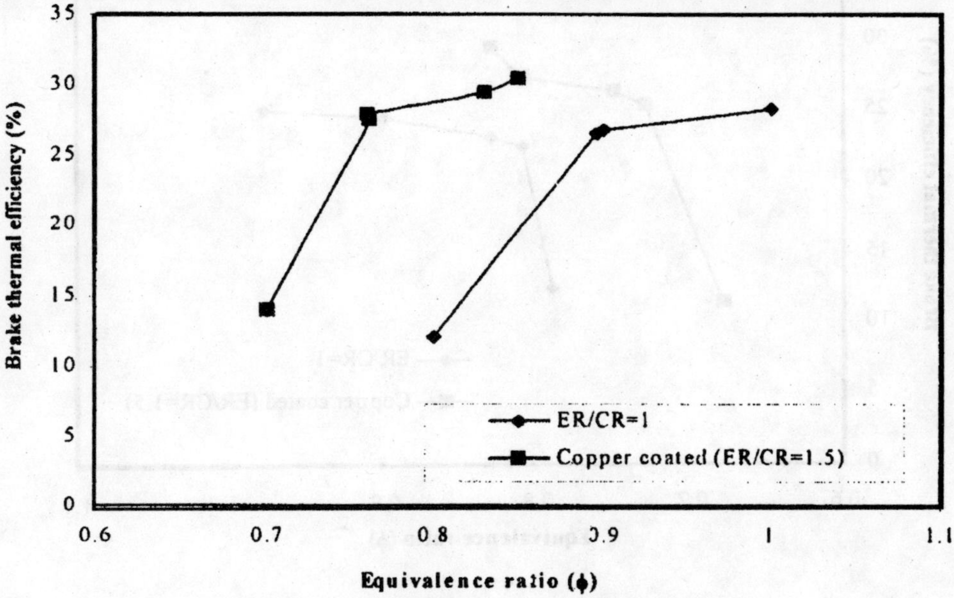

(a) ϕ Vs Efficiency (CR = 8, Speed = 1500 rpm)

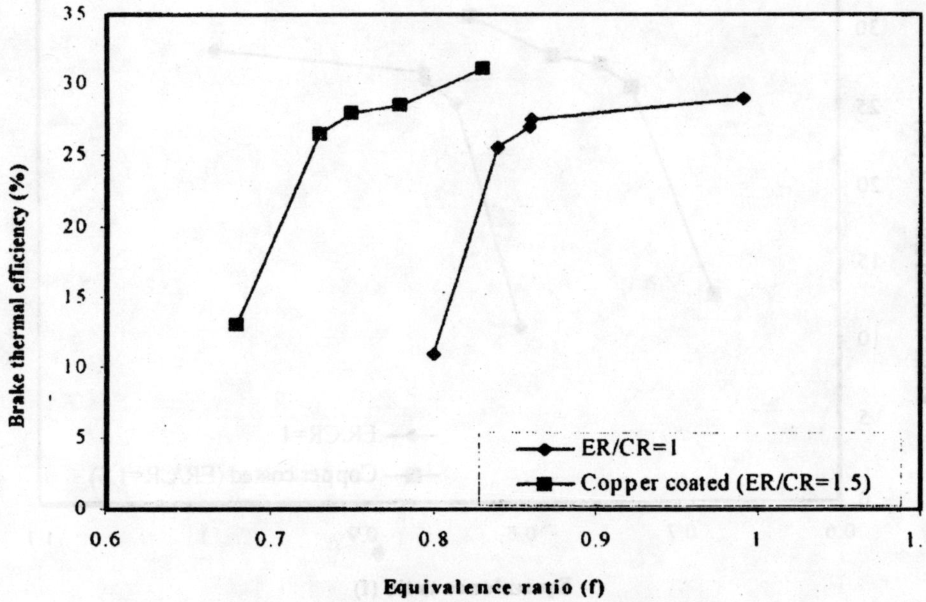

(b) ϕ Vs Efficiency (CR = 9, Speed = 1500 rpm)

Contd...

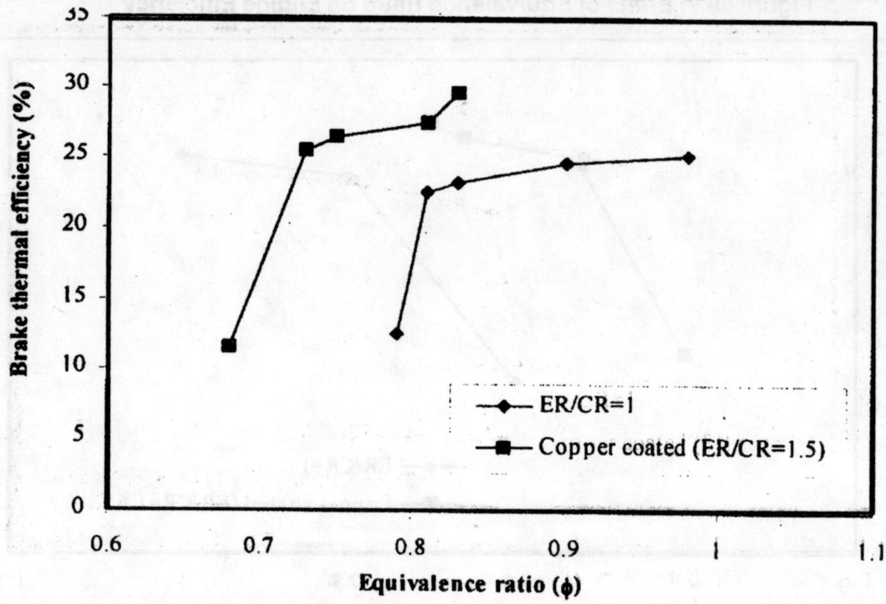

(c) ϕ Vs Efficiency (CR = 8, Speed = 1600 rpm)

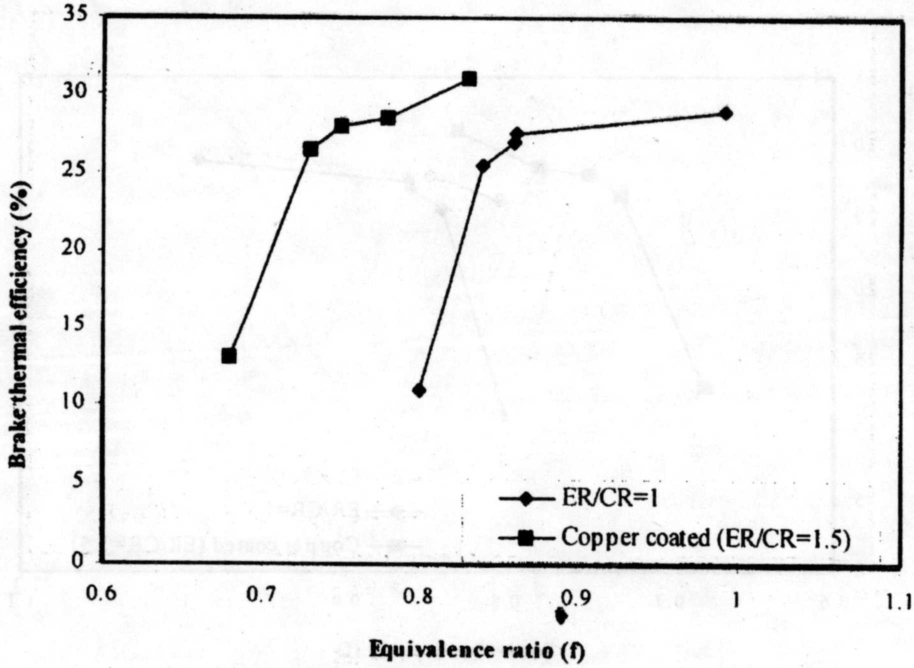

(d) ϕ Vs Efficiency (CR = 9, Speed = 1500 rpm)

Contd...

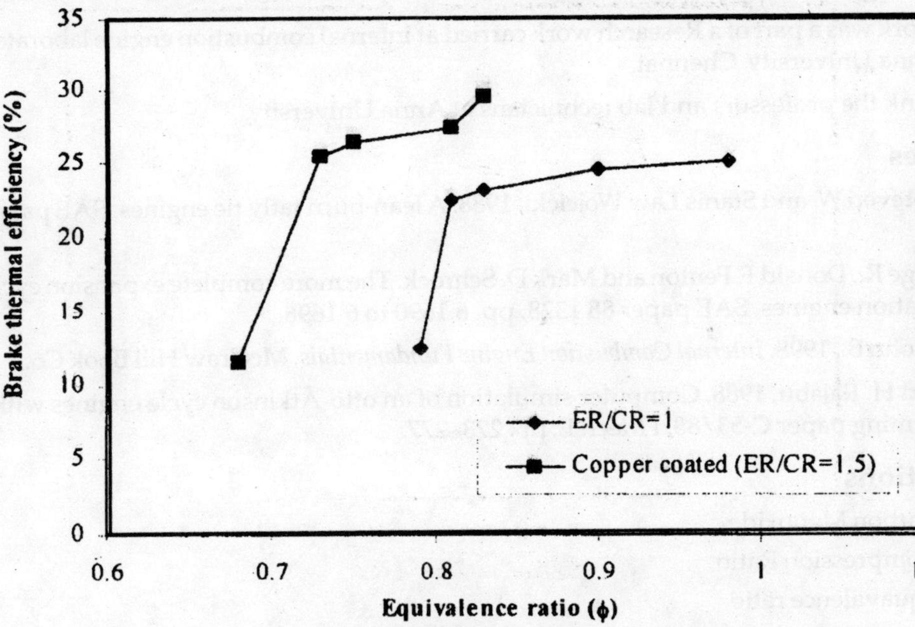

(e) ϕ Vs Efficiency (CR = 8, Speed = 1600 rpm)

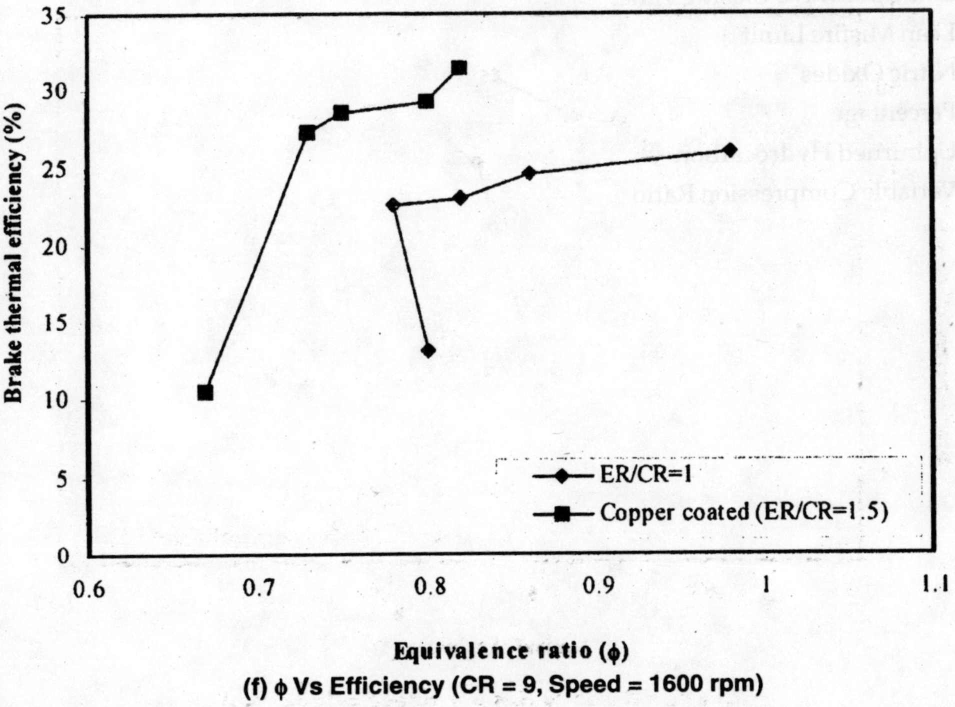

(f) ϕ Vs Efficiency (CR = 9, Speed = 1600 rpm)

Acknowledgement

This work was a part of a Research work carried at internal combustion engine laboratory at MIT campus, Anna University, Chennai.

We thank the professors and lab technicians at Anna University.

References

Beyerlein, Steven W. and Stanis Law Wojcicki, 1988. A lean-burn fatly tic engines. SAE paper No. 880 574.

Eakin, George R., Donald F. Fenton and Mark D. Schrock. The more complete expansion cycle applied to irrigation engines, SAE paper 88 1328, pp. 6.1690 to 6.1696.

Heywood, John B., 1998. *Internal Combustion Engine Fundamentals*. McGraw Hill Book Co.

Ma, T.H. and H. Rajabu, 1988. Computer simulation of an otto-Atkinson cycle engines with variable valve timing paper C-53/88, I Mech E, pp. 273–277.

Abbreviations

CO: Carbon Monoxide
CR: Compression Ratio
ϕ: Equavalence ratio
ER: Expansion Ratio
EE: Extended Expansion
LIVCT: Late Inlet Valve Closing Time
LML: Lean Misfire Limit
NO_x: Nitric Oxides
%: Percentage
UBHC: Unburned Hydrocarbon
VCR: Variable Compression Ratio

Chapter 67

Microbial Quality of Fresh and Smoked *Channa punctatus* Available in Imphal Market with Reference to Salted and Smoked in Laboratory

Ch. Sarojnalini and N. Jaya Devi
Department of Life Sciences, Manipur University,
Canchipur – 795 003, Manipur, India

ABSTRACT

Fresh (FF) and market smoked (MS) *Channa punctatus* available in Imphal during November and December 2005 were analysed for microbiological quality. Some of the fresh fishes were smoked in the laboratory after treating with different salt concentration in 3 per cent, 4 per cent and 5 per cent. Total plate counts of bacteria (TPC) in fresh and market smoked *C. punctatus* were 10^3 and 10^5 respectively. No fungus was detected in FF but it was 10^3 in MS. Count of *Staphylococcus aureus* and faecal *Streptococci* and *Salmonella* were higher in MS than FF. Coliform was not detected in MS however, these were found in FF. *E. coli* was not detected in both the samples. Total plate count of bacteria and fungi were very low in the laboratory smoked (LS) *C. punctatus* with different concentrations of salt.

Keywords: *Channa punctatus, Market smoked, Laboratory smoked.*

Introduction

Fish is an important source of protein in the diet of the people of Manipur. The fishes are consumed after frying or roasting or as an ingredient in vegetable curry preparation to add flavour and taste

(Singh *et al.*, 1990). Smoking not only imparts colour and flavour, but also has a preservative effect since formaldehyde, phenol and other substances evolved from smoke are deposited on the fish during the process (Stansby, 1963). Though there are reports on smoke curing of fish from India as well as from abroad (Mariappan *et al.*, 2004; Lu *et al.*, 1991), the age old technique adopted in Manipur is different from others as salting is not involved in the processing.

Fresh and smoked *Channa punctatus* in one of the most important fish which are popular in Manipur. Although the fishes are cured in such condition that the basic concern is the status of the fish in relation to health and hygiene. The hygienic quality of a fish is judge by its microbiological analysis. So the present work is undertaken. This article reports the microbiological quality of *Cllanna punctatus* of fresh (FF), market smoked (MS) in traditional Manipuri style, which are available in Imphal market and smoked in laboratory (LS) after treating with salt in a specially designed kiln.

Materials and Methods

Fresh *Channa punctatus* and smoked *Channa punctatus* of length 15.7–17.6 cm and 31.33–46.62 g weight were purchased from different fish sellers and brought to the laboratory.

Enumeration of total plate count (TPC), total fungal count (TFC), and most probable number (MPN) of coliforms and detection of pathogenic bacteria *viz.*, *Staphylococcus aureus.* faecal *Streptococci*, *Salmonella* and *E. coli* were done as per APHA (1976). TCP was enumerated on trypton agar; TFC on acidified potato dextrose agar; MPN of coliforms an brilliant green lactose bile broth (BGLB); *E. coli* on eosine methylene blue (EMB) agar after enrichment in BGLB; *Salmonella* on brilliant green agar (BGA) after enrichment in selenite cysteine broth (SCB) and faecal *Streptococci* on KF Streptococcal agar. The suspected pathogenic bacterial colonies were further tested using methods of APHA (1976) and Fungal colonies on PDA were picked up, stained with cotton blue in lactophenol and identified following the methods of Gilman (1957) and Ellis (1971, 1976).

Results and Discussion

Table 67.1 shows the bacterial and fungal count of fresh and smoked *Channa punctatus*, MS had the highest TPC (10^5 g^{-1}) followed by FF (10^3 g^{-1}) and LS (10 g^{-1}) *Staphylococcus aureus*, faecal *Streptococci*, Coliforms, *Salmonella* and *E. coli* were not detected in LS except very low count of *S. aureus* and faecal *Streptococci* in 2 per cent salt smoked fish. These lower values might be due to the incorporation of salt in smoking. Coliform was absent in the MS however TFC, *S. aureus*, faecal *Streptococci* and *Salmonella* were high in the MS than the FF. In general smoked fishes available in market are not salted and this may be the reason for higher microbial count. Contamination during processing, storage, handling and selling in the market might also be the reason for the higher count of bacteria. Microfloral count is related to the moisture content of the sample. Lowering of moisture retards the spoilage of fish (Stansby, 1963). In this experiment both MS and LS have high moisture content (49.15–52.57 per cent). However microfloral count is much higher in MS. Reduce in microfloral count in LS might be due to the treatment of salt before smoking as well as taking care during processing, storage and handling.

Fungal flora isolated from MS was *Cladosporium* sp. and *Candida* sp. The possibility of the incidence of toxic fungi fungal or metabolites leading to food poisoning cannot be ruled out unless proper care is taken. Small numbers of *Staphylococci* is not a serious problem but food poisoning may occur if the product is handled carelessly during processing resulting in multiplication of the organism (Iyer, 1979).

Table 67.1: Bacterial and Fungal Count of Fresh and Smoked *Channa pultctatus* Available in Imphal Market

Parameters	*Fresh C. punctatus*	*Smoked C. punctatus*
Total plate count (TPC) g^{-1}	$8 \times 10^3 - 9.9 \times 10^3$	$5.0 \times 10^5 - 7.6 \times 10^5$
Total fungal count (TFC) g^{-1}	ND	$3.2 \times 10^3 - 4.7 \times 10^3$
Staphylococcus aureus g^{-1}	$4.4 \times 10^3 - 5 \times 10^3$	$3.0 \times 10^4 - 4 \times 10^4$
Faecal *Streptococci* g^{-1}	$6.0 \times 10^2 - 7.2 \times 10^2$	$2.5 \times 10^5 - 3.6 \times 10^5$
Coliform (MNP) g^{-1}	210	ND
Salmonella g^{-1}	$1.7 \times 10^4 - 2.2 \times 10^4$	$1.22 \times 10^6 - 1.3 \times 10^6$

ND: Not detected; Mean value of 6 samples.

Table 67.2: Bacterial and Fungal Count of Laboratory Smoked *Channa punctatus* with Different Salt Concentrations

Parameters	*3% Salt Smoked Fish*	*4% Salt Smoked Fish*	*5% Salt Smoked Fish*
Total plate count (TPC) g^{-1}	$2.2 \times 10^2 - 3.3 \times 10^2$	$7.9 \times 10 - 9.0 \times 10$	$5.0 \times 10^2 - 7.3 \times 10^2$
Total fungal count (TFC) $^{-1}$	ND	ND	ND
Staphylococcus aureus g^{-1}	2.0×10	ND	ND
Faecal *Streptococci* g^{-1}	3.0×10	ND	ND
Coliform (MNP) g^{-1}	ND	ND	ND
Salmonella g^{-1}	ND	ND	ND

ND: Not detected.

Table 67.3: Fungal Flora Present in Market Smoked *Channa punctatus* (in per cent of total flora)

Fungi	*Market Smoked Channa punctatus*
Cladosporum sp.	15.625
Candida sp.	84.375

Faecal contamination is evidenced by the presence of faecal *Streptococci*, coliforms and *E. coli*. Their presence in foods possibly indicates the presence of enteric pathogens (Frazier and Westhoff, 1983). *E. coli* was totally absent in all the samples and *Salmonella* was detected in FF and MS.

From the results of the present experiment high fungal and bacterial count in MS is a mater of great concern for the health and hygiene of consumers. Attempts may be made to reduce the moisture content of MS by extending the time of smoking or by exposing to sunshine. More importantly proper care should be taken during processing, storage, handling and selling. The fishes should be properly fried, roasted or cooked before consumption.

Acknowledgement

Authors are thankful to Associate Prof. M. Shyamkesho Singh, Department of Life sciences, Manipur University, Canchipur for his guidance in the identification of fungus.

References

APHA, 1976. *Compendium of Method for Microbiological Examination of Foods*, (Ed.) M.L. Speak. American Public Health Association, Washington.

Ellis, M.B., 1971. *Dermatiaceous Hypmycetes*. CMI, Kew, Surrey, U.K., 608 pp.

Ellis, M.B., 1976. *More Dermatiaceous Hypmycetes*. CMI, Kew, Surrey, UK., 507 pp.

Frazier, W.C. and Westhoff, D.C., 1983. *Fundamentals of Food Microbiology*. Tata McGraw Hill, New York.

Gilman, J.C., 1957. *A Manual of Soil Fungi*. The Iowa State University Press, Iowa, USA, 450 pp.

Iyer, T.S.G., 1979. Pathogens and faecal organisms in fish and fishery product. In: *Quality Control in Fish Processing*. Central Institute of Fisheries Technology, Cochin (ICAR), pp. 45–59.

Joseph, A.C., Prabhu, P.V. and Balachandra, K., 1987. Studies on dired/smoked Barracuda. *Fish Technol.*, 24: 96–98.

Lilabati, H. and Vishwanath, W., 1996. Biochemical, mutritional and microbiological quality of three smoked fishes (*Wallago attu, Tor putitora, Puntius jayarami*) sold in the Imphal market, Manipur. *J. Freshwater Biol.*, 8(4): 217–223.

Lilabati, H. and Vishwanath, W., 1997. Biochemical, nutritional and microbiological quality of smoked *Anabas testudineus* brought to Manipur. *J. Freshwater Biol.*, 9(1): 23–27.

Lilabati, H. and Vishwanath, W., 2000. Biochemical, nutritional and microbiological quality of six species of smoked fishes from Manipur. *Fishery Technology*, 37: 89–94.

Lilabati, H. and Vishwanath, W., 2000. Changes in the microbial population during storage of the smoked fish, *Amblypharyngodon mola* of Manipur. *Indian J. Fish.*, 47(4): 349–354.

Lilabati, H., Vishwanath, W. and Shyamkesho Singh, M., 1999. Changes in bacterial and fungal quality during storage of smoked, *Esolnus danricus* of Manipur. *Fishery Technology*, 36(1): 36–39.

Lu, Y.J., Pace, D.R. and Plahar, D.W., 1991. Storage conditions and microbial quality of smoked dry herring in Ghana. *J. Food Protect*, 54: 557–559.

Mariappan, S., Sugumar, G. and Sukumar, D., 2004. Survival of *Salmonella, Staphylococcus aureus* and *Vibrio parahaemolyticus* in hot-smoked fish (*Nelmipterus* sp.) fillets under low temperature storage. *Indian Journal of Microbiology*, 44: 117–120.

Sing, M.B., Sarojnalini, C. and Vishwanath, W., 1990. Nutritive value of sundried *Eromus danricus* and smoked *Lepidocephalus guntea*. *Food Chemistry*, 36: 89–96.

Stansby, M.E., 1963. Cured fishery products. In: *Industrial Fishery Technology*, (Eds.) M.E. Stansby and E. Robert. Krieger Publ. Co., Huntington, New York, 415, pp.

Vishwanath, W. and Lilabati, H., 1998. Microbiological quality of fresh and smoked *Clarias batracllus* of Manipur. *Nat. Acad. Sci. Letters*, 21(3 and 4): 62–64.

Vishwanath, W., Lilabati, H. and Bijen, M., 1998. Biochemical, nutritional and microbiological quality of fresh and smoked mud eel fish *Monopterus albus*: A comparative study. *Food Chemistry*, 61: 153–156.

Chapter 68

Impact of Tsunami on Physico-chemical Characteristics of Coastal Water in Kulachel Area of Kanyakumari District

V. Ramasubramanian, R. Jeyaprakash and S. Jeyakumar

Department of Plant Biology and Plant Biotechnology,
Ayya Nadar Janaki Ammal College (Autonomous),
Sivakasi – 626 124, Tamil Nadu, India

ABSTRACT

Comprising over 70 per cent of the earth's surface, water is undoubtly the most precious natural resources on our planet. The global and regional climatic changes, influence life process even at local level. The recent tsunami in the Indian Ocean has influenced the quality of water in coastal areas of Tamil Nadu. Thus, we programmed to assess the quality of water in tsunami severely hitted kulachel area in Kanyakumari District.

Introduction

On 26th December 2004, an earthquake of *M* 9.0 shook the entire coast of India and Andaman Nicobar Islands. Centred off the coast of Sumatra in the Indian Ocean, the earthquake caused extensive damage in the Andaman and Nicobar Island regions of India and was felt strongly all along the east coast of India from Srikakulam in Andra Pradesh (AP) to Kanyakumari in Tamil Nadu (TN). The earthquake set off hugs tsunami, which travelled throughout the world oceans and severly affect the several countries in the Indian Ocean. The present study was made to assess the important water quality parameters due to unusual changes in the coastal region of Kulachel area in Kanyakumari District.

Material and Methods

The water samples were collected from the coastal region of Kulachel area in Kanyakumari District during 10th January 2005. The water samples were collected at four different part of the area, *viz.*, bottom of the sea (20 m depth), estuary region, Poomparai valkal area and middle of the sea (for away 1.5 km) from sea shore). Samples were bought to the laboratory and the water quality considering the parameters such as, pH, alkalinity, electrical conductivity, total dissolved solids, turbidity, total hardness, calcium, magnesium, nitrite, nitrate, chloride, fluoride, sulphate and phosphate contents were analysed. Sample taken in that area before the tsunami was maintained as control for comparison.

Results and Discussion

In the experimental samples, collected from different areas, turbidity, total dissolved solids and electrical conductivity were more compare to the control samples (Table 68.1). pH of the experimental samples ranged between 7.5–8.2 in all the stations, which is marginally higher than the control samples. The dissolved oxygen level in the water sample ranged from 1.5 to 6.1 mg/l during January 2005. While it is 6.8 mg/l control samples indicating a drastic reduction after the tsunami. The alkalinity in, experimental samples ranged between 98–188 ppt, but in control, the alkalinity was 94 ppt, the calcium, magnesium and nitrate concentration in experimental samples showed higher values than the control water samples. Sulphate and phosphate content of water samples were also increased after tsunami. The same result was also proposed by chadha *et al.* (2005) and Reddy *et al.* (2005). Along the west coast of India, the presence of high salinity, alkalinity, low dissolved oxygen and nutrien trich water has been reported during south-west monsoon seasons, when subsurface water reaches the coastal zone due to upwelling (Sankaranarayanan and Jayaraman, 1972).

Table 68.1

Sl.No.	*Parameters*	*Control*	*Inside the Sea*	*Estuary*	*Poomparai Valkal*	*Away from 1.5 km Seashore*
1.	Appearance	C and C	C and C	C and C	C and C	Sl. Brownish
2.	pH	7.2	8.2	7.5	7.5	7.6
3.	Electrical Conductivity (mS/cm)	1451	34720	12885	28800	2100
4.	Turbidity	0.84	1	2	4	1.2
5.	Total dissolved solids (mg/l)	916	24304	9006	20160	1147
6.	Dissolved oxygen	6.8	5.80	3.7	6.1	1.5
7.	Alkalinity	94	116	120	188	98
8.	Hardness	1050	5500	2100	4650	1480
9.	Calcium	115	400	160	300	171
10.	Magnesium	46	1080	408	936	54
11.	Nitrite	0.01	0.07	1.38	0.58	0.05
12.	Nitrate	5	1	4	2	2
13.	Sulphate	146	2440	774	1845	184
14.	Phosphate	0.01	0.07	0.02	0.07	0.1

All the values are expressed in mg/l, except pH and Electrical conductivity

C and C: Clear and Colourless.

However water samples during January 2005 from the coastal region of Kulachel area exhibited high alkalinity, hardness and low dissolved oxygen indicating the impact of tsunami on the characteristics of coastal water. This study reveals that, drastic change in the coastal water quality due to the tsunami has brought about a change in the coastal biodiversity.

References

Chadha, R.K., Latha, G., Harry Yeh, Curt Peterson and Toshitama Katada, 2005. The tsunami of the great Sumatra earthquake of *M.* 9.0 on 26th December 2004. Impact on the east coast of India. *Curr. Sci.*, 88(8): 1297–1301.

Reddy, H.R.V., Katti, R.J., Raveesha, K.P., Chandarasekara, Vikas, S.J., Nagendra Babu, R.S. and K.S. Shiv Kumar, 2005. Coastal water quality off Dakshina Kannada before and after tsunami. *Curr. Sci.*, 88(7): 1026–1027.

Sankaranarayanan, V.M. and N.R. Jayaraman, 1972. *Curr. Sci.*, 46: 204–205.

Index